Environmentally Sustainable Livestock Production

Environmentally Sustainable Livestock Production

Special Issue Editor

Ilkka Leinonen

MDPI • Basel • Beijing • Wuhan • Barcelona • Belgrade

Special Issue Editor
Ilkka Leinonen
Scotland's Rural College
UK

Editorial Office
MDPI
St. Alban-Anlage 66
4052 Basel, Switzerland

This is a reprint of articles from the Special Issue published online in the open access journal *Sustainability* (ISSN 2071-1050) from 2018 to 2019 (available at: https://www.mdpi.com/journal/sustainability/special_issues/Environmentally_Sustainable_Livestock_Production)

For citation purposes, cite each article independently as indicated on the article page online and as indicated below:

LastName, A.A.; LastName, B.B.; LastName, C.C. Article Title. *Journal Name* **Year**, *Article Number*, Page Range.

ISBN 978-3-03897-552-6 (Pbk)
ISBN 978-3-03897-553-3 (PDF)

Contents

About the Special Issue Editor vii

Preface to "Environmentally Sustainable Livestock Production" ix

Ilkka Leinonen
Achieving Environmentally Sustainable Livestock Production
Reprinted from: *Sustainability* **2019**, *11*, 246, doi:10.3390/su11010246 1

Nathan Pelletier, Maurice Doyon, Bruce Muirhead, Tina Widowski, Jodey Nurse-Gupta and Michelle Hunniford
Sustainability in the Canadian Egg Industry—Learning from the Past, Navigating the Present, Planning for the Future
Reprinted from: *Sustainability* **2018**, *10*, 3524, doi:10.3390/su10103524 6

Nathan Pelletier
Social Sustainability Assessment of Canadian Egg Production Facilities: Methods, Analysis, and Recommendations
Reprinted from: *Sustainability* **2018**, *10*, 1601, doi:10.3390/su10051601 30

Tianyi Cai, Degang Yang, Xinhuan Zhang, Fuqiang Xia and Rongwei Wu
Study on the Vertical Linkage of Greenhouse Gas Emission Intensity Change of the Animal Husbandry Sector between China and Its Provinces
Reprinted from: *Sustainability* **2018**, *10*, 2492, doi:10.3390/su10072492 47

Erasmus K.H.J. zu Ermgassen, Melquesedek Pereira de Alcântara, Andrew Balmford, Luis Barioni, Francisco Beduschi Neto, Murilo M. F. Bettarello, Genivaldo de Brito, Gabriel C. Carrero, Eduardo de A.S. Florence, Edenise Garcia, Eduardo Trevisan Gonçalves, Casio Trajano da Luz, Giovanni M. Mallman, Bernardo B.N. Strassburg, Judson F. Valentim and Agnieszka Latawiec
Results from On-The-Ground Efforts to Promote Sustainable Cattle Ranching in the Brazilian Amazon
Reprinted from: *Sustainability* **2018**, *10*, 1301, doi:10.3390/su10041301 65

Ana Beatriz Santos and Marcos Heil Costa
Do Large Slaughterhouses Promote Sustainable Intensification of Cattle Ranching in Amazonia and the Cerrado?
Reprinted from: *Sustainability* **2018**, *10*, 3266, doi:10.3390/su10093266 91

María I. Nieto, Olivia Barrantes, Liliana Privitello and Ramón Reiné
Greenhouse Gas Emissions from Beef Grazing Systems in Semi-Arid Rangelands of Central Argentina
Reprinted from: *Sustainability* **2018**, *10*, 4228, doi:10.3390/su10114228 119

Gwendolyn Rudolph, Stefan Hörtenhuber, Davide Bochicchio, Gillian Butler, Roland Brandhofer, Sabine Dippel, Jean Yves Dourmad, Sandra Edwards, Barbara Früh, Matthias Meier, Armelle Prunier, Christoph Winckler, Werner Zollitsch and Christine Leeb
Effect of Three Husbandry Systems on Environmental Impact of Organic Pigs
Reprinted from: *Sustainability* **2018**, *10*, 3796, doi:10.3390/su10103796 141

Ilkka Leinonen, Michael MacLeod and Julian Bell
Effects of Alternative Uses of Distillery By-Products on the Greenhouse Gas Emissions of Scottish Malt Whisky Production: A System Expansion Approach
Reprinted from: *Sustainability* **2018**, *10*, 1473, doi:10.3390/su10051473 . . . . **161**

Cristina Ullrich, Marion Langeheine, Ralph Brehm, Venja Taube, Diana Siebert and Christian Visscher
Influence of Reduced Protein Content in Complete Diets with a Consistent Arginine–Lysine Ratio on Performance and Nitrogen Excretion in Broilers
Reprinted from: *Sustainability* **2018**, *10*, 3827, doi:10.3390/su10113827 . . . . **179**

Henrik Saxe, Lorie Hamelin, Torben Hinrichsen and Henrik Wenzel
Production of Pig Feed under Future Atmospheric CO_2 Concentrations: Changes in Crop Content and Chemical Composition, Land Use, Environmental Impact, and Socio-Economic Consequences
Reprinted from: *Sustainability* **2018**, *10*, 3184, doi:10.3390/su10093184 . . . . **192**

Katrin Gerlach, Alexander J. Schmithausen, Ansgar C. H. Sommer, Manfred Trimborn, Wolfgang Büscher and Karl-Heinz Südekum
Cattle Diets Strongly Affect Nitrous Oxide in the Rumen
Reprinted from: *Sustainability* **2018**, *10*, 3679, doi:10.3390/su10103679 . . . . **210**

Michael MacLeod, Vera Eory, William Wint, Alexandra Shaw, Pierre J. Gerber, Giuliano Cecchi, Raffaele Mattioli, Alasdair Sykes and Timothy Robinson
Assessing the Greenhouse Gas Mitigation Effect of Removing Bovine Trypanosomiasis in Eastern Africa
Reprinted from: *Sustainability* **2018**, *10*, 1633, doi:10.3390/su10051633 . . . . **223**

Ricarda Maria Schmithausen, Sophia Veronika Schulze-Geisthoevel, Céline Heinemann, Gabriele Bierbaum, Martin Exner, Brigitte Petersen and Julia Steinhoff-Wagner
Reservoirs and Transmission Pathways of Resistant Indicator Bacteria in the Biotope Pig Stable and along the Food Chain: A Review from a One Health Perspective
Reprinted from: *Sustainability* **2018**, *10*, 3967, doi:10.3390/su10113967 . . . . **238**

Marco Baldi, Maria Cristina Collivignarelli, Alessandro Abbà and Ilaria Benigna
The Valorization of Ammonia in Manure Digestate by Means of Alternative Stripping Reactors
Reprinted from: *Sustainability* **2018**, *10*, 3073, doi:10.3390/su10093073 . . . . **264**

About the Special Issue Editor

Ilkka Leinonen (Dr) works at the Department of Rural Economy, Environment and Society at Scotland's Rural College (SRUC), United Kingdom. He obtained his PhD from the University of Joensuu (Finland) in 1997 and has since then carried out agricultural, ecological, and environmental research, with a main interest in quantitative methodologies, especially process-based modelling of ecosystem carbon, water, and nutrient dynamics. During recent years, he has mainly focused on the environmental sustainability of agricultural production systems, using and further developing methods of life cycle assessment (LCA) to quantify greenhouse gas emissions and other environmental impacts of agriculture and to investigate mitigation strategies in the livestock sector, for example, through changes in feeding, breeding, housing, and manure management. His other expertise includes ecophysiological plant and animal modelling, physical energy balance modelling, and methods for image analysis and remote sensing.

Preface to "Environmentally Sustainable Livestock Production"

Livestock production is a major global source of greenhouse gas emissions and it is also associated with other environmental issues, such as ammonia emissions and regional nutrient imbalances. This Special Issue presents 14 scientific papers assessing measures that aim to improve the environmental sustainability of livestock production and to mitigate its environmental impacts. Globally, the most important livestock species (beef and dairy cattle, pigs, broiler chicken and laying hens) are covered in the papers. The scope of the papers ranges from farm level mitigation methods to national level system changes. In general, most of the papers identify the efficiency of production as a key factor affecting the emissions arising from the livestock sector. In many studies, holistic approaches, such as environmental life cycle assessment, are used to assess the possible improvements in emission intensity, and the required links to other dimensions of sustainability, for example, using the methodology of social life cycle assessment, are also demonstrated in this Special Issue.

Ilkka Leinonen
Special Issue Editor

sustainability

MDPI

Editorial

Achieving Environmentally Sustainable Livestock Production

Ilkka Leinonen

Department of Rural Economy, Environment and Society, Scotland's Rural College (SRUC), Peter Wilson Building, Kings Buildings, West Mains Road, Edinburgh EH9 3JG, UK; ilkka.leinonen@sruc.ac.uk; Tel.: +44-131-5354044

Received: 2 January 2019; Accepted: 4 January 2019; Published: 7 January 2019

Livestock production is a major global source of greenhouse gas emissions [1] and high density of livestock in certain areas can also create local environmental issues such as harmful levels of ammonia emissions and regional nutrient imbalances [2,3]. However, future improvements in the global livestock sector can also be seen as a potential opportunity for delivering a significant share of the necessary mitigation of global warming and other environmental problems [1]. In fact, it has been demonstrated that significant reductions in various environmental impacts and especially in the emissions intensities (i.e., the amount of emissions per unit of product) have been already achieved in livestock production during the past decades, for example through breeding, optimized feeding, improved health status of animals, and improved manure management [4–6].

This Special Issue presents results from studies on different measures aiming to improve the environmental sustainability of livestock production and to mitigate environmental impacts, including emissions of greenhouse gases (carbon dioxide, methane, and nitrous oxide), nitrogen and phosphorus excretion, ammonia emissions, land use and use of energy and other resources. Globally the most important livestock species (beef and dairy cattle, pigs, broiler chicken and laying hens) are covered in the papers published here. The mitigation methods assessed in the papers include general system changes (together with novel approaches to comparison of alternative production systems), changes in feeding, improvement of animal health, and new technologies of manure management.

The scope of the papers included in this Special Issue range from farm level mitigation methods to national level system changes. One of the national-level articles, written by Pelletier et al. [7], presents an overview of the sustainability of the Canadian egg production industry. The authors provide an interdisciplinary perspective to this industry, considering its past, present, and potential futures. Their analysis covers environmental, institutional, and socio-economic sustainability with a special emphasis on animal welfare. The analysis identifies major challenges for sustainable egg industry, including shifting consumer and other stakeholder preferences and expectations, and conflicts between the expectations and scientific evidence. A thorough discussion is provided on possible strategies to resolve these issues.

The methodology of Life Cycle Assessment (LCA) is becoming a standard in evaluating the environmental sustainability of livestock systems, but LCA also has applications in the areas of social (S-LCA) and economic (life cycle costing) sustainability assessment. In contrast to relatively well established methods of environmental Life Cycle Assessment, the methods of S-LCA are still under development. In his second paper in this Special Issue, Nathan Pelletier [8] develops a suite of context-appropriate indicators and metrics to characterize the social risks and benefits specific to activities at Canadian egg production facilities and then applies those indicators to perform a detailed assessment of the "gate-to-gate" social risks and benefits of Canadian egg production facilities. The analysis presented here provides a starting point for expanding the methodology to cover the whole production chain and thus enable a full social life cycle assessment.

In their paper, Cai et al. [9] carry out a national level study on a reduction of the emission intensities (per unit of the monetary value of output) of non-CO_2 greenhouse gases arising from the livestock sector in China, through an analysis of the contribution of each province to the overall national emissions. They especially concentrate on the role of three "driving factors," i.e., environmental efficiency, productive efficiency and economic share, in determining the national greenhouse gas emission intensity. Their findings suggest that the productive efficiency (i.e., the input of production factors per unit of output value) is the main contributor to the changes in the emission intensities, and improvement in this factor has resulted in a considerable reduction of livestock sector emission intensity at the national level during the period of 1997–2016.

The effect of system changes on the productivity of cattle production in the Amazon biome was evaluated by zu Ermgassen et al. [10]. The key question in their study is how to handle the increasing agricultural production in Brazil (e.g., currently the World's second largest cattle herd) and at the same time protect remaining natural vegetation. The authors found that currently the cattle productivity in the Amazon biome is very low, and intensification of the cattle production systems would be the key factor in achieving higher environmental sustainability. In their paper, they present results from six initiatives in the Brazilian Amazon, which have successfully improved the productivity in beef and dairy systems.

Brazilian livestock production is also considered in the paper by Santos and Costa [11]. They tested the hypothesis according to which large slaughterhouses are potential leverage points for promoting sustainable intensification in the beef supply chain in Amazonia and the Cerrado, due to their interactions with ranchers, their location at the agricultural frontier, and their ability to control access to the market. The authors' conclusion was that although cattle-ranching intensification (with positive effect on environmental indicators) has occurred in the Cerrado, this development is independent of the presence of large slaughterhouses. Instead, the authors suggest that conservation measures such as a strong monitoring systems and more restrictive environmental policies would be the key promotes of environmental sustainability, especially at the Amazonia region.

In their article, Nieto et al. [12] assess the on-farm greenhouse gas emissions from beef production in semi-arid rangelands in Argentina and apply statistical analyses to identify the relationship between emissions and current farm management practices. Their results highlight the importance of efficient production in achieving environmental sustainability of livestock production. Their findings indicate that the emissions per product were low on farms that had improved livestock care management, applied rotational grazing, and had access to technical advice. The authors suggest that "implementation of realistic, relatively easy-to-adopt farming management practices has considerable potential for mitigating the GHG emissions in the semi-arid rangelands of central Argentina."

Livestock production in Europe has been considered to be highly intensive, and thus relatively efficient. However, many low-intensity systems exist as well, especially in organic production. In the study by Rudolph et al. [13] a system comparison using environmental LCA was carried for three European organic pig production systems, namely indoor, partly outdoor, and outdoor. The authors found a great between-farm variation in three environmental indicators: global warming, acidification and eutrophication potentials. The differences between the farms were mainly affected by feed production and to some extent also by housing. There were no between-system differences in global warming potential, but acidification potential was highest in the indoor system (as a result of ammonia emissions) and the eutrophication potential highest in the outdoor system (as a result of nutrient leaching). The authors conclude that the occurrence of organic farms with low environmental impacts indicates that it is possible to manage organic pig production in an environmentally friendly way.

Using agricultural by-products (that are not suitable for human consumption) as part of livestock feed has been considered to be one method to improve the environmental sustainability of livestock production. In their study, Leinonen et al. [14] assess the environmental consequences of using distillery by-products as a protein source for beef cattle in Scotland. Their study highlights the complexity of livestock feed production chains. This was demonstrated by the alternative uses of agricultural by-products (in this case either as a livestock feed or as a source of renewable energy), and the environmental impacts arising from those options were analyzed through the system expansion-based LCA approach.

Another option to reduce the environmental impacts through livestock feeding is to apply resource efficient feed formulation. Ullrich et al. [15] evaluated the potential of reducing the crude protein level of the broiler diet by using supplementation of single amino acids to achieve an optimum amino acid balance of the feed. Their experimental results confirm some earlier modelling studies [16] according to which a balanced diet with lowered crude protein concentration can reduce the amount of nitrogen excreted, which has multiple environmental benefits. It is also demonstrated that such an improvement can be achieved without compromising animal performance.

The link between livestock feeding and climate change is not only a one direction process. The feed formulation in the future may also be affected by changes in the availability and quality of certain feed ingredients, and such changes can be induced by global warming and increased atmospheric CO_2 concentration. Saxe et al. [17] applied a consequential life cycle assessment to quantify the environmental impacts and socio-economic effects that altered crop yields and chemical composition of the crops at elevated CO_2 levels in the future can have on pig feed formulation. They predict that the elevated CO_2 reduces the land use demand for pig feed production, but at the same time increases the demand for protein crops (soya), due to reduced protein concentration of feed crops. This will have considerable environmental and economic consequences.

In cattle production, methane from enteric fermentation and manure management is generally considered to be the most significant greenhouse gas. However, nitrous oxide emissions related to ruminant feeding have also their own role in the total emissions from this livestock sector. In their article, Gerlach et al. [18] present a new method for determining the concentrations of CO_2, CH_4, and N_2O in the ruminal gas phase of steers after ingestion of different forage types. Depending on the diet, high concentration of N_2O were found in the rumen, indicating that that fermented forages rich in nitrogen can be a significant source of greenhouse gases.

Improving the health status of animals is one option to maintain high production efficiency of livestock and in this way keep the emission intensity at the minimal level. In their paper, MacLeod et al. [19] apply the FAO livestock model GLEAM to quantify the greenhouse gas emissions from East African cattle production systems and the effects of an endemic disease trypanosomiasis on the emissions. The authors found that removing that disease could lead to a reduction in the emissions intensity per unit of protein produced, as a result of increases in milk yields and higher cow fertility rates. Another major issue related to animal health is antibiotic resistance, and this has also links to the environmental impacts of livestock production. In their comprehensive review, Schmithausen et al. [20] highlight knowledge gaps and various factors that contribute to the transmission of antibiotic-resistant bacteria between animals, humans, and the environment in pig production, following a holistic "One Health" approach.

Although most papers in this special issue focus on livestock husbandry and its effect on animal performance when considering possible methods for reducing the environmental impacts of livestock system (and their potential effects on human health), different manure management options can potentially also control such impacts. Reduction of ammonia emission through improved manure management has direct consequences on human and animal health, and it also affects numerous environmental issues such as eutrophication, acidification, and global warming. In this Special Issue, novel technologies of manure management are considered in the paper by Baldi et al. [21], who show results of a comparison of ammonia stripping methods aiming to reduce the emissions arising from digestate derived from anaerobic digestion of livestock manure and corn silage.

In summary, this Special Issue demonstrates a range of opportunities that would help to reduce the environmental impacts of global livestock production. Most of the papers identify the efficiency of the production as a key factor to affect the emission intensity of the livestock products. To assess the possible improvements in efficiency, holistic approaches such as Live Cycle Assessment would be necessarily needed, and further methodological development in this area is still required [22]. This would be especially a challenge when linking together the three pillars of sustainability (environmental, social, and economic) in sustainability assessments of livestock production.

Funding: This research was partly funded by the Scottish Government Rural Affairs and the Environment Portfolio Strategic Research Programme 2016-2021, WP 1.4 'Integrated and Sustainable Management of Natural Assets' and 2.4 'Rural Industries'.

Conflicts of Interest: The author declares no conflict of interest.

References

1. Gerber, P.J.; Steinfeld, H.; Henderson, B.; Mottet, A.; Opio, C.; Dijkman, J.; Falcucci, A.; Tempio, G. *Tackling Climate Change through Livestock. A Global Assessment of Emissions and Mitigation Opportunities*; Food and Agriculture Organization of the United Nations (FAO): Rome, Italy, 2013; ISBN 978-92-5-107921-8.
2. Vitousek, P.M.; Naylor, R.; Crews, T.; David, M.B.; Drinkwater, L.E.; Holland, E.; Johnes, P.J.; Katzenberger, J.; Martinelli, L.A.; Matson, P.A.; et al. Nutrient imbalances in agricultural development. *Science* **2009**, *324*, 1519–1520. [CrossRef]
3. Leinonen, I.; Eory, V.; MacLeod, M. Applying a process-based livestock model to predict spatial variation in agricultural nutrient flows in Scotland. *J. Clean. Prod.* **2018**, *209*, 180–189. [CrossRef]
4. Pelletier, N.; Ibarburu, M.; Xin, H. Comparison of the environmental footprint of the egg industry in the United States in 1960 and 2010. *Poult. Sci.* **2014**, *93*, 241–255. [CrossRef] [PubMed]
5. Tallentire, C.W.; Leinonen, I.; Kyriazakis, I. Breeding for efficiency in the broiler chicken. A review. *Agron. Sustain. Dev.* **2016**, *36*, 66. [CrossRef]
6. Tallentire, C.W.; Leinonen, I.; Kyriazakis, I. Artificial selection for improved energy efficiency is reaching its limits in broiler chickens. *Sci. Rep.* **2018**, *8*, 116. [CrossRef] [PubMed]
7. Pelletier, N.; Doyon, M.; Muirhead, B.; Widowski, T.; Nurse-Gupta, J.; Hunniford, M. Sustainability in the Canadian Egg Industry—Learning from the Past, Navigating the Present, Planning for the Future. *Sustainability* **2018**, *10*, 3524. [CrossRef]
8. Pelletier, N. Social Sustainability Assessment of Canadian Egg Production Facilities: Methods, Analysis, and Recommendations. *Sustainability* **2018**, *10*, 1601. [CrossRef]
9. Cai, T.; Yang, D.; Zhang, X.; Xia, F.; Wu, R. Study on the Vertical Linkage of Greenhouse Gas Emission Intensity Change of the Animal Husbandry Sector between China and Its Provinces. *Sustainability* **2018**, *10*, 2492. [CrossRef]
10. Zu Ermgassen, E.; Alcântara, M.; Balmford, A.; Barioni, L.; Neto, F.; Bettarello, M.; Brito, G.; Carrero, G.; Florence, E.; Garcia, E.; et al. Results from On-The-Ground Efforts to Promote Sustainable Cattle Ranching in the Brazilian Amazon. *Sustainability* **2018**, *10*, 1301. [CrossRef]
11. Santos, A.; Costa, M. Do Large Slaughterhouses Promote Sustainable Intensification of Cattle Ranching in Amazonia and the Cerrado? *Sustainability* **2018**, *10*, 3266. [CrossRef]
12. Nieto, M.; Barrantes, O.; Privitello, L.; Reiné, R. Greenhouse Gas Emissions from Beef Grazing Systems in Semi-Arid Rangelands of Central Argentina. *Sustainability* **2018**, *10*, 4228. [CrossRef]
13. Rudolph, G.; Hörtenhuber, S.; Bochicchio, D.; Butler, G.; Brandhofer, R.; Dippel, S.; Dourmad, J.; Edwards, S.; Früh, B.; Meier, M.; et al. Effect of Three Husbandry Systems on Environmental Impact of Organic Pigs. *Sustainability* **2018**, *10*, 3796. [CrossRef]
14. Leinonen, I.; MacLeod, M.; Bell, J. Effects of Alternative Uses of Distillery By-Products on the Greenhouse Gas Emissions of Scottish Malt Whisky Production: A System Expansion Approach. *Sustainability* **2018**, *10*, 1473. [CrossRef]
15. Ullrich, C.; Langeheine, M.; Brehm, R.; Taube, V.; Siebert, D.; Visscher, C. Influence of Reduced Protein Content in Complete Diets with a Consistent Arginine–Lysine Ratio on Performance and Nitrogen Excretion in Broilers. *Sustainability* **2018**, *10*, 3827. [CrossRef]
16. Leinonen, I.; Williams, A.G. Effects of dietary protease on nitrogen emissions from broiler production: A holistic comparison using Life Cycle Assessment. *J. Sci. Food Agric.* **2015**, *95*, 3041–3046. [CrossRef]

17. Saxe, H.; Hamelin, L.; Hinrichsen, T.; Wenzel, H. Production of Pig Feed under Future Atmospheric CO2 Concentrations: Changes in Crop Content and Chemical Composition, Land Use, Environmental Impact, and Socio-Economic Consequences. *Sustainability* **2018**, *10*, 3184. [CrossRef]
18. Gerlach, K.; Schmithausen, A.; Sommer, A.; Trimborn, M.; Büscher, W.; Südekum, K. Cattle Diets Strongly Affect Nitrous Oxide in the Rumen. *Sustainability* **2018**, *10*, 3679. [CrossRef]
19. MacLeod, M.; Eory, V.; Wint, W.; Shaw, A.; Gerber, P.; Cecchi, G.; Mattioli, R.; Sykes, A.; Robinson, T. Assessing the Greenhouse Gas Mitigation Effect of Removing Bovine Trypanosomiasis in Eastern Africa. *Sustainability* **2018**, *10*, 1633. [CrossRef]
20. Schmithausen, R.; Schulze-Geisthoevel, S.; Heinemann, C.; Bierbaum, G.; Exner, M.; Petersen, B.; Steinhoff-Wagner, J. Reservoirs and Transmission Pathways of Resistant Indicator Bacteria in the Biotope Pig Stable and along the Food Chain: A Review from a One Health Perspective. *Sustainability* **2018**, *10*, 3967. [CrossRef]
21. Baldi, M.; Collivignarelli, M.; Abbà, A.; Benigna, I. The Valorization of Ammonia in Manure Digestate by Means of Alternative Stripping Reactors. *Sustainability* **2018**, *10*, 3073. [CrossRef]
22. Mackenzie, S.G.; Leinonen, I.; Kyriazakis, I. The need for co-product allocation in the Life Cycle Assessment of agricultural systems—Is "biophysical" allocation progress? *Int. J. Life Cycle Assess.* **2017**, *22*, 128–137. [CrossRef]

sustainability

MDPI

Article

Sustainability in the Canadian Egg Industry—Learning from the Past, Navigating the Present, Planning for the Future

Nathan Pelletier [1,*], Maurice Doyon [2], Bruce Muirhead [3], Tina Widowski [4], Jodey Nurse-Gupta [5] and Michelle Hunniford [6]

1 340 Charles Fipke Centre for Innovative Research, 3247 University Way, University of British Columbia, Kelowna, BC V1V1V7, Canada
2 Department of Agricultural Economics and Consumer Science, Laval University, Quebec City, QC G1V 0A6, Canada; Maurice.Doyon@eac.ulaval.ca
3 Research Oversight and Analysis and Department of History, University of Waterloo, Waterloo, ON N2L 3G1, Canada; muirhead@uwaterloo.ca
4 Department of Animal Biosciences, University of Guelph, Guelph, ON N1G2W1, Canada; twidowsk@uoguelph.ca
5 Department of History, University of Waterloo, Waterloo, ON N2L 3G1, Canada; jnursegu@uwaterloo.ca
6 Burnbrae Farms Ltd., Lyn, ON K0E1M0, Canada; mhunniford@burnbraefarms.com
* Correspondence: nathan.pelletier@ubc.ca; Tel.: +1-250-807-8245

Received: 30 August 2018; Accepted: 27 September 2018; Published: 30 September 2018

Abstract: Like other livestock sectors, the Canadian egg industry has evolved substantially over time and will likely experience similarly significant change looking forward, with many of these changes determining the sustainability implications of and for the industry. Influencing factors include: technological and management changes at farm level and along the value chain resulting in greater production efficiencies and improved life cycle resource efficiency and environmental performance; a changing policy/regulatory environment; and shifts in societal expectations and associated market dynamics, including increased attention to animal welfare outcomes—especially in regard to changes in housing systems for laying hens. In the face of this change, effective decision-making is needed to ensure the sustainability of the Canadian egg industry. Attention both to lessons from the past and to the emerging challenges that will shape its future is required and multi- and interdisciplinary perspectives are needed to understand synergies and potential trade-offs between alternative courses of action across multiple aspects of sustainability. Here, we consider the past, present and potential futures for this industry through the lenses of environmental, institutional (i.e., regulatory), and socio-economic sustainability, with an emphasis on animal welfare as an important emergent social consideration. Our analysis identifies preferred pathways, potential pitfalls, and outstanding cross-disciplinary research questions.

Keywords: Canada; eggs; sustainability; animal welfare; economics; supply management

1. Introduction

Food systems are at the center of human well-being. In addition to satisfying a basic human need (i.e., regular access to food in sufficient quantity and of sufficient quality), food is also often a central contributor to our economies and cultures, and often even to our individual identities. However, activities in the agri-food system are also at the center of many of our most pressing sustainability challenges. The production of food—in particular, in the livestock sector—contributes a large fraction of current anthropogenic resource demands and environmental pressures [1–3]. For this reason, livestock industries are naturally the focus of a growing body of sustainability research and management initiatives [4].

Taken together with projected increases in food production globally and a trend towards diets higher in livestock products, this has spurred considerable interest in the concept and practice of "sustainable intensification" in the livestock sector [4]. According to Pretty et al. [5], sustainable intensification is defined as "producing more output from the same area of land while reducing negative environmental impacts and at the same time increasing contributions to natural capital and the flow of environmental services." Clearly, however, sustainable intensification efforts may also have potential benefits and trade-offs across socio-economic, institutional, and other aspects of sustainability that must be carefully considered.

Life cycle thinking (LCT) has emerged as a core concept in sustainability science [6]. LCT refers to adopting a systems-level perspective on industrial activities. This perspective enables us to understand how different kinds of potential sustainability benefits and impacts are distributed along agri-food supply chains, as well as trade-offs that may occur with respect to different valued outcomes when particular changes are implemented. Environmental life cycle assessment (e-LCA) is a commonly used tool, based on LCT, for studying and managing the resource/environmental dimensions of food supply chains [4]. In recent years, a rich body of research has applied this tool to evaluate a variety of livestock production systems and technologies in different contexts, and as a basis for understanding the respective merits of potential sustainable intensification technologies (for a review of 173 recent papers, see McClelland et al. [7]).

While such research is clearly of considerable value, it is by itself insufficient to support sustainability decision-making for the livestock sector since it considers only a subset of important sustainability criteria that inform our decisions [8]. In reality, the varied forces that influence how we produce and consume livestock products along with the associated benefits and impacts are complex, often interacting, and variable over time. They include changes in technology and management practices, evolving societal expectations and consumer preferences, and the regulatory context in which specific industries operate. Efforts to understand current sustainability challenges in the livestock sector and to identify preferred paths forward can benefit from interdisciplinary approaches that evaluate these forces with respect to historical trends, current conditions, and possible futures.

Egg Farmers of Canada, the industry body governing the production and marketing of eggs within the supply-managed Canadian egg industry, provides research monies to support four Research Chairs at Canadian universities. These Research Chairs respectively undertake independent research in the fields of economics (Doyon), public policy (Muirhead), animal welfare (Widowski), and sustainability (Pelletier) of broad or direct relevance to the egg industry. The current analysis brings together the expertise, research, and perspectives of each of these Chairs to present an integrated study of the past, present and possible futures for this industry. Specifically, the purpose of the analysis is to identify: (1) the key factors that have shaped the modern Canadian egg industry; (2) the issues and opportunities it currently faces; and (3) potential synergies and trade-offs across the multiple dimensions of sustainability that should be considered on an interdisciplinary basis in choosing among viable paths looking forward.

2. Methods

The analysis is presented in three sections. The first section provides a historical perspective, describing the emergence and evolution of the Canadian egg industry over the past century until the present. It is organized into subsections respectively addressing: key technological and management changes, including their influence on the efficiency and environmental sustainability impacts of egg production; the development and implications of the supply management system that governs the industry; the factors that have influenced the economics of egg production, including changing consumer preferences and social expectations; and, as an important aspect of the latter, the emergence of animal welfare as both a societal concern and field of study/application for the egg industry.

The second section, also organized into four subsections, describes challenges that the industry currently faces in each of these domains that may undermine its sustainability, as well as potential solutions. Possible trade-offs across sustainability domains that such solutions may imply are identified. On this basis, the final section summarizes some of the key areas for interdisciplinary research and collaboration that are necessary to support choosing among alternative courses of action to enable a sustainable egg industry in Canada into the future.

3. Discussion

3.1. Canadian Egg Industry Retrospective (Circa 1920 to Present)

3.1.1. Technology, Management, and Resource Efficiency

Although a small number of specialized, commercial egg farms in Canada existed in the early part of the 20th century, egg production was generally one among a series of activities undertaken on the mixed-farming operations that were characteristic of Canadian agriculture at that time. Beginning in the 1920s, however, the egg industry entered a period of significant and sustained industrialization. Two major developments that were particularly important to the specialization and intensification of egg production were the adoption of cage systems for housing laying hens and improved genetic selection for egg production and feed conversion efficiency.

From 1923–1924, D.C. Kennard performed the first experiments keeping laying hens in confinement at the Ohio Agricultural Research Station, using livability and production as the main measures of success [9]. However, cage systems were not immediately adopted for commercial use. This was because indoor confinement for longer periods of time was only possible once the complete ration, which was developed in 1924, was made available on the wider market in 1929 and the Rural Electrification Act of 1936 in the US enabled barns to be lit artificially. The ability to keep laying hens indoors revolutionized the egg industry [10,11]. Whereas eggs were previously a seasonal (spring) food in temperate climates, they could now be produced continuously year-round [12].

According to Lee, 1938 marked the year when farmers had "satisfactorily solved the many problems of management which were responsible for the failure of earlier battery plants (so called because of the methodical (militaristic) nature of organizing the cages in stacked groups or batteries)" [10]. The need for steel during World War II, however, meant that the production of cages was halted (except for a small number of chick batteries) until after the war when the battery cage boom started on the Pacific coast of the US and in the UK [10].

Individual cages were initially used because they eliminated issues with cannibalism and allowed for the practice of "positive culling," or removing birds from the flock that are "laying at a slow unprofitable rate or [have] quit laying altogether" [13]. However, due to the cost advantages of housing multiple birds in the same cage, the battery cage was adapted to house small groups of hens [14]. Beginning in the 1960s egg production in Canada transitioned from "free-run" (i.e., indoor non-cage) to cage-based production.

Genetic selection of laying hens for egg production and feed use efficiency in cage systems also began in earnest following World War II. Canadian Donald Shaver built a global breeding organization for layer and broiler strains [15]. The availability of electricity-powered incubators and the relatively short incubation cycle for eggs facilitated rapid progress in selection for production efficiency [16]. At the same time, development of improved management practices for disease prevention such as biosecurity protocols, along with the advent of poultry vaccines, served to improve bird health and reduce losses due to mortality [17,18]. Over time, advancements in housing technology including artificial ventilation systems and climate control, automated feeding, egg collection and manure removal were implemented, thereby reducing labor and allowing for thousands or tens of thousands of birds to be housed in a single barn.

In combination, these technology and management changes have enabled considerable improvements in resource efficiency and the reduction of environmental impacts associated with producing eggs in Canada. For example, annual rate of lay among Canadian laying hens has increased from less than 100 eggs per year in the early 20th century to over 300 at present [19]. Between 1962 and 2012, an interval of 50 years, rate of lay increased by more than 50%. This same 50-year interval was also marked by declining mortality rates (falling from roughly 13% in the early 1960s to 3.2% at present for pullets and laying hens combined), and by much improved feed conversion efficiencies. With respect to the latter, producing 1 kg of eggs in 1962 required over 3 kg of feed compared to the current average of 2 kg on contemporary egg farms [20].

Efficiency changes have been equally pronounced along the supply chains that ultimately support Canadian egg production. Among these, the most influential in terms of life cycle resource use/emissions-related sustainability impacts have been: (a) a 50% reduction in the energy intensity of ammonia production for nitrogen fertilizers (one of the most energy and emissions intensive aspects of modern agriculture); (b) improved yield-to-fertilizer and energy input ratios for the production of agricultural feed inputs (for example, corn yields in the province of Ontario increased 96% over this interval while nitrogen inputs per ton of crop declined 44%); and (c) improved efficiencies in freight transport, which connect activities all along the Canadian egg supply chain [20].

As a result, of these changes, the overall environmental footprint (i.e., including all supply chain activities) of producing eggs in Canada has, on average, declined 61%, 68%, and 72% for acidifying, eutrophying, and greenhouse gas (GHG) emissions while energy, land and water use decreased by 41%, 81% and 69% respectively per unit production. Moreover, despite that egg production volumes roughly doubled in Canada since the early 1960s, the absolute resource and environmental impacts for the industry as a whole were estimated to be 41%, 51% and 57% lower for acidifying, eutrophying, and greenhouse gas emissions, respectively. Supply chain energy, land and water use are 10%, 71% and 53% lower in aggregate [20]. These changes reflect a combination of farm-level efficiency gains (15–40%, depending on impact category) associated with improved management practices, superior bird genetics, and vaccine developments; changes in feed composition (29–60%); and changing efficiencies in activities along the supply chains that ultimately support egg production (0–56%) [20].

Life cycle assessment research has hence enabled a nuanced understanding of both the magnitude and distribution of a variety of environmental sustainability impacts along the contemporary Canadian egg supply chain, the relative importance of specific inputs to and activities associated with egg production as sources of these impacts, and the comparative impacts of egg production in alternative housing systems [21]. A relatively small number of variables explain the characteristic sources and distribution of life cycle resource use and emissions for eggs and egg product supply chains in Canada. Among these, feed composition and feed conversion efficiency in pullet and (in particular) layer facilities emerge as the strongest explanatory variables. Manure management is the second critical determinant of life cycle resource efficiency and emissions. Manure-related emissions are influenced by several factors, including feed composition (i.e., N and P content of feed inputs), feed conversion efficiencies, and manure handling strategies. Although more strongly influenced by supply chain feed inputs, direct water and energy use in facilities also make non-trivial contributions to the overall water and energy resource requirements for egg production. The contributions of egg processing and packaging, as well as egg breaking and further processing to overall supply chain resource use and emissions for eggs and egg products are relatively small [21]. These insights are largely consistent with similar research of intensive egg production in other countries (for example, [18,22–27]).

Pelletier [21] reported resource efficiencies and life cycle environmental performance by housing system type for the Canadian egg industry in 2012. Among the five housing technologies considered (i.e., conventional cages, enriched colony cages (which provide all of the equipment found in conventional cages with the addition of equipment that is intended to allow hens to express some of their behavioral priorities), free-run, free-range, and organic), both the life cycle inventory and impact assessment results suggested fairly similar levels of performance between systems for most variables. Feed conversion efficiencies were slightly higher in cage-based production, and mortality rates were substantially lower compared to non-cage systems. Of note was the higher variability in performance levels observed between reporting facilities for non-cage production systems, likely reflecting the substantial research and development investments and management experience gained for cage-based production over time relative to the emerging cage-free sector. Only for organic production were life cycle resource use and emissions significantly different from the other housing systems [21]. Here, the lower observed resource use and emissions intensity of organic eggs was attributable to the lower impacts of feed production rather than differences in farm-level efficiencies, where mortality rates were highest among the housing systems considered. At present, over 80% of laying hens in Canada are kept in conventional cages, and the remainder in either enriched cages, free-run barns, or free-range systems.

3.1.2. Shifting Regulatory Conditions for the Production and Marketing of Eggs

The intensification of the egg industry from the 1920s onward meant that increasing numbers of eggs were finding their way onto the market. While the consumption of eggs rose steadily over time, the growth in egg production often outstripped demand, leading to overproduction and hardship for egg farmers [28,29]. Critically, farmers only received prices that allowed a fair return for short periods of time after low prices had pushed out the most vulnerable producers. While a shortage of product briefly forced prices up, the cycle would start over as farmers re-entered the business or ramped up production and again prices would plummet due to an oversupply of eggs. This was a perennial problem and the Canadian Minister of Agriculture in 1959, Douglas Harkness, noted that "The only long-term solution to the current problem is a decrease in egg production to the point where there is a more realistic balance between supply and domestic requirements" [30].

An additional complication was cheap egg imports from the United States. This meant Canadian farmers often had to accept egg prices that not only reflected tough domestic competition, but also a cheaper international price. The US, with its more robust economies of scale, undercut Canadian prices. Furthermore, provinces with surplus production would seek markets outside their borders, often undercutting prices and causing local producers' incomes to drop [31].

The new realities of the price-cost squeeze borne by agricultural producers and trends towards more specialized, capitalized, and vertically integrated operations caused significant concern among family farmers who saw producer prices drop and their earnings decrease despite ever more costly farm investments. While some industry experts believed that the movement towards vertical integration was inevitable and that only a few of the largest Canadian egg producers would survive, others argued that egg producers needed to create a system that would permit them to succeed without the corporatization of the family farm [32,33].

The continued instability of egg prices galvanized discussions about the need for farmers to organize and institute measures that would secure fair returns. Foremost among these discussions were proposals to create provincial marketing boards to regulate the production and sale of eggs on a quota system basis. Critics of the idea of marketing boards cited infringements on their freedom to act independently. Some egg farmers also recognized that provincial production quotas would only work if a national production plan were instituted that addressed the issue of egg imports from the US [34].

In July 1971, provincial representatives reached an agreement in principle whereby production would be controlled by marketing boards in each province and the national market would be shared based on average production figures calculated from 1968 to 1970. This led to the passage of the Farm Products Marketing Agencies Act in January 1972, which allowed for the creation of the Canadian Egg Marketing Agency (CEMA) later that year. The system of "supply management" had come to the Canadian egg industry [35].

Despite some initial growing pains, by 1976 CEMA had turned the corner, becoming solvent and revising and consolidating the comprehensive egg marketing plan, which meant uniform pricing, quota planning, and overproduction penalties introduced in all provinces. The processes were coming into place that allowed for the successful operation of a national supply management system that promoted its social and economic sustainability. The system of supply management helped end market chaos and enabled farmers to earn a living wage and offer consumers a fair price, while demonstrating a commitment to improved egg farming practices. At present, CEMA, renamed Egg Farmers of Canada in 2008, oversees the quota-based production and marketing of eggs from over 1000 farms distributed across all ten Canadian provinces and one territory.

3.1.3. Shifting Social Preferences and Socio-Economic Conditions

Supply Management

Over the years, the economics of the egg industry has been driven by improvements in technology, genetics, feeds, and management; resulting in productivity gains in rate of lay, in feed conversion and in lower mortality rates. Supply management, which shaped the marketing of eggs in Canada since 1971, has been similarly important for the economics of egg production, largely through the prevention of egg surpluses—a primary rationale for its implementation. Production controls were intended to prevent the boom/bust cycle of commodities production. Whereas the huge swings so characteristic of commodity sales played havoc with planning and investment [36], supply management provided income stability for farmers and, for consumers, predictability in meal costing and planning [37].

Price fluctuations were also perceived to favor processors and supermarkets, who drove down the prices paid to farmers. Indeed, many farmers believed that they were being exploited by several ruthless buyers who were only interested in making money [38]. In July 1971, for example, a dozen eggs cost 17 cents retail in Toronto, while the cost of production was about 31 cents. Prices were, in the words of a Canadian Egg Producers Council document provided to the minister of agriculture, "disastrously low ... extending from 1970, through 1971 and the first half of 1972" [39].

As the document went on to note, "Not only this but the public interest is badly served by the economic waste and inefficiency that is an inevitable consequence of severe cyclical instability, and by the instability of consumer prices—sometimes extremely (and to the producer disastrously) low, but at others unnecessarily high as production inevitably falls to inadequate levels under the pressure of persistent losses." With the passage of Bill C-176, the *Farm Products Marketing Agencies Act* on 31 December 1971, the way was clear to establish a national system of egg production, based on provincial organization, that soon stabilized production, farmer incomes, and consumer prices. Significant improvement in farm income for egg farmers subsequently allowed them to invest in technology, as well as to respond to changes in consumer demand, such as for specialty eggs [40].

Although egg farmers' financial situation has been improved by supply management, eggs in Canada remain a cheap source of good quality protein. Figure 1 shows that evolution of the index price for eggs is similar to that of beef, relative to general inflation (CPI). One should note, however, that the real price (versus the index) of eggs is much cheaper than the price of beef, on a gram of protein basis.

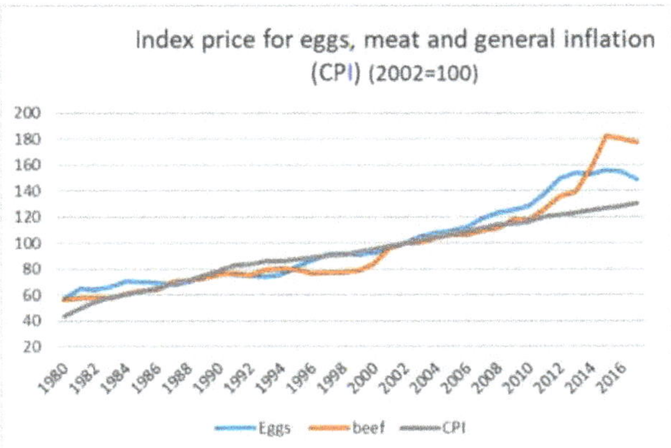

Figure 1. Consumer price index (CPI) for eggs, meat and general inflation in Canada from 1980-2016 (source: Agriculture and Agri-Food Canada [41]).

Shifting Consumer Preferences

The third major variable impacting the economics of egg production has been shifting consumer preferences. For instance, in 1980, Canadians were consuming roughly 22 dozen eggs a year. However, due to cholesterol concerns in the 80s and 90s, per capita consumption reached a low of 14.5 dozen in 1995, as illustrated in Figure 2. New research and a better comprehension of the various types of fats has rehabilitated egg consumption, which has been steadily increasing to the current annual rate of more than 20 dozen eggs per capita. Thus, Canadian egg farmers have seen steady market growth for over ten years, after a decade of significant cuts in production quota in the 80s.

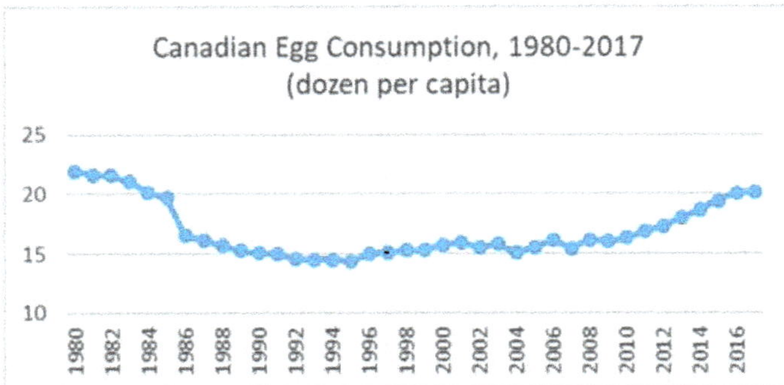

Figure 2. Canadian egg consumption (dozen per capita), 1980-2017 (source: Agriculture and Agri-Food Canada [42]).

Another important change in consumer demand in the 2000's is seen in the larger share of the specialty eggs market, which was roughly 12% in Canada in 2017. Specialty eggs are a value-added product, with differentiation based on either egg composition (omega-3, vitamin D eggs), the perceived quality (brown eggs), or the conditions under which eggs are produced (organic, free-run, and free-range eggs), as illustrated in Figure 3.

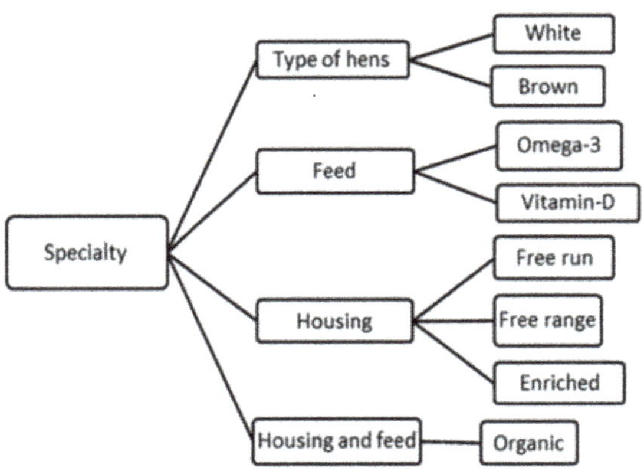

Figure 3. Specialty eggs in the Canadian market, by attribute.

The emergence of the specialty eggs market reflects the fact that buying food has evolved from a purely survival focus to include more nuanced nutritional as well as social preferences and environmental considerations. Animal welfare is an important issue that has been partially addressed through the offering of specialty eggs with specific animal welfare attributes (Figure 3 under housing). Those eggs command a higher retail price to reflect higher cost of production [43]. However, a disconnect between consumer willingness-to-pay and the desire for animal welfare [44] is likely to result when internalization of animal welfare costs is imposed on producers [45].

3.1.4. Animal Welfare

Public Concern for the Welfare of Laying Hens

Concurrent with the adoption of battery cages came growing public debate regarding the welfare of laying hens. Newspaper articles as early as 1953 advanced the same arguments that are still articulated today to either criticize or justify the use of cages. Key criticisms include lack of space, the inability of hens to perform natural behaviors, and that "the system makes the hen 'a mere egg laying machine'" [46]. In defense of cages, arguments are that they decrease disease and facilitate the provision of clean food and water.

Terms such as laying hen "plants" [10], "factories" or "egg machines" [47] were initially used to describe production efficiencies and were lauded because of the amount of control given to egg farmers. In the post-war period, however, such terms were given a negative connotation, and emotive terminology, for example "concentration camp," was increasingly used [48]. The book "Animal Machines" [48], and a subsequent UK government report addressing the welfare of farm animals [14], criticized animal production systems that severely restrict animal movement and behavior. The Brambell Report also clarified that a high production rate cannot alone be used as an indicator of good animal welfare. Nonetheless, by 1970, most hens in Canada (and elsewhere in the developed world) were kept in battery cages [49].

Beginning in 1976, regulations for animal housing systems in Europe were gradually established, beginning with the Council of Europe Convention on the Protection of Animals Kept for Farming Purposes [49]. Over the next few decades, a few individual European countries enacted minimum space allowances (or outright bans) for hens in cages, and a 1986 EU Directive set a minimum size for cages. This culminated in the EU Council Directive 99/74 for laying hens, which prohibited the use of non-enriched cage systems in all EU countries as of January 2012 [50]. According to the Directive, hens in cages must have a minimum of 750 cm^2 per hen and hens in non-cage systems must be stocked no greater than nine hens per m^2. All hens, regardless of housing system, must have a nest, perches, and litter to allow pecking and scratching [50]. In 2016, 55.6% of hens in the EU were housed in enriched cages with the remainder in free-run, free-range, and organic non-cage systems.

In Canada, the process for setting animal welfare standards for farm animals, coordinated by the Canadian Federation for Humane Societies (CFHS), was initiated in 1980 [51]. This process aimed to develop voluntary codes of practices for all livestock species. The first Code of Practice for Care and Handling of Chickens was published in 1983, followed by the Code of Practice for the Care and Handling of Pullets and Laying hens in 2003. These codes laid out recommendations for space allowances and other aspects of animal care, primarily for hens housed in battery cages [51]. In 2005, the National Farm Animal Care Council was established and developed a new and more rigorous process for developing science-informed codes. Code committees currently include multi-stakeholder representation from farmers, government agencies, scientists, CFHS, and national food retailer, restaurant, and food service associations. The process for code development includes establishment of a scientific committee that drafts a scientific report and review of the literature on key welfare issues, as well as regular re-evaluation of each code in response to developments in scientific and production knowledge [51].

In 2012, Egg Farmers of Canada (EFC) initiated a review of the 2003 Code under the guidance of the National Farm Animal Care Council. The new Code of Practice for laying hens was released in early 2017. The code requires all hens to be provided nests, perches, scratch mats or foraging material by 2036, similar to the EU Directive. While allowing for continued adoption of enriched cages, the code also sets more detailed standards for non-cage systems than are in place in the EU. Although the codes are not legislated, "when included as part of an assessment program, those who fail to implement requirements may be compelled by industry associations to undertake corrective measures or risk a loss of market options. Requirements also may be enforceable under federal and provincial regulation" [52].

Food Retailers as Drivers of Change

In Canada (and North America more broadly), changes to hen housing based on animal welfare concerns are coming much later compared to Europe and have been primarily market or producer driven (in response to pressure from customers and animal protection groups) rather than regulatory. In 2000, the United Egg Producers (a US trade organization) developed animal husbandry guidelines and an auditing program that set a minimum space allowance for laying hens in conventional battery cages [53]. That same year, McDonald's set a higher minimum space requirement for hens in their egg supply chain in the USA and Canada [54], followed soon after by Burger King and Wendy's. Over the next decade, several US states passed laws that essentially banned conventional cages [53]. However, a federal bill (The Egg Products Inspection Act Amendment "The Egg Bill)" introduced to US Congress in 2012 and 2013 that aimed to eliminate conventional battery cages but still allow for the adoption of enriched cages subsequently failed [55]. Corporate campaigns soon followed to obtain pledges from major food retailers, food service industries, and food manufacturers to purchase eggs solely from hens kept in cage-free systems.

In 2015, the first few North American restaurants, grocers and manufacturers made pledges to purchase eggs only from non-cage systems. In February 2016, EFC announced that no new conventional cage systems were to be installed after July 2016, but that both enriched cages and non-cage systems would be allowed. Shortly thereafter, the Retail Council of Canada, which comprises most large grocery chains in Canada, committed to sourcing only cage-free eggs by the end of 2025 [53]. By the end of 2016, an unprecedented number of North American corporations made commitments to cage-free housing which will require new or retrofitted barns for over 200 million hens in the United States and Canada by 2026 [55].

3.2. *Canadian Egg Industry Prospective*

3.2.1. Current Challenges and Opportunities for Improved Technology, Management, and Resource Efficiency

Leveraging continued efficiency gains and emissions reductions in the egg industry is important from the perspective of resource and environmental sustainability and may be supported by four separate but complementary foci. However, any recommended management or technology initiatives need necessarily be considered taking into account potential trade-offs with respect to other aspects of sustainability—for example, costs, animal welfare impacts, and acceptability to consumers.

Sustainability Management Best Practices

The first necessary focal area involves identification and dissemination of environmental sustainability best management practices in the context of current production norms, specific to housing system type. The 2012 benchmark LCA of the Canadian egg industry [21] demonstrated considerable variability in efficiencies among egg farms, with highest variability among producers using non-cage systems. For example, non-trivial differences in feed conversion efficiencies, rate of lay, and mortality rates are observed between industry leaders and laggers. Scenarios to assess the life cycle resource use and emissions mitigation potential of achieving best reported performance for feed conversion efficiency and rate of lay, as well as sourcing low-impact feeds suggest that further reducing the resource and emissions intensity of Canadian egg products by over 50% may be possible [21]. Concerted efforts to identify the factors that enable some farms to outperform others—whether related to management practices, in-place technologies, or other variables (for example, breed of hen), and programs to mainstream these factors industry-wide may enable substantial improvements in industry-average performance. Promotion and achievement of best practices would likely improve the profitability of egg production for individual farmers—in particular, through reducing use of costly feed and energy inputs. It is unclear, however, what positive or negative animal welfare impacts may arise, hence any recommended strategies must be carefully assessed on this basis.

Sustainable Feed Sourcing and Formulation

The second priority focal area is feed composition. The largest share of supply chain resource use and emissions associated with contemporary egg production is attributable to the feeds supplied to pullets and layers [18,21]. Data collected for the 2012 national benchmark LCA study indicated considerable diversity in the range and geographical origin of feed inputs from agricultural, livestock, fisheries, and other production systems [21]. Each feed input that may potentially be sourced for use in poultry feeds is also characterized by a distinctive life cycle resource use and emissions profile, with considerable variability between specific feed materials. Each material is similarly distinct in terms of its nutritional value for poultry, as well as its cost. Feed formulation is currently informed primarily by nutrition and cost considerations. However, "least environmental cost" feed input sourcing is the most critical lever for supply chain environmental sustainability management for egg production. Development of a regionally resolved feed formulation decision support tool that will integrate nutritional, cost, and environmental impact data for major feed input supply chains for the Canadian poultry industry is therefore desirable. Integration of nutritional and cost criteria with environmental impact data is essential, since feed input sourcing recommendations based on environmental criteria alone may result in feeds that are uneconomical or that result in poor feed conversion efficiency (hence negating any gains associated with lower environmental cost feed inputs).

Nitrogen Use Efficiency

Nitrogen use efficiency (included, but not limited to, manure management) is the third priority focal area for environmental sustainability management in this industry [21,56,57]. Nitrogen use efficiency and loss is important in terms of the net energy, nitrogen, and carbon footprint balance of egg production, and has important implications for air quality, human and animal health [58]. For farms producing their own feeds and cycling manure nitrogen on-farm, minimizing N loss is also economically important. Several variables are influential in nitrogen use efficiency and cycling, such as feed composition, feed conversion efficiency, moisture content of hen excreta, and manure management strategies [57]. The latter includes collection and handling technologies, residency time in storage, storage cover, and land application and incorporation methods. Manure belt systems and manure drying have been proposed as one strategy to reduce losses of nitrogen as ammonia from layer hen manure, as well as to improve air quality [59]. Implementation of any such technology should be carefully assessed with respect to the estimated potential systems-level (i.e., life cycle) benefits and trade-offs, including costs. However, interventions that improve air quality will generally also improve hen welfare as well as worker health.

Sustainable Intensification Technologies

The fourth priority focal area for improving the environmental performance of egg production is the identification and implementation of sustainable intensification technologies at farm level. Some promising sustainable intensification technologies for the egg industry include, for example, those related to waste valorization, lighting, use of renewable energy sources, and energy-efficient housing. Each of these will have cost implications, hence consideration of payback time is important. Therefore, too, is attention to potential hen welfare impacts.

From a "life cycle" perspective, waste valorization represents a significant opportunity for improving both resource/environmental efficiencies and profitability in the egg industry. While prior research has underscored the potential limitations and benefits of a subset of relevant waste valorization opportunities (for example, biogas production from poultry manure) [60–64], current knowledge regarding the distribution and fate of key waste streams along Canadian egg supply chains, as well as the comparative efficacy of existing waste valorization technologies is underdeveloped. Moreover, further research and technology development for novel waste valorization strategies is needed in support of increased diversion of under-used waste streams including egg shells, mortalities, and end-of-lay hens. With respect to the latter, the depopulation strategies for hens at end of lay, as well as potential hen transport requirements if a larger fraction of spent hens are to be processed for human consumption, will be particularly important from an animal welfare perspective.

Direct energy inputs to layer facilities for lighting, heating, ventilation, and other processes make a non-trivial contribution to life cycle resource use and emissions for egg production and are also an important cost of production consideration [18,21]. Similar energy inputs are also required upstream along egg supply chains for breeder facilities, hatcheries, and pullet facilities. Integration of renewable energy systems both for layer facilities and along egg supply chains may therefore provide significant opportunities for improving the life cycle environmental sustainability performance of the Canadian egg industry. A variety of renewable energy technologies are currently being employed at a subset of egg production facilities across the country. To date, however, there has been no systematic accounting of the distribution, scale, feasibility, mitigation potential and scalability of these technologies for egg production supply chains. With respect to the latter considerations, any such accounting must necessarily consider geographical and climatic factors including the spatial and temporal distribution of solar and wind resources to advance regionally appropriate, renewable energy technology deployment recommendations. Economic costs and payback time need also be considered.

Another area for green technology development and deployment in the egg industry is with respect to hen housing and other building infrastructure. As much as 30% of greenhouse gas (GHG) emissions are attributable to the building sector, largely due to energy use over the lifespan of buildings (UNEP 2009). Net zero energy building technologies aim to create buildings that produce at least as much renewable energy on site as they consume on an annual basis. Such technologies hence have the potential to substantially mitigate anthropogenic GHG emissions [65]. Little work has been advanced to date to evaluate the feasibility and mitigation potential of net zero energy building technologies in the intensive animal agriculture sector (also a key GHG emitter), where housing is typically employed for confined poultry, as well as for pork and dairy production. Such production facilities require energy inputs for lighting, climate control, ventilation, feed delivery, egg collection, manure management, and sanitation activities. Direct, farm-level energy use for egg production may account for as much as 25% of cradle-to-farm gate life cycle energy use and GHG emissions [18,21]—depending on farm location, on-farm efficiencies, and energy sources. Changes to ventilation systems may, however, have negative impacts on air quality, in turn impacting both worker and hen welfare, and short-term technology costs must be weighed against long-term returns resulting from energy savings.

Lighting systems for livestock production, in particular for poultry, are influential for animal health and productivity [66,67]. Diverse lighting systems have been used in the poultry industry. Most recently, light emitting diode (LED) lighting systems have been developed for poultry housing. These systems are primarily marketed based on their energy efficiency compared to competing lighting systems, which can effect significant cost savings for producers. However, several researchers have reported differences in egg weight, shell strength, rate of lay, bird behavior and feed conversion efficiency under different single and combined monochromatic LED light regimes [68–70]. Carefully selected LED lighting regimes may therefore have important implications for environmental sustainability performance which go far beyond direct, farm-level energy savings. This is particularly true with respect to changes in feed use efficiency, since feed inputs are the largest contributor to both costs and to supply chain resource use and emissions for egg production [18,21], as well as rate of lay and mortality rates—both of which influence feed use efficiency. Optimizing LED lighting systems for sustainability objectives therefore presents an important research area that must necessarily bridge resource/environmental, economic, and animal welfare considerations.

The current environmental performance profile of the industry largely reflects the production of eggs by hens in conventional cages. As previously described, slightly superior efficiencies are currently achieved in cage and enriched cage facilities, reflecting optimization of both management strategies and genetics for cage-based production in the industry over time. Efficiencies are slightly lower, and more variable between farms for non-cage systems (for example, with respect to mortality rates) [20]. As the industry transitions away from conventional cage-based production, both farm management and genetics optimization efforts will be required to close the currently modest efficiency gap between production in conventional cages and alternative housing systems. This will be similarly important to maintaining the profitability of egg production. It should be noted, however, that selective breeding for increased rate of lay looking forward may result in unacceptable welfare trade-offs as a result of further compromising the skeletal integrity of hens.

3.2.2. Current Challenges and Opportunities with Respect to Regulatory Conditions

Stakeholders Support for Supply Management

If Canadian egg farmers were not to show strong support for supply management, it would likely be dismantled in a relatively short period of time—in particular, given persistent pressure from international trading partners such as the United States. The former Canadian Wheat Board is a good example of the potential consequences of non-unanimous support among farmers—the institution lost its ability to be the sole purveyor of Western Canadian wheat and barley. Although egg farmers have and continue to demonstrate strong support for the system, how growth in consumption is allocated between Canadian regions has caused tensions. For instance, some regions have argued in the past that they should get greater quota allocation based on their lower cost of production. However, others argue that since those regions already produce more than they consume, their cost advantage is cancelled by the cost of transporting eggs long distance. Those tensions have been solved through negotiations and are currently low.

While challenges exist, supply management continues to find support among Canadians because they recognize that the system allows farmers to receive adequate compensation for their products, while also providing consumers with a fair price, and encourages a strong and stable domestic supply. In addition, unlike the United States and the European Union, which subsidize their producers heavily and have not been able to control periods of intense overproduction (especially in the dairy sector, which has led to needless producer suffering and wasted production as well as harmful effects for the environment) Canada only produces what is required to meet demand [71–73]. Similar to the dairy industry, the egg industry also suffers when farmers cannot cover their cost of production. In 2013, for example, French farmers destroyed hundreds of thousands of eggs to protest overproduction and the subsequent drop in prices [74]. Canada's system of supply management is not only more responsible, it is also a more sustainable model because it does not lead to resource wastage and pollution associated with overproduction. It also fairly compensates workers, who are then able to support their rural economies and create more vibrant and sustainable rural communities [75]. Doyon and Bergeron [76] found that investments on farms under supply management in Canada are more important (excluding quota investments) than those on non-supply-managed farms, because of the confidence and stability of revenue associated with supply management. Those on-farm investments also generated more jobs for farms under supply management and the economic impact was mostly directed towards rural areas.

Supply Management and Animal Welfare

A supply-managed industry is also better equipped to facilitate an orderly transition to non-conventional cage-based production and may actually support greater innovation given that producers get predictable and fair returns. This, in turn, encourages experimentation without the same degree of risk experienced elsewhere [40]. Beyond product innovation and consumer choice, Canada's supply-managed egg sector may allow for a smoother transition to alternative housing options in egg farming because it will, as it becomes mainstream, compensate farmers for the additional costs of production associated with these more expensive housing alternatives to conventional cages. The turmoil that was caused when the European Union demanded new housing standards without fair farmer compensation should serve as a warning against unsupported and uncompensated mandatory changes in farming practices [74].

3.2.3. Current Challenges and Opportunities with Respect to Social Preferences and Socio-Economic Conditions

Specialty Eggs and Collective Pricing

The first important economic issue currently facing the Canadian egg industry relates to pricing along the value chain. The current cost of production formula that is used by EFC to determine producer egg prices reflects egg production in battery cages. Farmers receive a premium from egg graders to produce specialty eggs and premiums are negotiated on an individual basis, sometimes with written contract or sometimes with a verbal agreement. Moreover, retail prices are sometimes at odds with the cost of production. For example, vitamin D eggs have a premium at retail of roughly 30% while the cost of production is barely affected. On the other hand, graders take important market risks while supplying organic eggs, given that they must guarantee a premium to egg farmers to motivate the capital cost involved, while having little to no guarantees on volume from retailers. Thus, organic eggs might end up in the regular supply chain, at the sole expense of the grader. As the percentage of specialty eggs on the market increases, the private nature of premiums might conflict with the collective principles of supply management. It might also cause problems in gathering sufficient information for determining the cost of production. How price will be transmitted along the value chain should also be a concern. Different options are being explored to address this issue. One possible option is to put at the disposal of farmers contract types for specialty eggs with price information. This would insure more fairness and reduce the asymmetry of information between graders and egg farmers. Another option would be, assuming that the growth of specialty eggs will continue, to have add-on to the cost of production for various types of specialty eggs. For example, organic dairy producers receive the cost of production plus 22 $ per hectoliter in Québec. This add-on can be revised on a regular basis using cost items.

Consumer Confidence

The second issue relates to the confidence of consumers in specialty eggs. This has not been an issue so far in Canada. However, as the variety of specialty eggs increase, how can the consumer be assured that, for example, the vitamin D and Omega-3 free-range brown eggs purchased are really what was paid for? Certification is likely to become important in this regard. As an example, Australia recently experienced a confidence crisis for consumers regarding what free-range implies in the absence of certification. The situation has also created uncertainty for egg farmers leading to a significant underproduction of free-range eggs. The increasing prevalence of sustainable sourcing initiatives in the food sector may also place new burdens on producers with respect to measuring and communicating around sustainability performance, goals, and progress.

The emergence of specialty eggs implies that attached to the private value of an egg is a social attribute that can be associated with a public good dynamic. However, even though specific social attributes are becoming more important for some buyers, preferences are currently heterogeneous among consumers. This raises interesting issues regarding achieving an appropriate balance between private and public choice, and whether and to what extent this is best influenced by market forces and/or regulations. For example, which type of production is best with respect to animal welfare, the environment, or for workers [22]? At what cost? Interdisciplinary research is required to identify the appropriate balance between competing attributes, the extent to which stakeholder incentives are aligned and, in turn, the preferred regulatory or market-based strategy.

Citizens Versus Consumers

The third issue relates to the potential disconnect between the pressure to transition to cage-free production (largely resulting from lobbying of retailers and fast-food chains in Canada by animal welfare organizations) and consumer willingness-to-pay for cage-free eggs. As previously mentioned, cage-free production may actually engender trade-offs with respect to animal welfare, the environment, workers' welfare, and economic efficiency. This nuance may be lost on most consumers, who tend to react adversely to the very idea of cage-based production.

For instance, Doyon et al. [44] found that when presenting consumers with fictive names for housing systems, the three out of twelves names containing the word cage were by far the ones considered as the least likely to promote hen welfare. On the other hand, the researchers also found that when consumers were provided with additional information on the enriched cage system, they showed increased preference for these eggs relative to regular eggs.

Given that cage-free eggs are readily available but represent less than 10% of the table eggs sold in Canada, one must naturally wonder about the impact on consumers and on demand for moving completely to cage-free egg production by 2025. Norwood and Lusk [77] found that consumers were willing to pay, on average, between 53% and 100% more for cage-free eggs over battery cage eggs. These results come with important standard deviation, meaning that some participants are not willing to pay more for cage-free eggs, and in some instances, some would pay less.

California is an interesting example. Cage-free eggs represented less than 10% of consumption in 2008 when a referendum (Proposition 2) to ban battery cages was voted on by 64% of eligible voters. Malone and Lusk [45] estimate that California's law banning battery cage increased the retail cost differential between California and the national egg price average by 51%, while that differential was historically around 15%. The cost for consumer surplus is estimated by the authors to be between $400 and $850 million.

3.2.4. Current Challenges and Opportunities for Improving Hen Welfare

As previously described, the Canadian egg industry is in the early stages of transitioning away from housing laying hens in conventional cages. However, to inform the selection of alternative housing systems, it is critical to understand the animal welfare trade-offs associated with different housing systems, as well as how these might impact on resource/environmental and socio-economic sustainability considerations. Rational decision-making must ultimately reflect consideration and accommodation across these domains.

Ethical Concerns and Scientific Measures

Concerns about animal welfare can be divided into three areas, which form the bases for measures used in science-based evaluations [78]. These comprise measures of biological function that include health, mortality, physical condition and production performance and measures of emotional or affective states of animals that include pain, fear, discomfort, reward, and pleasure. Naturalness, a third concept of what constitutes a good life for animals, is less amenable to scientific evaluation. However, it is often defined as the ability to perform natural or species-typical behavior. By quantifying and comparing the innate drive of animals to perform specific behavior patterns we can determine what animals want (find rewarding), lending objective measures of affective states to the concept of naturalness. All three categories of criteria are articulated in most formal definitions of animal welfare, but it is important to recognize that different stakeholders vary in the degree of importance they place on the different measures. Farmers and veterinarians tend to value measures of biological function, animal welfare scientists tend to value measures of subjective states and members of the broader community value naturalness [78]. In the case of housing systems for laying hens, these criteria are often conflicting. Public acceptance can drive farming practices, but public perceptions do not always align with scientific evidence [79].

While most formal definitions of animal welfare include the ability to express "normal" or innate behavior, the scientific consensus is that it is neither practical nor necessary for hens to be able to perform all types of behavior [80]. Empirical research has focused on identifying the candidate behaviors that are most important for hens. Specifically, nesting, perching, foraging, and dustbathing are considered to be behavioral needs or behavioral priorities [80]. Thus, provision of enough space and resources to support these behavior patterns are included in most science-based standards.

Welfare Trade-Offs Related to Different Laying Hen Housing Systems

Several literature reviews and scientific reports have summarized findings from studies comparing the welfare of hens in different housing systems [81–84], including reports comparing various welfare indicators collected on large-scale commercial farms [85–89]. Generally, all these reports indicate significant trade-offs for different aspects of welfare for hens in different housing systems. Some observed differences likely reflect the lesser degree of experience among farmers housing hens in alternative systems, underscoring the desirability of identifying and disseminating best management practices with respect to improved welfare outcomes in these systems.

Conventional cages generally result in good health, hygiene, and low mortality, but the lack of nests, scratch areas and perches coupled with lack of space impose a high degree of behavioral restriction for hens. Even basic activities such as locomotion, stretching and wing-flapping are significantly constrained in conventional cages [84]. It is also well established that the lack of load bearing exercise reduces bone strength of hens.

Hens housed in cage-free systems with nests, perches, and litter (or free-range with access to the outdoors) have substantial opportunities to perform a greater range of behaviors, but also incur a significantly greater risk of mortality, injury, and poor health [83,84]. Feather pecking and cannibalism are major causes of feather loss, poor welfare, and higher mortality, which are mitigated by the highly controversial practice of beak trimming in many management systems. A recent analysis of data from 3500 commercial flocks in the EU indicated that mortality rates were significantly higher and considerably more variable in non-cage and especially in free-range systems compared to cage systems [89]. In that study, genetic strain of the hen was a significant factor and risks were considerably greater when hens were not beak trimmed. Smothering, another cause of mortality for hens housed in large group sizes, occurs when birds mass together and pile on top of one another [90]. Piling can occur when birds become frightened and panic or when they crowd together at different times of day to access different resources, for example in communal nests [91] or to dust bathe on litter [92].

Enriched (furnished) cages generally result in production, health, and mortality rates that are comparable to or better than conventional cages [21,85,86,88]. With more space and added furnishings, hens in enriched cages have greater opportunities to express motivated behavior than in conventional cages. Nests in cages are generally well-used and result in hens showing more "settled" (satisfied) nesting behavior [93]. Perching behavior and/or the increased space and locomotion in enriched cages results in stronger bones than in conventional cages [94,95]. However, the scratch mats provided in enriched cages do not fully support foraging or dust bathing behavior [85], and this poses one of the biggest challenges for full welfare benefits of enriched cages.

Trade-Offs among Welfare, Economics, Environmental Impact and Human Health

Providing hens with greater space allowances increases both building capital costs and operating costs per dozen eggs produced [96]. Labor costs can be substantially higher, particularly in non-cage systems. Housing hens in alternative systems can also result in reduced efficiencies from lost product (damaged eggs in enriched systems and eggs laid outside of the nest in non-cage systems), higher mortality and reduced feed efficiencies due to energy expenditure related to increased exercise. Poor feather condition from feather pecking increases bird heat loss which must be compensated by higher feed intake or supplemental heating of barns [97]. All these factors can increase cost of production and will also undermine resource efficiency/environmental objectives.

Air quality is also significantly affected by provision of foraging substrate. Higher levels of aerial dust and microbes compromise hen health and worker health and safety [98,99]. Emissions of ammonia and particulate matter from barns may also increase risks to public health [100]. Developing means for mitigating air quality issues inherent in non-cage systems is an important area for future research.

Genetic selection for increased egg production has also come at the expense of hen health and welfare. The demand for calcium to support shell formation of the large numbers of eggs that modern layers produce results in poor skeletal health manifested as osteoporosis, fragile bones, and subsequent risks for bone breakage [101]. Despite advances in nutrition to support the calcium and phosphorus requirements of hens, a major portion of the calcium required for egg formation comes from the hen's skeleton which progressively weakens over her lifetime. Osteoporosis is exacerbated by restricted housing and the lack of load bearing exercise. However, although opportunities for exercise in enriched and non-cage housing do result in stronger bones, the skeleton of the modern laying is still relatively weak and increased freedom of movement also increases risks for bone fractures from collisions with furnishings. The keel bone (sternum) is particularly susceptible to fractures and prevalence rates have been reported to range from 10 to 30% in conventional cages, 20 to 60% in enriched cages and greater than 85% in non-cage systems [102]. Genetic selection for improved bone strength may provide a solution, although there appears to be an inverse relationship between production traits (egg number and shell quality) and bone strength.

Huge gains can be made for not only improving hen welfare and but also reducing loss in resource efficiencies by refining system design (technology) and optimizing nutrition, genetics, and management of alternative housing systems. The large degree of variability found in the literature (e.g., [89]) highlights the potential for non-cage systems to perform well. Research aimed at improving system design, with regards to hens being able to navigate the system without injury, and to readily use nests is essential. Performance testing combined with genome-wide DNA marker analyses are proving valuable tools for genetic selection that balances production and feed efficiency, skeletal integrity and the behavioral traits (i.e., use of nests and reductions in feather pecking) necessary for improving efficiencies and hen welfare in non-cage systems [103].

4. Synthesis and Conclusions

Based on our analysis of historical trends and current regulatory and socio-economic conditions, we identify several major challenges for or threats to the sustainability of the egg industry, along with a variety of alternative strategies to resolve them. We further posit that resolution of these challenges must be supported not only by interdisciplinary research but also by knowledge transfer to enable reconciliation of potential trade-offs between alternative courses of action.

These identified challenges largely relate to shifting consumer and other stakeholder preferences and expectations, which create pressures on egg producers to adapt their practices to produce eggs in particular ways and with particular attributes [104]. It should be underscored that such preferences may not be sufficiently knowledge-based, nor sufficiently informed of potential trade-offs. Foremost among these is the current pressure to transition from cage to non-cage housing systems, based on animal welfare considerations. In a recent survey, US consumers were asked to indicate whether moving from conventional to cage-free housing would have none, positive or negative impact on hen health, hen behavior, natural resource use efficiency, worker health and safety, food safety and egg quality [79]. Well over 50% and upwards of 70% of respondents indicated positive impacts for all the various attributes, respectively, although scientific evidence indicates either no or negative impacts for all of them.

There are distinctive trade-offs with respect to welfare outcomes, with different housing systems variously providing for superior or poorer outcomes, depending on the specific measure considered. At the same time, alternative housing systems imply different resource efficiency/environmental sustainability outcomes. Performance in non-cage systems is currently more variable and somewhat poorer than conventional and enriched colony cage-based production, hence research and development of technology and management strategies to optimize production in alternative housing systems is highly recommended. Perhaps more important is the transfer of research and practical knowledge to farmers transitioning to these new housing and management systems. It is well established that the knowledge, skills, attitudes, and beliefs of the stock people caring for livestock and poultry have profound effects on welfare and production performance of the animals in their care. Training can be used to increase knowledge, change attitudes, and improve performance. While many farmers move from managing simpler to more complex housing systems, a steep learning curve can be expected, with performance and efficiencies continuously improving as they did for conventional cages. Since many farmers may view the transition as being forced upon them, their attitudes about the systems may be negative and may also need to evolve with experience [105,106].

One strategy that has been tested in the UK is the development of comprehensive animal welfare assessments and benchmarking tools combined with feedback and educational materials for farmers. This top-down approach can be delivered within quality assurance schemes or as industry-wide initiatives and involve scientists, veterinarians, government extension specialists and various industry stakeholders. One example is Assurewel [107], a 6-year collaboration among the University of Bristol, RSPCA, and the Soil Association of the UK. As part of this, the FeatherWel project [108] specifically aimed to reduce injurious pecking on UK farms by providing an assessment tool that farmers could use to measure and track feather condition in their flocks together with advice on practical strategies to prevent the problem. Mullan et al. [109] reported that 59% of the 662 UK farmers involved in the project made management changes to improve welfare during the first year of the program, and there was a significant reduction in feather loss from year 1 to year 2. Another approach tested more broadly in the EU as part of the Horizon 2020 EU Research and Innovation program was the Hennovation project [110]. Hennovation also targeted solutions to feather pecking but by developing and disseminating technical innovations using "practice-driven innovation networks" comprising farmers, scientists, veterinarians and farm advisors. Management practices were developed and tested on farms by farmers and the results shared through on-line training, web-based tools and facilitated sessions [111].

In Canada, improvements in both animal welfare and resource use efficiency could be realized by combining efforts for development, knowledge transfer and implementation of best practices for both aspects of sustainability together. The infrastructure and resources of the supply management system could support the framework for such an approach. Currently, a feather scoring system for Canadian egg producers has been distributed through the provincial boards as part of an epidemiological study and benchmarking exercise [112].

The transition to alternative housing systems also presents challenges for Canadian producers from an economic stand-point, which can be partially remedied through the development of cost of production formulas that ensure a fair return to producers within the supply-managed industry. Such a development will serve to reduce risk for both producers and graders, and provide more predictable prices for consumers.

The continued rise of sustainable sourcing as a management consideration for agri-food supply chains, along with generally increasing expectations for accountability and transparency with respect to sustainability management, reporting, and demonstration of improvement, will also likely challenge the egg industry to respond accordingly. From a logistical perspective, it is imperative that farmers be enabled to participate in related initiatives in a non-burdensome manner. This will require development of rigorous sustainability measurement and reporting tools that are both transparent and easy to use. It must be anticipated that multi-criteria sustainability reporting tools that incorporate a combination of environmental, animal welfare, economic, and other indicators will likely highlight the inevitability of trade-offs associated with different management or technology alternatives for the industry. With respect to improvement opportunities, a variety of technology and management options are available that may improve environmental sustainability outcomes, but each must be simultaneously evaluated with respect to potential negative impacts on animal welfare and cost of production.

Another persistent threat to industry sustainability is pressure to dismantle the supply management system that currently governs egg production in Canada. This would likely precipitate consolidation and vertical integration in the industry, with many/most of the currently 1000+ farms disappearing. The opening up of the Canadian market to US egg imports would also likely considerably reduce domestic production.

Loss of supply management would also potentially undermine the ability to orchestrate a smooth transition to alternative housing systems in the Canadian egg industry. There is need for further research to understand potential welfare and sustainability trade-offs of production in supply-managed versus non-supply-managed contexts. Although conventional economic logic would predict higher efficiencies under free market conditions, data suggest similar feed conversion efficiency between US and Canadian flocks but higher mortality rates for US layers (6.7% mortality rate reported for the US for 2010 for conventional cage production, compared to 3.2% for Canada in 2012). However, the concentration of egg production in the US in grain-producing areas creates higher transport-related efficiencies compared to the nationally distributed production (which often necessitates more transportation of feed inputs) characteristic of the Canadian supply-managed industry. At the same time, distributed production reduces the need for more higher impact, refrigerated transport and, potentially, food waste.

There are also a variety of existential threats to the industry that merit consideration. Managing for the future is clearly fraught with uncertainties. Nonetheless, a subset of additional challenges that the egg industry will almost certainly grapple with can be identified with some measure of confidence. First, the growing awareness of the centrality of the agri-food system (in particular, the livestock sector) to many of our most pressing sustainability concerns points towards increased competition for legitimacy in the food space looking forward. Indeed, this is underscored by the rapid emergence and proliferation of sustainable sourcing schemes for agri-food products, largely driven by food processors, retailers, and fast-food chains. Grace of the nutritional value of eggs (currently the reference product for nutritional quality in broadly accepted measures such as Protein Digestibility Corrected Amino Acid Scores (PDCAAS) and Dietary Indispensable Amino Acid Scores (DIAAS)), as well as the relative efficiency of producing poultry compared to swine and ruminants, the egg industry is relatively advantaged at the outset. Nonetheless, it would be both prudent and strategic for the industry to be proactive with respect to positioning in this regard—in particular with respect to actively communicating around the coupled nutritional and sustainability benefits of eggs. This may also prove important considering growing markets for alternatives to animal products, including egg replacement products.

Second, projected growth in food production globally will only serve to exacerbate competition for land, energy, water, and other resources, as well as further disrupt global biogeochemical cycles and systems including the nitrogen and phosphorus cycles and the global climate system. The egg industry will likely be impacted by these changes at multiple levels. This includes emerging regulatory responses which may require operational and technological changes within the industry, potential cost increases for inputs such as feed and energy, greater uncertainty associated with yields along feed input supply chains and increased extreme weather events (and associated heating and ventilation challenges). These phenomena will create sustainability risks and opportunities. It is incumbent on the industry, at the leadership level, to remain attentive and to respond nimbly and effectively to such risks and opportunities as they emerge.

A third wild card for the egg industry is the extent to which disruptive technologies may alter both perceptions and norms regarding the production and consumption of food. For example, emerging technologies such as 3D printing of food will enable precision nutrient delivery, tailored to individual dietary needs and preferences, as well as incorporation on non-traditional protein alternatives such as insect protein. Against this backdrop, positioning of traditional foods such as eggs in the food space based on nutritional attributes may enjoy diminishing returns.

More directly relevant to the egg industry will be growth of the in-vitro biomass and animal product replacements sectors. Since the widely publicized creation of the first lab-grown beef burger in 2013, research, investment, and commercial development in this emerging disruptor sector has burgeoned. Cultured animal products may redefine how we think about, produce, and consume animal protein in the future. While clearly directly competitive with beef, pork, and chicken in the near to medium term, comparable advances in producing cultured egg and dairy substitutes are also likely over time. More directly relevant to the egg industry are vegan and vegetarian egg replacement products, which may have lower resource and environmental impacts as well as eliminate animal welfare-related concerns.

In short, the egg industry faces multiple risks and opportunities from a sustainability perspective in the coming years. Some are immediate and tangible, others less certain with respect to probability, magnitude, and consequence. Supporting the egg industry in successfully navigating this future will require effective research on multiple fronts and, in many cases, interdisciplinary research and knowledge mobilization that draws on the perspectives, tools, and competencies of multiple research areas to recognize and navigate inevitable trade-offs.

Author Contributions: All authors contributed to the conception, research and writing of this manuscript.

Funding: Funds for publication costs are provided by the Natural Sciences and Engineering Research Council of Canada/Egg Farmers of Canada Industrial Research Chair in Sustainability (Pelletier). This research received no external funding.

Acknowledgments: The research chairs held by Pelletier, Doyon, Muirhead and Widowski are each supported by Egg Farmers of Canada.

Conflicts of Interest: The authors declare no conflict of interest.

References

1. Pelletier, N.; Tyedmers, P. Forecasting potential global environmental costs of livestock production 2000–2050. *Proc. Natl. Acad. Sci. USA* **2010**, *107*, 18371–18374. [CrossRef] [PubMed]
2. Foley, J.; Ramankutty, N.; Brauman, K.; Cassidy, E.; Gerber, J.; Johnston, M.; Mueller, N.; O'Connel, C.; Ray, D.; West, P.; et al. Solutions for a cultivated planet. *Nature* **2011**, *478*, 337–342. [CrossRef] [PubMed]
3. Rockström, J.; Williams, J.; Daily, G.; Noble, A.; Matthews, N.; Gordon, L.; Wetterstrand, H.; DeClerck, F.; Shah, M.; Steduto, P.; et al. Sustainable intensification of agriculture for human prosperity and global sustainability. *Ambio* **2017**, *46*, 4–17. [CrossRef] [PubMed]
4. Pelletier, N. Life cycle thinking, measurement and management for food system sustainability. *Environ. Sci. Technol.* **2015**, *49*, 7515–7519. [CrossRef] [PubMed]
5. Pretty, J.; Toulmin, C.; Williams, S. Sustainable intensification in African agriculture. *Int. J. Agric. Sustain.* **2011**, *9*, 5–24. [CrossRef]
6. Sala, S.; Farioli, F.; Zamagni, A. Progress in sustainability science: Lessons learnt from current methodologies for sustainability assessment: Part 1. *Int. J. Life Cycle Assess.* **2013**, *18*, 1653–1672. [CrossRef]
7. McClelland, S.; Arndt, C.; Gordon, D.; Thoma, G. Type and number of environmental impact categories used in livestock life cycle assessment: A systematic review. *Livest. Sci.* **2018**, *209*, 39–45. [CrossRef]
8. Garnett, T. Three perspectives on sustainable food security: Efficiency, demand restraint, food system transformation. what role for life cycle assessment? *J. Clean. Prod.* **2014**, *73*, 10–18. [CrossRef]
9. Kennard, D.C. The trend toward confinement in poultry production. *Poult. Sci.* **1928**, *8*, 23–28.
10. Lee, C.E. *Profitable Broiler Battery and Layer Cage Management*, 6th ed.; Beacon Milling Co., Inc.: Cayuga, NY, USA, 1951.
11. Hanke, O.A.; Skinner, J.L.; Florea, J.H. Chronology: American Poultry History. In *American Poultry History (1823–1973)*; American Printing and Publishing Inc.: Madison, WI, USA, 1974; pp. 680–741.
12. Freidberg, S.E. The Triumph of the Egg. *Comp. Stud. Soc. Hist.* **2008**, *50*, 400–423. [CrossRef]

13. Hartman, R. *Keeping Chickens in Cages: A Description of the Outdoor Individual Cage System of Poultry Management as Developed Mainly in Southern California*; Pacific Poultryman: Palo Alto, CA, USA, 1950; p. 11.
14. Brambell, F. *Report of the Technical Committee to Enquire into the Welfare of Animals Kept under Intensive Livestock Husbandry Systems*; Her Majesty's Stationery Office: London, UK, 1965.
15. Shaver Poultry. The Shaver Heritage Story. Available online: https://www.shaver-poultry.com/en/about-us/history/ (accessed on 12 August 2018).
16. Siegel, P.; Dodgson, J.; Andersson, L. Progress from chicken genetics to the chicken genome. *Poult. Sci.* **2006**, *85*, 2050–2060. [CrossRef] [PubMed]
17. Sharma, B. Poultry production, management, and bio-security measures. *J. Agric. Environ.* **2010**, *11*, 120–125. [CrossRef]
18. Pelletier, N.; Ibarbaru-Blanc, M.; Xin, H. Comparison of the U.S. egg industry's environmental footprint in 1960 and 2010. *Poult. Sci.* **2014**, *93*, 243–255.
19. StatsCan. Production and Disposition of Eggs, Annual. Table: 32-10-0119-01 (Formerly CANSIM 003-0020). Available online: https://www150.statcan.gc.ca/t1/tbl1/en/cv.action?pid=3210011901 (accessed on 12 August 2018).
20. Pelletier, N. Changes in the life cycle environmental footprint of egg production in Canada from 1962–2012. *J. Clean. Prod.* **2018**, *176*, 1144–1153. [CrossRef]
21. Pelletier, N. Life cycle assessment of Canadian egg products, with differentiation by hen housing system type. *J. Clean. Prod.* **2017**, *152*, 167–180. [CrossRef]
22. Mollenhorst, H.; Berentsen, P.; de Boer, I. On-farm quantification of sustainability indicators: An application to egg production systems. *Br. Poult. Sci.* **2006**, *47*, 405–417. [CrossRef] [PubMed]
23. Cederberg, C.; Sonesson, U.; Henriksson, M.; Sund, V.; Davis, J. *Greenhouse Gas Emissions from Swedish Production of Meat, Milk and Eggs 1990 and 2005*; SIK Report 793; Swedish Institute for Food and Biotechnology: Gothenberg, Sweden, 2009.
24. Wiedemann, S.; McGahan, E. *Environmental Assessment of an Egg Production Supply Chain Using Life Cycle Assessment*; Australian Egg Corporation Limited: North Sydney, Australia, 2011.
25. Leinonen, I.; Willaims, A.; Wiseman, J.; Guy, J.; Kyriazakis, I. Predicting the environmental impacts of chicken systems in the United Kingdom through a life cycle assessment: Egg production systems. *Poult. Sci.* **2012**, *91*, 26–40. [CrossRef] [PubMed]
26. Pelletier, N.; Ibarbaru-Blanc, M.; Xin, H. Life cycle assessment of greenhouse gas emissions for contemporary intensive egg production systems in the Upper Midwestern United States. *J. Clean. Prod.* **2013**, *54*, 108–114. [CrossRef]
27. Ghasempour, A.; Ahmadi, E. Assessment of environmental impacts of egg production chain using life cycle assessment. *J. Environ. Manag.* **2016**, *183*, 980–987. [CrossRef] [PubMed]
28. Poultry Keeping on the Modern Farm. *Can. Poult. J.* **1931**, *6*, 85.
29. Decades: A look back through the last 100 years of Canadian Poultry. *Can. Poult. Mag.* **2013**, *100*, 32. Available online: https://www.canadianpoultrymag.com/100th-anniversary/key-developments/decades-13025 (accessed on 7 June 2018).
30. Harkness, D. Memorandum to Cabinet re Eggs. LAC, MG32 B 40 46, B 20S, File 6-3-8 Eggs (General) 1957–1963; 1959. Available online: http://data2.archives.ca/pdf/pdf001/p000000590.pdf (accessed on 30 September 2018).
31. Skogstad, G. *The Politics of Agricultural Policy-Making in Canada*; University of Toronto Press: Toronto, ON, Canada, 1987.
32. Ferries, C. Integration: The basic differences between vertical integration and contractual farming. *Can. Poultrym.* **1961**, *8*, 26–27.
33. CDA's Poultry Director Looks at Marketing Boards. *Can. Poultrym.* **1966**, *1*, 31.
34. Bushfield, E. Should We Socialize the Poultry Industry. *Can. Poultrym.* **1941**, *2*, 15.
35. Canadian Egg Marketing Agency. *First Annual Report of the Canadian Egg Marketing Agency for Presentation to Minister of Agriculture Canada, The Honourable E. F. Whelan, National Farm Products Marketing Council and First Annual Public Meeting*; Canadian Egg Marketing Agency: Ottawa, ON, Canada, 1974; pp. 2–3.
36. Scott, J. The more they lay, the poorer you get. *The Globe and Mail*, 5 January 1972; 25.
37. Elizabeth Thompson Advises: Budget, don't boycott, woman says. *The Globe and Mail*, 2 January 1967; p. 12.

38. Stephens, D. Processors, retailers advise marketing boards: Farmers sow cooperation to reap increased profits. *The Globe and Mail*, 7 March 1967; p. B5.
39. Egg Marketing Agency. *Proposal for a National Egg Marketing Plan for Canada*; RG 17, Vol 3765, File 611.9E1, Part 3; Agric Poultry Others; Egg Marketing Agency: Ottawa, ON, Canada, 1972.
40. Tamini, L.; Doyon, M.; Zan, M. Investment behavior of Canada egg producers: Analyzing the impacts of change in risk aversion and in the variability of eggs prices and production costs. *Br. Food J.* **2018**, *120*, 96–107. [CrossRef]
41. Consumer Price Index. Agriculture and Agri-Food Canada. Available online: http://www.agr.gc.ca/eng/industry-markets-and-trade/canadian-agri-food-sector-intelligence/poultry-and-eggs/poultry-and-egg-market-information/industry-indicators/consumer-price-index/?id=1384971854414 (accessed on 28 September 2018).
42. Per capita Disappearance. Agriculture and Agri-Food Canada. Available online: http://www.agr.gc.ca/eng/industry-markets-and-trade/canadian-agri-food-sector-intelligence/poultry-and-eggs/poultry-and-egg-market-information/industry-indicators/per-capita-disappearance/?id=1384971854413 (accessed on 28 September 2018).
43. Doyon, M.; Bergeron, S. Understanding strategic behaviour and its contribution to hypothetical bias when eliciting values for a private good. *Can. J. Agric. Econ.* **2016**, *64*, 653–666. [CrossRef]
44. Doyon, M.; Bergeron, S.; Crandfield, J.; Tamini, L.; Criner, G. Consumer preferences for improved hen housing: Is a cage a cage? *Can. J. Agric. Econ.* **2016**, *64*, 739–751. [CrossRef]
45. Malone, T.; Lusk, J. Putting the chicken before the egg price: An ex post analysis of California's battery cage ban. *J. Agric. Resour. Econ.* **2016**, *41*, 518–532.
46. Brady, T. Britons Debate Which Comes First: The Egg or the Hen's Happiness. *The New York Times*, 5 August 1953; p. 17.
47. Lewis, H.R. America's Debt to the Hen. *Natl. Geogr.* **1927**, *51*, 453–467.
48. Harrison, R. *Animal Machines: The New Factory Farming Industry*; Vincent Stuart Publishers Ltd.: London, UK, 1964; p. 186.
49. Appleby, M.C. The European Union ban on conventional cages for laying hens: History and prospects. *J. Appl. Anim. Welf. Sci.* **2003**, *6*, 103–121. [CrossRef] [PubMed]
50. Council Directive 1999/74/EC Laying Down Minimum Standards for the Protection of Laying Hens. *Off. J. Eur. Community* **1999**, *L203*, 53–57. Available online: https://eur-lex.europa.eu/LexUriServ/LexUriServ.do?uri=OJ:L:1999:203:0053:0057:EN:PDF (accessed on 15 August 2018).
51. National Farm Animal Care Council. *Brief History of the Codes*; National Farm Animal Care Council (NFACC): Lacombe, AB, Canada. Available online: http://www.nfacc.ca/brief-history-of-the-codes (accessed on 15 August 2018).
52. National Farm Animal Care Council. *Code of Practice for the Care and Handling of Pullets and Laying Hens*; National Farm Animal Care Council (NFACC): Lacombe, AB, Canada, 2017. Available online: http://www.nfacc.ca/poultry-layers-code-of-practice (accessed on 15 August 2018).
53. Mench, J.; Sumner, D.; Rosen-Molina, J. Sustainability of egg production in the United States—The policy and market context. *Poult. Sci.* **2011**, *90*, 229–240. [CrossRef] [PubMed]
54. Kaufman, M. McDonald's tells farmers to treat chickens better. *The Washington Post*. 23 August 2000. Available online: https://www.washingtonpost.com/archive/politics/2000/08/23/mcdonalds-tells-farmers-to-treat-chickens-better/e03b291a-d563-4321-b14e-57ec73338952/?noredirect=on&utm_term=.8fa99c27445b (accessed on 15 August 2018).
55. Shields, S.; Shapiro, P.; Rowan, A. A decade of progress toward ending the confinement of farm animals in the United States. *Animals* **2017**, *7*, 40. [CrossRef] [PubMed]
56. *Retail Council of Canada Grocery Members Voluntarily Commit to Course Cage-Free*; Retail Council of Canada: Toronto, ON, Canada, 2016; Available online: https://www.retailcouncil.org/media/newsreleases/retail-council-of-canada-grocery-members-voluntarily-commit-to-source-cagefree (accessed on 15 August 2018).
57. Gerber, P.; Uwizeye, A.; Shulte, R.; Opio, C.; de Boer, I. Nutrient use efficiency: A valuable approach to benchmark the sustainability of nutrient use in global livestock production? *Curr. Opin. Environ. Sustain.* **2014**, *9–10*, 122–130. [CrossRef]

58. Westhoek, H.; Lesschen, J.; Leip, A.; Rood, T.; Wagner, S.; De Marco, A.; Murphy-Bokern, D.; Pallière, C.; Howard, C.; Oenema, O.; et al. *Nitrogen on the Table: The Influence of Food Choices on Nitrogen Emissions and the European Environment*; Centre for Ecology & Hydrology, European Nitrogen Assessment Special Report on Nitrogen and Food: Edinburgh, UK, 2015.
59. Ghaly, A.; Alhattab, M. Drying poultry manure for pollution potential reduction and production of organic fertilizer. *Am. J. Environ. Sci.* **2013**, *9*, 88–102. [CrossRef]
60. De Meester, S.; Demeyer, J.; Velge, F.; Peene, A.; Van Lanenhove, H.; Dewulf, J. The environmental sustainability of anaerobic digestion as a biomass valorization technology. *Bioresour. Technol.* **2012**, *121*, 396–403. [CrossRef] [PubMed]
61. Billen, P.; Costa, J.; Van der Aa, L.; Van Caneghem, J.; Vandecasteele, C. Electricity from poultry manure: A cleaner alternative to direct land application. *J. Clean. Prod.* **2015**, *96*, 467–475. [CrossRef]
62. Fernandez-Lopez, M.; Puig-Gamero, M.; Lopez-Gonzalez, D.; Avalos-Ramirez, A.; Valverde, J.; Sanchez-Silva, L. Life cycle assessment of swine and dairy manure: Prolysis and combustion processes. *Bioresour. Technol.* **2015**, *182*, 184–192. [CrossRef] [PubMed]
63. Hajjaji, N.; Houas, A.; Pons, M.-N. Thermodynamic feasibility and life cycle assessment of hydrogen production via reforming poultry fat. *J. Clean. Prod.* **2016**, *134*, 600–612. [CrossRef]
64. Molins, G.; Alvarez, M.; Garrido, N.; Macanas, J.; Carrillo, F. Environmental impact assessment of polylactide (PLA)/chicken feathers biocomposite materials. *J. Polym. Environ.* **2018**, *26*, 873–884. [CrossRef]
65. United Nations Environment Program. *Buildings and Climate Change: Summary for Decision-Makers*; United Nations Environment Program Sustainable Buildings Initiative, United Nations Environment Program: Nairobi, Kenya, 2009.
66. Er, D.; Wang, Z.; Cao, J.; Chen, Y. Effect of monochromatic light on the egg quality of laying hens. *J. Appl. Poult. Res.* **2007**, *16*, 605–612. [CrossRef]
67. Hassan, M.; Sultana, S.; Choe, H.; Ryu, K. Effect of combinations of monochromatic LED light color on the performance and behavior of laying hens. *J. Poult. Sci.* **2014**, *51*, 321–326. [CrossRef]
68. Karakaya, M.; Parlat, S.; Yilmaz, M.; Yildirm, I.; Ozalp, B. Growth performance and quality properties of meat from broiler chickens reared under different monochromatic light sources. *Br. Poult. Sci.* **2009**, *50*, 76–82. [CrossRef] [PubMed]
69. Huber-Eicher, B.; Suter, A.; Spring-Stahli, P. Effects of colored light-emitting diode illumination on behavior and performance of laying hens. *Poult. Sci.* **2013**, *92*, 869–873. [CrossRef] [PubMed]
70. Mendes, A.; Paixao, S.; Restelatto, R.; Morello, G.; de Moura, D.; Possenti, J. Performance and preference of broiler chickens exposed to different lighting sources. *J. Appl. Poult. Res.* **2013**, *22*, 62–70. [CrossRef]
71. Hornby, C.; Felix, B. Europe's farmers dump milk to protest low prices. *Reuters*. 16 September 2009. Available online: https://www.reuters.com/article/us-europe-dairy-protests/europes-farmers-dump-milk-to-protest-low-prices-idUSTRE58F4KG20090916 (accessed on 5 August 2018).
72. Hadden, H. Got milk? Too much of it, say US dairy producers. *The Wall Street Journal*. 20 May 2017. Available online: https://www.wsj.com/articles/got-milk-too-much-of-it-say-u-s-dairy-farmers-1495278002 (accessed on 5 August 2018).
73. Dillon, J. Got too much milk? dairy dumping highlights production bottlenecks, Northeast surplus. *Vermont Public Radio*. 9 May 2018. Available online: http://digital.vpr.net/post/got-too-much-milk-dairy-dumping-highlights-production-bottlenecks-northeast-surplus#stream/0 (accessed on 5 August 2018).
74. EU blamed for French Egg Crisis. *Euractiv*. 13 August 2013. Available online: https://www.euractiv.com/section/agriculture-food/news/eu-blamed-for-french-egg-crisis/ (accessed on 5 August 2018).
75. Supply management benefits farmers, rural economy. *Western Producer*. 9 July 2015. Available online: https://www.producer.com/2015/07/supply-management-benefits-farmers-rural-economy/ (accessed on 5 August 2018).
76. Doyon, M.; Bergeron, S. *Economic Impact from Farm Investments in Canada*; Project Report, CIRANO, 2018RP-12; CIRANO: Quebec, QC, Canada, 2018.
77. Norwood, B.; Lusk, J. A calibrated auction-conjoint valuation method: Valuing pork and egg produced under differing animal welfare conditions. *J. Environ. Econ. Manag.* **2011**, *62*, 80–94. [CrossRef]
78. Fraser, D.; Weary, D.; Pajor, E.; Milligan, B. A scientific conception of animal welfare that reflects ethical concerns. *Animal Welfare* **1997**, *6*, 187–205.

79. Ochs, D.; Wolf, C.; Widmar, N.; Bir, C. Consumer perceptions of egg-laying hen housing systems. *Poult. Sci.* **2018**, *97*, 3390–3396. [CrossRef] [PubMed]
80. Weeks, C.; Nicol, C. Behavioural needs, priorities and preferences of laying hens. *World's Poult. Sci. J.* **2006**, *62*, 296–307. [CrossRef]
81. Scientific Panel on Animal Health and Welfare. Welfare aspects of various systems for keeping laying hens. *EFSA J.* **2005**, *197*, 1–23. Available online: https://efsa.onlinelibrary.wiley.com/doi/pdf/10.2903/j.efsa.2005.197 (accessed on 15 August 2018).
82. LayWel. Welfare Implications of Changes in Production Systems for Laying Hens: Deliverable 7.1 Overall Strengths and Weaknesses of Each Defined Housing System for Laying Hens, and Detailing the Overall Welfare Impact of Each Housing System. 2006. Available online: http://www.laywel.eu/web/pdf/deliverable%2071%20welfare%20assessment.pdf (accessed on 15 August 2018).
83. Lay, D.; Fulton, R.; Hester, P.; Karcher, D.; Kjaer, J.; Mench, J.A.; Porter, R. Hen welfare in different housing systems. *Poult. Sci.* **2011**, *90*, 278–294. [CrossRef] [PubMed]
84. Widowski, T.; Classen, H.; Newberry, R.; Petrik, M.; Schwean-Lardner, K.; Yue Cottee, S.; Cox, B. *Code of Practice for the Care and Handling of Pullets, Layers and Spent Fowl: Review of Scientific Research on Priority Issues*; National Farm Animal Council: Lacombe, AB, Canada, 2013. Available online: https://www.nfacc.ca/resources/codes-of-practice/poultry-layers/Layer_SCReport_2013.pdf (accessed on 25 September 2018).
85. Rodenburg, T.; Tuyttens, F.; De Reu, K.; Herman, L.; Zoons, J.; Sonck, B. Welfare assessment of laying hens in furnished cages and non-cage systems: An on-farm comparison. *Anim. Welf.* **2008**, *17*, 363–373.
86. Sherwin, C.; Richards, G.; Nicol, C. Comparison of the welfare of layer hens in 4 housing systems in the UK. *Br. Poult. Sci.* **2010**, *51*, 488–499. [CrossRef] [PubMed]
87. Petrik, M.; Guerin, M.; Widowski, T. On-farm comparison of keel fracture prevalence and other welfare indicators in conventional cage and floor-housed laying hens in Ontario, Canada. *Poult. Sci.* **2015**, *94*, 579–585. [CrossRef] [PubMed]
88. Blatchford, R.; Fulton, R.; Mench, J. The utilization of the Welfare Quality® assessment for determining laying hen condition across three housing systems. *Poult. Sci.* **2016**, *95*, 154–163. [CrossRef] [PubMed]
89. Weeks, C.; Lambton, S.; Williams, A. Implications for welfare, productivity and sustainability of the variation in reported levels of mortality for laying hen flocks kept in different housing systems: A meta-analysis of ten studies. *PLoS ONE* **2016**, *11*, e0146394. [CrossRef] [PubMed]
90. Barrett, J.; Rayner, A.; Gill, R.; Willings, T.; Bright, A. Smothering in UK free-range flocks. Part 1: Incidence, location, timing and management. *Vet. Rec.* **2014**, *175*, 19. [CrossRef] [PubMed]
91. Bright, A.; Johnson, E. Smothering in commercial free-range laying hens: A preliminary investigation. *Vet. Rec.* **2011**, *168*, 512. [CrossRef] [PubMed]
92. Campbell, D.; Makagon, M.; Swanson, J.; Siegford, J. Litter use by laying hens in a commercial aviary: Dust bathing and piling. *Poult. Sci.* **2016**, *95*, 164–175. [CrossRef] [PubMed]
93. Hunniford, M.; Widowski, T. Curtained nests facilitate settled nesting behaviour of laying hens in furnished cages. *Appl. Anim. Behav. Sci.* **2018**, *202*, 39–45. [CrossRef]
94. Jendral, M.; Korver, D.; Church, J.; Feddes, J. Bone mineral density and breaking strength of white leghorns housed in conventional, modified, and commercially available colony battery cages. *Poult. Sci.* **2008**, *87*, 828–837. [CrossRef] [PubMed]
95. Casey-Trott, T.; Korver, D.; Guerin, M.; Sandilands, V.; Torrey, S.; Widowski, T. Opportunities for exercise during pullet rearing, Part II: Long-term effects on bone characteristics of adult laying hens at the end-of-lay. *Poult. Sci.* **2017**, *96*, 2518–2527. [CrossRef] [PubMed]
96. Matthews, W.; Sumner, D. Effects of housing system on the costs of commercial egg production. *Poult. Sci.* **2014**, *94*, 552–557. [CrossRef] [PubMed]
97. Peguri, A.; Coon, C. Effect of feather coverage and temperature on layer performance. *Poult. Sci.* **1993**, *72*, 1318–1329. [CrossRef]
98. David, B.; Mejdell, C.; Michel, V.; Lund, V.; Moe, R. Air quality in alternative housing systems may have an impact on laying hen welfare. Part II—Ammonia. *Animals* **2015**, *5*, 886–896. [CrossRef] [PubMed]
99. Zhao, Y.; Shepherd, T.; Li, H.; Xin, H. Environmental assessment of three egg production systems–Part I: Monitoring system and indoor air quality. *Poult. Sci.* **2015**, *94*, 518–533. [CrossRef] [PubMed]

100. Shepherd, T.; Zhao, Y.; Li, H.; Stinn, J.; Hayes, M.; Xin, H. Environmental assessment of three egg production systems—Part II. Ammonia, greenhouse gas, and particulate matter emissions. *Poult. Sci.* **2015**, *94*, 534–543. [CrossRef] [PubMed]
101. Widowski, T.; Casey-Trott, T.; Hunniford, M.; Morrissey, K. Welfare of laying hens: An overview. In *Achieving Sustainable Production of Eggs Volume 2: Animal Welfare and Sustainability*; Roberts, J.R., Ed.; Burleigh Dodds Series in Agricultural Science (Book 17); Burleigh Dodds Science Publishing: Cambridge, UK, 2017.
102. Harlander-Matauschek, A.; Rodenburg, T.; Sandilands, V.; Tobalske, B.W.; Toscano, M. Causes of keel bone damage and their solutions in laying hens. *World's Poult. Sci. J.* **2015**, *71*, 461–472. [CrossRef]
103. Preisinger, R. Innovative layer genetics to handle global challenges in egg production. *Br. Poult. Sci.* **2018**, *59*, 1–6. [CrossRef] [PubMed]
104. Swanson, J.; Lee, Y.; Thompson, P.; Bawden, R.; Mench, J. Integration: Valuing stakeholder input in setting priorities for socially sustainable egg production. *Poult. Sci.* **2011**, *90*, 2110–2121. [CrossRef] [PubMed]
105. Hemsworth, P. The human factor: Influence on livestock performance and welfare. *Proc. N. Z. Soc. Anim. Prod.* **2000**, *60*, 237–240.
106. Sinclair, M.; Zito, S.; Phillips, C. The Impact of Stakeholders' Roles within the Livestock Industry on Their Attitudes to Livestock Welfare in Southeast and East Asia. *Animals* **2017**, *7*, 6. [CrossRef] [PubMed]
107. Assurewel. Available online: http://www.assurewel.org/aboutassurewel (accessed on 22 August 2018).
108. FeatherWel. Available online: http://www.featherwel.org (accessed on 22 August 2018).
109. Mullan, S.; Szmaragd, C.; Cooper, M.; Wrathall, J.; Jamieson, J.; Bond, A.; Atkinson, C.; Main, D. Animal welfare initiatives improve feather cover of cage-free laying hens in the UK. *Anim. Welf.* **2016**, *25*, 243–253. [CrossRef]
110. Hennovation. Available online: http://hennovation.eu/index.html (accessed on 15 September 2018).
111. Van Dijk, L.; Buller, H.; MacAllister, L.; Main, D. Facilitating practice-led co-innovation for the improvement in animal welfare. *Outlook Agric.* **2017**, *46*, 131–137. [CrossRef]
112. Van Staaveren, N.; Decina, C.; Baes, C.; Widowski, T.; Berke, O.; Harlander-Matauschek, A. A description of laying hen husbandry and management practices in Canada. *Animals* **2018**, *8*, 114. [CrossRef] [PubMed]

sustainability

MDPI

Article

Social Sustainability Assessment of Canadian Egg Production Facilities: Methods, Analysis, and Recommendations

Nathan Pelletier *

Natural Science and Engineering Research Council of Canada (NSERC)/Egg Farmers of Canada (EFC) Industrial Research Chair in Sustainability, Department of Biology/Faculty of Management, University of British Columbia Okanagan, Kelowna, BC V1V 1V7, Canada

* Correspondence: nathan.pelletier@ubc.ca; Tel.: +1-250-807-8245

Received: 19 March 2018; Accepted: 14 May 2018; Published: 16 May 2018

Abstract: A detailed assessment of the "gate-to-gate" social risks and benefits of Canadian egg production facilities was undertaken based on the United Nations Environment Programme/Society of Environmental Toxicology and Chemistry (UNEP/SETAC) Guidelines for Social Life Cycle Assessment. Data were collected via survey from a representative subset of Canadian egg farms, and evaluated against a novel suite of indicators and performance reference points developed for relevance in the Canadian context. The evaluation focused on interactions with four stakeholder groups (Workers; Local Communities; Value Chain Partners; and Society) in eighteen thematic areas. This assessment resulted in a rich and highly nuanced characterization of the potential social risks and benefits attributable to contemporary egg production facilities in Canada. Overall, risks were low and benefits were identified for Local Communities, Value Chain Partners, and Society stakeholder groups, but mixed for the Workers stakeholder group. With respect to the latter, identified areas of higher risk are related, in particular, to a subset of indicators for Working Hours, Equal Opportunities and Fair Salary. As such, the results suggest opportunities and strategies for the Canadian egg industry both to capitalize on its current successes as well as to proactively engage in improving its social sustainability profile. The study also contributes a novel set of social sustainability metrics for use and continued development in the Canadian egg sector as well as other agri-food sectors in Canada and beyond. The inevitable challenge in social life cycle assessment (LCA) of developing non-arbitrary performance reference points for social indicators for which clear norms do not exist, and similarly for establishing non-arbitrary scales and thresholds for differentiating between performance levels, is underscored. A necessary next step with respect to the methods presented herein is for stakeholder groups to carefully consider and refine the performance reference points and characterization thresholds that have been developed, in order to assess their alignment with context-specific social sustainability priorities for this industry, and also to extend the analysis to encompass other value chain stages to enable a full social life cycle assessment.

Keywords: social life cycle assessment; social impact assessment methods; social sustainability; eggs; Canada

1. Introduction

Social life cycle assessment (S-LCA) is the least developed of the three, complementary strands of life cycle assessment, which provide analytical frameworks for life cycle-based sustainability measurement and management [1]. In complement to environmental LCA and life cycle costing, S-LCA is intended to improve decision support through understanding and identifying measures to reduce the social impacts associated with product life cycles [2–5].

The "Guidelines for Social Life Cycle Assessment of Products" [6] provide the first major step towards consensus methodologies for S-LCA. These guidelines provide "a map, a skeleton, and a flashlight" [7] for implementation of S-LCA studies. In this context, "map" refers to the broad nature of the guidelines and their preliminary nature. "Skeleton" reflects that the guidelines provide a foundation on which it is envisaged that stakeholders will engage in fleshing out more specific methodological requirements. "Flashlight" highlights that the guidelines illuminate future research needs [7]. In short, considerable work remains—in particular with respect to developing widely accepted social indicators and impact assessment methods, which may vary by sector/context. Uncertainty with respect to scoring and weighting social sustainability performance for specific indicators remains a widely recognized challenge [8]. Despite current limitations, the Guidelines are nonetheless the most widely employed reference document for S-LCA studies [5,9,10].

Among published studies, researchers have developed and applied disparate indicators and impact assessment methods. Variability in approach to indicator development and usage is reflected in the range of qualitative through semi-quantitative and quantitative indicators that have been employed [11–13]. With respect to impact assessment methods, the range of approaches may be in part attributable to alternative paradigmatic bases for approaching S-LCA, as well as the necessity of identifying context-appropriate reference points against which to assess social sustainability performance [8,10,14,15]. The development of "Methodological Sheets of Subcategories of Impact for a Social LCA" [16] has provided a first reference point for improving consistency in current practice.

Research has also varied in terms of bottom-up versus top-down approaches to social life cycle assessment. Some authors have stressed the importance of context and company-specific assessments (for example, see [17,18]), since social impacts may vary widely between companies producing otherwise similar products. Others have emphasized the desirability of full supply chain social life cycle inventory modelling and impact assessment, which is much more feasible when generic social life cycle inventory data are employed [1]. In light of the general lack of detailed, process-level social life cycle inventory databases to support full supply chain models in many contexts, gate-to-gate studies are a common and necessary step to making such data available.

Use of social life cycle assessment in agri-food contexts has been limited to date, hence availability of published case studies (for example, see [5,13–15,19]) is also limited. Among these, approaches to indicator development and scoring are also varied—reflecting, in part, the highly diverse nature of such contexts [8,13,15].

Eggs and egg products are an important part of the Canadian diet, and the egg industry makes a significant contribution to national agricultural production in Canada [20]. Activities in this industry have potential social risks and benefits for a wide range of stakeholders. To date, research efforts have not been brought to bear in order to understand these potential risks and benefits in support of improved decision making for social sustainability in this industry, nor have appropriate metrics for assessing risks and benefits been advanced.

The general objective of the current analysis was to use the Guidelines for Social LCA [6] and the Methodological Sheets of Subcategories of Impact for a Social LCA [16] as a basis for developing and applying a suite of context-appropriate indicators and metrics to characterize the social risks and benefits specific to activities at Canadian egg production facilities (i.e., what would be the egg production stage in a full S-LCA of egg production). Taking the work of Reveret et al. [19] as inspiration, the assessment used directly collected data from egg farmers in Canada to characterize social risks and benefits for four stakeholder groups (Workers, Local Communities, Value Chain Partners, and Society). Additional work will be necessary to similarly map social risks and benefits that may exist elsewhere along egg value chains in support of a full social life cycle assessment of this industry.

The results of the study include a novel array of social indicators and performance reference points appropriate to the Canadian egg farm context (but also adaptable for application in egg industries elsewhere as well as other agri-food contexts), which should be further refined by egg industry stakeholders looking forward. The results also include a first set of social science-based benchmarks of the social benefits and risks attributable to Canadian egg production for the base year 2012. As such, these results offer insights as to key leverage points for further improving the social sustainability performance of egg production in Canada.

The format of the manuscript broadly mirrors the International Organization for Standardization (ISO) 14044 norm in terms of the four stages of LCA, and the parallel UNEP/SETAC [6] guidelines for social LCA. Section 2 (Methods) describes the Goal and Scope of the study, including details regarding the collection of life cycle inventory (LCI) data, life cycle impact assessment (LCIA) categories and methods, and objectives regarding the interpretation of the study. Section 3 provides detailed life cycle impact assessment (LCIA) results, and Section 4 presents the interpretation and discussion of results.

2. Materials and Methods

2.1. *Goal and Scope of the Study*

This analysis was undertaken in order to develop and apply methods to characterize the social risks and benefits for the stakeholder groups Workers, Local Communities, Value Chain Partners, and Society associated with activities at contemporary Canadian egg facilities in 2012 (i.e., a "gate-to-gate analysis"). The goal was to arrive at a nuanced understanding of the current social sustainability profile of egg facilities in Canada, supported by development of an appropriate suite of social sustainability metrics, using data directly collected from industry participants and in correspondence with the Guidelines for Social Life Cycle Assessment [6].

The system boundary for the assessment of Canadian egg production facilities is specific to the activities of the participating facilities (i.e., a "gate-to-gate" analysis) and attendant interactions with their stakeholders in 2012. The activity variable is 1000 egg facility worker hours (worker hours is commonly used for this purpose in S-LCA), which results in 82.6 tonnes of egg production. Worker hours was chosen because this will enable integration of future work examining other supply chain stages, using this same aggregating activity variable, for a full social life cycle assessment. Moreover, this will also allow expressing supply chain sustainability impacts per unit of eggs produced in parallel to results from E-LCA studies. It is important to note here that, due to the focus on gate-to-gate interactions only, it was not necessary to define aspects such as allocation principles, cut-offs, etc., nor to develop a flow diagram.

2.1.1. Life Cycle Inventory Analysis

All data used for this detailed assessment of Canadian egg facilities were collected directly via survey from industry participants. Supplementary Information A (SI A) presents the survey that was used for data collection purposes. The survey was designed based on the stakeholder groups and indicator categories described in the Guidelines for Social Life Cycle Assessment. Efforts were made to ensure a representative sample, taking into account region of production (province-level) and farm size. Surveys were administered by provincial egg boards and their field officers.

2.1.2. Life Cycle Impact Assessment Methods

According to the Guidelines for S-LCA, social life cycle impact assessment (S-LCIA) involves:

(a) aggregating some inventory data within subcategories and categories; and
(b) making use of additional information, such as internationally accepted levels of minimum performance, to help understand the magnitude and the significance of the data collected in the Inventory phase.

However, in contrast to environmental life cycle impact assessment, standardized, widely accepted social life cycle impact assessment methods that allow aggregation of different kinds of social risk are not yet available for social life cycle assessment. The Guidelines do, however, refer to Methodological

Sheets of Subcategories of Impact for a Social LCA [16], which were consulted in order to identify relevant sub-categories for the current analysis.

The S-LCIA stage of this facility-specific analysis focuses on interpretation and presentation of the social life cycle inventory data that were collected from Canadian egg facilities and associated impact assessment results, with results organized by stakeholder group and social theme following the Guideline recommendations. Four stakeholder groups were considered: Workers; Local Communities; Value Chain Partners; and Society (Consumers were not considered, since the analysis did not extend beyond the egg facility gate). Data and results were further organized within each stakeholder group into subcategories. Table 1 describes the stakeholder groups and subcategory combinations that were evaluated.

Table 1. Stakeholder groups, social themes considered in the social life cycle assessment of Canadian egg facilities.

STAKEHOLDER: WORKERS
Freedom of association and collective bargaining rights
Fair salary
Working hours
Equal opportunities
Health and safety
Social benefits, social security, and job security
STAKEHOLDER: LOCAL COMMUNITY
Access to resources
Safe and healthy living conditions
Respect for indigenous rights
Community engagement
Local employment
STAKEHOLDER: SOCIETY
Public commitment to sustainability issues
Contribution to economic development
Employee training
Corruption
STAKEHOLDER: VALUE CHAIN PARTNERS
Fair competition
Promoting social responsibility
Supplier relationships

Establishing Performance Reference Points

In order to interpret the social sustainability performance of Canadian egg production facilities, it was necessary to identify performance reference points (PRPs) for either ideal or minimum acceptable performance along a spectrum of possible performance levels for the subcategory indicators for each social theme. These reference points were chosen, to the extent possible, relative to relevant international and/or Canadian norms. In many cases, however, clear norms were not identifiable as a basis for assessing performance. In such cases, thresholds for differentiating between performance levels were instead based either on ideal performance (for example, a zero percent incidence rate of an undesirable outcome, or a one hundred percent incidence rate of a desirable outcome) or based on % incidence rate along a continuum from undesirable to desirable outcomes. In some cases, more than one norm was identified as potentially relevant. Here, results are presented for each norm. Supplementary Information B (SI B) provides a series of 17 tables describing impact subcategories and subcategory indicators, and reporting the performance reference points and thresholds used to assess performance for each indicator.

It is important to note here that, where clear norms do not exist, the determination of performance reference points is subjective and somewhat arbitrary. Ideally, these should be defined in stakeholder-specific contexts via deliberative democratic procedures that reflect a shared value structure and objectives. For example, within the egg industry, stakeholders could potentially engage in dialogue and processes aimed at determining thresholds that correspond with shared industry values and objectives regarding social sustainability outcomes. For the purpose of the current analysis, many of the performance reference points were researcher-defined and should hence be viewed as placeholder values only. They might also be viewed, however, as a basis for benchmarking, tracking changes, and goal setting with respect to the sustainability performance of the industry over time relative to the 2012 benchmark results.

To assist in the presentation of performance levels, the S-LCIA results are colour coded so as to indicate the performance levels achieved by the industry in 2012. The system of color codes developed by Reveret et al. [19] to represent performance levels for a social life cycle assessment of the Canadian dairy industry is adopted. This system utilizes a scale spanning four performance levels (Figure 1). These are "risky behaviour," "compliant behaviour," "proactive behaviour," and "committed behaviour".

A **risky behaviour** is a practice that may potentially result in a serious, undesirable consequence for stakeholders. This includes illegal behaviours as well as behaviours that, although not illegal, may generally be viewed as negative.

A **compliant behaviour** is one that meets minimum requirements, norms or expectations. This performance level signifies that the organization is not acting in a risky manner, nor is it making any proactive efforts to achieve a socially desirable outcome.

A **proactive behaviour** is one that indicates some level of initiative towards achieving a more socially desirable outcome than may be legally required.

A **committed behaviour** is one that reflects leadership or clear striving to achieve socially desirable outcomes. This level of behaviour goes beyond marginal improvements by demonstrating considerably better outcomes than would be associated with compliant behaviour.

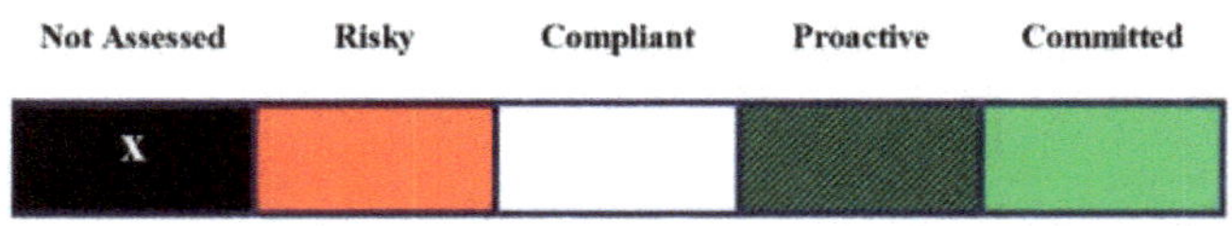

Figure 1. Colour codes employed to characterize performance levels relative to performance reference points.

Performance levels were assigned based on specific thresholds for each indicator as described in SI B. A fixed scale for thresholds was not feasible—rather, thresholds deemed appropriate to differentiating performance levels for each indicator were developed. As with the assignment of PRPs in the absence of clear norms, the assignment of threshold levels is inevitably somewhat arbitrary unless defined in stakeholder-specific contexts via deliberative democratic procedures that reflect a shared value structure and objectives. This is a widely recognized challenge in S-LCA [6,16].

The methodology for defining the performance reference points and thresholds for each stakeholder group/social theme/sub-category indicator are schematically described in Figure 2. Detailed descriptions of the category indicator-specific methods and rationales are provided in SI B (Tables S1–S17).

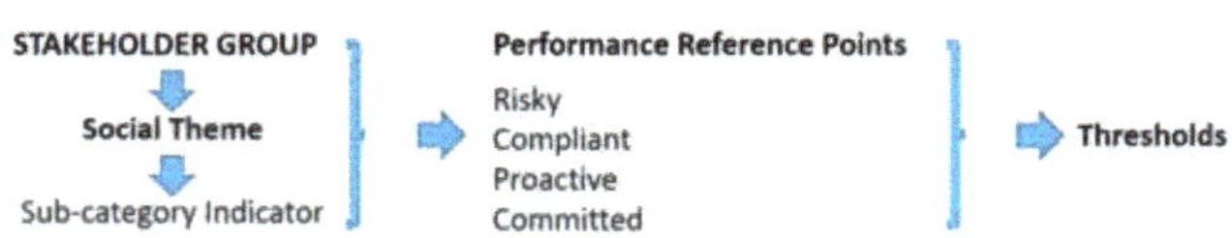

Figure 2. Schematic representation of the relationships between performance reference points/thresholds for each stakeholder group/social theme/sub-category indicator combination utilized in the social life cycle impact assessment.

It should also be noted that not all performance levels are relevant for all indicators. For example, if no clear norm for an indicator can be identified, then a "risky" behavior cannot be assessed. Nonetheless, it is still useful (for example, for the purpose of benchmarking or sectoral goal setting) to differentiate between the performance of companies, which may engage in socially desirable behaviours to varying degrees, despite the lack of any clear norm or requirement.

To facilitate an accessible presentation of results in Figures 3–7, in some cases, indicator results are aggregated if the same performance score/colour code is assigned for multiple related indicators for a subcategory that can be expressed under a common indicator heading. Aggregation and weighting are not otherwise applied.

2.1.3. Interpretation and Recommendations

The Life Cycle Interpretation phase requires systematically reviewing the results of the Life Cycle Impact Assessment phase with respect to the research questions. Here, the interpretation phase focuses in particular on:

- elucidating the key social risks and benefits for stakeholders that are specifically associated with activities that occur at Canadian egg production facilities.
- identifying priority areas for interventions to improve the social license of Canadian egg producers, either via communications regarding the social benefits associated with the egg industry, or commitments to monitor and seek to improve the social sustainability profile of Canadian egg production with respect to specific social risk areas.
- identifying priority areas for further data collection and research.
- highlighting weaknesses of the current study and recommending areas for further research and methods development.

3. Results

3.1. Life Cycle Inventory Results

In total, usable surveys were collected from 59 egg facilities together representing 357,286 worker hours in Canada in 2012. The number of part- and full-time workers employed in these facilities was 248, with an average of 1440.7 h worked per worker per year. For detailed social life cycle inventory results for each indicator, see SI C Tables S18–S35, which present the social life cycle inventory data, expressed as worker hour-weighted average results, as well as low and high reported values where relevant.

3.2. Life Cycle Impact Assessment Results

Figures 3–7 present summary social life cycle impact assessment results for each stakeholder group, social theme and sub-category indicator. For detailed social life cycle impact assessment results for each sub-category indicator, please see SI D Tables S36–S53.

STAKEHOLDER: WORKERS

Figures 3 and 4 provide a summary of LCIA results for the Workers stakeholder category.

Freedom of Association and Collective Bargaining Rights

No employers indicated that employment at their facilities is conditioned by any restrictions on the right of employees to collective bargaining (compliant). However, for only 28% of worker hours did respondents indicate that employees are free to join unions of their choice. Moreover, a 0% union representation rate was reported and it would also appear that participation of employees in organizational planning is not currently the industry norm. Only for 10% of worker hours was it indicated that employees have access to neutral, third-party dispute resolution. Based on the current

Canadian average rate of union representation a risky behaviour is assessed. In light of current lack of norms but low levels of provision, compliant behaviours are assessed with respect to employee participation, codified minimum notice periods regarding operational changes, and employee access to neutral, binding, and independent dispute resolution procedures. The Canadian egg industry might, however, consider options for improving employee representation and participation (SI D Table S36).

Fair Salary

Minimum wage standards were exceeded across the Canadian egg facilities surveyed for both the lowest paid and average employee, hence, on this basis, a "proactive" score is assigned. The majority (77%) of workers receive at least 10% more than the minimum wage, 41% earn at least 50% more, and 26% earn at least 100% more than the minimum wage. Relative to norms for Canadian agricultural workers, wages for the lowest paid employees at some Canadian egg facilities might be considered risky in that they are below this average. Average employee wages are, however, similar to the average for Canadian agricultural workers and hence a compliant score is assessed for this indicator.

Another area for improvement is with respect to the regularity and documentation of pay. For only 57% of worker hours did respondents indicate that regular documentation of pay was provided. Here, a risky behaviour is assessed. There were no reported complaints regarding deductions on wages, hence a compliant behaviour is assessed for this indicator (SI D Table S37).

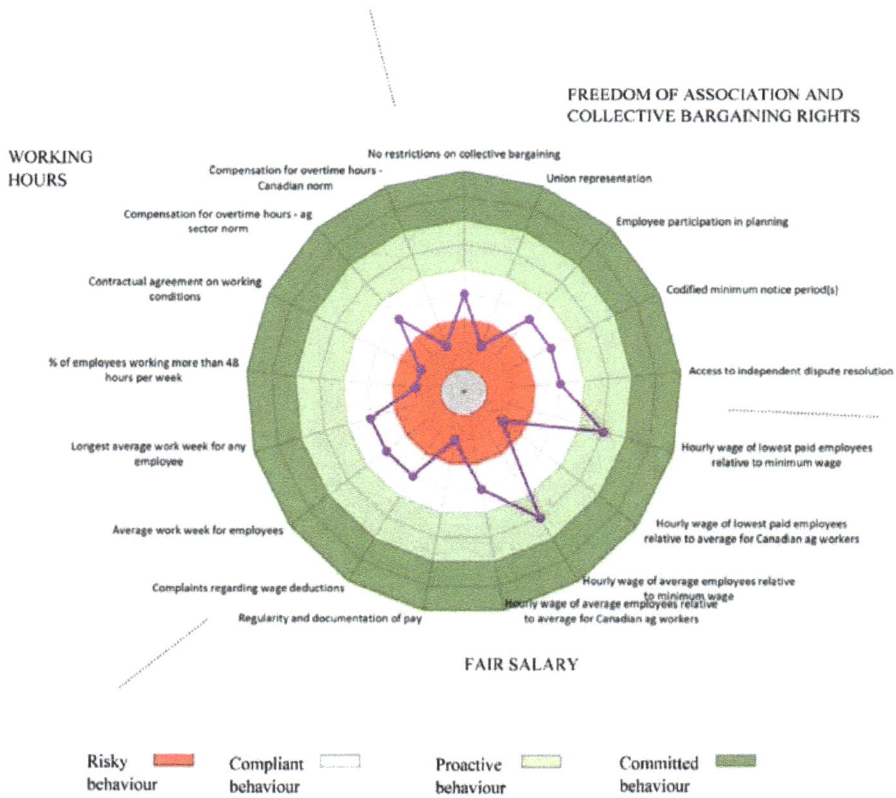

Figure 3. Social life cycle impact assessment scores for Canadian egg facilities in 2012 with respect to the stakeholder group "Employees" (freedom of association and collective bargaining rights, fair salary, working hours).

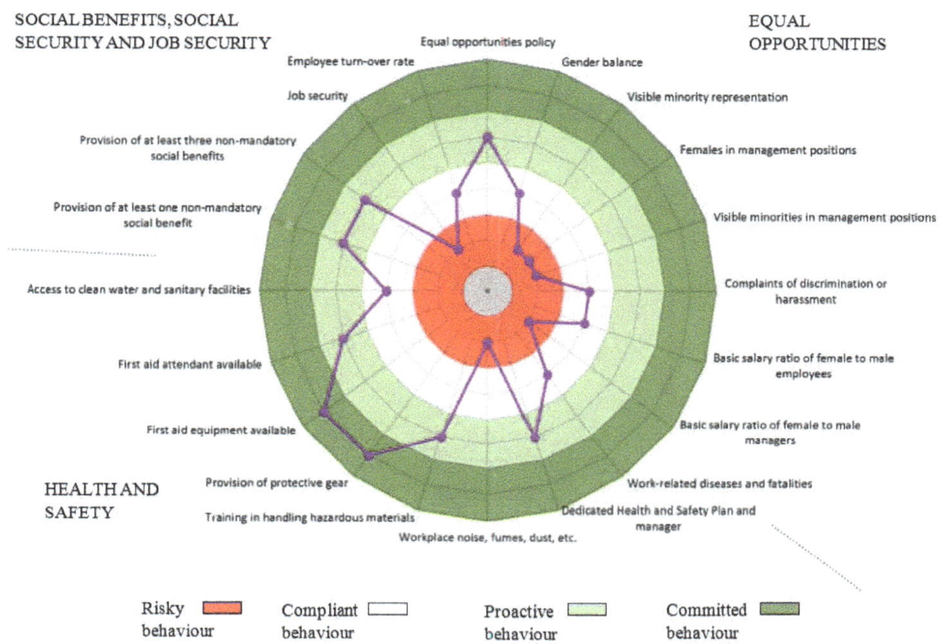

Figure 4. Social life cycle impact assessment scores for Canadian egg facilities in 2012 with respect to the stakeholder group "Employees" (equal opportunities, health and safety, social benefits, social security and job security).

Working Hours

The weighted average work week for employees in Canadian egg facilities in 2012 was 38.83 h, which is slightly higher than the Canadian average but much lower than the ILO maximum of 48 h (compliant behaviour). This is, however, likely consistent with working hours elsewhere in the agricultural sector. The longest average work week is 47 h (compliant behaviour). Respondents also indicated, however, that more than 10% of working hours correspond to work weeks in excess of 48 h. Here, a risky behaviour is assessed. There is also a very low incidence of contractual agreements regarding working hours and overtime compensation (risky behavior).

Although some employers report paying their employees a 50% overtime premium, overall, the survey data indicate a low rate of overtime pay (5.8% on average) in Canadian egg facilities. This seemingly corresponds to the norm for the Canadian agriculture sector (compliant behaviour). However, if the norm for Canadian workers as a whole is taken as the reference point, then a risky behaviour is assessed (SI D Table S38).

Equal Opportunities

Based on reported data, employees on Canadian egg farms are largely Caucasian (97%) and male (63%), with an even age class distribution. Twenty-eight percent of respondents reported having a formal equal opportunities policy in place for their facilities (proactive behaviour). In many cases, lack of such a policy may simply reflect that the facilities are small family/owner operated facilities with no other employees. However, if equality of representation of males and females is taken as the performance reference point (compliant behaviour), or a proportionate degree of inclusion of visible minorities (risky behaviour), then there may, indeed, be space for improvement in this regard. This is similarly true for representation of females and visible minorities in management positions (risky behaviour).

Considering salary equality, it would appear that male and female employees receive equal remuneration (compliant behaviour). However, female managers apparently receive lower salaries (85%) relative to male managers (risky behaviour) (SI D Table S39).

Health and Safety

According to Employment and Social Development Canada (ESDC) [21], in Canada "one in every 68 employed workers in 2010 was injured or harmed on the job and received workers compensation as a result". Although 2.7 works days were lost, on average, due to occupational accidents, injuries or illness in Canadian egg production facilities, only 0.14 WCB claims were filed per facility industry-wide, which is well below the Canadian average. No medically-diagnosed, work-related diseases or fatalities were reported (compliant behaviour).

Respondents indicated that 53% of worker hours occurred in workplaces characterized by high levels of noise, fumes, or dust (risky behaviour). While all employees are provided with appropriate protective gear when required to handle hazardous materials (committed), only half are reportedly trained in handling such materials (proactive).

Employees generally have access to clean water and sanitary facilities (compliant behavior) and, in most cases (82% of worker hours—committed behaviour) first aid equipment is maintained and available to employees. A trained, designated first aid attendant was available for 62% of the reported worker hours (proactive behaviour). A dedicated Health and Safety plan is in place for 45% of the worker hours (proactive behavior), and a dedicated Health and Safety manager for 31% of the worker hours (proactive behaviour).

Overall, it would appear that health and safety incidents were low in 2012, but that the Canadian egg industry would benefit from development and implementation of industry-wide standards with respect to health and safety plans, management, and training. The average surveyed Canadian egg facility invested $1762 in health and safety measures in 2012 (SI D Table S40).

Social Benefits, Social Security, and Job Security

It would appear that the provision of social benefits is not consistent within the Canadian egg industry. Health insurance benefits are provided to less than half of employees, dental and retirement benefits to roughly one third, and child care and/or maternity/paternity benefits to less than one employee in seven. Across these five categories, the average social benefit provision level is 30%. It should be noted here that some employers in Canadian egg facilities do offer full social benefits to their employees, while others offer none.

With respect to employment contracts, only 1/4 of worker hours were subject to formal employment contracts in 2012. Of these, roughly 2/3 were full time contracts, and 1/3 were part time contracts. The majority of reported worker hours were not subject to an employment contract. This indicates a low level of job security. Based on a reference level of a 100% formal contract rate, a risky behavior is assessed.

According to the Conference Board of Canada (CBC) [22], the voluntary employee turn-over rate in Canada in 2012–2013 was 7.3%, and the involuntary turn-over rate was 3.7%. Employee turn-over rates in Canadian egg facilities were comparatively low (4.1% overall). Half of employees who left jobs at Canadian egg facilities in 2012 quit, while one quarter were dismissed. Compared to the Canadian average, a compliant behaviour is assessed (SI D Table S41).

STAKEHOLDER: LOCAL COMMUNITY

Figure 5 provides a summary of LCIA results for the Local Community stakeholder category.

Access to Resources

The total expenditures of egg facilities on infrastructure with mutual community access and benefit in 2012 were very low (compliant behaviour assessed). However, some survey respondents spent as much as $26,000 on voluntary, charitable donations and investments in their community, with an average of $2205 per facility (proactive behaviour assessed) (SI D Table S42).

Safe and Healthy Living Conditions

Only one accident in a local community was reported across surveyed facilities for 2012, hence the industry average accident rate is very low. This single accident is taken as an outlier, hence a compliant score is assigned. However, while many facilities did not receive any complaints from the local community regarding nuisance issues, some facilities received as many as two complaints. Overall, 0.4 complaints were received per facility in 2012. Relative to an ideal scenario performance reference point of 0 complaints, a risky behaviour is assessed.

In light of the complaint rate, it would seem prudent to having in place formal protocols to maintain living conditions, minimize risk, and respond to grievances in the community. Twenty-seven percent of facilities reported having formal protocols in place to maintain safe and healthy living conditions, while 32% have formal protocols to minimize risk and 17% have formal protocols to respond to grievances (proactive behavior). No fines for infringements at the local level were reported (compliant behavior).

One current approach to ensuring safe and healthy living conditions in local communities and minimizing risk is the implementation of an environmental farm management plan. Two thirds of worker hours occurred at facilities having an environmental farm management plan in 2012. Here, proactive behavior is assessed (SI D Table S43).

Respect for Indigenous Rights

While the survey data indicate that no complaints were received by Canadian egg facilities from First Nations community members in 2012, it is notable that no meetings were held with First Nations members. Also notable is that only 2% of worker hours were reported to occur in areas where land rights conflicts exist (risky behavior), which is unlikely to be accurate, and that only 2% of worker hours occurred in facilities having formal policies in place to protect the rights of indigenous community members. As no norms exist in this regard, compliant behaviors are assessed. It is nonetheless recommended that proactive behavior be encouraged in this domain (SI D Table S44).

Community Engagement

Only 7% percent of worker hours in 2012 occurred in egg facilities having formal policies on community engagement in place (compliant behavior). Reported attendance of meetings with stakeholder groups was variable, with some facilities reporting having attended as many as six meetings. On average, reporting facilities devoted 9 volunteer hours to community initiatives in 2012. Here, proactive behaviours are assessed (SI D Table S45).

Local Employment

Although less than 1/3 of Canadian egg facilities report having a formal policy regarding local hiring (proactive behavior), the local hiring rate was 87% in 2010 (committed behavior). Similarly, the majority share (64%) of goods and services purchased by egg facilities was locally sourced (proactive behavior) (SI D Table S46).

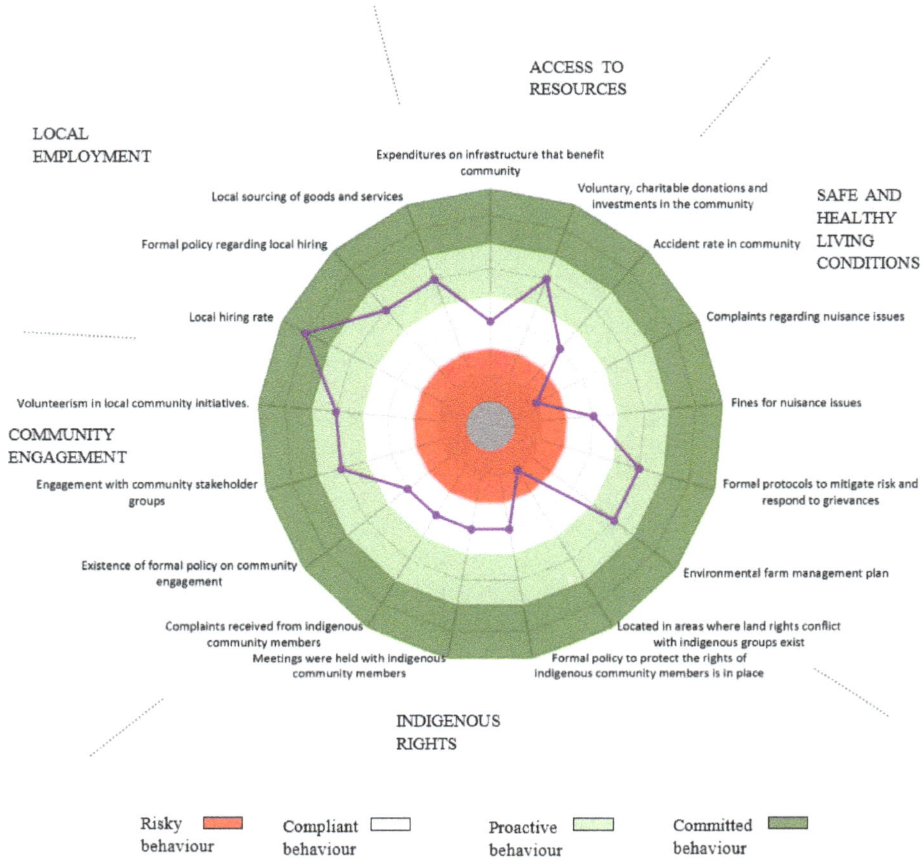

Figure 5. Social life cycle impact assessment scores for Canadian egg facilities in 2012 with respect to the stakeholder group "Local Community".

STAKEHOLDER: VALUE CHAIN PARTNERS

Figure 6 provides a summary of LCIA results for the Value Chain Partners stakeholder category.

Fair Competition

Although no facilities reported having documentation of procedures to prevent engaging in or being complicit in anti-competitive behaviours (compliant), no legal actions, fines or complaints were reported for 2012 (compliant) (SI D Table S47).

Promoting Social Responsibility

Three percent of worker hours occurred in facilities that report adhering to a specific code of conduct regarding the protection of human rights (compliant behavior). No supplier audits for environmental or social responsibility were reported (compliant behavior). Canadian egg facilities surveyed for this study participated in a weighted average of 0.16 initiatives to promote value chain environmental or social responsibility, with a range from 0 to 5 initiatives per facility (proactive behavior). In light of this low activity level, it would seem that there is opportunity here for initiatives to enhance the promotion of social responsibility in the Canadian egg industry (SI D Table S48).

Supplier Relationships

Overall, it appears that Canadian egg facilities maintain good relationships with their value chain partners. No complaints were received from suppliers with respect to coercive communications or insufficient lead time in 2012, and a weighted average of 0.01 complaints were received (representing one facility with two complaints, which is taken as an outlier) with respect to timeliness of payments. On the basis of this very low level, a compliant behavior is assessed for the industry as a whole (SI D Table S49).

STAKEHOLDER: SOCIETY

Figure 7 provides a summary of LCIA results for the Society stakeholder category.

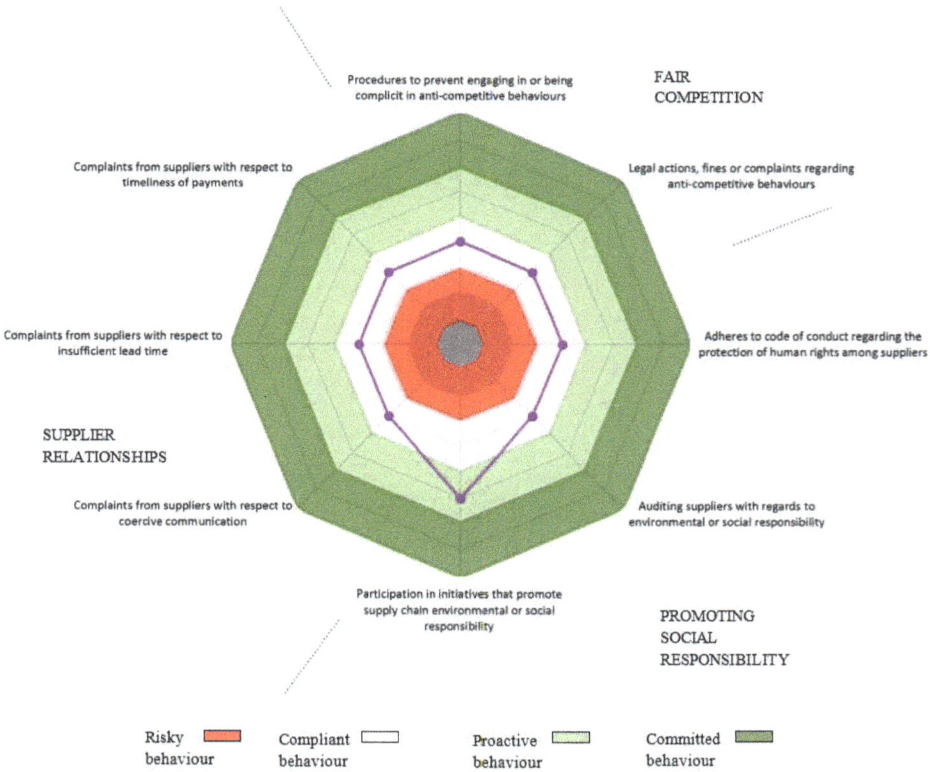

Figure 6. Social life cycle impact assessment scores for Canadian egg facilities in 2012 with respect to the stakeholder group "value chain partners".

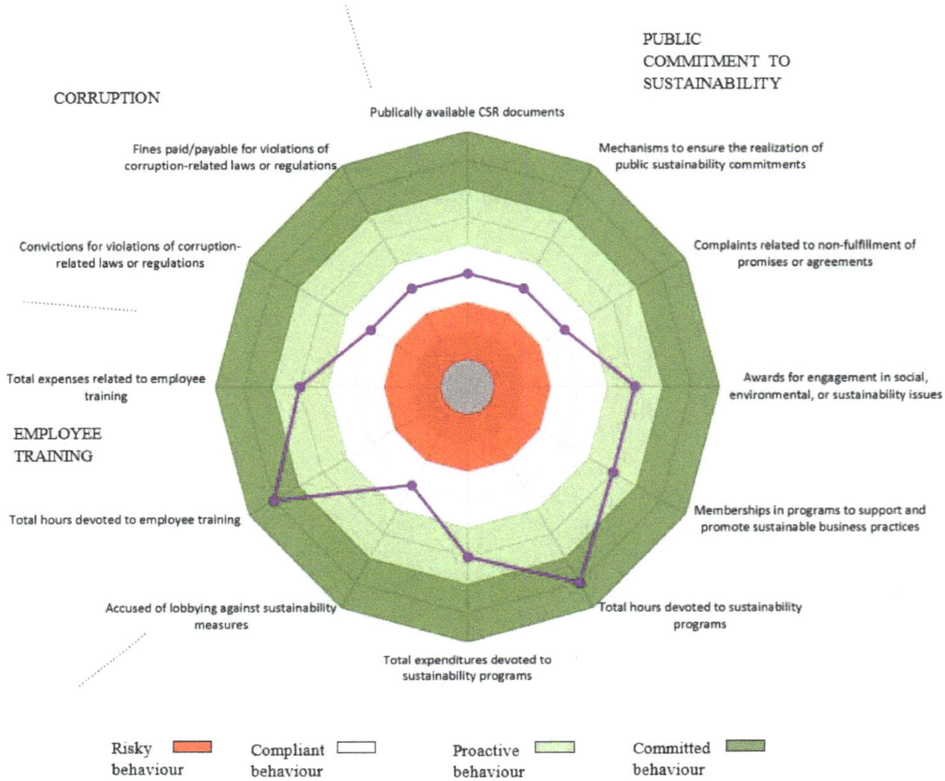

Figure 7. Social life cycle impact assessment scores for Canadian egg facilities in 2012 with respect to the stakeholder group "Society".

Public Commitment to Sustainability

Among those surveyed, only 8% of worker hours occurred in facilities having publicly available CSR or other documents that communicate commitments to sustainability issues (compliant behaviour). For 1%, specific mechanisms are in place to ensure the realization of such commitments (compliant behaviour). On this basis, it would appear that documentation and communication of sustainability commitments is not currently widespread in the Canadian egg industry.

Outside of formal commitments and communications, however, some egg facilities are, indeed, engaging in sustainability initiatives. Facilities report an average of over 10 h devoted to sustainability programs in 2012 (committed behavior). Some facilities were active members in programs to support sustainable business practices (proactive behavior), and others received awards for engagement in sustainability issues (proactive behavior). The average expenditure on sustainability programs was $803 (proactive behavior). No complaints were reported with respect to non-fulfillment of sustainability commitments or activities to undermine sustainability measures (compliant behaviour).

In sum, it would appear that monitoring and communicating with respect to sustainability issues is not currently the norm in the Canadian egg industry, but that some facilities are already actively engaged in this domain. A significant opportunity may hence exist to enhance the social license of Canadian egg facilities through implementation and documentation of formalized sustainability commitments (SI D Table S50).

Contribution to Economic Development

The scale of these contributions varies widely between egg facilities, reflecting the commensurately broad range in size of operations. For example, total revenue ranges from $260,000 to $9,464,000, with an industry average of $2,508,000 among reporting facilities. On average, participating Canadian egg facilities paid $248,000 of wages in 2012, along with $78,000 in taxes and reported an average of $1,758,000 in total expenses (SI D Table S51).

Employee Training

Significant resources were dedicated to employee training in Canadian egg facilities in 2012. On average, workers were provided with 53 h of training in 2012 (committed behaviour), and total expenses related to employee training averaged $529 per facility (proactive behaviour) (SI D Table S52).

Corruption

Two convictions for violations of corruption-related laws or regulations were reported by a single facility, with an accompanying fine of $500. This single facility report is therefore taken as an outlier, as it would appear that such violations are not commonplace within the Canadian egg industry (hence compliant scores are assigned) (SI D Table S53).

4. Discussion

The results of this assessment provide a rich and highly nuanced characterization of the potential social risks and benefits attributable to contemporary egg production facilities in Canada. They provide a series of first benchmarks for the Canadian egg industry. They also highlight where and to what extent stakeholders are currently benefiting from positive social contributions made by the industry—for example, through local job creation, high levels of employee training, provision of protective gear and first aid equipment for employees, and time devoted to sustainability programs (areas where facilities demonstrated "committed" behaviors). These contributions go beyond what is otherwise required by law. The results also point towards areas of potential social risk for some stakeholders (such as the low incidence of formal employment contracts and union representation for workers, low levels of overtime compensation, or sub-optimal workplace diversity and equality). As such, the results suggest opportunities and strategies for the Canadian egg industry both to capitalize on its current successes as well as to proactively engage in improving its social sustainability profile.

For the "Workers" stakeholder category, the directly collected data confirmed the very low level of representation and collective bargaining for employees of egg facilities. The representation rate of 0% was much lower than the Canadian average of 31%, but likely in-line with current norms for the Canadian agricultural sector. Similarly, employee salaries appear to be consistent with Canadian agricultural worker norms, but lower than those of the average Canadian. Most employees are, in fact, remunerated at levels significantly higher than the minimum wage, but a small fraction of employees receive full- or part-time minimum wage salaries. Province-by-province assessments of employee wages relative to regional living wage estimates are recommended in order to ensure that the lowest paid employees are not at risk. Compensation for overtime pay should also be monitored, and industry-wide norms developed. On average, worker hours are within acceptable norms although a small fraction of employees work in access of 48 hours per week. Overall, improvements with respect to formal, contractual agreements for employees regarding working conditions are recommended.

With respect to equal opportunities in the Canadian egg facility workplace, the majority of employees are currently Caucasian males. Visible minorities have a very low level of representation in this workforce. For general employees, salaries are equitable for males and females. However, the data collected for 2012 suggest a disparity in pay for female compared to male managers of roughly 15%. Additional research and possible mitigation strategies may be desirable to increase diversity and ensure equality for employees.

Health and safety conditions for workers in Canadian egg facilities appear to be very good. Although facilities are often characterized by high levels of dust and/or noise, employers report high levels of worker training, use of protective gear, and access to first aid equipment and sanitary facilities. Many facilities report the use of dedicated health and safety plans, as well as designated first aid attendants. No medically-diagnosed, workplace-related diseases or fatalities were reported for 2012.

Provision of social benefits to workers varies widely between egg facilities. Many facilities report providing at least one non-mandatory social benefit and some more than three. Others provide no additional benefits to employees. This suggests opportunities for the development of norms within the Canadian egg industry to promote consistent provision of non-mandatory social benefits in order to improve conditions for employees. Improving job security through the implementation of formal employment contracts is also recommended, as current usage of contractual agreements is low.

For the stakeholder group "Local Community," it would appear that Canadian egg facilities are often proactively engaged in making meaningful contributions to the communities in which they operate. Egg facilities largely employ local community members, and source a large fraction of inputs from local businesses. Most facilities reported some level of voluntary, charitable donations or investments in their communities, with an average of $2205 spent on local community initiatives per facility in 2012. Participation in community initiatives is also high. In addition, many facilities report implementation of an Environmental Farm Management Plan, which serves to reduce risk of nuisance issues, as well as the existence of formal protocols to mitigate risk for local communities and respond to grievances. The overall accident rate in local communities as a result of egg facility activities was very low. One area of potential social risk relates to interactions with indigenous communities. Although many egg facilities are situated in communities proximate to First Nations communities, only 2% reported operating in areas where land rights conflicts exist, and reported levels of proactive engagement with First Nations communities were low. Formalizing "good neighbour" policies and outreach may be desirable. This could potentially be paired with initiatives to increase workplace diversity.

Management decisions and activities similarly impact on society at large, and may result in either societal risks or benefits. Demonstrating commitment to and actions consistent with contributing positively to society is fundamental to the concept of corporate social responsibility (CSR). CSR reporting, including transparent mechanisms for measuring, monitoring, goal setting and follow-up, is increasingly important to maintaining social license, as well as accessing emerging market opportunities associated with sustainability objectives.

Many Canadian egg facilities are already taking steps to improve and promote the sustainability of their practices. Facilities reported an average of 10 h devoted to sustainability initiatives in 2012, and related expenses of roughly $800 per facility. Development and reporting of formal CSR documents is, however, relatively uncommon. An industry-wide initiative to support operators in developing and implementing CSR strategies is strongly encouraged. Indeed, in light of increased attention by both commodity groups and large retailers to developing and implementing sustainability measurement, reporting and certification requirements, it is strongly recommended that the Canadian egg industry become proactive in this respect. This will require the support and cooperation of both producers and their representative bodies in order to develop and implement industry-wide strategies and initiatives.

Social life cycle assessment is a relatively new field of research, and methodological development is on-going. To date, very few social LCA studies of food products have been reported (for example, see [11,13,14,19,23,24]). Varied approaches have often been employed [11,12] and, given the diversity of agri-food production contexts, the feasibility and desirability of consistency between studies is debatable.

The current analysis represents only a partial (i.e., gate-to-gate) evaluation of the potential social risks and benefits attributable to the Canadian egg industry. Future work should expand the analysis to encompass value chain stages and stakeholder interactions both upstream and downstream of Canadian egg facilities. While considerable efforts were invested to ensure appropriate representation across regions of production (provinces) and farm size, the sample nonetheless represents a relatively small subset of egg production facilities in Canada. In this respect, "scaling up" results is challenging, and collecting larger data samples wherever feasible in such analyses is clearly desirable. The survey questions (and related indicators that can be assessed) should also be periodically revisited so as to ensure that all relevant social sustainability issues are taken into consideration—in particular given that conditions may change over time, and the suite of social sustainability issues deemed appropriate for inclusion in such analyses will likely continue to evolve. Moreover, while the impact assessment methods that were developed here provide a valuable example of how social risks and benefits might be assessed in egg production or for other food production activities, it is important to note that defining performance reference points for social life cycle impact assessment where clear norms do not exist is challenging and inevitably somewhat arbitrary. This is similarly true with respect to identifying appropriate thresholds for distinguishing between levels of "sustainability performance" for each indicator considered, even where norms do exist. Multiple, competing bases are possible. For this reason, it is essential that methods and data are transparently presented so as to enable identification of potential real or perceived bias. Moreover, it may be preferable to present results at the indicator level (as in the current study) rather than aggregating results across indicators, since such aggregation requires weighting (introducing an additional level of uncertainty and potential bias) and may also bias interpretation if numerous "easy to achieve" norms outweigh a limited number of more important social sustainability performance indicators.

Clearly, a desirable next step, in the extension of this methods development and analysis exercise, is for the Canadian egg industry to engage in a multi-stakeholder dialogue in order to define mutually agreed upon, clear thresholds for performance levels as well as goals for social sustainability performance in this industry [13]. The novel set of social sustainability metrics developed here can therefore provide a useful starting point for continued development and application in the Canadian egg sector and may be usefully adapted for application in other agri-food sectors in Canada and beyond.

Supplementary Materials: The following are available online at http://www.mdpi.com/2071-1050/10/5/1601/s1, SI A—Social Life Cycle Inventory Survey, SI B—Social Life Cycle Impact Assessment Indicators and Performance Reference Points, SI C—Detailed Social Life Cycle Inventory Results, and SI D—Detailed Social Life Cycle Impact Assessment Results.

Funding: This research was funded by Egg Farmers of Canada. Publication was supported through the NSERC/Egg Farmers of Canada Industrial Research Chair in Sustainability.

Conflicts of Interest: The author declares no conflict of interest. The funding sponsors had no role in the design of the study; in the collection, analyses, or interpretation of data; in the writing of the manuscript, and in the decision to publish the results.

References

1. Pelletier, N.; Sala, S.; Ostaglu, E.; Rosenbaum, E.; Vasta, A.; Benoit-Norris, C.; Norris, G. Social sustainability in trade and development policy. *Int. J. Life Cycle Assess.* **2018**, *23*, 629–639. [CrossRef]
2. Jorgensen, A.; Le Bocq, A.; Nazarkina, L.; Hauschild, M. Methodologies for social life cycle assessment. *Int. J. Life Cycle Assess.* **2008**, *13*, 96–103. [CrossRef]
3. Jorgensen, A.; Lai, L.; Hauschild, M. Assessing the validity of impact pathways for child labour and wellbeing in social life cycle assessment. *Int. J. Life Cycle Assess.* **2010**, *15*, 5–16. [CrossRef]
4. Jorgensen, A. Social LCA—A way ahead? *Int. J. Life Cycle Assess.* **2013**, *18*, 296–299. [CrossRef]
5. Zamagni, A.; Feschet, P.; De Luca, A.; Iofrida, N.; Buttol, P. Social Life Cycle Assessment: Methodologies and Practice. In *Sustainability Assessment of Renewables-Based Products: Methods and Case Studies*; Dewulf, J., De Meester, S., Rodrigo, A., Eds.; Wiley: New York, NY, USA, 2015; Available online: http://ca.wiley.com/WileyCDA/WileyTitle/productCd-111893394X.html# (accessed on 5 January 2017).
6. UNEP/SETAC. Guidelines for Social Life Cycle Assessment of Products. In *The UNEP/SETAC Life Cycle Initiative*; UNEP/SETAC: Paris, France, 2009.

7. Benoit-Norris, C.; Norris, G.; Valdivia, S.; Ciroth, A.; Moberg, A.; Bos, U.; Prakash, S.; Ugaya, C.; Beck, T. The guidelines for social life cycle assessment of products: Just in time! *Int. J. Life Cycle Assess.* **2010**, *15*, 156–163. [CrossRef]
8. Do Carmo, B.; Margni, M.; Baptiste, P. Addressing uncertain scoring and weighting factors in social life cycle assessment. *Int. J. Life Cycle Assess.* **2017**, *22*, 1609–1617. [CrossRef]
9. Wu, R.; Yang, D.; Chen, J. Social life cycle assessment revisited. *Sustainability* **2014**, *6*, 4200–4226. [CrossRef]
10. Iofrida, N.; De Luca, A.; Strano, A.; Gulisano, G. Can social research paradigms justify the diversity of approaches to social life cycle assessment? *Int. J. Life Cycle Assess.* **2018**, *23*, 464–480. [CrossRef]
11. Paragahawewa, U.; Blackett, P.; Small, B. *Social Life Cycle Analysis (S-LCA): Some Methodological Issues and Potential Application to Cheese Production in New Zealand*; AgResearch Ltd.: Palmerston North, New Zealand, 2009.
12. Smith, J.; Barling, D. Social impact and life cycle assessment: Proposals for methodological development for SMEs in the European food and drink sector. *Int. J. Life Cycle Assess.* **2014**, *19*, 944–949. [CrossRef]
13. De Luca, A.; Iofrida, N.; Strano, A.; Falcone, G. Social life cycle assessment and participatory approaches: A methodological proposal applied to citrus farming in Southern Italy. *Integr. Environ. Assess.* **2015**, *11*, 383–396. [CrossRef] [PubMed]
14. Chen, W.; Holden, N. Social life cycle assessment of average Irish dairy farm. *Int. J. Life Cycle Assess.* **2017**, *22*, 1459–1472. [CrossRef]
15. Petti, L.; Ramirex, P.; Traverso, M.; Ugaya, C. An Italian tomato "Cuore di Bue" case study: Challenges and benefits of using subcategory assessment method for social life cycle assessment. *Int. J. Life Cycle Assess.* **2018**, *23*, 569–580. [CrossRef]
16. Benoît-Norris, C.; Vickery-Niederman, G.; Valdivia, S.; Franze, J.; Traverso, M.; Ciroth, A.; Mazijn, B. Introducing the UNEP/SETAC methodological sheets for subcategories of social LCA. *Int. J. Life Cycle Assess.* **2011**, *16*, 682–690. [CrossRef]
17. Spillemaecker, S.; Vanhoutte, G.; Traverniers, L.; Lavrysen, L.; van Braeckel, D.; Mazijin, B.; Rivera, J. *Integrated Product Assessment: The Development of the Label 'Sustainable Development' For Products, Ecological, Social, and Economical Aspects of Integrated Product Policy*; Belgian Science Policy: Brussels, Belgium, 2004.
18. Dreyer, L.; Hauschild, M.; Schierbeck, J. Characterisation of social impacts in LCA. *Int. J. Life Cycle Assess.* **2010**, *15*, 247–259. [CrossRef]
19. Reveret, J.P.; Couture, J.M.; Parent, P. Socio-economic LCA of milk production in Canada. In *Social Life Cycle Assessment*; Muthu, S., Ed.; Springer Science + Business Media: Singapore, 2015. [CrossRef]
20. AAFC. *Canada's Poultry and Egg Industry Profile at a Glance*; Agriculture and Agri-Food Canada: Ottawa, ON, Canada, 2013.
21. ESDC. *Work—Work-Related Injuries. Indicators of Well-Being in Canada*; Employment and Social Development Canada: Toronto, Canada, 2015. Available online: http://www4.hrsdc.gc.ca/.3ndic.1t.4r@-eng.jsp?iid=20 (accessed on 10 June 2016).
22. CBC. *What Are the Overall Voluntary and Involuntary Turnover Rates for Canadian Organizations? HR Questions and Answers*; Conference Board of Canada: Ottawa, ON, Canada, 2015; Available online: http://www.conferenceboard.ca/topics/humanresource/questions.aspx#Q1 (accessed on 10 June 2016).
23. Kruse, S.; Flysjo, A.; Kasperczyk, N.; Scholz, A. Socioeconomic indicators as a complement to life cycle assessment—An application to salmon production systems. *Int. J. Life Cycle Assess.* **2009**, *14*, 8–18. [CrossRef]
24. Manik, Y.; Leahy, J.; Halog, A. Social life cycle assessment of palm oil biodiesel: A case study in Jambi Province of Indonesia. *Int. J. Life Cycle Assess.* **2013**, *18*, 1386–1392. [CrossRef]

sustainability

MDPI

Article

Study on the Vertical Linkage of Greenhouse Gas Emission Intensity Change of the Animal Husbandry Sector between China and Its Provinces

Tianyi Cai [1,2], Degang Yang [1,*], Xinhuan Zhang [1], Fuqiang Xia [1] and Rongwei Wu [1,2]

1 Xinjiang Institute of Ecology and Geography, Chinese Academy of Sciences, Urumqi 830011, China; caitianyi14@mails.ucas.ac.cn (T.C.); zhangxh@ms.xjb.ac.cn (X.Z.); xiafq@ms.xjb.ac.cn (F.X.); wurongwei16@mails.ucas.ac.cn (R.W.)

2 University of Chinese Academy of Sciences, Beijing 100049, China

* Correspondence: dgyang@ms.xjb.ac.cn; Tel.: +86-0991-7827319

Received: 30 May 2018; Accepted: 13 July 2018; Published: 16 July 2018

Abstract: China's carbon intensity (*CI*) reduction target in 2030 needs to be allocated to each province in order to be achieved. Thus, it is of great significance to study the vertical linkage of *CI* change between China and its provinces. The existing research on the vertical linkage focuses more on energy-related economic sectors in China; however, attention has not been paid to China's animal husbandry (*AH*) sector, although the role of the China's AH sector in greenhouse gas (*GHG*) reduction is increasingly important. This study firstly established a vertical linkage of change in greenhouse gas emission intensity of the animal husbandry sector (*AHGI*) between China and its 31 provinces based on the logarithmic mean Divisia index (*LMDI*) decomposing method from the perspective of combining emission reduction with economic development, and quantified the contributions of each province and its three driving factors of environmental efficiency (*AHEE*), productive efficiency (*AHPE*), and economic share (*AHES*) to reducing China's *AHGI* during the period of 1997–2016. The main results are: (1) The *AHGI* of China decreased from 5.49 $tCO_2eq/10^4$ yuan in 1997 to 2.59 $tCO_2eq/10^4$ in 2016, showing a 75.25% reduction. The *AHGI* in 31 provinces also declined and played a positive role in promoting the reduction of national *AHGI*, but there were significant inter-provincial differences in the extent of the contribution. Overall, the provinces with higher emission levels contributed the most to the reduction of China's *AHGI*; (2) The *AHPE* and *AHEE* factors in 31 provinces cumulatively contributed to the respective 68.17% and 11.78% reduction of China's *AHGI*, while the *AHES* factors of 31 provinces cumulatively inhibited the 4.70% reduction. Overall, the *AHPE* factor was the main driving factor contributing to the reduction of China's *AHGI*. In the future, improving the level of *AHEE* through *GHG* emissions reduction technology and narrowing the inter-provincial gap of the level of *AHPE* are two important paths for promoting the reduction of China's *AHGI*.

Keywords: *LMDI*; provincial contribution; emission reduction; decomposition; driving factor

1. Introduction

In 2015, at the "Paris Climate Conference", the international community reached a legally-binding target of limiting the global temperature increase to no more than 2 °C [1]. To attain this target, global greenhouse gas (*GHG*) emissions must be reduced by 70% by 2050 [2]. Moreover, the implementation of effective emission reduction depends on the emission reduction policies and cooperation of different countries and their economic sectors [3]. China is the world's largest emitter of *GHG* and largest developing country [4]. The Chinese government had integrated emission reduction with economic development and promised the world that by 2030 the CO_2 emission intensity of its GDP would be reduced by 60–65% compared to 2005 levels [5]. Meanwhile, this carbon intensity (*CI*) reduction target was allocated to each province in order to be achieved [6–8].

However, China's daunting *CI* reduction target is only intended for CO_2 emissions, with no clear quantitative target for non-CO_2 *GHG* emission reduction [4]. In fact, non-CO_2 *GHG* emissions are an important contributor to global warming [9], and the animal husbandry (*AH*) sector has been universally recognized as the primary source of global non-CO_2 *GHG* (CH_4 and N_2O) emissions [9–13]. Non-CO_2 emissions directly from the *AH* sector contribute approximately 19% to the current climate warming [14], but there is also very large potential for emission reduction [10,15–18]. China's *AH* sector emitted 445 million tons (2005) of non-CO_2 *GHG* (CO_2 equivalents), accounting for approximately 6% of China's total *GHG* emissions and 30% of the total non-CO_2 *GHG* emissions [19]. With the continuous rise in the per capita income level and the continuous progress in urbanization in China, the demand for livestock food in the residents' diet will continue to grow [20–24]. Thus, the *AH* sector will play a more important role in reducing *GHG* emissions in China. China's *CI* reduction targets need to be allocated to each province to achieve. Therefore, it is of great significance to conduct a study from the perspective of combining emissions reduction with economic development to uncover the vertical linkage of change in greenhouse gas emission intensity of animal husbandry sector (*AHGI*) between China and its provinces, and evaluate the driving factors of change in *AHGI*, which will provide an important basis for formulating provincial *AHGI* reduction targets for China.

Macro studies on global *GHG* emission intensity of *AH* sector have been reported. Caro's study found that it was vastly different among countries of the intensity of *GHG* emission per unit of economic output value in *AH* sector in 2010, with the highest intensity occurring in developing countries, and particularly Africa [11]. Bennetzen's study showed that *GHG* emissions per unit livestock product has decreased by approximately 44% since 1970 in global. The decoupling of livestock production and emissions had shown very large regional differences. While developing countries have contributed to the doubling of global livestock production, they have also contributed more *GHG* emissions [25,26]. However, more literature has focused on the study of the carbon footprint of animal products from a small scale perspective [27–32].

Existing studies on the vertical linkage and its drivers of CI between China and its provinces have mainly focused on energy-related CO_2 emission intensity [7,33–35], with the logarithmic mean Divisia index (*LMDI*) approach being the most commonly used decomposition method. Wang et al. [7] used the *LMDI* method to investigate the contribution given by China's 30 provinces to the decline of national *CI* from 1997 to 2008. The results showed that Liaoning, Heilongjiang, and Hebei have made greater contributions to the decline of national *CI*, the contribution of a province to the decline of national *CI* laid mainly in the promotion of its energy efficiency. Zhang et al. [33] applied the *LMDI* technique to explore the *CI* drivers in 29 Chinese provinces from 1995 to 2012. The results indicated that the overall *CI* of China decreased rapidly and energy intensity is the most significant driver for the decrease of *CI*. Tan et al. [34] examined the role of activities related to the electric power industry in reducing China's *CI* from 1998 to 2008 utilizing the *LMDI* technique. The results revealed that the provinces with higher emission levels contributed the most to China's improvements in *CI*. Wang et al. [35] established panel data models to explore the influencing factors of *CI* in China. The results showed that the economic level of activity was negatively correlated to *CI* on both national and regional levels. However, attention has not been paid to China's *AH* sector. The existing research literature related to China's *AH* sector have either focused more on the macro horizontal perspective on the regional characteristics of emissions and intensity in various provinces [36–39] and the decomposition of the drivers for carbon emissions [39–43], or on the micro perspective on the differences in the carbon footprint between different feeding methods [44,45] and different animal products [46–48]. Meng et al. [36] estimated the *AH*'s *GHG* emissions of China's different livestock division by life circle assessment method. The results revealed that there were higher emissions of CO_2 equivalent of livestock in agricultural areas than in pastoral areas, but the emission intensity was lower than in pastoral areas. Yao et al. [37] used the same method to obtain a conclusion that high livestock carbon emissions areas were mostly located in prairie areas or major grain producing areas of China. Luo et al. [38] explored the spatial and temporal heterogeneity of CO_2 emission intensity in China's agricultural sector and found that Central China had the highest agricultural CO_2 emission intensity than Western and Eastern China. Tian et al. [39,41] studied regional characteristics and driving factor of agricultural carbon emissions in China with *LMDI* model. The results showed that the traditional agricultural provinces, especially the major crop production areas were the main emission source regions. The efficiency factor, labor factor, and industry structure factor had strong inhibitory effects on China's agricultural carbon emissions, while the economy factor had a strong positive effect. Yao et al. [43] conducted a decomposition study on the factors affecting the changes of *AH* carbon emissions based on the *LMDI* method and found that *AH* production efficiency improvement was the most important factor to restrain the sustained growth of the *AH* carbon emissions. Xiong et al. [40,42] used the same decomposition method to draw conclusions that the economy factor was the critical factor to promote the increase in agricultural carbon emissions in Xinjiang province of China, while the main inhibiting factor was the efficiency factor. These research results have enriched the body of work on *GHG* emissions reduction in China's *AH* sector from different perspectives. However, the above-mentioned studies have mostly treated both the *AH* sector and crop farming sector as part of the agricultural sector and neglected the particularity of the *AH* sector. In fact, first of all, the *AH* sector differs from the crop farming sector in its production methods, the economic benefits, the generation mechanism of *GHG*, and the emission reduction strategies; secondly, studying the *GHG* emissions and intensity of the *AH* sector from the horizontal perspective of provinces cannot reveal the vertical linkage between the partial and overall changes. More importantly, vertical linkage is an important basis for the Chinese government to formulate *AHGI* reduction targets for all provinces.

Our work is different from the above research in the following respects: First, this study estimated the *GHG* emissions (CO_2 equivalent) per unit of *AH* economic output value in China and its 31 provinces from 1997 to 2016. Second, we established a vertical linkage of change in *AHGI* between china and its 31 provinces from the perspective of combining emissions reduction with economic development, and separated the factors affecting the change of *AHGI* into environmental efficiency, productive efficiency and economic share. We used the *LMDI* decomposition method to support our study, because it was proven to be a feasible tool in achieving complete decomposition without residuals, allowing subgroup estimations to be aggregated in a consistent manner and performing attribution analyses on the impact of subcategory estimates [49]. Third, we quantified the contributions of each province and its three driving factors to the change of China's *AHGI* during the period of 1997–2016. This study aims to provide a reference for formulating provincial *AHGI* reduction targets for the Chinese government in a more scientific and reasonable way.

2. Materials and Methods

2.1. Estimation of AHGI

This study adopted the *GHG* emissions (CO_2 equivalent) per unit of the *AH* economic output value as the *AHGI* indicator. The formula is shown as follows:

$$AHGI = AHGE/AHGDP \tag{1}$$

where *AHGI*, *AHGE*, and *AHGDP* denote greenhouse gas emission intensity of animal husbandry sector (in $tCO_2eq/10^4$ yuan), the sum of greenhouse gas emissions from animal husbandry sector (in tCO_2eq) and economic output value of animal husbandry sector (10^4 yuan), respectively. The *AHGI* of each province is calculated similarly.

According to the "Guidelines for Compiling Low-carbon Development and Provincial Greenhouse Gas Inventory" issued by the Department of Climate Change of the National Development and Reform Commission of China (*NDRCC*) in 2013 [50], the provincial *GHG* inventory for *AH* includes two parts: CH_4 and N_2O emissions in animal manure management and CH_4 emissions from animal enteric fermentation in this study. Due to the vast territory of China, the climate and soil conditions in different geographical regions have different effects on CH_4 and N_2O emissions during the storage and management of animal manure, so the emissions factors of different emission sources also adopt the recommended values of the sub-regions in China by the guidelines. The formula is shown as follows:

$$AHE_{CH_4,manure} = \left(\sum AP_i \times EF_{CH_4,manure,i}\right) \times 10^{-3} \quad (2)$$

$$AHE_{N_2O,manure} = \left(\sum AP_i \times EF_{N_2O,manure,i}\right) \times 10^{-3} \quad (3)$$

$$AHE_{CH_4,enteric} = \left(\sum AP_i \times EF_{CH_4,enteric,i}\right) \times 10^{-3} \quad (4)$$

$$AHGE = \left[\left(AHE_{CH_4,manure} + AHE_{CH_4,enteric}\right) \times 25 + AHE_{N_2O,manure} \times 298\right] \quad (5)$$

where $AHE_{CH_4,manure}$ represent the sum of the CH_4 caused by the manure management, in *t* CH_4/year; $AHE_{N_2O,manure}$ represent the sum of the N_2O emissions caused by the manure management, in *t* N_2O/year; $AHE_{CH_4,enteric}$ represents the sum of the CH_4 emissions caused by enteric fermentation, in t CH_4/year; AP_i is the amount of annual feeding of animal *i* in head; *i* include cattle, buffalo, dairy cows, horses, donkeys, mules, camels, goats, sheep, pigs and poultry in China; Non-ruminant animals, especially poultry, which have small body weights, produce low CH_4 emissions. Therefore, CH_4 emissions from non-ruminant animals are not included in animal enteric fermentation emissions; $EF_{CH_4,manure,i}$ is the CH_4 emission factor of manure management of animal *i* in kg CH_4/head/year, as shown in Table A1; and $EF_{N_2O,manure,i}$ is the N_2O emission factor of manure management of animal *i* in kg N_2O/head/year, as shown in Table A2; $EF_{CH_4,enteric,i}$ is the CH_4 emission factor of enteric fermentation of animal *i*, in kg CH_4/head/year. The guidelines give the methane emission factors for enteric fermentation of different animals in different feeding modes (large-scale farming, farmer raising, and grazing) in China. However, it is difficult to obtain the provincial data of livestock activity by distinguishing different feeding methods, so the CH_4 emission factor for animal enteric fermentation in this study uniformly adopted *IPCC* recommendations [51] as shown in Table 1, in kg CH_4/head/year. In order to facilitate the summation of total greenhouse gas emissions and comparative analysis in each province, we converted CH_4, N_2O to CO_2 equivalent. 25 and 298 are the relative molecular warming forcing of CH_4 and N_2O in a 100-year horizon, respectively [52]. *AHGE* represent the sum of greenhouse gas emissions from animal husbandry sector, in tCO_2eq. The *AHGE* of each province is calculated similarly. The total amount of *AHGE* in China is equal to the sum of total *AHGE* in 31 provinces.

Table 1. The CH_4 emission factor for animal enteric fermentation (kg CH_4/head/year).

Livestock	Emission Factor	Livestock	Emission Factor	Livestock	Emission Factor
Buffalo	55	Mule	10	Sheep	5
Dairy cow	61	Camel	46	Goat	5
Cattle	47	Horse	18	Pig	1
Donkey	10				

2.2. Decomposition of AHGI

2.2.1. Decomposing Process

As the total *AHGE* and *AHGDP* from China are generated by the *AH* sectors of the 31 provinces across the country, this study firstly established a vertical linkage of the *AHGI* calculation between the national and provincial *AH* sectors. The calculation is shown as follows:

$$AHGI = \frac{AHGE}{AHGDP} = \sum_{i=1}^{31} AHGE_i / \sum_{i=1}^{31} AHGDP_i \quad (6)$$

The existing studies have shown that changes in China's *CI* have been impacted not only by the changes of *CI* of provinces but also by changes of the *GDP* share of the provinces in the national *GDP* [7,34]. As China's *AH* sector is one of the economic production sectors, changes in its intensity should also comply with the above assertion. Therefore, the factors that drive the changes in *AHGI* in the country are divided into two categories: the *AHGI* of each province and its share of *AHGDP* in the national *AHGDP*. Formula (6) is decomposed as follows:

$$AHGI = \sum_{i=1}^{31} \frac{AHGE_i}{AHGDP_i} \frac{AHGDP_i}{AHGDP} \tag{7}$$

However, the changes of *AHGI* in various provinces are also impacted by different factors. The differences in resource endowments, development methods, technical levels, and supporting policies for the development of *AH* in various provinces in China determine the inter-provincial differences in the level of *AH* production [53]. However, this regional difference can be summarized as the difference in the level of *GHG* emissions and economic benefits of each province's *AH* sector from the perspective of the *AHGI* calculation formula. The most perfect scenario for the development of *AH* sector is to reduce the amount of *GHG* emissions while increasing the economic benefits of livestock. This is also a sustainable path for *AH* sector to mitigate and adapt to climate change [54,55]. Therefore, this study classifies the driving factors that drive the changes of *AHGI* in 31 provinces into two categories: the animal husbandry environmental efficiency (*AHEE*) factor (*GHG* emissions per unit of production factors; the lower the former is, the higher the environmental efficiency) and the animal husbandry productive efficiency (*AHPE*) factor (the input of production factors per unit of *AH* output value; the lower the former is, the higher the productive efficiency). Formula (7) is, therefore, further broken down as follows:

$$AHGI = \sum_{i=1}^{31} \frac{AHGE_i}{AHBS_i} \frac{AHBS_i}{AHGDP_i} \frac{AHGDP_i}{AHGDP} = \sum_{i=1}^{31} AHEE_i AHPE_i AHES_i \tag{8}$$

where $AHBS_i$ denotes the input of production factors in the *AH* sector (it can be replaced with the breeding scale of livestock) in the ith province of China. To make it easier for comparing different types of livestock in different provinces, the equivalent standardized cattle index is introduced to estimate the total provincial livestock population. The estimation method and parameters are cited from [56], one cattle is a standard unit, and the reference parameters for converting other livestock to standard cattle units are shown in Table 2. $AHEE_i$, $AHPE_i$ and $AHES_i$ are the three drivers of *AHGI* in each province, indicating the ith province's *AHEE* (*GHG* emissions per unit of livestock), *AHPE* (livestock input per unit of *AHGDP*), and *AHES* (the provincial proportion of the *AHGDP* in the national *AHGDP*).

Table 2. The reference parameters for converting other livestock to standard cattle units.

Livestock	Parameters	Livestock	Parameters	Livestock	Parameters
Buffalo	1.3	Mule	1	Pig	0.3
Dairy cow	2.6	Camel	1.75	Poultry	0.01
Horse	0.8	Goat	0.2		
Donkey	0.6	Sheep	0.25		

2.2.2. Calculating Method

After decomposing the *AHGI* into the above three driving factors, as shown in Equation (8), next we will present the method used to calculate. Referring to Ang [49], the change of *AHGI* can be decomposed using an additive *LMDI* method as follows:

$$\ln\frac{AHGI_T}{AHGI_0} = \Delta AHGI = AHGI_T - AHGI_0 = \Delta AHEE + \Delta AHPE + \Delta AHES \tag{9}$$

where T represents the end of the period; 0, the start of the period; $\Delta AHGI$, the change of *AHGI* in China; $\Delta AHEE$, the contribution of *AHEE* factor; $\Delta AHPE$, the contribution of *AHPE* factor; and $\Delta AHES$, the contribution of *AHES* factor. $\Delta AHEE$, $\Delta AHPE$ and $\Delta AHES$ can be calculated by:

$$\Delta AHEE = \sum_{i=1}^{31} w_i^{AH^*} \ln(AHEE_{i,T}/AHEE_{i,0}) \tag{10}$$

$$\Delta AHPE = \sum_{i=1}^{31} w_i^{AH^*} \ln(AHPE_{i,T}/AHPE_{i,0}) \tag{11}$$

$$\Delta AHES = \sum_{i=1}^{31} w_i^{AH^*} \ln(AHES_{i,T}/AHES_{i,0}) \tag{12}$$

where $w_i^{AH^*}$ in the above equations is the logarithmic weighting scheme specified in the following:

$$w_i^{AH^*} = L\left(w_{i,0}^{AH}, w_{i,T}^{AH}\right) / \sum_{i=1}^{31} L\left(w_{i,0}^{AH}, w_{i,T}^{AH}\right) \tag{13}$$

$$L\left(w_{i,0}^{AH}, w_{i,T}^{AH}\right) = \left(w_{i,T}^{AH} - w_{i,0}^{AH}\right) / \ln\left(w_{i,T}^{AH} / w_{i,0}^{AH}\right) \tag{14}$$

$$w_i^{AH} = AHEE_i AHPE_i AHES_i / \sum_{i=1}^{31} AHEE_i AHPE_i AHES_i \tag{15}$$

2.3. *Data Collection and Processing*

This study involves 31 provinces in mainland China and the study period spans from 1997 to 2016. Considering the feeding cycles which are more than one year of cattle, buffalo, dairy cows, horses, donkeys, mules, goats, and sheep [51], the data for the annual total livestock raised is denoted by the year-end inventory data. The feeding cycle for poultry (55 days) and pig (200 days) is less than one year [37], and the data for the annual total is expressed by the current year's slaughter volume [53]. The time-series livestock activity data of each province used in this paper are obtained from the China Rural Statistical Yearbook (1998–2017). The time-series livestock economic output data of every province are obtained from the China Statistical Yearbook (1998–2017) and then converted to outputs in constant 1997 price. All abbreviations used in this study are described in Table A3.

As China's *CI* reduction target is a long-term target, until 2030, this study took 1997 as the baseline period and 2016 as the comparison period as the basis for this research.

3. Results and Discussion

3.1. *Changes of AHGI in China and Its Provinces*

From a national perspective, the *AHGI* of China decreased from 5.49 $tCO_2eq/10^4$ yuan in 1997 to 2.59 $tCO_2eq/10^4$ in 2016, showing a 75.25% reduction within this period, as shown in Figure 1. This is in line with the finding by Caro [11] that the *AHGI* in global developing countries has been on the decline.

From a provincial perspective, compared 2016 with 1997, the *AHGI* were all decreasing in 31 provinces, but the extent of the decrease varied significantly among the provinces. Figure 2 shows that the *AHGI* in 15 provinces (including Hainan, Heilongjiang, Yunnan, Anhui, Henan, Guizhou, Shanxi, Hebei, Shaanxi, Guangxi, Xinjiang, Inner Mongolia, Jilin, Shandong and Sichuan) declined faster than 75.25% (the aggregate reduction in *AHGI* of China), while the *AHGI* in other 16 provinces declined slower than 75.25%. Hainan's *AHGI* declined from 6.68 $tCO_2eq/10^4$ yuan to 1.76, which was the largest decline (133.28%) of all the provinces. Liaoning's *AHGI* declined from 3.11 $tCO_2eq/10^4$ yuan to 2.54, which was the smallest decline (19.91%) of all the provinces.

According to the above changing trends of the *AHGI* in China and its provinces during the period of 1997–2016, this paper argues that these reductions of the *AHGI* are due to two reasons. On the one hand, the *AHGE* has been growing slowly with the transformation of China's *AH* production from an extensive mode to an intensive mode during the study period. On the other hand, the *AHGDP* has been growing rapidly due to advancement of livestock feeding and management techniques, and standardization, scale, and organizational level of *AH* production. However, there are significant inter-provincial differences in resource endowments, development methods, technical levels, and supporting policies for the development of *AH*, which made the extent of reduction different among the 31 provinces.

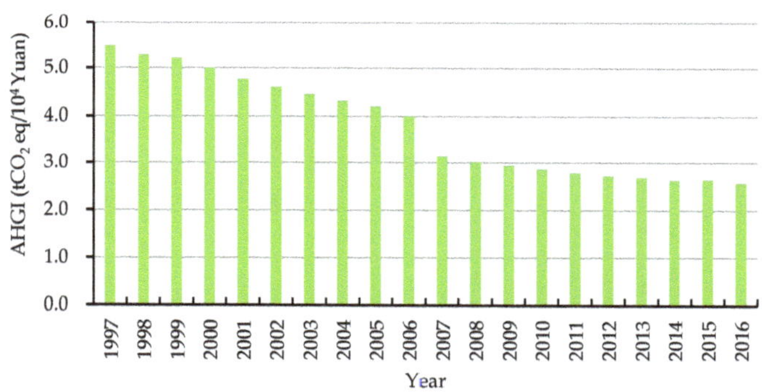

Figure 1. Changes of *AHGI* in China during the period 1997–2016.

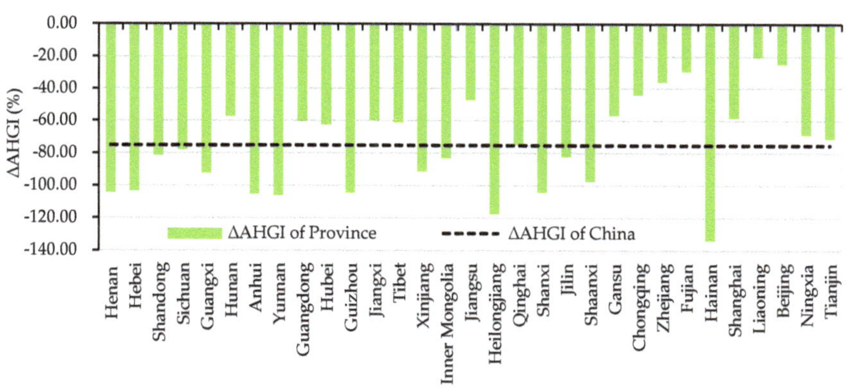

Figure 2. Changes of *AHGI* in 31 provinces, comparing 2016 with 1997.

It was reported that 445 million tCO_2eq were emitted from animal enteric fermentation and manure management in the Second National Communication on Climate Change (*NCCC*) issued by the Chinese government in 2005, using 21 and 310 as the values for Global Warming Potential (*GWP*) of CH_4 and N_2O [19]. Our estimates of 503 million tCO_2eq emissions for the corresponding year are comparable to these data and differences can be attributed to different *GWP* (this paper used *IPCC* Fourth Assessment Report value (25 and 298 of CH_4 and N_2O)) [52]. However, this will not affect the overall research results and, thus, the estimated value of *AHGI* used in this study is reliable.

3.2. Analysis of Provincial Contributions to the Reduction of China's AHGI

Compared 2016 with 1997, China's *AHGI* showed a 75.25% reduction. The decomposition results (Table 3, Figure 3) of provincial contributions from the *LMDI* model showed that all provinces played a positive role in promoting the 75.25% reduction of China's *AHGI*, but there were significant differences among provinces in the extent of the contribution. The top 10 contributing provinces were Henan (−7.11%), Hebei (−5.58%), Shandong (−5.55%), Sichuan (−5.29%), Guangxi (−4.81%), Hunan (−4.29%), Anhui (−4.24%), Yunnan (−3.31%), Guangdong (−3.23%), and Hubei (−2.82%), while the bottom ten contributing provinces, in descending order, were Tianjin (−0.06%), Ningxia (−0.14%), Beijing (−0.34%), Liaoning (−0.35%), Shanghai (−0.58%), Hainan (−0.59%), Fujian (−1.01%), Zhejiang (−1.01%), Chongqing (−1.12%), and Gansu (−1.25%). Considering the largest and smallest contributors, Henan contributed approximately 119 times as much as Tianjin.

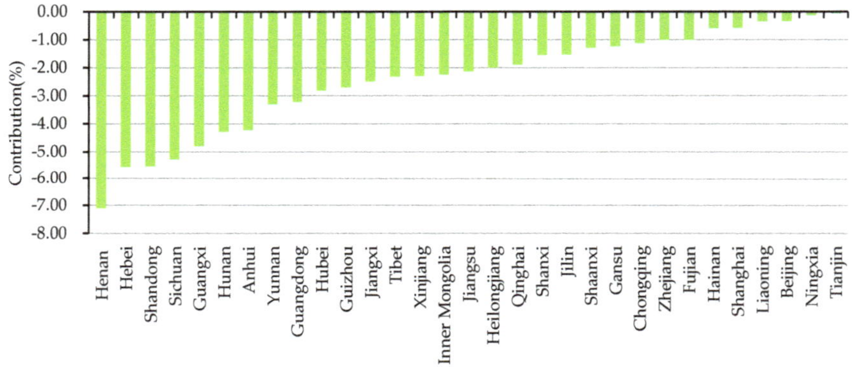

Figure 3. Provincial contributions to the reduction of China's *AHGI*, comparing 2016 with 1997.

The study period witnessed a decline of the *AHGI* in all provinces, which contributed positively to the reduction in national *AHGI*. However, the results of linear regression between the reduction of *AHGI* in 31 provinces and the contributions of 31 provinces to the country showed that the R^2 value was only 0.1543 (Figure 4). This showed that there was an inconsistency between the extent of the actual reduction of *AHGI* in each province and its contributions to the country. Taking Hainan and Hunan as an example, the actual decline in *AHGI* in Hainan (133.28%) is the highest in 31 provinces, but Hainan's contribution (0.59%) to the reduction of China's *AHGI* is ranked the reciprocal sixth position among all provinces. The decline in *AHGI* in Hunan (56.74%) is only ranked 24th, but its contribution (4.29%) is ranked sixth among all provinces. The reason is that the proportion of *AH* scale, *GHG* emissions, and share of the output value in Hainan are relatively small relative to the entire country (0.75%, 0.75%, and 1.11% in 2016, respectively). This is reflected in the *LMDI* decomposition model of this study as a smaller weighting factor. Consequently, the actual decline in *AHGI* in Hainan has less impact on the decline in national *AHGI*. Likewise, the proportion of the above-mentioned three factors in Hunan are relatively large in the country (5.21%, 5.68%, and 6.27% in 2016), which makes Hunan have a greater impact on the decline in national *AHGI*. Similar provinces include Sichuan, Guangdong, Shaanxi, and Shanxi. This indicates that the proportion of the provincial *AH* scale, *GHG* emissions, and share of economic output value in the country are important for measuring their contributions.

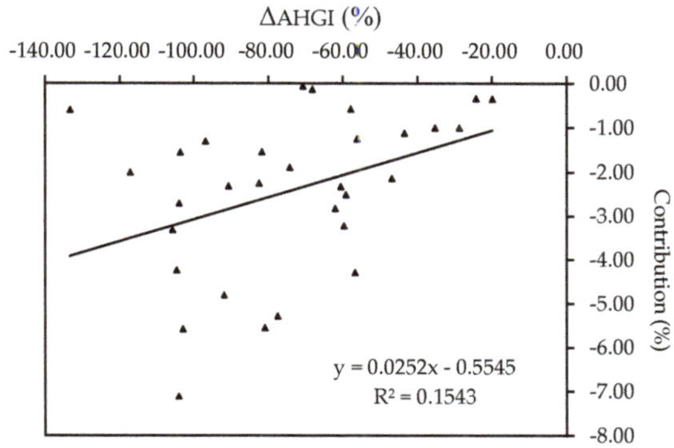

Figure 4. The scatter diagram of the regression variables between the change of *AHGI* in 31 provinces and the contributions of the 31 provinces to the country.

In terms of the regional characteristics of animal husbandry production, among the top 10 contributing provinces, Henan, Hebei, Shandong, Sichuan, Hunan, Anhui, and Hubei are the major grain-producing regions in China, and they are also the major livestock breeding provinces in the agricultural areas of China. There are abundant feeding resources from crops, and the standardization, scale, and organizational level of *AH* production are high in these provinces. They are currently China's main base for producing livestock such as live pigs, beef cattle, and poultry [53,57], and shoulder the important function of supplying animal-based food to the country; the top ten provinces cumulatively contributed to a reduction of 46.22% in China's *AHGI*, accounting for 61.42% of the total contributions by 31 provinces. Meanwhile, for these provinces in 2016, it accounted for approximately 55.16%, 55.01%, and 61.05% of the scale of livestock breeding, *AHGE*, and *AHGDP*, respectively, of the corresponding total amount in the country.

Among the bottom ten contributing provinces, Beijing, Tianjin, Shanghai, and Chongqing are the four municipalities directly governed by the Chinese government. Zhejiang, Fujian, Hainan, and Liaoning are the coastal provinces of Eastern China. The above eight provinces and municipalities are the major consumers of animal-based food in China. They are densely populated, with relatively high levels of urbanization and industrialization, and their *AH* production is dominated by modern urban *AH* [37] and the proportion of *AHGDP* in the *GDP* is relatively low; in contrast, Gansu and Ningxia are in the northwest agro-pastoral ecotone of China, with a relatively fragile ecological environment, and strong resource and environment constraints for the development of *AH*, and the total output from *AH* is not large [24,58]. The bottom ten provinces cumulatively contributed to a reduction of only 6.45% in China's *AHGI*, accounting for 8.57% of total provincial contributions. Meanwhile, for these provinces in 2016, it accounted for approximately 14.39%, 12.97% and 13.07% of the scale of livestock breeding, *AHGE*, and *AHGDP*, respectively, of the corresponding total amount in the country.

The remaining 11 provinces cumulatively contributed to a reduction of 22.58% in China's *AHGI*, accounting for 30.01% of the provincial total contributions. Among these provinces, Inner Mongolia, Xinjiang, Qinghai, and Tibet are the traditional pastoral areas, with a long development history of *AH*. With rich resources in natural grassland, the focus is primarily on the development of grassland *AH*. They are currently China's major breeding base for cattle, sheep, horses, and other herbivorous livestock [58]; on the other hand, Jiangxi, Heilongjiang, Jiangsu, and Jilin are the main grain-producing regions in China and are rich in feed resources, the focus is primarily on the development of *AH* in agricultural areas. For these provinces in 2016, it accounted for approximately 30.45%, 32.02%, and 25.88% of the scale of livestock breeding, *AHGE*, and *AHGDP*, respectively, of the corresponding total amount in the country.

The discussion above showed that there was an inconsistency between the extent of the actual reduction of *AHGI* in each province and their contributions to the country, China's progress in reducing *AHGI* was mainly made by provinces with a large *GHG* emissions from *AH*. The conclusion drawn by this study that provinces contributed to the reduction in national *CI* is consistent with the conclusion drawn by Tan's research [34] on the Chinese power industry and Wang's research [7] on China's economic development. The revelation for us is that the government should comprehensively consider the actual decline of *AHGI* in all provinces and its contributions to the reduction in national *AHGI* in formulating provincial *AHGI* reduction targets for China. Each province should not be required to have the same decline in *AHGI* as in the national *AHGI*. In addition, priority should be given to assessing the emission reduction responsibilities and impact in large *AH* provinces, because while these provinces emit a larger share of *AHGE*, they also supply a larger share of animal-based food and create a larger share of *AHGDP* for the country. More importantly, these provinces are also major contributors in reducing the *AHGI* of China.

3.3. *Analysis of Driving Factors' Contributions to the Reduction of China's AHGI*

The decomposition results of three driving factors' contributions from the *LMDI* model (Table 3, Figure 5) indicate that there is a complex linkage among the three driving factors: *AHEE*, *AHPE*, and *AHES*. Three driving factors comprehensively determine the reduction of China's *AHGI* through two different contributions of positive promotion and negative inhibition, but the way the three driving factors exert impact and the extent of their contributions vary significantly among provinces. Overall, the *AHPE* factors of all provinces are the main factors contributing to the reduction of China's *AHGI*, cumulatively contributing to 68.17%, followed by the *AHEE* factors (11.78%). The *AHES* factors of various provinces have the nature of a "double-edged sword", having cumulatively inhibited the 4.70% reduction of China's *AHGI*.

Table 3. The decomposition results of vertical linkage of changes in *AHGI* between China and its provinces based on the *LMDI* model.

Province	The Decomposition Results of Three Driving Factors (%)				Province	The Decomposition Results of Three Driving Factors (%)			
	Δ*AHEE*	Δ*AHPE*	Δ*AHES*	Total		Δ*AHEE*	Δ*AHPE*	Δ*AHES*	Total
China	−11.78	−68.17	4.70	−75.25	Jiangsu	−0.15	−1.05	−0.94	−2.15
Henan	−2.17	−6.93	1.99	−7.11	Heilongjiang	−0.34	−3.29	1.63	−2.00
Hebei	−0.52	−4.96	−0.09	−5.58	Qinghai	0.07	−1.82	−0.15	−1.90
Shandong	−1.25	−4.62	0.32	−5.55	Shanxi	−0.29	−1.25	−0.01	−1.55
Sichuan	−0.47	−5.67	0.85	−5.29	Jilin	−0.15	−1.97	0.57	−1.55
Guangxi	−1.17	−3.40	−0.23	−4.81	Shaanxi	−0.23	−1.13	0.06	−1.31
Hunan	−0.27	−2.96	−1.05	−4.29	Gansu	0.02	−1.36	0.10	−1.25
Anhui	−1.43	−2.54	−0.27	−4.24	Chongqing	−0.16	−0.58	−0.38	−1.12
Yunnan	−0.53	−4.74	1.96	−3.31	Zhejiang	−0.08	−0.26	−0.66	−1.01
Guangdong	−0.26	−2.08	−0.88	−3.23	Fujian	−0.44	−0.03	−0.54	−1.01
Hubei	−0.52	−2.11	−0.19	−2.83	Hainan	−0.30	−0.72	0.43	−0.59
Guizhou	−0.44	−2.90	0.63	−2.71	Shanghai	0.07	−0.24	−0.41	−0.58
Jiangxi	−0.66	−1.29	−0.56	−2.51	Liaoning	−0.07	−0.47	0.19	−0.35
Tibet	0.10	−1.90	−0.53	−2.33	Beijing	0.04	−0.11	−0.27	−0.34
Xinjiang	−0.22	−3.00	0.91	−2.31	Ningxia	0.06	−0.44	0.25	−0.14
Inner Mongolia	0.02	−4.14	1.86	−2.26	Tianjin	−0.02	−0.20	0.16	−0.06

Note: Positive and negative values in the table represent the contribution of two different properties, the positive value is the contribution to inhibit the reduction of China's *AHGI*, while negative value is the contribution to promote the reduction of China's *AHGI*.

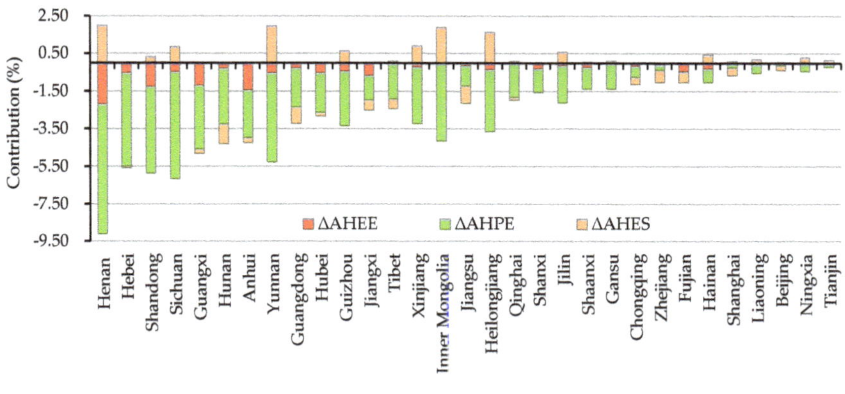

Figure 5. Driving factors' contributions to the reduction of *AHGI* in each province, comparing 2016 with 1997.

3.3.1. AHPE Factor

With the transformation of China's *AH* production from an extensive mode to an intensive mode and the continuous improvement in the productivity of *AH*, the livestock input for China's 10,000 yuan of *AHGDP* dropped from the 5.98 standard cattle in 1997 to 3.17 standard cattle in 2016, and the level of *AHPE* increased about 63.47%. The livestock input for 10,000 yuan of *AHGDP* in all provinces of China also decreased (Figure 6). As a result, the *AHPE* factors of all provinces contributed positively to the reduction of *AHGI*. Among them, Henan's contribution (6.93%) was the highest, and Tianjin's contribution (0.20%) was the lowest. Overall, the *AHPE* factors of all provinces cumulatively contributed to a reduction of 68.17% in *AHGI* of China. However, due to the vast territory of China, there are significant regional differences among provinces in terms of the economic benefits of livestock products. Taking 2016 as an example, the livestock input for 10,000 yuan of *AHGDP* for China's frontier grassland pastoral areas (Tibet (28.59), Qinghai (15.20), Gansu (7.72), Xinjiang (7.20), Inner Mongolia (4.75), Ningxia (4.63), Guizhou (4.18), Yunnan (4.14), and Guangxi (3.73)) was higher than the national average (3.17) due to the comprehensive impact of the productivity level of *AH*, the distance to the consumer market of livestock products, and the national dietary habits [24]. On the other hand, the livestock input for 10,000 yuan of *AHGDP* for China's central and eastern regions (Henan (2.55), Anhui (2.47), Hubei (2.62), (Jiangsu (2.40), Zhejiang (2.35), and Tianjin (1.46)), which are close to the consumer market, have a larger scale of *AH*, and a higher intensity, was lower than the national average. This finding shows that the *AHPE* factor of each province is the leading factor in promoting the reduction of China's *AHGI*, but there are still large gaps among provinces. Overall, the productive efficiency level of China's frontier grassland pastoral areas is lower than that of central and eastern provinces.

3.3.2. AHEE Factor

The *GHG* emissions per unit standard cattle in China dropped from 0.92 tCO_2eq in 1997 to 0.82 tCO_2eq in 2016, and the level of *AHEE* increased about 11.51%. However, the changes of *AHEE* levels were different in 31 provinces (Figure 6), as a result, the *AHEE* factors of each province have both positive and negative contributions in reducing the China's *AHGI*. Among 31 provinces, the *GHG* emissions per unit standard cattle in these seven provinces (Tibet (1.16 increased to 1.20), Qinghai (1.05 increased to 1.08), Inner Mongolia (0.895 increased to 0.899), Ningxia (0.81 increased to 0.89), Gansu (0.89 increased to 0.90), Beijing (0.62 increased to 0.70), and Shanghai (0.59 increased to 0.75)) increased slightly, which cumulatively inhibited the 0.38% reduction of China's *AHGI*. On the one hand, Shanghai and Beijing are currently China's cities with relatively developed economy and urbanization among the seven provinces. As the demand of urban residents for milk consumption continues to grow, the proportion of dairy cows in their livestock breeding structure has been increasing, resulting in an increase in emissions. On the other hand, Tibet, Qinghai, Inner Mongolia, Ningxia, and Gansu are China's traditional grassland pastoral regions, so their livestock breeding focuses mainly on cattle, sheep, horses, and other ruminant livestock. The methane emissions generated by the enteric fermentation of these animals are relatively high [12,59]. As these livestock breeding scales increase, so will the amount of emissions. The *GHG* emissions per unit standard cattle in the remaining 24 provinces have been on the decline, which cumulatively promoted the 12.16% reduction of China's *AHGI*. Overall, the *AHEE* factors of all provinces only cumulatively promoted the 11.78% reduction of China's *AHGI*. This indicates that the reduction of China's *AHGI* during the study period depended more on the substantial increase in the *AHPE* of all the provinces than on the improvement of *AHEE*.

3.3.3. AHES Factor

The *AHGDP* of China continued to grow from 713.33 billion yuan in 1997 to 1707.90 billion yuan in 2016, which is a relative increase of 87.31%. The 15 provinces of Henan, Yunnan, Inner Mongolia, Heilongjiang, Xinjiang, Sichuan, Guizhou, Jilin, Hainan, Shandong, Ningxia, Liaoning, Tianjin, Gansu, and Shaanxi all saw an increase in their *AHES* (Figure 6). However, the increase in *AHES* of these provinces was significantly faster than the increase in *AHEE*, which inevitably led to a rapid increase in carbon emissions when they expanded the scale of *AH* production. Therefore, the *AHES* factors of these provinces all inhibited the reduction of China's *AHGI*. However, the *AHES* in the remaining 16 provinces was on the decline. As the *AHES* declined, so did the *GHG* emissions. Therefore, the *AHES* factors of these provinces all promoted the reduction of China's *AHGI*. However, due to the difference in the base of *AHGDP* in each province, the increase and decrease based on the *AHGDP* were also different. As a result, the contribution from the *AHES* factors of different provinces varied. Overall, the cumulative negative contribution from the *AHES* factors of 15 provinces, including Henan, was 11.88%, while the *AHES* factors of the other 16 provinces, including Anhui, cumulatively made a positive contribution of 7.18%. In all, the *AHES* factors of all provinces cumulatively inhibited a 4.70% reduction of China's *AHGI*. This means that, in terms of contributions by each province to the reduction in China's *AHGI*, the *AHES* factor is a weight with the nature of a "double-edged sword", and can decrease or increase the contribution value of each province. If a province can expand its scale of *AH* production and also improve the level of *AHEE*, there will be a greater increase in the *AHGDP* of the province, and a greater positive contribution to the reduction in national *AHGI*. If not, then the result will be the opposite.

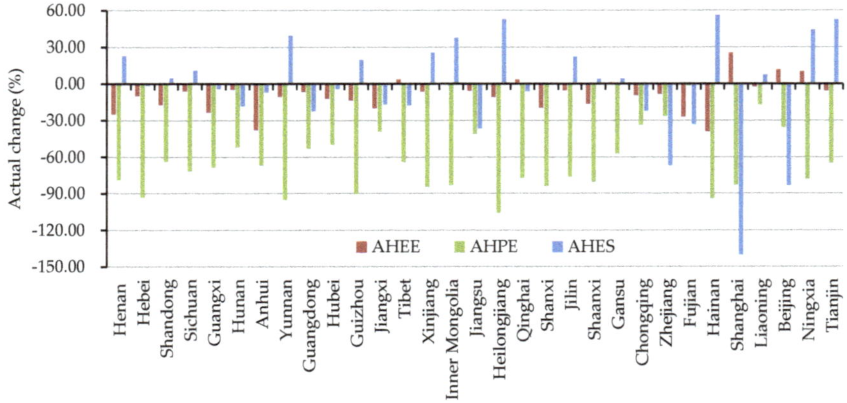

Figure 6. Actual change of three driving factors in each province of China, comparing 2016 with 1997.

Currently, China is the foremost producer of *GHG* emissions and largest developing country in the world. China is also a major producer of livestock products in the world. China's *AH* sector not only shoulders responsibility for supplying livestock products to more than 1.3 billion people, but also takes care of the livelihoods of more than 200 million farmers and herdsmen. With the continuous rise in the per capita income level and the continuous progress of urbanization in China, the demand for livestock food in the residents' diet will continue to increase. Therefore, it is an objective need and an inevitable trend for various provinces in China to expand the scale of breeding and pursue economic benefits of *AH*. In terms of the driving factors that contribute to the reduction of China's *AHGI* during the research period, *AHPE* factors played a major role. However, in terms of the reduction responsibilities of the *AH* sector in China, the development of *AH* should take into account both economic and environmental benefits. The most perfect scenario for the development of *AH* sector is to reduce the amount of *GHG* emissions while increasing the economic benefits of livestock [54,55]. This is also a sustainable path for China's *AH* sector to mitigate and adapt to climate change. With reference to Caro's results [11], the *AHGI* of China in 2010 (2.87 $tCO_2eq/10^4$ yuan) was much lower than that of Vietnam (17.97 $tCO_2eq/10^4$ yuan) and India (16.16 tCO_2 $eq/10^4$ yuan), both of which have higher emission intensities in Asia, but it is 3.78 times that of the United States (0.76 $tCO_2eq/10_4$ yuan). This shows that although the *AHGI* has been continuously on the decline, there is still a large gap compared with the developed countries, and there is still more room for greater emissions reduction in the future. On the one hand, Wang's research shows that the greatest technical emission reduction potential for China's *AH* sector will be 253 Mt of CO_2 emission by 2020 [60], accounting for 56.85% of total *GHG* emissions in 2005 from *AH*. This shows that the technical means will be important measures to reduce *GHG* emissions for China's *AH* sector in the future. On the other hand, Herrero's research indicates that *GHG* emissions can be effectively reduced in *AH* sector by using feed additives to improve the animal's forage digestibility and strengthening the storage and management of animal manure and other technical means [15]. Therefore, in the future, all provinces should focus more on improving the level of *AHEE*, and further reduce the *GHG* emissions of livestock by improving the raising of livestock and manure management technologies so as to promote the conversion from advantage in production scale to advantage in contribution to emissions reduction. In addition, there is a large gap among all provinces in China in terms of *AHPE*. Therefore, narrowing the inter-provincial gap of *AHPE* is also an important path for promoting the reduction of China's *AHGI*.

3.4. Advantages and Limitations of This Study

This study focused on *AHGI* reduction in China and its provinces. Compared with previous studies, this study went deeper in the following three aspects. First, on the decomposition scale, it was different from the existing decomposition studies on the *GHG* emissions from *AH* [39,41–43], which have been done only from the horizontal provincial scales. This study firstly established a vertical linkage of change in *AHGI* between china and its 31 provinces and clearly revealed the linkage between partial changes and overall changes. This is more in line with China's current need to allocate *GHG* reduction tasks from top to bottom. Second, in terms of driving factors, the existing studies have mostly explained the impact on *GHG* emissions of *AH* [39,41–43]. Based on the *AHGI* calculation formula, we separated the factors affecting the change of China's *AHGI* into environmental efficiency, productive efficiency, and economic share. Our research effectively combined emissions reduction and economic development, thus, it can provide a more specific basis for each province to formulate *GHG* emission reduction measures for the *AH* sector. Third, in terms of research findings, previous studies [36–39] found that provinces such as Henan, Shandong, Hebei, and Sichuan have always been major contributors to *GHG* emissions from the *AH* sector in China. Our findings indicated that these provinces were not only major contributors to *GHG* emissions from the *AH* sector in China, they were also the main contributors to the reduction of China's *AHGI*. This is of great significance for a comprehensive and objective understanding of the responsibilities of the provinces with large *GHG* emissions for their emissions reduction tasks.

However, although there are some advantages in the study, it also needs for further development. First, we only discussed the decomposition results of provincial contributions to the reduction of China's *AHGI* comparing 2016 with 1997. However, we did not further divide the study period into different time periods for detailed analysis and discussion. Future studies should strengthen the comparative analysis at different time periods. In addition, we only revealed the linkage of *AHGI* change between china and its 31 provinces, but did not delve more deeply into the issues of fairness and efficiency of contribution values in each province. In future research, the *LMDI* decomposition method and the performance evaluation method should be combined to answer the issues of fairness and efficiency of each province's contribution value. Three major factors, which are the reduction of greenhouse gas emissions, economic growth and the supply of livestock food, should be included in the performance evaluation system together.

4. Conclusions

This study established a vertical linkage of *GHG* emission intensity change of the animal husbandry sector between China and its 31 provinces based on *LMDI* decomposing method from the perspective of combining emission reductions with economic development. In addition, the study quantified the contributions of each province and its three driving factors of environmental efficiency, productive efficiency, and economic share to reducing China's animal husbandry *GHG* emission intensity during the period of 1997–2016. The final conclusions are as follows.

(1) The *AHGI* of China decreased from 5.49 $tCO_2eq/10^4$ yuan in 1997 to 2.59 $tCO_2eq/10^4$ in 2016, showing a 75.25% reduction. Compared 2016 with 1997, the *AHGI* in 31 provinces also declined and played a positive role in promoting the reduction of the national *AHGI*, but there were significant differences among provinces in the extent of contribution. Henan, the largest contributor, contributed to a 7.11% reduction of China's *AHGI*, and Tianjin was the smallest contributor (0.06%). The top ten provinces (Henan, Hebei, Shandong, Sichuan, Guangxi, Hunan, Anhui, Yunnan, Guangdong and Hubei) cumulatively contributed to a reduction of 46.22% in China's *AHGI*, accounting for 61.42% of the total contributions by 31 provinces; while the bottom ten provinces (Tianjin, Ningxia, Beijing, Liaoning, Shanghai, Hainan, Fujian, Zhejiang, Chongqing and Gansu) cumulatively contributed to a reduction of only 6.45% in China's *AHGI*, accounting for 8.57% of total provincial contributions. Overall, there was an inconsistency between the extent of the actual reduction of *AHGI* in each province and its contributions to the country. China's progress in reducing *AHGI* was mainly made by provinces with a large *GHG* emissions from *AH*.

(2) Three driving factors (environmental efficiency, productive efficiency, and economic share) comprehensively determine the reduction of China's *AHGI* through two different contributions of positive promotion and negative inhibition, but the way in which the three driving factors exert their impact and the extent of their contributions vary significantly among provinces. The productive efficiency and environmental efficiency factors in 31 provinces cumulatively contributed to the respective 68.17% and 11.78% reduction of China's *AHGI*, while the economic share factors of 31 provinces cumulatively inhibited the 4.70% reduction of China's *AHGI*. Overall, the productive efficiency factors are the main driving factors contributing to the reduction of China's *AHGI*. The reduction of China's *AHGI* during the study period depended more on the substantial increase in the *AH*'s productive efficiency than on the improvement of *AH*'s environmental efficiency. The economic share factor was a weight with the nature of a "double-edged sword", which can decrease or increase the contribution value of each province to the reduction of China's *AHGI*. In the future, improving the level of *AHEE* through *GHG* emission reduction technology and narrowing the inter-provincial gap of the level of *AHPE* are two important paths for promoting the reduction of China's *AHGI*. In terms of improving the level of *AHEE*, all provinces need to do this. However this is even more urgent in the seven provinces, including Tibet, Qinghai, Inner Mongolia, Ningxia, Gansu, Beijing and Shanghai. In terms of narrowing the inter-provincial gap of the level of *AHPE*, the gap is mainly reflected between the frontier grassland pastoral areas in the western and the agricultural

areas in Central and Eastern China. The key provinces that need to improve the level of *AHPE* are located in grassland pastoral areas, including Tibet, Qinghai, Gansu, Xinjiang, Inner Mongolia, Ningxia, Guizhou, Yunnan and Guangxi.

Author Contributions: T.C. and D.Y. designed the research; T.C., D.Y., and F.X. performed the research; R.W. analyzed the data; T.C. drafted the manuscript, which was revised by D.Y. and X.Z. All authors have read and approved the final manuscript.

Funding: This research was funded by Grant Program of Clean Development Mechanism Fund of China (CDMF) grant number [No. 2014092].

Acknowledgments: We acknowledge the Grant Program of Clean Development Mechanism Fund of China (CDMF) (No. 2014092).

Conflicts of Interest: The authors declare no conflict of interest.

Appendix A

Table A1. The CH_4 emission factor for animal manure management (kg CH_4/head/year).

Region	Dairy Cow	Cattle	Buffalo	Horse	Donkey	Mule
North China	7.46	2.82	-	1.09	0.60	0.60
Northeast China	2.23	1.02	-	1.09	0.60	0.60
East China	8.33	3.31	5.55	1.64	0.90	0.90
Central and Southern China	8.45	4.72	8.24	1.64	0.90	0.90
Southwest China	6.51	3.21	1.53	1.64	0.90	0.90
Northwest China	5.93	1.86	-	1.09	0.60	0.60
Region	**Camel**	**Sheep**	**Goat**	**Pig**	**Poultry**	
North China	1.28	0.15	0.17	3.12	0.01	
Northeast China	1.28	0.15	0.16	1.12	0.01	
East China	1.92	0.26	0.28	5.08	0.02	
Central and Southern China	1.92	0.34	0.31	5.85	0.02	
Southwest China	1.92	0.48	0.53	4.18	0.02	
Northwest China	1.28	0.28	0.32	1.38	0.01	

Note: North China: Beijing, Tianjin, Hebei, Shanxi, Inner Mongolia; Northeast China: Heilongjiang, Jilin, Liaoning; East China: Shanghai, Jiangsu, Zhejiang, Anhui, Fujian, Jiangxi, Shandong; Central and Southern China: Henan, Hubei, Hunan, Guangxi, Guangdong, Hainan; Southwest China: Chongqing, Sichuan, Guizhou, Yunnan, Tibet; Northwest China: Shaanxi, Gansu, Ningxia, Qinghai, Xinjiang.

Table A2. The N_2O emission factor for animal manure management (kg N_2O/head/year).

Region	Dairy Cow	Cattle	Buffalo	Horse	Donkey	Mule
North China	1.846	0.794	-	0.330	0.188	0.188
Northeast China	1.096	0.931	-	0.330	0.188	0.188
East China	2.065	0.846	0.875	0.330	0.188	0.188
Central and Southern China	1.710	0.805	0.860	0.330	0.188	0.188
Southwest China	1.884	0.691	1.197	0.330	0.188	0.188
Northwest China	1.447	0.545	-	0.330	0.188	0.188
Region	**Camel**	**Sheep**	**Goat**	**Pig**	**Poultry**	
North China	0.330	0.093	0.093	0.227	0.007	
Northeast China	0.330	0.057	0.057	0.266	0.007	
East China	0.330	0.113	0.113	0.175	0.007	
Central and Southern China	0.330	0.106	0.106	0.157	0.007	
Southwest China	0.330	0.064	0.064	0.159	0.007	
Northwest China	0.330	0.074	0.074	0.195	0.007	

Note: The provinces included in different regions are the same as Table A1.

Table A3. Abbreviations.

Item	Descriptions
GHG	Greenhouse gas
AH	Animal husbandry
CI	Carbon intensity
LMDI	The logarithmic mean Divisia index
AHGI	Greenhouse gas emission intensity of animal husbandry sector (in tCO_2 eq/10^4 yuan)
AHGE	The sum of greenhouse gas emissions from animal husbandry sector (in tCO_2 eq)
AHGDP	Economic output value of animal husbandry sector (10^4 yuan)
AHBS	The input of production factors in the animal husbandry sector (it can be replaced with the breeding scale of livestock by converting to standard cattle units)
AHEE	The animal husbandry environmental efficiency (GHG emissions per unit of livestock)
AHPE	The animal husbandry productive efficiency (livestock input per unit of *AHGDP*)
AHES	The animal husbandry economic share (the provincial proportion of the *AHGDP* in the national *AHGDP*)

References

1. Vandyck, T.; Keramidas, K.; Saveyn, B.; Kitous, A.; Vrontisi, Z. A global stocktake of the Paris pledges: Implications for energy systems and economy. *Glob. Environ. Chang.* **2016**, *41*, 46–63. [CrossRef]
2. Food and Agriculture Organization (FAO). *The State of Food and Agriculture Climate Change, Agriculture and Food Security*; FAO: Rome, Italy, 2016.
3. Intergovernmental Panel on Climate Change (IPCC). *Clim. Chang. 2014: Synthesis Report. Contribution of Working Groups I, II and III to the Fifth Assessment Report of the Intergovernmental Panel on Climate Change*; Core Writing Team, Pachauri, R.K., Meyer, L.A., Eds.; IPCC: Geneva, Switzerland, 2014; 151p.
4. Liu, L.-C.; Wu, G. The effects of carbon dioxide, methane and nitrous oxide emission taxes: An empirical study in China. *J. Clean. Prod.* **2017**, *142*, 1044–1054. [CrossRef]
5. National Development and Reform Commission of China (NDRCC). *Enhanced Actions on Climate Change: China's Intended Nationally Determined Contributions*; NDRCC: Beijing, China, 2015.
6. Chen, Y.; Lin, S. Decomposition and allocation of energy-related carbon dioxide emission allowance over provinces of China. *Nat. Hazards* **2014**, *76*, 1893–1909. [CrossRef]
7. Wang, F.; Feng, G.; Wu, L. Regional Coutribution to the Decline of National Carbon Intensity in China's Economic Growth. *Econ. Res.* **2013**, *8*, 143–155.
8. Wang, K.; Zhang, X.; Wei, Y.-M.; Yu, S. Regional allocation of CO_2 emissions allowance over provinces in China by 2020. *Energy Policy* **2013**, *54*, 214–229. [CrossRef]
9. Smith, P.M.B.; Ahammad, H.; Clark, H.; Dong, H.; Elsiddig, E.A.; Haberl, H.; Harper, R.; House, J.; Jafari, M.; Masera, O.; et al. Agriculture, Forestry and Other Land Use (AFOLU). In *Climate Change 2014: Mitigation of Climate Change. Contribution of Working Group III to the Fifth Assessment Report of the Intergovernmental Panel on Climate Change*; Edenhofer, O., Pichs-Madruga, R., Sokona, Y., Farahani, E., Kadner, S., Seyboth, K., Adler, A., Baum, I., Brunner, S., Eickemeier, P., et al., Eds.; Cambridge University Press: Cambridge, UK; New York, NY, USA, 2014; pp. 811–922.
10. Smith, P.; Martino, D.; Cai, Z.; Gwary, D.; Janzen, H.; Kumar, P.; McCarl, B.; Ogle, S.; O'Mara, F.; Rice, C.; et al. Greenhouse gas mitigation in agriculture. *Philos. Trans. R. Soc. Lond. Ser. B Boil. Sci.* **2008**, *363*, 789–813. [CrossRef] [PubMed]
11. Caro, D.; Davis, S.J.; Bastianoni, S.; Caldeira, K. Global and regional trends in greenhouse gas emissions from livestock. *Clim. Chang.* **2014**, *126*, 203–216. [CrossRef]
12. Ripple, W.J.; Smith, P.; Haberl, H.; Montzka, S.A.; McAlpine, C.; Boucher, D.H. Ruminants, climate change and climate policy. *Nat. Clim. Chang.* **2013**, *4*, 2–5. [CrossRef]
13. O'Mara, F.P. The significance of livestock as a contributor to global greenhouse gas emissions today and in the near future. *Anim. Feed. Sci. Technol.* **2011**, *166–167*, 7–15. [CrossRef]
14. Reisinger, A.; Clark, H. How much do direct livestock emissions actually contribute to global warming? *Glob. Chang. Boil.* **2017**, *24*, 1749–1761. [CrossRef] [PubMed]
15. Herrero, M.; Henderson, B.; Havlík, P.; Thornton, P.K.; Conant, R.T.; Smith, P.; Wirsenius, S.; Hristov, A.N.; Gerber, P.; Gill, M.; et al. Greenhouse gas mitigation potentials in the livestock sector. *Nat. Clim. Chang.* **2016**, *6*, 452–461. [CrossRef]

16. Havlik, P.; Valin, H.; Herrero, M.; Obersteiner, M.; Schmid, E.; Rufino, M.C.; Mosnier, A.; Thornton, P.K.; Bottcher, H.; Conant, R.T.; et al. Climate change mitigation through livestock system transitions. *Proc. Natl. Acad. Sci. USA* **2014**, *111*, 3709–3714. [CrossRef] [PubMed]
17. Gerber, P.J.S.H.; Henderson, B.; Mottet, A.; Opio, C.; Dijkman, J.; Falcucci, A.; Tempio, G. *Tackling Climate Change through Livestock—A Global Assessment of Emissions and Mitigation Opportunities*; Food and Agriculture Organization of the United Nations (FAO): Rome, Italy, 2013.
18. Frank, S.; Beach, R.; Havlík, P.; Valin, H.; Herrero, M.; Mosnier, A.; Hasegawa, T.; Creason, J.; Ragnauth, S.; Obersteiner, M. Structural change as a key component for agricultural non-CO_2 mitigation efforts. *Nat. Commun.* **2018**, *9*, 1–8. [CrossRef] [PubMed]
19. National Coordination Committee on Climate Change (NCCC). *Second National Communication on Climate Change of The People's Republic of China*; NCCC: Beijing, China, 2012.
20. Tilman, D.; Clark, M. Global diets link environmental sustainability and human health. *Nature* **2014**, *515*, 518–522. [CrossRef] [PubMed]
21. Lin, J.; Hu, Y.; Cui, S.; Kang, J.; Xu, L. Carbon footprints of food production in China (1979–2009). *J. Clean. Prod.* **2015**, *90*, 97–103.
22. Hawkins, J.; Ma, C.; Schilizzi, S.; Zhang, F. China's changing diet and its impacts on greenhouse gas emissions: An index decomposition analysis. *Aust. J. Agric. Resour. Econ.* **2018**, *62*, 45–64. [CrossRef]
23. Popp, A.; Calvin, K.; Fujimori, S.; Havlik, P.; Humpenöder, F.; Stehfest, E.; Bodirsky, B.L.; Dietrich, J.P.; Doelmann, J.C.; Gusti, M.; et al. Land-use futures in the shared socio-economic pathways. *Glob. Environ. Chang.* **2017**, *42*, 331–345. [CrossRef]
24. Li, X.L.; Yuan, Q.H.; Wan, L.Q.; He, F. Perspectives on livestock production systems in China. *Rangel. J.* **2008**, *30*, 211. [CrossRef]
25. Bennetzen, E.H.; Smith, P.; Porter, J.R. Decoupling of greenhouse gas emissions from global agricultural production: 1970–2050. *Glob. Chang. Biol.* **2016**, *22*, 763–781. [CrossRef] [PubMed]
26. Bennetzen, E.H.; Smith, P.; Porter, J.R. Agricultural production and greenhouse gas emissions from world regions—The major trends over 40 years. *Glob. Environ. Chang.* **2016**, *37*, 43–55. [CrossRef]
27. Tichenor, N.E.; Peters, C.J.; Norris, G.A.; Thoma, G.; Griffin, T.S. Life cycle environmental consequences of grass-fed and dairy beef production systems in the Northeastern United States. *J. Clean. Prod.* **2017**, *142*, 1619–1628. [CrossRef]
28. Lesschen, J.P.; van den Berg, M.; Westhoek, H.J.; Witzke, H.P.; Oenema, O. Greenhouse gas emission profiles of European livestock sectors. *Anim. Feed. Sci. Technol.* **2011**, *166–167*, 16–28. [CrossRef]
29. Reckmann, K.; Traulsen, I.; Krieter, J. Life Cycle Assessment of pork production: A data inventory for the case of Germany. *Livest. Sci.* **2013**, *157*, 586–596. [CrossRef]
30. Gollnow, S.; Lundie, S.; Moore, A.D.; McLaren, J.; van Buuren, N.; Stahle, P.; Christie, K.; Thylmann, D.; Rehl, T. Carbon footprint of milk production from dairy cows in Australia. *Int. Dairy J.* **2014**, *37*, 31–38. [CrossRef]
31. Patra, A.K. Accounting methane and nitrous oxide emissions, and carbon footprints of livestock food products in different states of India. *J. Clean. Prod.* **2017**, *162*, 678–686. [CrossRef]
32. Desjardins, R.; Worth, D.; Vergé, X.; Maxime, D.; Dyer, J.; Cerkowniak, D. Carbon Footprint of Beef Cattle. *Sustainability* **2012**, *4*, 3279–3301. [CrossRef]
33. Zhang, W.; Li, K.; Zhou, D.; Zhang, W.; Gao, H. Decomposition of intensity of energy-related CO_2 emission in Chinese provinces using the LMDI method. *Energy Policy* **2016**, *92*, 369–381. [CrossRef]
34. Tan, Z.; Li, L.; Wang, J.; Wang, J. Examining the driving forces for improving China's CO_2 emission intensity using the decomposing method. *Appl. Energy* **2011**, *88*, 4496–4504. [CrossRef]
35. Wang, Z.; Zhang, B.; Liu, T. Empirical analysis on the factors influencing national and regional carbon intensity in China. *Renew. Sustain. Energy Rev.* **2016**, *55*, 34–42. [CrossRef]
36. Meng, X.; Cheng, G.; Zhang, J.; Wang, Y.; Zhou, H. Analyze on the spatialtemporal characteristics of GHG estimation of livestock's by life cycle assessment in China. *China Environ. Sci.* **2014**, *34*, 2167–2176.
37. Yao, C.; Qian, S.; Li, Z.; Liang, L. Provincial animal husbandry carbon emissions in China and temporal-spatial evolution mechanism. *Resour. Sci.* **2017**, *39*, 698–712.
38. Luo, Y.; Long, X.; Wu, C.; Zhang, J. Decoupling CO_2 emissions from economic growth in agricultural sector across 30 Chinese provinces from 1997 to 2014. *J. Clean. Prod.* **2017**, *159*, 220–228. [CrossRef]

39. Tian, Y.; Zhang, J.-b.; He, Y.-y. Research on Spatial-Temporal Characteristics and Driving Factor of Agricultural Carbon Emissions in China. *J. Integr. Agric.* **2014**, *13*, 1393–1403. [CrossRef]
40. Xiong, C.; Yang, D.; Xia, F.; Huo, J. Changes in agricultural carbon emissions and factors that influence agricultural carbon emissions based on different stages in Xinjiang, China. *Sci. Rep.* **2016**, *6*, 36912. [CrossRef] [PubMed]
41. Tian, J.; Yang, H.; Xiang, P.; Liu, D.; Li, L. Drivers of agricultural carbon emissions in Hunan Province, China. *Environ. Earth Sci.* **2016**, *75*, 1–17. [CrossRef]
42. Xiong, C.; Yang, D.; Huo, J. Spatial-Temporal Characteristics and LMDI-Based Impact Factor Decomposition of Agricultural Carbon Emissions in Hotan Prefecture, China. *Sustainability* **2016**, *8*, 262. [CrossRef]
43. Yao, C.; Qian, S.; Mao, Y.; Li, Z. Decomposition of impacting factors of animal husbandry carbon emissions change and its spatial differences in China. *Trans. Chin. Soc. Agric. Eng.* **2017**, *33*, 10–19.
44. Zhuang, M.; Gongbuzeren, L.W. Greenhouse gas emission of pastoralism is lower than combined extensive/intensive livestock husbandry: A case study on the Qinghai-Tibet Plateau of China. *J. Clean. Prod.* **2017**, *147*, 514–522. [CrossRef]
45. Luo, T.; Yue, Q.; Yan, M.; Cheng, K.; Pan, G. Carbon footprint of China's livestock system—A case study of farm survey in Sichuan province, China. *J. Clean. Prod.* **2015**, *102*, 136–143. [CrossRef]
46. Xue, B.; Wang, L.Z.; Yan, T. Methane emission inventories for enteric fermentation and manure management of yak, buffalo and dairy and beef cattle in China from 1988 to 2009. *Agric. Ecosyst. Environ.* **2014**, *195*, 202–210. [CrossRef]
47. Xu, X.; Lan, Y. A comparative study on carbon footprints between plant- and animal-based foods in China. *J. Clean. Prod.* **2016**, *112*, 2581–2592. [CrossRef]
48. Wang, L.-Z.; Xue, B.; Yan, T. Greenhouse gas emissions from pig and poultry production sectors in China from 1960 to 2010. *J. Integr. Agric.* **2017**, *16*, 221–228. [CrossRef]
49. Ang, B.W. LMDI decomposition approach: A guide for implementation. *Energy Policy* **2015**, *86*, 233–238. [CrossRef]
50. National Development and Reform Commission of China (NDRCC). *Guidelines for the Development of Low Carbon and Provincial Greenhouse Gas Inventories*; NDRCC: Beijing, China, 2013.
51. IPCC. *IPCC Guidelines for National Greenhouse Gas Inventories Volume 4: Agriculture, Forestry and Other Land Use*; IPCC: Geneva, Switzerland, 2006.
52. IPCC. *Clim. Chang. 2007: Synthesis Report. Contribution of Working Groups I, II and III to the Fourth Assessment Report of the Intergovernmental Panel on Climate Change*; Core Writing Team, Pachauri, R.K., Reisinger, A., Eds.; IPCC: Geneva, Switzerland, 2007.
53. Wang, G.G.; Wang, M.L.; Yang, C. Characteristics and Mechanism of Animal Husbandry's Geographical Agglomeration in China. *J. Nat. Resour.* **2014**, *29*, 2137–2146.
54. Thornton, P.K. Livestock production: Recent trends, future prospects. *Philos. Trans. R. Soc. B Boil. Sci.* **2010**, *365*, 2853–2867. [CrossRef] [PubMed]
55. Herrero, M.; Thornton, P.K. Livestock and global change: Emerging issues for sustainable food systems. *Proc. Natl. Acad. Sci. USA* **2013**, *110*, 20878–20881. [CrossRef] [PubMed]
56. Ren, J. *Grassland Science*; China Agriculture Press: Beijing, China, 1961.
57. Fu, Q.; Zhu, Y.; Kong, Y.; Sun, J. Spatial analysis and districting of the livestock and poultry breeding in China. *J. Geogr. Sci.* **2012**, *22*, 1079–1100. [CrossRef]
58. Wang, G.; Wang, M.; Wang, J.; Yang, C.; Liu, Y. Characteristics and influencing factors of grass-feeding livestock breeding in China: An economic geographical perspective. *J. Geogr. Sci.* **2016**, *26*, 501–512. [CrossRef]
59. Key, N.; Tallard, G. Mitigating methane emissions from livestock: A global analysis of sectoral policies. *Clim. Chang.* **2011**, *112*, 387–414. [CrossRef]
60. Wang, W.; Koslowski, F.; Nayak, D.R.; Smith, P.; Saetnan, E.; Ju, X.; Guo, L.; Han, G.; de Perthuis, C.; Lin, E.; et al. Greenhouse gas mitigation in Chinese agriculture: Distinguishing technical and economic potentials. *Glob. Environ. Chang.* **2014**, *26*, 53–62. [CrossRef]

sustainability

MDPI

Article

Results from On-The-Ground Efforts to Promote Sustainable Cattle Ranching in the Brazilian Amazon

Erasmus K.H.J. zu Ermgassen [1,*], Melquesedek Pereira de Alcântara [2], Andrew Balmford [1], Luis Barioni [3], Francisco Beduschi Neto [4,5], Murilo M. F. Bettarello [6], Genivaldo de Brito [7], Gabriel C. Carrero [2,8], Eduardo de A.S. Florence [4], Edenise Garcia [7], Eduardo Trevisan Gonçalves [9], Casio Trajano da Luz [9], Giovanni M. Mallman [7], Bernardo B.N. Strassburg [10,11], Judson F. Valentim [12] and Agnieszka Latawiec [10,11,13,14]

1 Conservation Science Group, Department of Zoology, University of Cambridge, David Attenborough Building, Pembroke Street, Cambridge CB2 3EQ, UK; apb12@cam.ac.uk
2 Institute for Conservation and Sustainable Development of the Amazon, Rua Barão de Solimões 12, Manaus, AM 69058-250, Brazil; melk.alcantara@idesam.org.br (M.P.d.A.); gabriel.carrero@idesam.org.br (G.C.C.)
3 Embrapa Agriculture Informatics, Av. André Tosello, nº 209, Campus da Unicamp, Campinas, SP 13083-886, Brazil; luis.barioni@embrapa.br
4 Instituto Centro de Vida, Av. Ariosto da Riva, 3473- St E, Alta Floresta, MT 78580-000, Brazil; chicobeduschineto@gmail.com (F.B.N.); florenceaf@hotmail.com (E.d.A.S.F.)
5 Grupo de Trabalho da Pecuária Sustentável, Av. Paulista 1754 -Cj. 146, Bela Vista, SP 01310-920, Brazil
6 Via Verde Consultoria Agropecuária em Sistemas Tropicais, Rua Mariana Amaral 138, São Sebastião do Paraíso, MG 37950-000, Brazil; mbettarello@viaverde.agr.br
7 The Nature Conservancy, Belém, PA 66035-115, Brazil; gdebrito@tnc.org (G.d.B.); egarcia@tnc.org (E.G.); gmallmann@tnc.org (G.M.M.)
8 Department of Geography, University of Florida, 3141Turlington Hall, Gainesville, FL 32611-7315, USA
9 Instituto de Manejo e Certificação Florestal e Agrícola, Estrada Chico Mendes 185, Piracicaba, SP 13426-420, Brazil; eduardo@imaflora.org (E.T.G.); casio@imaflora.org (C.T.d.L.)
10 International Institute for Sustainability, Estrada Dona Castorina 124, Rio de Janeiro, RJ 22460-320, Brazil; b.strassburg@iis-rio.org (B.B.N.S.); a.latawiec@iis-rio.org (A.L.)
11 Rio Conservation and Sustainability Science Centre, Department of Geography and the Environment, Pontifical Catholic University of Rio de Janeiro, Rio de Janeiro, RJ 22453-900, Brazil
12 Embrapa Acre. Rodovia BR-364, km 14, Caixa Postal 321., Rio Branco, SC 69908-970, Brazil; judson.valentim@embrapa.br
13 School of Environmental Sciences, University of East Anglia, Norwich NR4 7TJ, UK
14 Institute of Agricultural Engineering and Informatics, Faculty of Production and Power Engineering, University of Agriculture in Krakow, Balicka 116B, 30-149 Krakow, Poland
* Correspondence: erasmus.zu.ermgassen@gmail.com or ekhjz2@cam.ac.uk; Tel.: +44-01223-765398

Received: 12 March 2018; Accepted: 16 April 2018; Published: 23 April 2018

Abstract: Agriculture in Brazil is booming. Brazil has the world's second largest cattle herd and is the second largest producer of soybeans, with the production of beef, soybeans, and bioethanol forecast to increase further. Questions remain, however, about how Brazil can reconcile increases in agricultural production with protection of its remaining natural vegetation. While high hopes have been placed on the potential for intensification of low-productivity cattle ranching to spare land for other agricultural uses, cattle productivity in the Amazon biome (29% of the Brazilian cattle herd) remains stubbornly low, and it is not clear how to realize theoretical productivity gains in practice. We provide results from six initiatives in the Brazilian Amazon, which are successfully improving cattle productivity in beef and dairy production on more than 500,000 hectares of pastureland, while supporting compliance with the Brazilian Forest Code. Spread across diverse geographies, and using a wide range of technologies, participating farms have improved productivity by 30–490%. High-productivity cattle ranching requires some initial investment (R$1300–6900/ha or US$410–2180/ha), with average pay-back times of 2.5–8.5 years. We conclude by reflecting on the challenges that must be overcome to scale up these young initiatives, avoid rebound increases in deforestation, and mainstream sustainable cattle ranching in the Amazon.

Keywords: livestock; Amazon; beef; dairy; sustainable intensification; land sparing

1. Introduction

There is growing competition for land use in Brazil. Beef, soy, and bioethanol production are forecast to grow 24%, 39%, and 27%, respectively, in the next decade [1], even as the government has committed to reforest 12 million hectares of land and reduce deforestation—with zero illegal deforestation by 2030 [2]. As pasture makes up the majority of agricultural land, high hopes are placed on the potential for increases in cattle productivity to spare land and accommodate the expansion of other land uses.

The productivity of Brazilian beef production is currently low; only one-third of its sustainable potential [3]. Brazil could in theory meet demand for beef, crops, and timber until 2040 without further conversion of natural ecosystems, by increasing cattle productivity to half of that potential [3]. Since livestock make up 37% of Brazil's greenhouse gas emissions [4] and extensive cattle ranching has historically been associated with deforestation, cattle productivity improvements are also key to Brazil's climate goals [5]. It is hoped that cattle intensification will reduce greenhouse gas emissions through land sparing [6], increased soil carbon sequestration [7], and increased greenhouse gas intensity [8]. The Brazilian contribution to the United Nations Framework Convention on Climate Change (UNFCC, New York, NY, USA), includes commitments to reduce deforestation and increase cattle productivity through the restoration of 15 million hectares of degraded pasture [2].

In this study, we report the results from six on-the-ground initiatives which have been working to turn theory into practice by increasing the productivity of cattle ranching in the Brazilian Amazon, a region with low productivity and high potential [3]. First, we describe the current state of beef and dairy production in the Brazilian Amazon, before we summarize the results from six initiatives which are raising cattle productivity in the region. We show that there are many ways for cattle ranching production to be increased on existing pastureland; these initiatives are diverse in geography and the technologies adopted, and we summarize common successes and challenges faced by all. We then finish by reflecting on the risks and mechanisms for achieving wide-scale higher-productivity cattle ranching in the region.

1.1. Beef Production in the Brazilian Amazon

Nearly one third (29%) of the Brazilian cattle herd, the second largest in the world, is found in the Amazon biome (Supplementary Material). Beef production in the region is characterized by extensive, pasture-based systems. Farmers traditionally keep zebu cattle breeds—80% of cattle are Nelore *Bos indicus* [9]—and use few chemical inputs (e.g., fertilizers) and little active pasture management, leading to gradual soil degradation and loss of productivity [10,11]. By some estimates, 40% of pastures are in a moderate or advanced state of degradation [12], and cattle stocking rates are well below their potential [3], with little increase seen since the early 2000s [13]. These systems are typically only marginally profitable [14].

The cycle of pasture degradation and low profitability has meant that cattle ranching has historically been associated with deforestation; pasture makes up 60% of deforested land in the Legal Amazon region [15]. Recently, beef production and deforestation have uncoupled (Figure 1) and there is growing acknowledgement of the complex mix of drivers underlying deforestation. From the mid-2000s onwards, deforestation fell 70% through a combination of improvements in enforcement on private land [16], expansion of protected areas [17], market-initiatives [18], and an economic slowdown [19]. As deforestation again creeps upwards [20], debate continues about the relative importance of beef production, land speculation, and the rapid expansion of cropland as underlying drivers of deforestation in the Amazon [21–24].

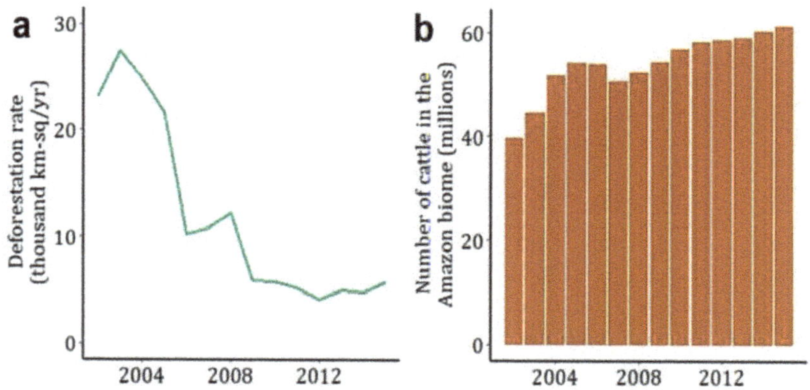

Figure 1. (**a**) Deforestation fell 70% from the mid-2000s onwards; (**b**) even as the cattle herd in the Amazon biome continued to grow. Data from: [25,26].

Finally, Amazonian beef is becoming increasingly integrated into the global economy. Improved animal health control, such as expanding the zone of eradication of foot-and-mouth disease, has facilitated a growth in exports [27]. While most beef from the Legal Amazon region is still consumed domestically, exports have more than doubled from <5% of production in early 2000s to 13.5–17.4% of production by 2011 (Figure S1, Supplementary Material).

1.2. Dairy Production in the Brazilian Amazon

Dairy production in the Amazon is a smaller scale operation than beef ranching. Dairy cattle make up only 3.9% of all cattle in the Amazon biome [25], which is responsible for 6.3–8.7% of Brazilian milk production [25]. Dairy farming is dominated by family farms (Figure 2), producing milk for subsistence or the local market. These farms have up to 70 cattle per farm, with low use of chemical inputs and a strong reliance on family labor [28,29]. Milk production is pasture-based, with some farms providing supplementary feed (e.g., sugar cane silage or concentrates) in the dry season or at the milking parlor.

Dairy productivity is therefore low and can be improved. Most dairy cattle are dual-purpose zebu breeds, though the use of dairy breeds and cross-breeds is increasing, for example, the number of registered Gir cattle (a specialized dairy breed) increased 70% (to more than 300,000 cattle) from 2007–2012 [30], though they still make up only a small proportion of the 22 million milked cows in the country [25]. Amazon municipalities have a median productivity of 689 L/cow/yr, which is lower than the median for the rest of Brazil (1224 L/cow/yr), and lags behind other international milk producers, such as New Zealand and the European Union, which produce 3500–4200 and 4000–8000 L/cow/yr, respectively [25,31,32].

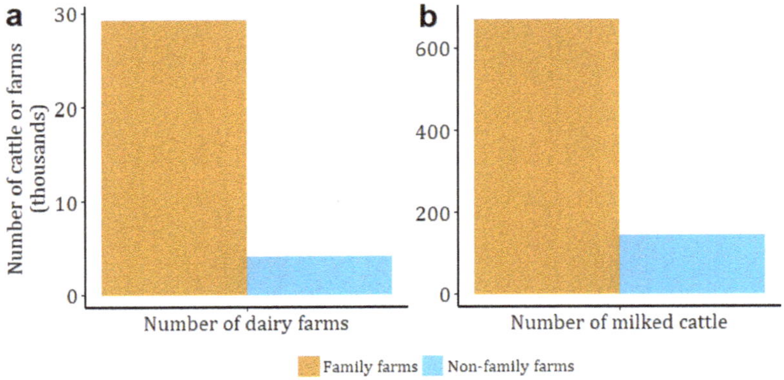

Figure 2. Dairy farming is dominated by family farms, both in terms of (**a**) the number of properties, and (**b**) number of dairy cattle. Data from: [33]. Family farming in Brazil is legally defined by a maximum farm size (ranging from 20–440 ha, dependent on the region), the number of permanent employees, and the proportion of non-agricultural income.

2. Materials and Methods

A questionnaire about the financial and production performance of sustainable cattle ranching initiatives (available in the Supplementary Material) was shared with the representatives of four organizations (The Nature Conservancy, Instituto Centro da Vida, Embrapa, and Idesam) who attended a conference on sustainable cattle ranching initiatives in Rio de Janeiro in September 2015. To increase our sample size, a further three initiatives operating in the Amazon region were then contacted (either directly or through the Brazilian Roundtable for Sustainable Beef, Portuguese acronym, GTPS); one of which (Florestas de Valor) provided sufficient data to participate. In total, six initiatives (led by five organizations) participated, as outlined in Table 1.

Two versions of the survey were circulated, one for beef and one for dairy intensification initiatives, structured as follows. Questions were grouped into eight sections on the (i) overview of the project (name of the initiative, and institutions involved); (ii) characteristics of the initiative (the number and types of farm participating, number of cattle and area of pasture intensified, and the year the initiative began); (iii) details of the package of technologies implemented on participating farms (farm and pasture management, forage species, use of supplementary feed, etc.); (iv) the costs involved in the implementation of improved farm management; (v) the costs involved in maintenance of improved pasture; (vi) the productivity achieved on the farm, in terms of stocking rates (animal units/ha, where one animal unit is equivalent to a 450 kg cow), beef production (in arroba/hectare/yr, where one arroba, abbreviated as "@", is a common Brazilian livestock unit, equivalent to 15 kg of carcass deadweight), or milk production (liters of milk per cow and per hectare per year); (vii) details of other measures of performance (e.g., environmental compliance, greenhouse gas emissions); and (viii) details of how farmers were recruited to each initiative and the respondent's reflections on the barriers and opportunities for improved cattle ranching.

The surveys were completed by project managers and field technicians for each initiative, who are co-authors of this review. Where survey responses were not clear, they were clarified via email by the first author. Survey data was complemented with published results from initiatives where available (e.g., [34–41], and all authors provided substantial revision of the manuscript text to ensure it accurately describes each intervention.

Table 1. Characteristics of the cattle intensification initiatives surveyed.

Name of Initiative	Lead Organization	Location	Beef or Dairy	Most Important Management Features	Year Project Started	Number of Farms	Hectares of Land under Intensification	Number of Cattle	Mean Farm Size, Hectares (Range)	Forest Code Compliance Required?
Intensification of beef cattle production systems with the use of mixed grass-legume pastures in Acre	Embrapa	State of Acre	Beef	Vegetative planting of mixed legume-grass pastures, persistent legume supply of symbiotically fixed nitrogen	*Pueraria phaseoloides* introduced in 1976;	5400 [a]	480,000 [a]	No data available	No data available	NA
					Arachis pintoi introduced in 1999	2000	137,600			
Novo Campo Program	ICV	Alta Floresta, Nova Canaã do Norte, Paranaíta e Cotriguaçu (MT)	Beef	Pasture rotation, pasture fertilization, application of GAP	2012	23	14,300	23,800	200 (30–900)	Yes
Do Campo à Mesa	TNC	São Félix do Xingu (PA)	Beef	Pasture rotation, pasture fertilization, application of GAP	2013	13	20,208	34,043	3077 (100–6900)	Yes
Silvopastoral system with rotational grazing for beef	Idesam	Apuí (AM)	Beef	Pasture rotation, agroforestry with timber and leguminous trees, improved book-keeping	2011	10	236	566 [b]	570 (53–3020)	Yes
Silvopastoral system with rotational grazing fo dairy	Idesam	Apuí, Manicoré, Novo Aripuanã (AM)	Dairy	Pasture rotation, agroforestry with leguminous trees, improved book-keeping, and drinking water system	2014	11	95	332 [b]	188 (83–340)	Yes
Florestas de Valor	IMAFLORA	São Félix do Xingu (PA)	Dairy	Rotational grazing, leguminous trees lining fenced plots	2015	6	50	145	83 (25–200)	Yes

[a] Figures from 2004, the last year that production practices in the region were surveyed; [b] Estimate based on mean stocking rates and pasture area of farms. Ranges are listed in brackets, where provided. State abbreviations: Embrapa = the Brazilian Corporation for Agriculture Research; ICV = Instituto Centra da Vida; TNC = The Nature Conservancy; Idesam = Institute for Conservation and Sustainable Development of the Amazon; IMAFLORA = Institute of Forestry and Agricultural Management and Certification; MT = Mato Grosso; AM = Amazonas; PA = Pará.

3. Results

We provide results from six sustainable cattle intensification initiatives in the Amazon biome, four working with beef producers and two with dairy producers (Table 1). While one of these initiatives was launched in 1976 and introduced legume pasture technologies which have since been adopted on more than 5000 farms, the remaining initiatives are more recent (established post-2011). These latter initiatives operate on 63 farms raising 59,000 cattle on 35,000 hectares of pasture in three states (Figure 3). The technologies deployed are diverse, ranging from relatively low-input leguminous systems to more input-intensive rotational grazing systems.

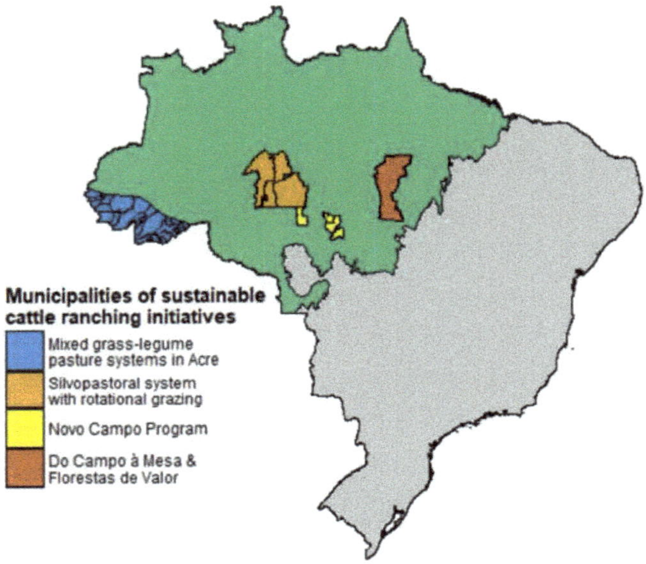

Figure 3. Map of Brazil, with the Amazon biome colored in green, and the municipalities where the sustainable cattle initiatives reported in this article are present shown in other colors.

Each initiative has achieved higher farm productivity, boosting meat production per hectare by 30–270% and dairy production per hectare up to 490% (Tables 2 and 3). While the use of higher-yielding technologies is profitable in most cases, it requires initial investment to improve farm productivity, with payback times ranging from 1.5–12 years. The specifics and results of each initiative are described in more detail below.

3.1. *Beef Case Study #1—Intensification of Cattle Production Systems with the Use of Mixed Grass-Legume Pastures in Acre*

In 1976, the Brazilian Corporation for Agriculture Research (Embrapa, Brasília, Brazil) established the Program for Reclamation, Improvement and Management of Pastures in the Brazilian Amazon (PROPASTO) which included a series of on-farm experiments to promote the adoption of mixed legume pastures in the state of Acre [40]. A number of cultivars were launched, of which one legume, *Pueraria phaseoloides* (tropical kudzu), was the first to be adopted at scale. By 2004, tropical kudzu was present in over 30% (480,000 ha) of the total pasture area in Acre and has been successfully planted in combination with a variety of grass species (Table S1, Supplementary Material) [40].

Table 2. Productivity and profitability of initiatives increasing productivity of beef production. Ranges listed in brackets, where provided.

Name of Initiative		Baseline Stocking Rate (AU/ha)	Stocking Rate (AU/ha)	Increase in Stocking Rate over Baseline	Baseline Productivity (@/ha/yr)	Productivity (@/ha/yr)	Average Increase in Productivity	Years to Break Even on Investment	Years to Achieve Max Productivity	Typical Profit/ Hectare/Year (R$)	Additional References
Intensification of beef cattle production systems with the use of mixed grass-legume pastures in Acre	Mixed grass-*Pueraria phaseoloides* pastures	1 *	1.5 (1–2)	1.5x	8 (4–10) *	10 (4.9–12.5)	1.3x	3 (2–4)	2 (1.5–3)	149–271	[39,42–44]
	Mixed grass-*Arachis pintoi* pastures		2.2 (1.5–3.59)	2.2x		12 (13–35)	1.5x	4 (3–5)	3 (2–5)	296–381	
Novo Campo Program		1.22 *	2.8 (1.5–3.5)	2.3x	4.7 *	10.8 (7–27)	2.1x	2.5 (1.5–4)	Data not provided	602 (173–1140)	[4,38]
Do Campo à Mesa		0.87 (0–2.81)	1.06 (0.27–3.05)	1.2x	4.5 (0.9–10.5)	5.87 (1.42–19.2)	1.3x	8.5 (7–12)	~6	432 (−546–1103)	[37]
Silvopastoral System with Rotational Grazing for Beef		0.60 (0.45–0.7)	2.4 (2.0–2.7)	4.0x	5.5 (4–7)	15 (12–20)	2.7x	5 (4–6)	5	~263	-

AU = animal stocking unit, equivalent to a 450 kg cow; @ = 15 kg of carcass (deadweight). * productivity data are not available from participating farms pre-adoption, and so the baseline data are estimates of the regional average productivity without the adopted technology. Estimates of profitability do not include revenues from farm activities not directly related with cattle production (e.g., sale of timber trees or crops), and costs are representative of the interventions made on participating farms (they do not, for example, consider the cost of acquiring land or purchasing cattle, as participating farms used on-farm resources for intensification).

Embrapa began by introducing mixed legume pastures on properties belonging to three farmers, who were identified as innovators [40]. Knowledge of these novel technologies then spread through word of mouth and trained agricultural extension officers. Legumes were promoted because of their ability to fix nitrogen, which reduces pasture maintenance costs and produces a protein-rich sward (Figure 4); a pasture sown with 20–45% Tropical kudzu produces nitrogen equivalent to approximately 60–120 kg of N/ha/yr [42]. Grass-legume associations cost between R$1350–2000/hectare to implement, and around R$100/ha/yr to maintain (Table 4) and are therefore a relatively low-cost intensification technology for pasture restoration and intensification. Tropical kudzu pastures produce modest productivity improvements, supporting 1.5 animal units/ha and producing 4.9–12.5 @/ha/yr (Table 2).

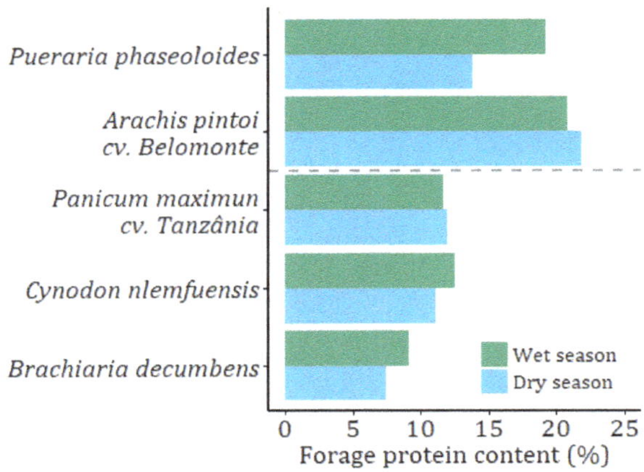

Figure 4. Leguminous pasture plants (above the dashed line) have higher protein content than conventional grasses (below dashed line). Source: [45].

Since a peak in the early 2000s, the popularity of Tropical kudzu has declined as it showed poor compatibility with some of the newer grass species being planted by farmers, such as African stargrass (*Cynodon nlemfuensis*), and also failed to persist when managed in more intensive production systems using mixed pastures with other grasses and rotational grazing at stocking rates above 1.5 animal units per hectare. For these situations, Embrapa promoted forage peanut cultivar Belomonte (*Arachis pintoi*) [41].

This new cultivar was released in 1999 in Bahia, Brazil. First planted by a single farmer in Acre in 2000, in April 2001 20 farmers planted this legume together with a variety of grasses (Figure 5 and Table S1, Supplementary Material). Adoption was rapid. By March 2004, close to 1000 small, medium and large farmers of Acre had already introduced forage peanut into 65,000 ha of pasture [41], and by 2015 forage peanut was planted across 2000 farms and 137,000 ha in Acre [39], approximately nine percent of the state's pasture area [46].

Forage peanut can be either planted along with other grasses during pasture restoration (i.e., replanting of a degraded pasture), or introduced onto existent pasture during the rainy season. Since forage peanut cultivar Belomonte does not produce seeds (it instead reproduces vegetatively), it must be planted using stolon cuttings. Farmers usually set aside an area (<1 ha) where forage peanut grows in dense stands (i.e., without competing grasses), from which the vegetative stolons are then harvested for planting in pasture. Embrapa have successfully developed a number of techniques for establishing grass-legume pastures using vegetative propagation of forage peanut and stoloniferous grasses, depending on the farmer's technology level, ranging from semi- to fully-mechanized and either conventional or no-till agriculture [34]. African stargrass-forage peanut pastures managed under rotational grazing can support up to 3 animal units/ha (Table 2), producing Nelore × Angus crossbred steers ready for slaughter within 24 months (Table S2, Supplementary Material), compared with the 36+ months typical of extensive systems [41]. These productivity improvements also improve the farm bottom line, increasing profitability from around R$41.10/ha/yr in traditional systems up to R$381.28/ha/yr (Table 2).

Figure 5. Nelore cattle (Bos indicus) grazing mixed pasture: forage peanut (Arachis pintoi cv. Belomonte) (the yellow flowering plants) with *Brachiara* spp.

Grass-legume pastures can mitigate greenhouse gas emissions by substituting for fossil-fuel dependent nitrogen fertilizers, by reducing cattle slaughter ages [42,47], and by increasing soil carbon sequestration [48]. Additionally, Costa et al. [49] reported that mixed pastures of *Brachiaria humidicola* and forage peanut cv. Mandobi in Acre had 24% lower N_2O emission (2.38 kg N ha^{-1} yr^{-1}) than pure pasture of the same grass (3.13 kg N ha^{-1} yr^{-1}) and similar emissions to native forest (2.47 kg N ha^{-1} yr^{-1}).

3.2. *Beef Case Study #2—Novo Campo Program*

The Novo Campo Program ("New Field" Program in English) has involved 23 farms in the Alta Floresta region of Mato Grosso since 2012. Led by the Instituto Centro de Vida (ICV), with collaboration with stakeholders from across the cattle supply chain (see Table 4 for a complete list of participating organizations), cattle productivity has been increased through a package of farm management changes. These include the introduction of pasture rotation, the adoption of so called "good agricultural practices" (GAP), correction of soil imbalances (e.g., by liming), pasture fertilization, and improved farm record keeping. The GAP is a voluntary set of "gold-standard" guidelines for sustainable production adopted across Brazil, which includes a check-list of 125 points of guidance across 11 areas of farm management, spanning farm economic management, social and environmental responsibilities, to pasture and herd management [50].

Together, these interventions have improved farm productivity and profitability, and reduced greenhouse gas emissions (Figure 6). Beef production per hectare has increased from ~4.7 (the regional average) to 7–24 @/ha/yr, which has reduced the cost of production per arroba on intensified farms by one third (R$66.33/@ vs. R$95.80/@) [38]. Profit increased from less than R$100/ha/yr to more than R$600/ha/yr of pasture (Table 2). These yield-raising technologies require an initial investment of R$1500–4000/ha, depending on the initial pasture condition, though these up-front costs are paid off after an average of 2.5 years (Table 2).

Table 3. Productivity of dairy intensification initiatives surveyed. Ranges are listed in brackets, where provided.

Name of Initiative	Baseline Stocking Rate (AU/ha)	Stocking Rate (AU/ha)	Increase in Stocking Rate over Baseline	Baseline Productivity (L/ha/yr)	Productivity (L/ha/yr)	Increase in Productivity Over Baseline	Baseline Productivity (L/cow/yr)	Productivity (L/cow/yr)	Increase in Productivity over Baseline	Years to Break Even on Investment	Years to Achieve Max Productivity	Typical profit/ha/yr (R$)
Silvopastoral System with Rotational Grazing for Dairy	0.75 (0.5–1.08)	3.5 (2.4–6.3)	4.7x	~1192	5794 (2969–9037)	4.9x	1551 (760–1825)	1954 (1642–2482)	1.26x	2.6 (1.8–6.8)	6 (5–7)	4425 (2176–8092)
Florestas de Valor	1.1 (0.9–1.2) *	3.1 (2.5–3.7)	2.8x	Data not provided	~3190	Data not provided	~760	~1100	1.4x	4 (3–5)	3 (2–4)	Data not provided

* productivity data are not available from participating farms pre-adoption, and so the baseline data are estimates of the regional average productivity without the adopted technology.

Table 4. Typical costs involved in each intensification initiative. Degraded pastures are pastures with declining pasture fertility; their restoration often requires soil correction, ploughing, and reseeding of grasses, whereas soil correction and ploughing may not be required for conventional pasture improvement. Ranges are listed in brackets, where provided.

Name of Initiative	Organizations Involved	Ranching Systems	Cost of Intensification (R$/ha)		Cost of Pasture Maintenance (R$/ha/yr)	Cost of Technical Assistance (R$/property/yr)
			Improvement of Degraded Pasture	Improvement of Conventional Pasture		
Intensification of beef cattle production systems with the use of mixed grass-legume pastures in Acre	Embrapa Acre, Federação de Agricultura do Estado do AcreFundo de Desenvolvimento da Pecuária do Estado do Acre, & Associação para o Fomento à Pesquisa de Melhoramento de Forrageiras	Cow-calf, calf raising & fattening, full cycle	Semi-mechanized conventional planting: 2011. Mechanized conventional planting: 1461–1920. Mechanized no-till planting: 1347–1806		~100	Data not collected
Novo Campo Program	ICV, International Institute for Sustainability (IIS), Embrapa, Solidariedad, Sindicatos Rurais de Alta Floresta e Cotriguaçu, JBS, McDonalds, Arcos Dourados, IMAFLORA, Althelia Ecosphere, Terras, GTPS, Fundo Vale, Norad, & the Moore Foundation	Calf raising & fattening	3500 (3000–4000)	2000 (1500–2000)	1800 (1500–2000)	8000 (6000–12000)
Do Campo à Mesa	TNC, Marfrig, Walmart, GTPS, & the Moore foundation	Calf raising & fattening	1890 (1750–1897)	1468 (1318–1571)	~680	Data not collected
Silvopastoral System with Rotational Grazing for Beef	Idesam, Centro para Investigación en Sistemas Sostenibles de Producción Agropecuária (CIPAV), Via Verde Consultoria Agropecuária, Fundo Vale, & Viveiro Santa Luzia	Cow-calf, calf raising & fattening	2666 (2412–3021)	All farms had degraded pasture	~216.25	~5480
Silvopastoral System with Rotational Grazing for Dairy	Idesam, CIPAV, & Via Verde Consultoria Agropecuária, Fundo Vale, & Viveiro Santa Luzia	Cow-calf	5355 (4900–6866)	All farms had degraded pasture	~275	~5480
Florestas de Valor	IMAFLORA, CAMPPAX (Cooperativa Alternativa Mista do Alto Xingu), ADAFAX (Associação Desenvolvimento da Agricultura Familiar do Alto Xingu), CFA (Casa Familiar Rural de São Felix do Xingu), Petrobras, Fundo Vale, & Fundo Amazônia	Cow-calf	~2500	All farms had degraded pasture	2000 (1200–2560)	2500 (2000–3000)

By improving animal growth rates and so achieving slaughter weight in fewer days, farmers reduce emissions from enteric fermentation across the animal's lifetime (enteric fermentation contributes 67–83% of emissions, excluding land use change—a topic we return to in Section 4 [8,51,52]). This is seen from the experience on Novo Campo Program farms. Emissions have been reduced by 36–59% (Figure 6), in large part through reductions in slaughter age down to 20–24 months (Table S2, Supplementary Material).

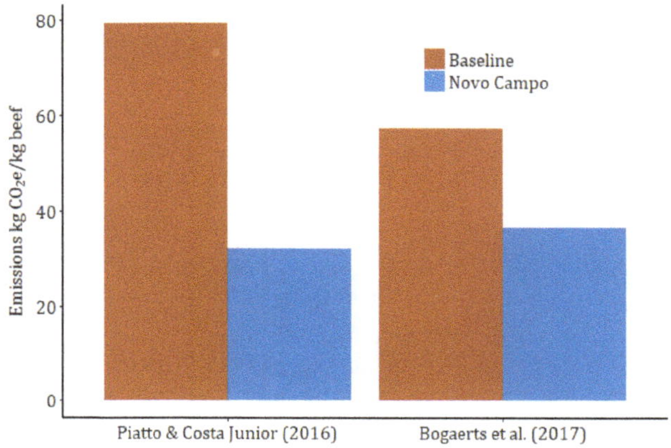

Figure 6. Two estimates for the emissions per kilogram of beef of conventional ranches and Novo Campo Program farms. Piatto and Costa Junior [53] compare emissions on pilot farms before pasture intensification (baseline), with emissions after two years of participating in the initiative. Bogaerts et al. [8] compare farms participating in the Novo Campo Program with neighboring non-participating farms. While both studies include emissions from enteric fermentation, manure management, pasture fertilization, and fossil fuels required for pasture restoration, Bogaerts et al. also include emissions from concentrate feed production, and Piatto and Costa Junior include carbon sequestration in improved pasture and soil carbon emissions from degraded pasture. No emissions from land use change are included, because no recent deforestation occurred on sampled farms.

Key changes in farm management are improved book-keeping and the introduction of rotational pasture management. Adequate book-keeping is fundamental to understanding and improving farm management processes, and yet is not done by a majority of farm managers or owners [4]. Farmers are therefore trained in the importance of recording the costs of all inputs and the quantity and value of beef produced, to allow the calculation of the income and profit per arroba of beef. Once the economic performance of the farm is established, rotational grazing is introduced.

Typically, 10–30% of the farms' pasture area is fenced off into ca. 5-hectare plots, which are targeted for pasture improvement. Pasture restoration begins with soil analysis to identify soil imbalances (e.g., pH). The pasture is then ploughed and limed (typically with 1500 kg/ha lime; Table S3, Supplementary Material), and the pasture is fertilized (400 kg/ha) and replanted, with *Panicum maximun* cv. Mombaça *or Panicum maximun* cv. Tanzânia grasses (Figure 7).

These fertilized plots have much higher productivity than conventionally managed pasture; in the first two years of the project, they produced 20.75 @/ha/yr compared with 10.75 @/ha/yr across the farm as a whole (Marcuzzo and de Lima, 2015). Cattle are moved through each fenced plot sequentially; the stocking rate and exact timing of the cattle rotation are based on the season, forage height, and planted species, manipulated to maximize cattle growth while maintaining pasture fertility. With *Panicum maximun* cv. Mombaça, cattle enter plots when the grass height is around 90 cm and are moved when it has been grazed down to around 40 cm (approximately every five days in the wet season, and less frequently during the dry season).

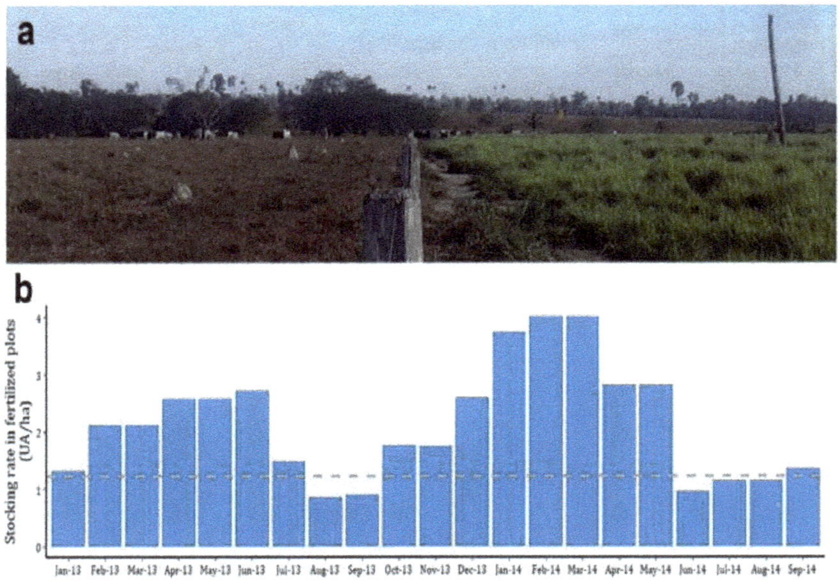

Figure 7. (**a**) High-yielding cattle pasture (right) on a Novo Campo Program farm, one month after replanting, compared with conventional, unreformed pasture (left); (**b**) Stocking rate in intensified pasture plots for the period January 2013–September 2014. The grey dashed line represents the mean stocking rate for farms in the region. Adapted from: [38].

While forage is in abundant supply in the rainy season (from approximately December–May), in the dry period, stocking rates in the intensified pasture areas are reduced (Figure 7), and supplementary feeding is necessary. Novo Campo Program farms have adopted a semi-feedlot feeding approach, where cattle are given supplementary concentrate feed in troughs in a confined area of pasture. One farm has also integrated soy and beef production to boost dry season feed availability. Soy is planted on 200 ha, which is seeded with *Brachiaria* spp. after the soy harvest. This additional pasture area then serves as an additional forage source during the dry period.

All Novo Campo participating farmers qualify for the GAP certificate, developed by Embrapa. The adoption of GAP requires training of farm personnel, and ultimately, approximately a 50% increase in on-farm labor. To support the dissemination of knowledge to staff on Novo Campo farms and beyond, ICV therefore linked up with a local university, UNEMAT Alta Floresta, to train an additional 40 agricultural extension officers in GAP, environmental licensing, farm financial analysis, and the use of farm management software [38].

Farms participating in the Novo Campo Program must also comply with Brazilian National Law No. 12.651, the so called 'Forest Code'. They must be registered in the rural environmental registry (Portuguese acronym CAR), cannot be blacklisted by the environmental police (IBAMA), and must have had no illegal deforestation post-2008. On joining the Novo Campo Program, many farms had degraded riparian areas, which legally must be reforested within 20 years. Properties received support from ICV in restoring these areas, with the restoration actions depending on the degradation status and location of streams. Where streams had some secondary regrowth and/or nearby forest, this might include only fencing-off streams from cattle to foster natural regeneration; where riparian areas were more degraded or isolated, they may have required direct seeding of trees, removal of grasses and the control of pests. In both cases, restoration is not cheap with costs varying from R$2360/ha for passive restoration, to R$9654/ha for active replanting [54].

To scale-up the results achieved in the Novo Campo Program, a commercial spin-off, PECSA, was launched in 2015. While management of the Novo Campo continues under ICV, PECSA applies the same package of technologies, in some cases intensifying more than the 30% of pasture area used in the Novo Campo Program to increase farm productivity 3–5 times above the regional average [55].

3.3. Beef Case Study #3 Do Campo à Mesa

Launched in 2013, as a collaboration between The Nature Conservancy (TNC) and several partners along the beef supply chain (Table 4), the do Campo à Mesa initiative ("From Field to Table" in English) operates on 13 farms in São Félix do Xingu, Pará, to boost productivity through the establishment of rotational pasture and training in GAP. Results after one year of the project are promising; stocking rates increased 20% and beef productivity 30% (Table 2), with substantially greater gains expected; beef productivity is forecast to increase more than 3.5-times to 17@/ha (10–27@/ha) within 12 years of the start of the project [37].

On joining the initiative, baseline data were collected by agricultural extension officers on the herd structure (e.g., number of animals, category, age, weight), farm operating costs, and soil condition. These are complemented with remote sensing analyses of the farm's land use. Many farms had considerable areas of degraded pasture, 23% and 18% of pasture was moderately or highly-degraded, respectively [37]. To combat this, management plans were drawn up to improve 20% of the farm's pasture area each year so that after five years all pasture would be in improved condition.

Goals for stocking rates were set, aiming to gradually increase stocking rates from 0.87 AU/ha to 3.0 AU/ha. To achieve this, the initiative has used pasture improvement (weeding and liming, and resowing of pasture if degraded), education of farm workers in GAP, and the establishment of rotational grazing. Livestock are also given 1.5 kg/head/day of protein supplementary feed in the dry season, to overcome the seasonal deficit in feed availability.

The main costs of improving farming practices come in three forms: pasture improvement and maintenance, adoption of GAP, and costs of environmental compliance; these costs vary strongly with farm size. Pasture improvement requires an initial investment of between R$1300–1900/ha/yr (Table 4), and adoption of GAP requires not only worker education, but also improvements in infrastructure and more farm labor. After 12 years of the project, the requirement for labor is forecast to increase on average 54%, with larger increases on the biggest farms [37]. On small farms labor is family-based and will be kept constant, while large farms plan to increase their number of employees three-fold. Do Campo à Mesa farms must also be compliant with the Forest Code, which also incurs substantial costs. Compulsory restoration of deforested areas added an extra 30–250% to the cost of adopting improved farm practices [37]. Taking all the costs of the transition to more sustainable farming practices together, pasture intensification and legal compliance generated better economic returns for large farms (>500 ha of pasture) ([37]. Costs per hectare for the three smallest properties were on average 2.3 times higher than for other farms, and the two smallest farms, with 44 and 126 ha of pasture area, were not projected to make a profit and subsequently elected not to continue with cattle intensification. To help overcome the initial cost barrier and support the growth of the program, TNC helps farmers apply for loans from the Brazilian government's low carbon fund (the "Plano ABC", in Portuguese).

3.4. Beef Case Study #4—Silvopastoral System with Rotational Grazing for Beef

In 2011, the Institute for Conservation and Sustainable Development of the Amazon (Portuguese acronym, Idesam) launched the Silvopastoral System with Rotational Grazing initiative ("Sistema Silvipastoril com Pastejo Rotacional", in Portuguese), on beef and dairy farms in Apuí, Amazonas. The initiative is working with 10 beef farms to boost productivity of smallholder beef production (results for dairy farms are listed in diary case study #1). While the planting of trees and shrubs, involves high up-front costs (Table 4), participating beef farms have improved productivity from 4–7@/ha/yr to 12–20@/ha/yr, and profitability from ~R$130/ha/yr to R$260/ha/yr (Table 2).

Farm improvement begins with a visit from an agricultural extension technician, collecting baseline farm information and drawing up management plans with the farmer. To introduce rotational grazing, an area of between 20–50 hectares is intensified on each farm by restoring pasture through the application of lime, where required. This area is then divided into six plots, sown with *Panicum maximum cv. Mombaça* or *Brachiaria brizantha*, fertilized with phosphorus, and managed in a rotational system. Cattle are moved through each plot approximately every 6–7 days according to the pasture condition.

These plots are divided by double electric fences (1.5 to 2 m in width) protecting a line of trees planted 3 m apart (Figure 8). The trees are mostly native species, half of which are planted for their timber or other economic value and the other half are a mix of leguminous tree species (20–30 trees/ha), including Inga-de-metro (*Inga edulis Mart.*) Leucaena (*Leucaena leucocephala var. cunningham*) Paricá (*Schizolobium amazonicum*) Gliricídia (*Gliricidia sepium*), Jatobá (*Hymenaea courbaril*), and *Parkia* spp. Among the trees, fodder shrubs are also planted, including *Tithonia diversifolia* and Cratília (*Cratylia argentea*). The principal benefits of planting leguminous trees and fodder shrubs in pasture are that the leaves provide a high-protein feed [56], increased shade which can reduce heat stress in cattle [57], and nitrogen-fixation which boosts grass growth and can improve soil condition [58].

Figure 8. (**a**) Producer standing with a tree line of 4-month old leguminous trees in an Idesam silvopastoral system; (**b**) silvopastoral system once the bushes and trees are 2 years old.

Silvopastoral systems do, however, require careful management and substantial initial investment. Trees need protection from heavy grazing for the first 12–24 months post-planting; thereafter they require occasional pruning if they get too broad in order to avoid excessive shade hindering grass growth [59]. Farmers must also closely monitor herd performance, including daily recording of stocking rates. These changes require on average a 20% increase in on-farm labor. Farmers are supported throughout the process by monthly visits from Idesam's agricultural extension staff. Costs of implementing silvopastoral systems are high, R$2400–3020/ha, though this is offset by low maintenance costs around R$220/ha/yr, in part because leguminous pastures do not require any nitrogen fertilizer application.

While it is hoped that productivity increases will reduce greenhouse gas emissions on participating farms, a recent analysis found that participating farms had higher greenhouse gas emissions than neighboring farms (47 vs. 40 kg CO_2e/kg beef) [8]. These results should, however, be treated with caution as the analysis used the Cool Farm Tool, an out-of-the-box greenhouse gas calculator which is not tailored for measuring emissions from integrated systems, and the input data were collected less than one year after the program's implementation. The environmental and economic impacts of integrated systems, such as silvopastoral systems, are difficult to model because the different parts of the management system interact [60], in this case, leguminous trees fertilize the pasture, supporting grass growth. The Cool Farm Tool, though comprehensive in many respects, does not consider these interactions, simplifying the farm's environmental footprint and potentially over-estimating emissions.

Similarly, while the Cool Farm Tool can calculate carbon sequestered in trees on-farm, this source of sequestration was not included in Bogaerts et al. [8]. Additionally, the farm-level data used were collected shortly after the implementation of rotational grazing, therefore, emissions associated with pasture improvement were counted before productivity gains had been realized. As it is expected to take five years for the systems to achieve full productivity (Table 2), using data from only the first year overestimates emissions from participating farms.

Finally, to participate in the initiative, farms must also be compliant with environmental legislation. They must be registered on the CAR, develop a PRAD (the "Projeto de Recomposição de Áreas Degradadas e/ou Alteradas", a plan for restoration if the property does not meet minimum legal requirements for forest cover), and agree to not clear any new areas of forest.

3.5. Dairy Case Study #1—Silvopastoral System with Rotational Grazing for Dairy

Idesam also work with 11 smallholder pilot farms (ranging from 83–340 ha in size; Table 1) in the state of Amazonas, to increase dairy productivity through the rotational management of pasture lined with timber and leguminous trees, and shrubs. As for Idesam's beef intensification in the region, the dairy initiative has seen productivity improvements, a 1.26-fold increase in milk production per cow and 4.9-fold increase in milk production per hectare (Table 3).

Plots of intensively managed pasture are divided by doubled electric fences protecting a line of trees and shrubs. Compared with Idesam's beef system, the dairy farms use a greater number of plots (~40) and trees (50 to 110/ha). Around 6 hectares is targeted for intensive management on each farm, with 0.1 to 0.9 hectares per plot. Forty-four percent of the tree species were planted to provide shade for cattle and timber as a source of long-term income for farmers; 56% were leguminous [61]. As often required in the region, the soil was supplemented with lime, before planting *Brachiaria brizantha*, *Panicum maximum cv. BRS zuri* or *cv. Massai* grasses, with phosphorus added as necessary. These grasses show high productivity in the shady conditions typical of silvopastoral systems [57,62]. Laboratory experiments have shown, for example, that shade can even increase the protein content of *Panicum maximum* grasses [63].

The system is managed in rotation, where cattle are moved through plots every 12 h to two days, depending on grass height. In the drier months, the deep-rooted leguminous trees continue to provide a source of fodder, and some farmers supplement feed with maize silage or "cut-and-carry" feeding of *Tithonia diversifolia*, *Inga edulis*, and *Cratylia argentea*. Lactating cattle on some farms also receive 1.5 kg of maize-based concentrate feed at milking each day. Water availability is crucial for high-productivity dairy production, and so drinking water is pumped into elevated water boxes, which distribute it by gravity through a system of buried hoses to each pasture plot (Figure 9).

Figure 9. (**a**) Installation of underground piping to deliver water to each plot in the Silvopastoral System with Rotational Grazing for Dairy initiative; (**b**) Dairy cattle drinking from one of the water troughs, with electric fencing in the foreground.

Systems with leguminous trees require approximately 15% more labor than conventional pasture-based systems. The trees require protection from grazing and insects for the first few months, as well as intermittent pruning during the first three years. The requirement for tree care reduces as the trees mature, and the rotational management of cattle in these systems requires no additional specific management, as cattle are moved twice a day for milking in any case.

The implementation of leguminous systems can be costly, ranging from R$4900–6900/ha to cover the costs of pasture reformation, tree planting, electric fencing (for managing rotational grazing), and construction of water sources in each plot (Table S3, Supplementary Material). Though these initial costs are paid off within 2–7 years (Table 3), as improved management boosts profitability from R$1281.15/ha/yr to around R$4425/ha/yr, Idesam has provided financial support to the first farmers of the program. Farmers paid 20% of the cost of implementation, with Idesam covering the remaining 80%. To access this financial support and participate in the initiative, farmers must commit to legal compliance with the Forest Code. Farms must be registered in the CAR, commit to not deforest further, and restore non-forested areas and degraded riparian strips, in line with the PRAD.

3.6. *Dairy Case Study #2—Florestas de Valor*

The dairy intensification project Florestas de Valor ("Forests of Value" in English) was launched by the Institute of Forestry and Agricultural Management and Certification (Portuguese acronym, IMAFLORA) in 2015, and operates on six farms in São Félix do Xingu in the state of Pará. By concentrating production on a small, intensively managed portion of pasture in each farm, they have increased stocking rates almost three-times above the regional average, with 85% higher productivity per cow (3240 L milk/cow/yr vs. 1750 L/cow/yr; Table 3).

Florestas de Valor operates on small properties, ranging from 25–200 hectares in size (Table 1). These farms rely almost entirely on family labor, and so it is important that the intensification does not increase the overall requirement for labor. This is achieved by focusing production on a small area in each farm, where 3.5–11 hectares are selected for intensification and divided into 10–15 fenced plots (Figure 10). The soil in each plot is analyzed before soil correction, and either direct resowing with *Brachiaria brizantha MG5*, *Panicum maximum cv. Mombaça*, *Brachiaria decumbens* or *Panicum maximum cv. Massai*, or pasture restoration through crop-livestock integration. On four properties, maize was planted on degraded pasture; once the maize was harvested, pasture grasses were then sown. The fences between pasture plots are planted with leguminous trees, including *Canavalia ensiformis*, *Inga edulis* and *Cajanus cajan*. Trees were planted three meters apart, with an average of 66 trees per hectare. Cattle remain approximately three days in each plot, thereby completing a cycle of each plot every 30–45 days. In the dry season, when grass growth is slower and over-grazing is more likely, less time is spent in the fenced plots, and cattle are instead put onto pasture that has been intentionally rested.

Overall, it costs around R$2500/ha to implement the rotational grazing and leguminous tree systems. These costs stem from costs of soil improvement, grass seeds, maize planting, fencing, solar panels, and in-pasture water sources (Table S3, Supplementary Material). Of the 50 hectares intensified, IMAFLORA funded 36 hectares, with farmers covering the costs of the remaining 14 hectares. In either case, because of improvements in productivity, the total initial cost is expected to be paid off within 3–5 years.

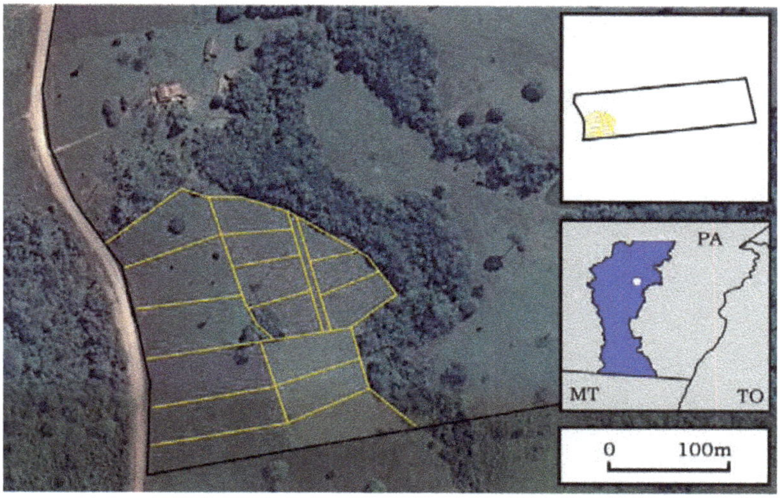

Figure 10. Around 10% of each property in the Florestas de Valor initiative is divided into small plots using fences lined with leguminous trees (yellow lines in the main image). Top right: intensified pasture area shown within the total farm boundary (black line). Lower right: The farm location is shown as a point within Sao Felix do Xingu (municipality colored in blue). State abbreviations: PA = Pará, MT = Mato Grosso, TO = Tocantins.

4. Discussion

The results from these initiatives suggest that there are a variety of available technologies that can increase cattle ranching productivity and profitability in the Amazon. Though diverse in the details, these initiatives share many similarities, including their focus on farmer training, farm record-keeping, and improved pasture management, in particular, the adoption of rotational grazing and pasture fertilization using chemical inputs or leguminous plants. These management changes require some initial investment (R$1300–6900/hectare), which is paid off within 2.5–8.5 years.

With the exception of the introduction of grass-legume pastures in Acre, the initiatives presented are young and further productivity gains are expected, with productivity expected to peak 1.5–7 years after implementation (Tables 2 and 3). Our study is, however, not a large-scale randomized controlled trial, and so these promising results come with a number of caveats. First, we do not claim that our review is exhaustive, though we present results from six of the thirteen initiatives that we are aware of which operate in the region (Table S4, Supplementary Material), and we believe our results are broadly representative of high-yielding cattle ranching in the Amazon. Second, these initiatives recruited farmers opportunistically, predominantly through farmer networks and open farm days (Figure 11). The farmers participating are therefore "early-adopters", who may differ from other farmers in systematic ways, for example by being less risk averse. We do not believe that the productivity gains that we observe result from these farmer differences or fundamental differences between these farms and their neighbors. For three of our six initiatives, we present productivity estimates from before and after the interventions showing clear productivity increases (for the other three, the Novo Campo Program, Florestas de Valor and mixed legume pastures in Acre, pre-intervention data were not available and our baseline data are estimates of the regional average productivity; Tables 2 and 3). Our data are also self-reported, though our productivity improvements are in line with previous literature on the productivity gains from improvements in farm management in Brazil [5,64–67], and our estimates of the costs of intensification are an important resource for accurately estimating the cost-effectiveness of cattle intensification in the Amazon. Previous work on the cost of cattle intensification has typically focused only on direct costs of pasture improvement, thereby underestimating the true cost of intensification for farmers. When modelling the pasture improvement included in Brazil's contribution to the UNFCC, De Oliveira Silva et al. [5], for example, estimate costs between R$365–1243/ha and maintenance costs from R$6.9–266.8/ha, estimates which are substantially lower than our figures. Their figures, however, do not include indirect costs such as costs from the transportation of inputs, increased labor, or costs from fencing and the implementation of rotational grazing.

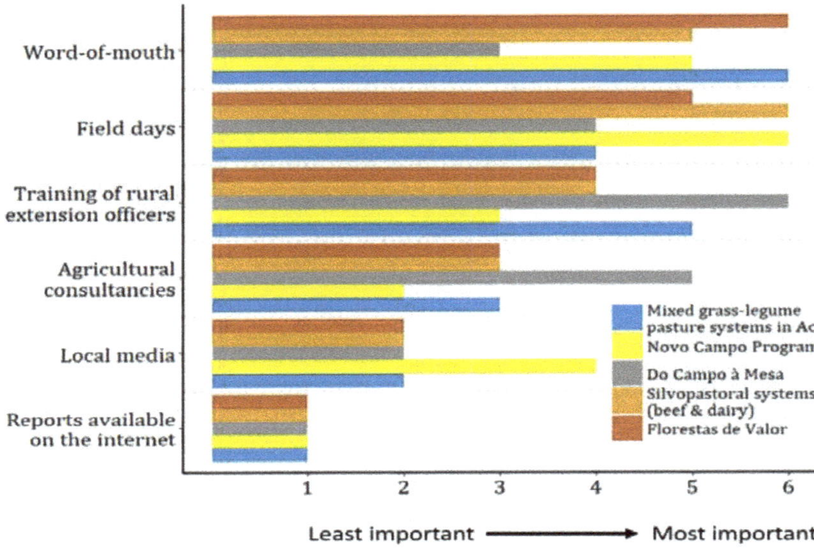

Figure 11. Relative importance (1 = least important, 6 = most important) of six different channels for recruiting farmers into the six initiatives described in this article. All initiatives recruited farmers opportunistically, as knowledge about the initiatives was spread for the most part through word-of-mouth, field days where neighboring farmers were invited to tour participating farms, and the training of rural extension officers in GAP. Idesam's beef and dairy initiatives took similar approaches to recruitment, and so the results are grouped for these two initiatives.

4.1. Risks of Cattle Intensification

Whatever the financial costs, increasing cattle ranching yields in the Amazon is not without risks [68]. Improving productivity can reduce greenhouse gas emissions from beef production, as seen both at the farm-scale in the Novo Campo Program (Figure 6) and other intensification initiatives in the region [8], and at the national-scale from coupled economic-environmental modelling [6]. Cohn et al. [6] find that the adoption of GAP in Brazil could halve greenhouse gas emissions from deforestation and agriculture, though in practice the land sparing effect of cattle productivity increases are likely to vary spatially. While in consolidated regions, economic theory suggests that intensification can help reduce deforestation through market and labor effects [69,70], in forested regions intensification risks rebound effects—the so-called "Jevon's paradox" where increases in the profitability of cattle ranching incentivize, rather than reduce, local deforestation. Reconciling targets for increased cattle production and zero illegal deforestation or even deforestation-free production will therefore require explicitly linking improvements in cattle ranching with habitat protection and efforts to reduce leakage in cattle supply chains [71,72]. It is also with this rebound effect in mind that the initiatives described in this article explicitly require participating farmers to comply with the Brazilian

Forest Code, have no recent illegal deforestation, and develop land use plans for reforestation where required. Further discussion of potential risks is included in the online Supplementary Material.

While the technologies discussed in this article can increase farm profitability, this is also not always the case; results from two of our initiatives show that pasture intensification may, in some cases, be more profitable on large, rather than small farms. In the do Campo à Mesa initiative, pasture intensification was not profitable within twelve years for the two smallest farms, suggesting that there is a tipping point in economic returns between 126 and 425 ha of pasture [37]. Similar economies of scale were found in a modelling study using data from the Novo Campo Program (which found that the introduction of GAP and rotational grazing intensification was only profitable on farms with >385 ha of pasture; IIS, 2015), and in other studies of cattle ranching economics [4,73]. On the other hand, these economies of scale appear to be technology and system dependent. Positive economic returns were seen for smallholder dairy producers, and silvopastoral beef systems in Apuí (Tables 2 and 3), which can turn a profit with as little as 20 hectares of pasture. Similarly, grass-legume pastures have been adopted by small- and large-farms alike in Acre [41,74]. Given that 78% of cattle-rearing farms (hosting 33% of cattle) in the Amazon biome have less than 200 hectares of pasture (Figure S2, Supplementary Material), it is important that efforts to improve profitability and farmer livelihoods in the cattle sector include both large- and small- landholders.

4.2. Barriers to Scaling up Sustainable Cattle Ranching

Cattle production in Brazil is set to grow; the Brazilian government recently set ambitious targets for increasing beef and dairy production by 40% [75]. The sustainable growth of the industry is not, however, guaranteed. To ensure that the cattle industry develops sustainably, improving farmer livelihoods while protecting the environment, will require a mix of the right financial incentives, efforts to support training of rural workers and agricultural extension services, and improved monitoring of cattle supply chains.

Most cattle ranchers adopt good agricultural practices because of the expected improvements in productivity and profitability [76]. However, implementation costs and difficulty in accessing credit are barriers for many producers. Among producers surveyed in Mato Grosso regarding adoption of good agricultural practices, 18% cited financial constraints as a barrier to adoption [76], and high implementation costs are also an important barrier for four of the six cattle initiatives described in this article (Figure 12).

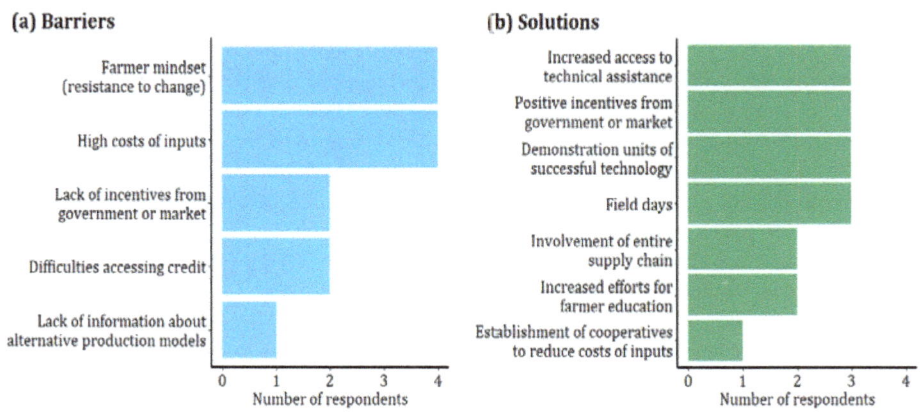

Figure 12. (**a**) The most important barriers to the implementation of sustainable cattle initiatives, and (**b**) the solutions to these barriers, as identified in the survey of the organizations running the six sustainable cattle ranching initiatives described in this article.

Sustainable growth in the cattle industry must also combine productivity improvements with the protection of native vegetation. In the Amazon biome, farmers are required to keep 80% of their land under forest (the threshold is set at 50% for small properties and properties in Ecological and Economic Zoning areas), and must reforest any land deforested above this threshold, or purchase certificates through the nascent forest trading scheme to compensate [77]. As the forest certificate market is not yet operating at scale, farmers currently rely on on-site reforestation for legal compliance, and the costs can be substantial. The do Campo à Mesa and Novo Campo Program initiatives report reforestation costs of R$868–6068/ha and R$2360–9654/ha, respectively. These figures are roughly equivalent to the costs of pasture intensification and implementation of GAP. Unfortunately, even where farmers can access credit, no credit lines currently support costs of compliance with the Forest Code [37].

The financial barriers to improved cattle production and compliance with the Forest Code can, however, be overcome by developing the right private and public incentives for producers (Figure 12). Currently, farmers receive the same price for their product, regardless of their environmental management. This could be fixed by the development of sustainable beef price premiums and certification schemes, such as the "Standard for Sustainable Cattle Production Systems" developed by the Sustainable Agricultural Network, which delivers a financial reward to producers implementing good practices [78]. Similarly, agricultural credit can be leveraged for sustainability, by making access to agricultural credit contingent on the adoption of sustainable ranching practices, and by supporting the costs of meeting the requirements of the Forest Code.

As an example of sustainable credit, in 2010 Brazil created the landmark ABC Program, one of the world's first credit lines for low carbon agriculture [79], which supports the costs of restoring degraded pasture and the implementation of integrated crop-livestock-forestry systems. The impact of the ABC Program has, however, been hampered by bureaucratic issues, unfavorable interest rates, and a lack of public awareness. Producers perceive the ABC program as being complex, slow, and overly bureaucratic [76]—posing a particular problem for small producers [80]. The ABC Program is often out-competed by other credit lines; its interest rates (7.5–8% per year) are double that of loans available through the National Rural Credit System, Brazil's main source of agricultural credit [79]. Public awareness is also a problem. Surveys in the Alta Floresta region of Mato Grosso show that most producers have not heard of the ABC Program and are not familiar with the concept of sustainable credit lines [80]. Only 10% of the ABC Program's budget is spent in the northern states, which make up the majority of the Amazon biome [81]. Overall, sustainable credit lines make up only 1.9% of all agricultural credit in Brazil [79], and there are currently no sustainable credit lines which are specifically aimed at smallholders, though these could be created within the existing National Program for Strengthening Family Agriculture (Portuguese acronym, PRONAF) [79].

The widespread adoption of sustainable cattle ranching will, of course, require more than just the correct mix of financial incentives. Barriers are also posed by a shortage of trained labor, farmer risk aversion, and the complexity of cattle supply chains. Improved farm performance cannot be achieved without the adequate training of farm staff. The lack of qualified labor is, however, acute in both beef and dairy production [76,82]. Sixty-five percent of ranchers surveyed in Alta Floresta, Mato Grosso, cited a shortage of qualified labor as the main barrier to the adoption of good agricultural practices [76]. Access to agricultural extension services is also limited [11]. Four-fifths of dairy farmers in Mato Grosso, for example, have never received technical assistance [28].

Farmer psychology also plays an important role (Figure 12). High-yielding cattle ranching costs more in the short term, though it generates positive returns in the longer-term (Tables 2 and 3). Many cattle ranchers are, however, risk averse [4] or not motivated by profit-maximization [83], and the transition from low-input, low-risk extensive systems to intensive pasture management requires to some degree a shift in mindset. Improved farm management begins with improved record-keeping, which is a foreign concept to most producers [4]. Rotational grazing systems also require that cattle are moved more frequently. Nelore cattle breeds have a reputation as being difficult to handle, though this is in large part because in extensive systems they are not used to contact with farm staff. While regular contact with farm laborers does improve their temperament [84], farmers can at first take some convincing about the feasibility of new management practices. The required shift in mindset is perhaps even greater for the adoption of silvopastoral systems and mixed grass-legume pastures. Farmers used to thinking of cattle as animals which graze grass may be initially reluctant to incorporate trees or herbaceous legumes (usually considered as undesirable species) into pasture as a source of forage and fertilizer.

These psychological barriers can perhaps be overcome by increasing familiarity with high-yielding systems, which remains low [80]. Awareness can be raised by establishing demonstration units on real farms, as in the six initiatives described in this article, and open-farm field days so that local farmers can witness and learn about new management options (Figure 11). As Brazilian farmers' receive most of their farming advice from other farmers [85], word-of-mouth dissemination of new technologies is critical, and can be effective, as seen in the experiences of legume pastures in Acre. The existence of local champions, long-term commitment of key players, and strategic partnerships among local stakeholders are also key to successful wide adoption of intensive cattle production systems [41].

Finally, there are structural barriers to sustainable cattle ranching. Cattle supply chains are complex, which means that deforestation is difficult to eradicate. While market initiatives (such as the "Terms of Adjustment of Conduct" and "G4" agreements) require meatpacking companies to block sales from properties with illegal deforestation (the G4 prohibits new deforestation altogether), this applies only to properties which supply cattle directly to slaughterhouses. As cattle may be born on one ranch, reared on a second, and fattened on a third, leakage is widespread. Though these agreements have reduced deforestation among the direct suppliers of slaughterhouses, it has not led to overall reductions in deforestation [71]. To permit growth of the Brazilian beef industry while reducing deforestation will therefore require efforts to reduce leakage. This could be achieved either by monitoring the movements of individual cattle, for example, using unique ear tags, or by monitoring farm-to-farm movement of batches of cattle. This information is already collected as part of the Guide to Animal Transport (GTA) used to track animal sanitation and health, but it is not used for monitoring environmental compliance.

While the barriers to scaling-up high-yielding cattle ranching in the Amazon are numerous, there is cause for optimism. First, cattle productivity is already increasing in most regions of Brazil [13,86]. Second, the example of leguminous pasture adoption in Acre shows that local demonstration farms can lead to technology diffusion at a regional scale in the Amazon. Third, though focused in southern Brazil, lessons can be learned from the dairy extension initiative, the Projeto Balde Cheio ("Full Bucket" project in English). The program began in 1999 in two municipalities in the states of São Paulo and Minas Gerais, where demonstration units were established on twelve farms. Operating on a budget of only R$5000–45,000 (US$5000–23,000) per year, agricultural extension officers from Embrapa worked with farmers to introduce a package of new practices, including improved farm book-keeping, soil conservation, pasture fertilization, and rotational management. On average, family farmers who joined the program increased milk production three-fold [82], with higher productivity arising from a combination of more lactating cows/area (31%), higher productivity/cow (24%), and better labor performance (37%), while using less land area (−7%). The initiative has since expanded, as the number of farmers assisted rose from 400 in 2010 to more than 3000 in 2012, and is now present in 483 municipalities nationwide, including farms in Rondônia, Pará and Amazonas, within the Amazon biome (Table S4, Supplementary Material).

5. Conclusions

As cattle ranching makes up the majority of agricultural land and productivity is still well below its sustainable potential, improvements in cattle productivity are key to the sustainable intensification of Brazilian agriculture. We present results from six cattle ranching initiatives which have achieved higher productivity and profitability in the Brazilian Amazon, while also supporting compliance with the Forest Code. These initiatives are, for the most part, still young and so we conclude by setting out three key conditions which are required to mainstream sustainable cattle ranching in the Amazon. If these conditions are met, we believe that the Brazilian beef industry can profitably produce more on less land and thereby facilitate growth in the agricultural sector while protecting Brazil's remaining native vegetation.

(1) Large-scale knowledge transfer—long-term funding and support is required for farmer-centered agricultural extension services, which increase awareness of high-yielding technologies and support small- and large-holders alike to adopt appropriate farming practices.
(2) Financial support for sustainable cattle ranching—farmers must be incentivized to adopt sustainable ranching practices, both through competitive, sustainable credit lines and through market signals. Rural credit lines should include sustainability criteria, and should help farmers not only increase agricultural production, but also meet the costs of Forest Code compliance. Market signals also matter, and just as some slaughterhouses offer price-premiums for high meat quality, price-premiums for GAP would encourage farmer uptake.
(3) Increase transparency in cattle supply chains—efforts by some slaughterhouses to monitor direct suppliers are a step in the right direction, but do not go far enough. All slaughterhouses should monitor both indirect and direct cattle suppliers, and monitoring efforts should be independently audited and publicly reported, so that deforestation may ultimately be eliminated from cattle supply chains.

Supplementary Materials: The following are available online at http://www.mdpi.com/2071-1050/10/4/1301/s1, supplementary text about the risks of cattle intensification; Tables S1–S4, and Figures S1 and S2.

Acknowledgments: We thank Fabio Ramos, Daniel Silva, André Novo, Artur Camago, Abdias Machado, Kemel Kalif, Márcio Rangel, Ezequiel Rodrigues do Valle, Marcelo Stabile, Mariano Cenamo, João Pedro Pio, Toby Gardner, Ben Phalan, Javier Godar, Livia Matyucheva, Priscila da Cunha Lessa, and Jose Guerrero for their input at various stages of the project. Erasmus K.H.J. zu Ermgassen was funded by BBSRC grant BB/J014540/1 and the Tim Whitmore Trust Fund. Funding for open access publishing come from BBSRC grant BB/J014540/1.

Author Contributions: Erasmus K.H.J. zu Ermgassen co-designed the study, wrote the survey, collected the data, analysed the data, and led on the writing of the manuscript. Bernardo B.N. Strassburg and Agnieszka Latawiec co-designed the study; Melquesedek Pereira de Alcântara, Francisco Beduschi Neto, Murilo M. F. Bettarello, Genivaldo de Brito, Gabriel C. Carrero, Eduardo de A.S. Florence, Edenise Garcia, Eduardo Trevisan Gonçalves, Casio Trajano da Luz, Giovanni M. Mallman, and Judson F. Valentim provided data, and all co-authors contributed substantially to the writing of the manuscript.

Conflicts of Interest: This manuscript reports the results of six initiatives which have increased the productivity of cattle ranching in the Brazilian Amazon. Many of the authors (Melquesedek Pereira de Alcântara, Francisco Beduschi Neto, Murilo M.F. Bettarello., Genivaldo de Brito, Gabriel C. Carrero, Eduardo de A.S. Florence, Edenise Garcia, Eduardo Trevisan Gonçalves, Casio Trajano da Luz, Giovanni M. Mallman, Bernardo B.N. Strassburg, Judson F. Valentim, and Agnieszka Latawiec) were directly involved in the development of these initiatives. The founding sponsors had no role in the design of the study; in the collection, analyses, or interpretation of data; in the writing of the manuscript, and in the decision to publish the results.

References

1. *FIESP Outlook FIESP 2026: Projections for Brazilian Agribusiness*; Federação das Indústrias do Estado de São Paulo: São Paulo, Brazil, 2016.
2. Brazil Intended Nationally Determined Contribution towards Achieving the Objective of the United Nations Framework Convention on Climate Change. Available online: http://www4.unfccc.int/submissions/INDC/Published%20Documents/Brazil/1/BRAZIL%20iNDC%20english%20FINAL.pdf (accessed on 16 April 2018).
3. Strassburg, B.B.N.; Latawiec, A.E.; Barioni, L.G.; Nobre, C.A.; da Silva, V.P.; Valentim, J.F.; Vianna, M.; Assad, E.D. When enough should be enough: Improving the use of current agricultural lands could meet production demands and spare natural habitats in Brazil. *Glob. Environ. Chang.* **2014**, *28*, 84–97. [CrossRef]
4. Barbosa, F.A.; Soares Filho, B.S.; Merry, F.D.; de Oliveira Azevedo, H.; Costa, W.L.S.; Coe, M.T.; da Silveira Batista, E.; Maciel, T.C.; Sheepers, L.C.; de Oliveira, A.R.; et al. *Cenários para a Pecuária de corte Amazônica;* Universidade Federal de Minas Gerais: Belo Horizonte, Brazil, 2015; p. 154.

5. De Oliveira Silva, R.; Barioni, L.G.; Queiroz Pellegrino, G.; Moran, D. The role of agricultural intensification in Brazil's Nationally Determined Contribution on emissions mitigation. *Agric. Syst.* **2018**, *161*, 102–112. [CrossRef]
6. Cohn, A.S.; Mosnier, A.; Havlik, P.; Valin, H.; Herrero, M.; Schmid, E.; O'Hare, M.; Obersteiner, M. Cattle ranching intensification in Brazil can reduce global greenhouse gas emissions by sparing land from deforestation. *Proc. Natl. Acad. Sci. USA* **2014**. [CrossRef] [PubMed]
7. De Oliveira Silva, R.; Barioni, L.G.; Hall, J.A.J.; Moretti, A.C.; Fonseca Veloso, R.; Alexander, P.; Crespolini, M.; Moran, D. Sustainable intensification of Brazilian livestock production through optimized pasture restoration. *Agric. Syst.* **2017**, *153*, 201–211. [CrossRef]
8. Bogaerts, M.; Cirhigiri, L.; Robinson, I.; Rodkin, M.; Hajjar, R.; Costa Junior, C.; Newton, P. Climate change mitigation through intensified pasture management: Estimating greenhouse gas emissions on cattle farms in the Brazilian Amazon. *J. Clean. Prod.* **2017**, *162*, 1539–1550. [CrossRef]
9. Piccoli, M.L.; Braccini Neto, J.; Brito, F.V.; Campos, L.T.; Bértoli, C.D.; Campos, G.S.; Cobuci, J.A.; McManus, C.M.; Barcellos, J.O.J.; Gama, L.T. Origins and genetic diversity of British cattle breeds in Brazil assessed by pedigree analyses. *J. Anim. Sci.* **2014**, *92*, 1920–1930. [CrossRef] [PubMed]
10. Townsend, C.R.; de Lucena Costa, N.; de Araújo Pereira, R.G. *Aspectos Econômicos da Recuperação de Pastagens no Bioma Amazônia*; EMBRAPA: Porto Velho, Brazil, 2009; p. 28.
11. Valentim, J.F. Desafios e estratégias para recuperação de pastagens degradadas e intensificação da pecuária a pasto na Amazônia Legal. In *Simpósio de Pecuária Integrada*; Pereira, D.H., Carneiro e Pedreira, B., Eds.; Recuperação de Pastagens: Anais; Fundação Uniselva: Cuiabá, Brazil, 2016; pp. 11–56.
12. Dias-Filho, M.B.; Andrade, C.M.S. *De Pastagens no Trópico Úmido*; Embrapa Amazônia Oriental: Belém, Brazil, 2006; p. 33.
13. Dias, L.C.P.; Pimenta, F.M.; Santos, A.B.; Costa, M.H.; Ladle, R.J. Patterns of land use, extensification, and intensification of Brazilian agriculture. *Glob. Chang. Biol.* **2016**, *22*, 2887–2903. [CrossRef] [PubMed]
14. Bowman, M.S.; Soares-Filho, B.S.; Merry, F.D.; Nepstad, D.C.; Rodrigues, H.; Almeida, O.T. Persistence of cattle ranching in the Brazilian Amazon: A spatial analysis of the rationale for beef production. *Land Use Policy* **2012**, *29*, 558–568. [CrossRef]
15. Ministério da Agricultura, Pecuária e Abastecimento; Ministério da Ciência, Tecnologia, Inovações e Comunicações; Ministério do Meio Ambiente. *Projeto Terraclass 2012*; Embrapa & INPE: Brasília, Brazil, 2014.
16. Börner, J.; Kis-Katos, K.; Hargrave, J.; König, K. Post-Crackdown Effectiveness of Field-Based Forest Law Enforcement in the Brazilian Amazon. *PLoS ONE* **2015**, *10*, e0121544. [CrossRef] [PubMed]
17. Soares-Filho, B.; Moutinho, P.; Nepstad, D.; Anderson, A.; Rodrigues, H.; Garcia, R.; Dietzsch, L.; Merry, F.; Bowman, M.; Hissa, L.; et al. Role of Brazilian Amazon protected areas in climate change mitigation. *Proc. Natl. Acad. Sci. USA* **2010**, *107*, 10821–10826. [CrossRef] [PubMed]
18. Gibbs, H.K.; Rausch, L.; Munger, J.; Schelly, I.; Morton, D.C.; Noojipady, P.; Soares-Filho, B.; Barreto, P.; Micol, L.; Walker, N.F. Brazil's Soy Moratorium. *Science* **2015**, *347*, 377–378. [CrossRef] [PubMed]
19. Assunção, J.; Gandour, C.; Rocha, R. Deforestation slowdown in the Brazilian Amazon: Prices or policies? *Environ. Dev. Econ.* **2015**, *20*, 697–722. [CrossRef]
20. Tollefson, J. Deforestation spikes in Brazilian Amazon. *Nature* **2016**, *540*. [CrossRef]
21. Carrero, G.; Fearnside, P. Forest Clearing Dynamics and the Expansion of Landholdings in Apuí, a Deforestation Hotspot on Brazil's Transamazon Highway. *Ecol. Soc.* **2011**, *16*. [CrossRef]
22. Datu Research. *Deforestation and the Brazilian beef Value Chain*; Datu Research: Durham, NC, USA, 2014.
23. Merry, F.; Soares-Filho, B. Will intensification of beef production deliver conservation outcomes in the Brazilian Amazon? *Elem. Sci. Anth.* **2017**, *5*, 24. [CrossRef]
24. Richards, P.D.; Walker, R.T.; Arima, E.Y. Spatially complex land change: The Indirect effect of Brazil's agricultural sector on land use in Amazonia. *Glob. Environ. Chang.* **2014**, *29*, 1–9. [CrossRef] [PubMed]
25. IBGE Pesquisa Pecuária Municipal (PPM). Available online: https://sidra.ibge.gov.br/pesquisa/ppm/quadros/brasil/2015 (accessed on 24 April 2017).
26. PRODES Desflorestamento nos Municipios. Available online: http://www.dpi.inpe.br/prodesdigital/prodesmunicipal.php (accessed on 26 April 2017).
27. Pacheco, P. Actor and Frontier Types in the Brazilian Amazon: Assessing Interactions and Outcomes Associated with Frontier Expansion. *Geoforum* **2012**, *43*, 864–874. [CrossRef]

28. Gomes, S.T.; Instituto Mato-Grossense de Economia Agropecuária. *Diagnóstico da Cadeia do Leite no Estado de Mato Grosso*; Famato, Senar-MT, & Sescoop-MT: Cuiabá, Brazil, 2012; p. 103.
29. Zoccal, R.; Alves, E.R.; Gasques, J.G. *Diagnóstico da Pecuária de Leite nacional: Estudo Preliminar. Contribuição para o Plano Pecuário 2012*; Embrapa Gado de Leite: Juiz de Fora, Brazil, 2011.
30. Santana, M.L.; Pereira, R.J.; Bignardi, A.B.; El Faro, L.; Tonhati, H.; Albuquerque, L.G. History, structure, and genetic diversity of Brazilian Gir cattle. *Livest. Sci.* **2014**, *163*, 26–33. [CrossRef]
31. Eurostat Database Agriculture Statistics. Available online: http://ec.europa.eu/eurostat (accessed on 9 January 2014).
32. Livestock Improvement Corporation (LIC); DairyNZ. *New Zealand Dairy Statistics 2013–14*; Livestock Improvement Corporation Limited and DairyNZ Limited: Hamilton, New Zealand, 2014; p. 55.
33. IBGE Censo Agropecuário. Available online: http://www2.sidra.ibge.gov.br/bda/acervo/acervo2.asp?e=v&p=CA&z=t&o=24 (accessed on 21 April 2017).
34. Andrade, C.M.S.; de Santos, D.M.; dos Ferreira, A.S.; Valentim, J.F. *Técnicas de Plantio Mecanizado de Forrageiras Estoloníferas por Mudas*; EMBRAPA: Rio Branco, Brazil, 2016; p. 22.
35. Carrero, G.C. Sistemas Silvipastoris com Pastejo Rotacional: Alternativas Sustentáveis para a Produção Pecuária na Amazônia. In *Gestão e Governança Local para a Amazônia Sustentável*; notas técnicas; PQGA/IBAM: Belém, Brasil, 2016; pp. 79–91.
36. Carrero, G.C.; Bettarello, M.; Alcântra, A.; Cenamo, M.C. Sistema extensivo versus sistema silvipastoril intensivo para pecuária de leite na Amazônia Brasileira: Produtividade, benefícios socioeconômicos e ecossistêmicos para mitigação e adaptação às mudanças climáticas. In *3° Congreso Nacional de Sistemas Silvopastoriles—VIII Congreso Internacional de Sistemas Agroforestales*; Peri, L., Ed.; Ediciones INTA: Santa Cruz, CA, USA, 2015; pp. 304–308.
37. Garcia, E.; Ramos Filho, F.S.V.; Mallmann, G.M.; Fonseca, F. Costs, Benefits and Challenges of Sustainable Livestock Intensification in a Major Deforestation Frontier in the Brazilian Amazon. *Sustainability* **2017**, *9*, 158. [CrossRef]
38. Marcuzzo, S.F.; de Lima, A. *A Programa Novo Campo: Estratégia de Pecuária Sustentável na Amazônia*; Instituto Centro de Vida: Alta Floresta, MT, Brazil, 2015.
39. Sá, C.P.d; Bayma, M.M.A.; Carvalho, B.P. *Avaliação dos Impactos da Tecnologia—Recomendação do Amendoim Forrageiro (Arachis Pintoi) em Pastagens no Acre*; EMBRAPA: Rio Branco, Brazil, 2016; p. 18.
40. Valentim, J.F.; Andrade, C.M.S. Tropical kudzu (*Pueraria phaseoloides*): Successful adoption in sustainable cattle production systems in the western Brazilian Amazon. In *Proceedings of the XX International Grassland Congress: Offered Papers*; Wageningen Academic Publishers: Wageningen, The Netherlands, 2005.
41. Valentim, J.F.; Andrade, C.M.S. Forage peanut (*Arachis pintoi*): A high yielding and high quality tropical legume for sustainable cattle production systems in the western Brazilian Amazon. In *Proceedings of the XX International Grassland Congress: Offered Papers*; Wageningen Academic Pub: Wageningen, The Netherlands, 2005.
42. Andrade, C.M.S. Produção de ruminantes em pastos consorciados. In Proceedings of the Simpósio SOBRE MAnejo Estratégico da Pastagem, Viçosa, MG, Brazil, 11–13 November 2010; pp. 171–214.
43. Panerai, A. Amendoim Forrageiro 2008. Available online: https://pt.slideshare.net/AlexandrePanerai/amendoim-forrageiro-embrapa (accessed on 8 May 2017).
44. De Sá, C.P.; Andrade, C.M.S.; Valentim, J.F. *Análise Econômica para a Pecuária de Corte em Pastagens Melhoradas no Acre*; EMBRAPA: Rio Branco, Brazil, 2010; p. 5.
45. Andrade, C.M.S. Importância das Leguminosas Forrageiras para a Sustentabilidade dos Sistemas de Produção de ruminantes. Available online: http://www.sbprc.famev.ufu.br/sites/sbprc.famev.ufu.br/files/Carlos%20Maur%C3%ADcio%20Andrade_As%20leguminosas%20e%20a%20sustentabilidade.pdf (accessed on 20 July 2017).
46. INPE Dados TerraClass 2014. Available online: http://www.inpe.br/cra/projetos_pesquisas/terraclass2014.php (accessed on 22 May 2017).
47. Valentim, J.F.; Andrade, C.M.S. Perspectives of grass-legume pastures for sustainable animal production in the tropics. *Reunião Annu. Soc. Bras. Zootec.* **2004**, *40*, 142–154.
48. Henderson, B.B.; Gerber, P.J.; Hilinski, T.E.; Falcucci, A.; Ojima, D.S.; Salvatore, M.; Conant, R.T. Greenhouse gas mitigation potential of the world's grazing lands: Modeling soil carbon and nitrogen fluxes of mitigation practices. *Agric. Ecosyst. Environ.* **2015**, *207*, 91–100. [CrossRef]

49. Costa, F.S.; Sales, M.F.L.; Freitas, A.C.R.; Souza, C.P. Modeling nitrous oxide emissions in grass and grass-legume pastures in the western Brazilian Amazon. In *International Symposium on Greenhouse Gases in Agriculture*; Embrapa: Brasília, DF, Brazil, 2016; Volume 2.
50. Do Valle, E.R. *Lista de verificação: Rebanhos Comerciais*; EMBRAPA: Brasília, DF, Brazil, 2013.
51. Cerri, C.C.; Moreira, C.S.; Alves, P.A.; Raucci, G.S.; Castigioni, B.d.A.; Mello, F.F.C.; Cerri, D.G.P.; Cerri, C.E.P. Assessing the carbon footprint of beef cattle in Brazil: A case study with 22 farms in the State of Mato Grosso. *J. Clean. Prod.* **2016**, *112*, 2593–2600. [CrossRef]
52. Siqueira, T.T.S.; Duru, M. Economics and environmental performance issues of a typical Amazonian beef farm: A case study. *J. Clean. Prod.* **2016**, *112*, 2485–2494. [CrossRef]
53. Piatto, M.; Costa Junior, C. *Good Livestock Practices Reduce GHG Emissions and Increase Meat Production in the Amazon*; Imaflora: Piracicaba, SP, Brazil, 2016.
54. International Institute for Sustainabilit (IIS). *Análise Econômica de Uma Pecuária Mais Sustentável*; International Institute for Sustainability: Rio de Janeiro, Brasil, 2015; p. 66.
55. PECSA Pecsa—Pecuária Sustentável da Amazônia. Available online: https://pecsa.com.br/pt/ (accessed on 21 July 2017).
56. Cardona, C.A.C.; Naranjo Ramírez, J.F.; Tarazona Morales, A.M.; Murgueitio Restrepo, E.; Chará Orozco, J.D.; Ku Vera, J.; Solorio Sánchez, F.J.; Flores Estrada, M.X.; Solorio Sánchez, B.; Barahona Rosales, R. Contribution of intensive silvopastoral systems to animal performance and to adaptation and mitigation of climate change. *Rev. Colomb. Cienc. Pecu.* **2014**, *27*, 76–94.
57. Broom, D.M.; Galindo, F.A.; Murgueitio, E. Sustainable, efficient livestock production with high biodiversity and good welfare for animals. *Proc. R. Soc. B Biol. Sci.* **2013**, *280*. [CrossRef] [PubMed]
58. Burlamaqui Bendahan, A. Système intégré culture-Elevage–Arbre (SILPF) dans l'État du Roraima, Amazonie Brésilienne. 2015. Available online: http://agritrop.cirad.fr/583402/ (accessed on 19 April 2017).
59. Dalzell, S.; Meat & Livestock Australia. *Leucaena: A Guide to Establishment and Management*; Meat and Livestock Australia: Sydney, Australia, 2006; ISBN 978-1-74191-013-1.
60. Marton, S.M.R.R.; Zimmermann, A.; Kreuzer, M.; Gaillard, G. Comparing the environmental performance of mixed and specialised dairy farms: The role of the system level analysed. *J. Clean. Prod.* **2016**, *124*, 73–83. [CrossRef]
61. Carrero, G.C.; Jacauna, M.A.; Sardinha, A.M.; Bettarello, M. Estabelecimento de árvores nativas de interesse ecológico e econômico para Sistemas Silvipastoris Intensivos em Apuí, Amazônia Brasileira. In *III Congresso Internacional de Sistemas Agroforestales—III Congresso Nacional de Sistemas Silvipastoriles, 2015, Iguazú, Argentina. 3o Congresso de Sistemas Silvipastoriles: VIII Congreso Internacional de Sistemas Agroforestales*; Ediciones INTA: Santa Cruz, CA, USA, 2015; Volume 1, pp. 21–25.
62. Alvim, M.J.; Paciullo, D.S.C.; de Carvalho, M.M.; Aroeira, L.J.M.; Carvalho, L.d.A.; Novaes, L.P.; Gomes, A.T.; de Miranda, J.E.C.; Ribeiro, A.C.C.L. *Sistema de Produção de Leite com Recria de Novilhas em Sistemas Silvipastoris.* EMBRAPA Gado de Leite. Available online: https://sistemasdeproducao.cnptia.embrapa.br/FontesHTML/Leite/LeiteRecriadeNovilhas/index.htm (accessed on 18 April 2017).
63. Silveira, S.R. *Valor Nutritivo de Panicum Maximum, Cultivares Tanzania y Massai, de Acuerdo a Niveles de Sombra y de Fertilizacio'n Nitrogenada*; INTA: Santiago del Estero, Argentina, 2012; pp. 289–293.
64. Crestani, S.; Mendonça, H.; Ribeiro, F.N.; Frederico, M.M.; Almeida, E.X.d; Portela, F.A. Steers performance in dwarf elephant grass pastures alone or mixed with Arachis pintoi. *Trop. Anim. Health Prod.* **2013**, *45*, 1369–1374. [CrossRef] [PubMed]
65. De Goes, R.H.d.T.e.B.; Mancio, A.B.; de Paula Lana, R.; Alves, D.D.; Leão, M.I.; Silva, A.T.S. Recria de novilhos mestiços em pastagens de Brachiaria brizantha, com diferentes níveis de suplementação, na região Amazônica. Desempenho animal. *Rev. Bras. Zootec.* **2005**, *34*, 1740–1750. [CrossRef]
66. Hoffmann, A.; Moraes, E.H.B.K.; Mousquer, C.J.; Simioni, T.A.; Gomes, F.J.; Ferreira, V.B.; Silva, H.M. Produção de Bovinos de Corte no Sistema de Pasto-Suplemento no Período Seco. *Nativa* **2014**, *2*, 119–130. [CrossRef]
67. Lesama, M.F.; Moojen, E.L. Produção animal em gramíneas de estação fria com fertilização nitrogenada ou associadas com leguminosa, com ou sem fertilização nitrogenada. *Ciênc. Rural* **1999**, *29*, 123–128. [CrossRef]
68. Latawiec, A.E.; Strassburg, B.B.N.; Valentim, J.F.; Ramos, F.; Alves-Pinto, H.N. Intensification of cattle ranching production systems: Socioeconomic and environmental synergies and risks in Brazil. *Animal* **2014**, *8*, 1255–1263. [CrossRef] [PubMed]
69. Angelsen, A. Policies for reduced deforestation and their impact on agricultural production. *Proc. Natl. Acad. Sci. USA* **2010**, *107*, 19639–19644. [CrossRef] [PubMed]

70. Vale, P.M. *The Conservation versus Production Trade-Off: Does Livestock Intensification Increase Deforestation? Evidence from the Brazilian Amazon*; Working Paper No. 197. 2014. Available online: http://www.lse.ac.uk/GranthamInstitute/wp-content/uploads/2014/12/Working-Paper-174-Vale-2014.pdf (accessed on 16 April 2018).
71. Alix-Garcia, J.; Gibbs, H.K. Forest conservation effects of Brazil's zero deforestation cattle agreements undermined by leakage. *Glob. Environ. Chang.* **2017**, *47*, 201–217. [CrossRef]
72. Phalan, B.; Green, R.E.; Dicks, L.V.; Dotta, G.; Feniuk, C.; Lamb, A.; Strassburg, B.B.N.; Williams, D.R.; zu Ermgassen, E.K.H.J.; Balmford, A. How can higher-yield farming help to spare nature? *Science* **2016**, *351*, 450–451. [CrossRef] [PubMed]
73. Harfuch, L.; Palauro, G.; Zambianco, W.M. *Economic Analysis of Investment for the Cattle Ranching Expansion*; INPUT & AGROICONE: São Paulo, Brazil, 2016; p. 8.
74. Valentim, J.F.; da Carneiro, J.C. *Quebra da Dormência e Plantio de Puerária em Sistemas de Produção Agropecuários e Agroflorestais*; Embrapa Acre-Sér. Anteriores INFOTECA-E: Rio Branco, Brazil, 1998.
75. Ministério da Agricultura, Pecuária e Abastecimento (MAPA). *Plano mais Pecuária*; Ministério da Agricultura, Pecuária e Abastecimento: Brasília, Brazil, 2014.
76. Latawiec, A.E.; Strassburg, B.B.N.; Silva, D.; Alves-Pinto, H.N.; Feltran-Barbieri, R.; Castro, A.; Iribarrem, A.; Rangel, M.C.; Kalif, K.A.B.; Gardner, T.; et al. Improving land management in Brazil: A perspective from producers. *Agric. Ecosyst. Environ.* **2017**, *240*, 276–286. [CrossRef]
77. Soares-Filho, B.; Rajão, R.; Merry, F.; Rodrigues, H.; Davis, J.; Lima, L.; Macedo, M.; Coe, M.; Carneiro, A.; Santiago, L. Brazil's Market for Trading Forest Certificates. *PLoS ONE* **2016**, *11*, e0152311. [CrossRef] [PubMed]
78. Newton, P.; Alves-Pinto, H.N.; Pinto, L.F.G. Certification, Forest Conservation, and Cattle: Theories and Evidence of Change in Brazil: Cattle certification in Brazil. *Conserv. Lett.* **2015**, *8*, 206–213. [CrossRef]
79. Lopes, D.; Lowery, S. *Rural Credit in Brazil*; 2017. Available online: http://www.forest-trends.org/documents/files/doc_5016.pdf (accessed on 7 February 2017).
80. Gurgel, A.G.; Costa, C.F.; Barbosa, L.B.C.; Piha, D. *Desafios e Restrições dos Produtores Rurais na Adoção de Tecnologias de Baixo Carbono ABC: Estudo de Caso em Alta Floresta, em Mato Grosso*. Fundação Getulio Vargas (FGV) Centro de Agronegócio (GV Agro), Escola de Economia de São Paulo (EESP), Observatório ABC. 2017. Available online: http://observatorioabc.com.br/wp-content/uploads/2017/05/Sumario_ABC_AltaFloresta_-1.pdf (accessed on 6 February 2017).
81. Gurgel, A.C.; Costa, C.F.; Belinky, A.; Monzoni, M.; Vendramini, A.; Rocha, F.C.; Martins, S.C. *Análise dos Recursos do Programa ABC*; Centro de Estudos em Sustentabilidade da Fundação Getulio Vargas, Centro de Agronegócio, Escola de Economia de São Paulo, & Aliança pelo Climate Uso da Terra. 2014. Available online: https://s3-sa-east-1.amazonaws.com/arquivos.gvces.com.br/arquivos_gvces/arquivos/340/ObsABC_Relatorio2_AnaliseRecursosRegional_Completo.pdf (accessed on 28 April 2015).
82. Novo, A.L.M. *Islands of Dairy in a Sea of Sugarcane: The Future of Family Dairy Farming in Brazil*; Wageningen University: Wageningen, The Netherlands, 2012.
83. Garrett, R.; Gardner, T.; Morello, T.; Marchand, S.; Barlow, J.; Ezzine de Blas, D.; Ferreira, J.; Lees, A.; Parry, L. Explaining the persistence of low income and environmentally degrading land uses in the Brazilian Amazon. *Ecol. Soc.* **2017**, *22*. [CrossRef]
84. Ceballos, M.C.; Góis, K.C.R.; Sant'Anna, A.C.; Paranhos da Costa, M.J.R. Frequent handling of grazing beef cattle maintained under the rotational stocking method improves temperament over time. *Anim. Prod. Sci.* **2016**. [CrossRef]
85. Pinto, L.F.G.; Hajjar, R.; Newton, P.; Agrawal, A.; Adshead, D.; Bini, D.; Bogaerts, M.; Cirhigiri, L.; Maguire-Rajpaul, V.A.; González-Chaves, A.; et al. *Transitioning to More Sustainable, Low-Emissions Agriculture in Brazil*; GGIAR, CIAT & FutureEarth. 2016. Available online: https://cgspace.cgiar.org/handle/10568/75631 (accessed on 31 August 2017).
86. Martha, G.B.; Alves, E.; Contini, E. Land-saving approaches and beef production growth in Brazil. *Agric. Syst.* **2012**, *110*, 173–177. [CrossRef]

sustainability

MDPI

Article

Do Large Slaughterhouses Promote Sustainable Intensification of Cattle Ranching in Amazonia and the Cerrado?

Ana Beatriz Santos * **and Marcos Heil Costa**

Department of Agricultural Engineering, Federal University of Viçosa, Viçosa-MG 36570-900, Brazil; mhcosta@ufv.br

* Correspondence: ana.bsantos15@gmail.com; Tel.: +55-31-3899-1902

Received: 9 August 2018; Accepted: 31 August 2018; Published: 13 September 2018

Abstract: This study investigated the influence of large slaughterhouses on five variables, two related to environment impact (land use change rate and greenhouse gases emissions (GE)), and three related to cattle-ranching intensification (protein from crops, calories from crops and stocking rate). In Amazonia, the results show a reduction of the land use change rate and GE in zones both with and without the influence of large slaughterhouses. The hypothesis that slaughterhouses are leverage points to reduce deforestation in the biome was not confirmed. The slaughterhouses also seem to have no effect on cattle ranching intensification, as protein and calories production increased significantly in both zones, while the stocking rates did not change in the influence zones. In the Cerrado, cattle-ranching intensification is a reality, and is occurring independently of the presence of large slaughterhouses. In conclusion, the results show no evidence that large slaughterhouses have promoted either cattle-ranching intensification or improvements in the sustainability of the cattle-ranching activity in Amazonia and the Cerrado.

Keywords: beef supply chain; beef cattle; sustainability assessment; land use change; greenhouse gases emissions

1. Introduction

In recent decades, the expansion of cattle ranching in Amazonia and the Cerrado has raised concerns regarding the increase of CO_2 emissions associated with beef production. Historically, Brazil's largest share of greenhouse gases (GHG) emissions comes from land use change, particularly the conversion of natural vegetation to pasturelands [1]. Despite the decrease in Brazilian CO_2 emissions between 2005 and 2010 (from 1.7 to 0.3 Mt-CO_2/year), the LULUCF (Land Use, Land Use Change and Forestry—see Appendix A for a full list of abbreviations) sector emissions still represented 45% of the total emissions in 2015 [2].

Sustainable intensification of cattle ranching has been proposed as a promising solution to reconcile the need for increased beef production and the need for reduction of GHG emissions [3,4]. This concept suggests that producing more beef on less land (referred to as intensification) may slow deforestation and suppression of native Cerrado vegetation and reduce GHG emissions. According to Strassburg et al. [4], increasing Brazilian pasture productivity to 49–52% of its potential would be sufficient to meet demands for beef until 2040. In addition, about 14.3 Gt-CO_{2e} could be mitigated; of this, 87% (12.5 Gt-CO_{2e}) would be due to the projected reduction in deforestation [4].

In addition to emissions from land use change, cattle ranching is the largest source of methane (CH_4) in the country. Together, the LULUCF sector and CH_4 emissions from enteric fermentation represented 58% of Brazilian GHG emissions in 2015 [2]. Several studies have demonstrated that investments in pasture management and animal feed are able to increase animal production and reduce the time cattle spend in pasture [5–8]. However, grass-feeding is the predominant management system in the country, and animal-feed supplementation with protein and calories is still uncommon [6]. The low rate of weight gain due to unsupplemented feeding makes the average slaughter age in Brazil about four years old, twice what it is in the United States [9].

Brazil's National Policy on Climate Change (PNMC—Política Nacional sobre Mudanças no Clima) has mandated a reduction of GHG emissions in several economic activities; in agriculture, it supports the adoption of techniques that make cattle ranching more productive on existing pasturelands [10]—i.e., intensification. According to Dias et al. [11], the average stocking rate grew from 0.70 to 1.48 head/ha in the Cerrado and 0.69 to 1.53 head/ha in the Amazon between 1990 and 2010. The adoption of technologies was responsible for a great part of this increase [12], but in various localities the pasture productivity remains low [11] and there is no evidence that cattle ranching is increasing in a sustainable way.

In the beef supply chain, slaughterhouses are potential leverage points for promoting sustainable intensification due to their interactions with ranchers, their location at the agricultural frontier, and their ability to restrict ranchers' access to the market [13]. In the last decade, international campaigns promoted by non-governmental organizations (NGOs) have linked illegal deforestation to the emergence of large slaughterhouses in Amazonia [14,15]. In July 2009, individual meatpacking companies in Pará signed the legally binding Terms of Adjustment of Conduct (TAC), which imposes penalties on companies purchasing from properties with recent illegal deforestation. These agreements have since been replicated in the states of Acre, Rondônia, Amazonas and Mato Grosso [13]. The four biggest meatpackers of the country (JBS, Bertín, Marfrig, and Minerva) also signed in 2009 an agreement with NGO Greenpeace. This agreement imposed that meatpackers would buy only from Brazilian Amazonia ranches with zero-deforestation and meet standards issued by international multi-stakeholder commodity roundtables [13,16].

The public concern about the contribution of beef production to forest loss and climate change demonstrates probable environmental benefits from slaughterhouse market domination as they have a direct influence on ranchers. Gibbs et al. [13] quantified the responses of four large JBS slaughterhouse units in southeastern Pará to zero-deforestation agreements signed in 2009. These units respected the agreement, avoiding trade with ranchers with illegal deforestation on their lands. Besides, there was a greater adherence to the Rural Environmental Registry (CAR—Cadastro Ambiental Rural) and a decrease of deforestation on the properties of JBS partners.

Despite the importance of the theme, previous studies have not directly evaluated the consequences of large slaughterhouses influence on the sustainable intensification of the cattle ranching activity. Until now, studies have evaluated cattle intensification from an economic point of view [17,18], as a source of GHG and potential mitigation strategy [3,4,6] and as an outcome of a sample of policies, certifications or agreements [13,19–21]. To evaluate the sustainable intensification promoted, the discussion of the role of large slaughterhouses should not be limited to the analysis of deforestation rates. In this context, it is also necessary to investigate changes in production—mainly the average of cattle herd per hectare and potential agricultural region—and in relevant environmental variables.

In this study, we evaluated whether large slaughterhouses have been able to promote changes in their supply areas to meet sustainable intensification. We analyzed five variables: two related to environmental impact (land use change rate and GHG emissions) and three related to intensification (protein and calories produced by crops, and stocking rate). For the environmental impact variables, we investigated whether the slaughterhouses presence promote a decrease of the land use change and GHG emissions. For the intensification variables, we investigated whether slaughterhouse presence help to promote improvements in ranching practices as indicated by the increase in calories and protein produced by crops—nutrients that might ultimately be used for animal supplementation or for other purposes—and in rangeland stocking rates.

2. Materials and Methods

This work was divided into four parts. First, we selected large slaughterhouses that started operation approximately midway between 2000 and 2013, and we delimited their influence zones. Second, we delimited control zones in regions that are far from slaughterhouse influence and outside both conservation units and indigenous lands. Third, in the influence zones, we tested for changes after the slaughterhouse started operation, looking specifically at rates of land use change, GHG emissions, protein from crops, calories from crops, and cattle stocking rates. Finally, we tested for changes in these variables in the control zones.

2.1. Study Area

The Amazon is the largest biome in Brazil, covering about 49% of the national territory (420 Mha). In recent decades, cattle ranching has dominated the process of occupation and exploration of this biome, following government-sponsored colonization projects and incentives [22]. Currently, about 38 million hectares of pasture is located in the Amazon (25% of the national total). Between 1980 and 2013, cattle herds destined for slaughter grew 800% (from 6.24 to 56.59 million head; Figure 1), which is 58% of the national increase for this period. In addition to the expansion of cattle ranching, a dramatic increase in the number of slaughterhouses registered at the Federal Inspection Service was also observed, from 1 in 1980 to 62 in 2016 (Figure 1).

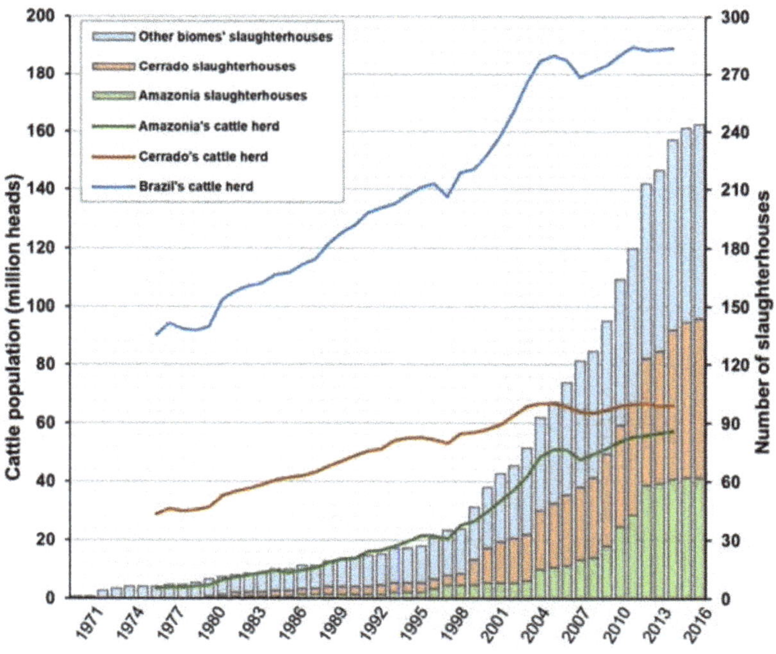

Figure 1. Number of slaughterhouses registered at the Federal Inspection Service and cattle population in Brazil, Amazonia and the Cerrado over time.

The Cerrado is the second largest biome in Brazil (200 Mha) and the most important region for cattle ranching, with 56 Mha of pasturelands. The biome contains the largest national herd (66 million head in 2014), representing 35% of the national total (Figure 1). As part of the new Brazilian agricultural frontier, the biome is credited as the driver of the country's ascendance in global agricultural commodity markets [6]. The number of slaughterhouses registered at the Federal Inspection Service in the Cerrado biome grew even more dramatically over the last few decades than the number in the Amazon: from 1 in 1980 to 82 in 2016 (Figure 1).

2.2. Period of Study and Datasets

We evaluated the following variables: land use change rate (ΔLU), GHG emissions (GE), protein from crops (PC), calories from crops (CC), and cattle stocking rate (SR). We obtained ΔLU from the forest cover dataset produced by Hansen et al. [23], while other agricultural variables were calculated from the dataset produced by Dias et al. [11]. Due to limitations of forest cover data availability, the period was 2000–2013.

The SR was obtained by dividing the number of beef cattle by the total pasture area. To construct the cattle maps, we used data provided by the Municipal Livestock Survey (PPM—Pesquisa Pecuária Municipal). We estimated the number of cattle destined for beef production by subtracting the number of dairy cows from the total number of cattle. To convert the tabular PPM data to a gridded cattle dataset, we calculated the ratio between the number of beef cattle and total pasture area in tabular form for each municipality in the Amazonian and Cerrado biomes. The total pasture for each municipality was extracted using Brazilian municipal boundaries polygons (spatial data) provided by IBGE. Due to the lack of data for certain years of the analysis, we replicated the available data in the missing years. Then, we constructed yearly maps for number of cattle by multiplying the municipality ratio (tabular data described above) and the amount of pasture for each grid cell of the municipality (map data). In the end, each municipality grid cell (i, j) was assigned a number of cattle proportional to that grid cell's total pasture area in that year (t).

GE is the sum of GHG emissions due to enteric fermentation and land use change. To estimate the CO_2 emissions due to land use change (E), we prepared a map of live below- and aboveground biomass (BGB and AGB) for the historic extent of the major vegetation physiognomies of the Amazon and the Cerrado. Starting from the BGB and AGB values from the LULUCF Reference Report from the Third National Communication of Brazil to the UNFCCC (United Nations Framework Convention on Climate Change) [24], we calculated total biomass values and then assigned these values to each grid cell in the vegetation map prepared by IBGE [25]. For non-forest vegetation physiognomies or anthropized areas (i.e., land areas transformed by human activity), we assigned biomass values corresponding to the average of subdivisions of the Brazilian classification system according to the dominant phytophysiognomy indicated on the vegetation map layer. The final biomass map of the historic vegetation expressed in Mg dry matter/ha is presented in Appendix B (Figure A1). Using these data, we obtained the total biomass (in Mg) for each grid cell (i, j) for each year t by multiplying the biomass values per area ($B_{(i,j)}$, in Mg dry matter/ha) by the amount of forest area ($F_{(i,j,t)}$, in ha) in the grid cell for that year. Then, we calculated the CO_2 emissions per pixel ($E_{(i,j,t)}$, in Tg-CO_2/year) by subtracting the total carbon in biomass of each grid cell (i, j) for each year (t + Δt) from the previous year's value (year t), according to Equation (1),

$$E_{(i,j,t)} = \frac{44}{12} \times 0.485 \times 10^{-6} \left(B_{(i,j)} \left(F_{(i,j,\,t)} - F_{(i,j,t+\Delta t)}\right)\right) \quad (1)$$

where 44/12 is used to convert g-C to g-CO_2, 0.485 to convert the dry matter biomass to carbon, and 10^{-6} to convert Mg to Tg.

We estimated CH_4 emissions by enteric fermentation (M) based on the Methane Emissions from Enteric Fermentation and Animal Manure Management Reference Report of the Third National Communication of Brazil to the UNFCCC [26]. Initially, we separated each grid cell's annual value for head of cattle ($C_{(i,j,t)}$) into three animal categories: adult males, adult females and young cattle. Using the Tier 2 approach described in IPCC [27], we identified the proportion of cattle in each of these three categories for each state by year ($R_{c,(i,j,t)}$, in percent, where c denotes animal category) and the corresponding emission factors by category ($f_{c(i,j,t)}$, in kg-CH_4 head 1/year^{-1}). As the emission factors and proportions are available only through 2010, we applied the 2010 values for the years 2011,

2012 and 2013. The total CH_4 emissions of each biome are presented in Appendix B and compared with other data. CH_4 emissions were converted to CO_2 equivalents (CO_{2e}) considering the GWP_{100} (Global Warming Potential over a 100-year time interval). The annual emissions per pixel due to enteric fermentation by cattle ($M_{(i,j,t)}$, in Tg-CO_{2e}) were then calculated according to Equation (2),

$$M_{(i,j,t)} = 28 \times 10^{-9} \sum_c C_{(i,j,t)} R_{(c,i,j,t)} f_{(c,i,j,t)} \quad (2)$$

where 28 is the GWP_{100} factor, and 10^{-9} is used to convert kg to Tg. Finally, we calculated the GE (Tg-CO_{2e}/year) emitted in year t as the sum of the M and E maps.

The CC and PC variables estimate the quantity of calories and protein produced in the region. These nutrients might be used for animal supplementation or for other purposes. We selected the three main feed crops used in the country for analysis: maize, soybean and sugarcane. To estimate the production (in metric tons) of each crop per pixel (i, j) in a year (t), we multiplied the crop productivity (in metric ton/ha) by the crop planted area (in ha) maps of Dias et al. [11]. Next, we multiplied the three production maps—soy (P^{so}), maize (P^{ma}) and sugarcane (P^{su})—by the dry matter fraction (d_c). The energy content (e_c) and protein content (p_c) were then used to convert dry matter values into calorie and protein values, respectively. The values of d_c, e_c, and p_c are given in Table 1 and are typical of Brazilian crops. Finally, the values for the protein (PC) and calorie (CC) maps were calculated according to Equations (3) and (4), respectively:

$$PC_{(i,j,t)} = 10^{-3}\left(P^{so}_{(i,j,t)} d_c^{so} p_c^{so} + P^{ma}_{(i,j,t)} d_c^{ma} p_c^{ma} + P^{su}_{(i,j,t)} d_c^{su} p_c^{su}\right) \quad (3)$$

$$CC_{(i,j,t)} = 0.239 \times 10^{-6}\left(P^{so}_{(i,j,t)} d_c^{so} e_c^{so} + P^{ma}_{(i,j,t)} d_c^{ma} e_c^{ma} + P^{su}_{(i,j,t)} d_c^{su} e_c^{su}\right) \quad (4)$$

Table 1. Values for dry matter fraction, energy content, and protein content of crops.

	Dry Matter (d_c) * (Fraction)	**Energy Content (e_c) * (MJ/kg of Dry Matter)**	**Protein Content (p_c) * (as a Fraction of Dry Matter)**
Maize	0.88	13.6	0.105
Soy	0.90	14.3	0.420
Sugarcane	0.23	9.10	0.0430

* Values obtained from Cardoso et al. [28].

In Equation (3), the conversion factor 10^{-3} is the result of multiplying 10^6 (used to convert tons to g) and 10^{-9} (used to convert g to Gg). In Equation (4), the factor 0.239 is used to convert joules (J) to calories (cal). The factor 10^{-6} is the result of multiplying 10^3 (used to convert tons to kg), 10^6 (used to convert MJ to J) and 10^{-15} (used to convert cal to Pcal).

2.3. Mapping of Large Slaughterhouses and Definition of Influence Zones

Beef slaughterhouse production data is usually classified information. To identify large slaughterhouses for the study, we first searched for those registered at the Federal Inspection Service (SIF—Sistema de Inspeção Federal). Registration is a condition for trading across states and exporting. Slaughterhouses not registered at SIF can sell only inside the state and thus are assumed to be small. To georeference the locations of slaughterhouses, we looked for each unit on Google Maps through the addresses reported to the Department for Inspection of Animal Products (DIPOA—Departamento de Inspeção de Produtos de Origem Animal) of the Brazilian Ministry of Agriculture, Livestock and Food Supply (MAPA—Ministério da Agricultura, Pecuária e Abastecimento). Other information, such as the opening or closing date, was collected from the National Register of Legal Entities (CNPJ—Cadastro Nacional de Pessoa Jurídica); registration with CNPJ is legally required to start business activities in Brazil. To restrict the analysis only to large units, we selected only slaughterhouses with slaughter capacity greater than 40 head/hour (classes MB1, MB2 and MB3, according to MAPA ordinance number 82 of 27 February 1976).

We found 144 slaughterhouse units with SIF registration in Amazonia and the Cerrado, including 61 that qualify as large units (42% of the total, Figure 2). As our analysis aims to determine the impact of the large slaughterhouses, ideally, the analyzed units should have been operating for close to half of the 2000–2013 study period, so that a "former" period can be compared to a "latter" period of similar duration. Thus, we selected slaughterhouses with a starting year for operations (y_{os}) between 2004 and 2008. Only 12 slaughterhouses satisfy this condition and could thus be used. The selected units are presented in Table 2, and their locations are shown in Figure 2.

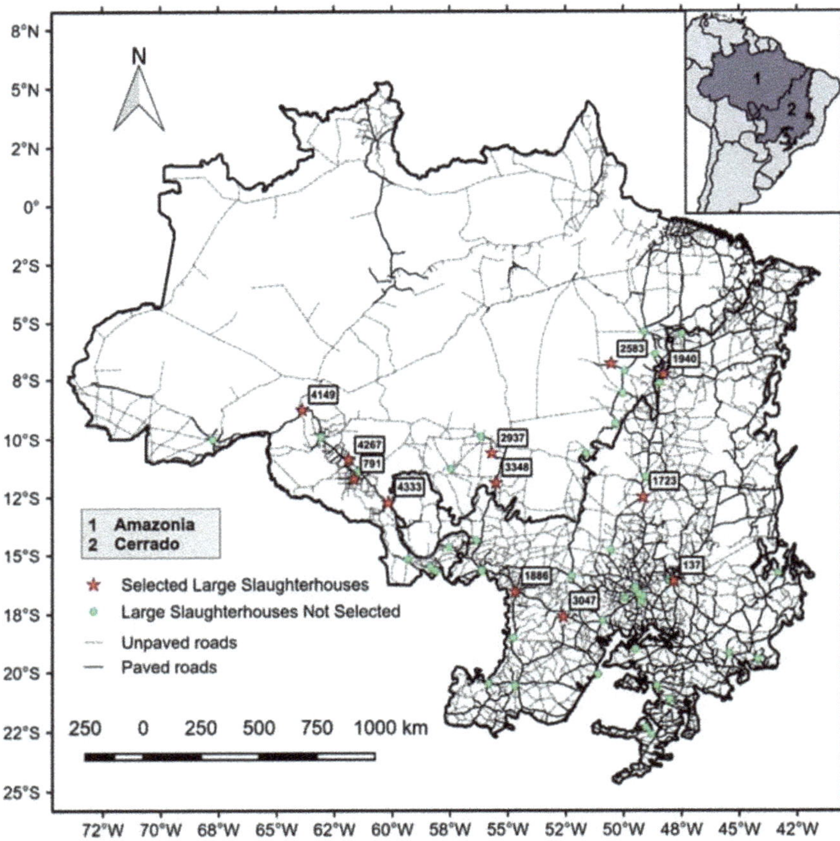

Figure 2. Locations of selected large slaughterhouses and large slaughterhouses that were not selected. Solid and dashed lines represent paved and unpaved roads, respectively.

Table 2. Characteristics of selected slaughterhouses.

SIF Code	Class *	Year of Operation Start (y_{os})	Latitude (°)	Longitude (°)	State	Biome
791	MB1	2006	11°43′ S	61°39′ W	Rondônia	Amazonia
3348	MB1	2004	11°54′ S	55°30′ W	Mato Grosso	Amazonia
3047	MB2	2006	17°36′ S	52°36′ W	Goiás	Cerrado
137	MB3	2008	16°06′ S	47°49′ W	Goiás	Cerrado
1723	MB3	2004	12°29′ S	49°08′ W	Tocantins	Cerrado
1886	MB3	2006	16°33′ S	54°40′ W	Mato Grosso	Cerrado
1940	MB3	2007	7°16′ S	48°16′ W	Tocantins	Amazonia/Cerrado
2583	MB3	2008	6°48′ S	50°31′ W	Pará	Amazonia
2937	MB3	2005	10°37′ S	55°41′ W	Mato Grosso	Amazonia
4149	MB3	2004	8°42′ S	63°55′ W	Rondônia	Amazonia
4267	MB3	2004	10°54′ S	61°53′ W	Rondônia	Amazonia
4333	MB3	2004	12°43′ S	60°10′ W	Rondônia	Amazonia

* MB1 are units with slaughter capacity greater than 80 head/hour and storage capacity greater than 20 t/day; MB2 are units with slaughter capacity greater than 80 head/hour that may or may not have storage capacity; and MB3 are units with slaughter capacity between 40 and 80 head/hour that may or may not have storage capacity.

We define the slaughterhouse influence zone as the likely cattle supply area around a slaughterhouse. We delimited the influence zone of each slaughterhouse unit by determining the distance that could realistically be traveled by a cattle truck. We assumed a maximum travel time of 8 h, which is the maximum travel time tolerated by cattle [29]. To select the truck routes, we used the Brazilian road network for 2010 prepared by the National Logistics and Transportation Plan (PNLT—Plano Nacional de Logística e Transporte). To account for vehicular speed limits, we assigned different velocities for each part of the route. In Brazil, the maximum permissible truck speeds are 90 km/h on paved roads and 60 km/h on unpaved roads (Law number 9503/1997 modified by Law number 13,281/2016).

However, it is not possible to adopt these speeds as the average. The high center of gravity of loaded trucks, the poor condition of Northern Brazilian roads [30] and the necessity for stops are some of the factors limiting driving speeds. Thus, we assumed an average speed of 10 km/h for distances traveled until reaching a paved or unpaved road, 20 km/h for distances traveled on unpaved roads and 40 km/h on paved roads. We also delimited intermediary zones spanning travel distances of 2 h, 4 h and 6 h to determine whether the influence on surrounding areas varies with distance from the slaughterhouse unit.

2.4. Definition of Control Zones

In this study, we also delimited control zones to determine whether the responses of the study variables occurred only in the influence zones. The control zones were chosen from areas outside the influence of any of the slaughterhouses selected for this study. The control zones could not be in areas around other slaughterhouses with slaughter capacity up to 40 head/hour that opened before 2000. We also excluded areas with indigenous lands and conservation units to avoid the effects of conservation measures. The control zones are of the same size as the average size of the 8 h-influence zones, and, in the absence of a y_{os}, we chose 2006 to separate the former and latter periods.

2.5. Data Analysis

We analyzed the changes in five variables, two related to environmental sustainability (land use change rate (ΔLU) and GHG emissions (GE)) and three related to cattle ranching intensification (protein from crops (PC), calories from crops (CC), and stocking rate (SR)). To determine whether the changes really were associated with the start of slaughterhouse operations, we performed two tests, T1 and T2 (Figure 3).

In the first test (T1), we tested for change inside the slaughterhouse influence zone (denoted by superscript S). We used a Wilcoxon paired test to compare the former period (denoted by subscript

F) with the latter period (denoted by subscript *L*), where the former period included the years from 2000 to y_{os}, and the latter period the years from y_{os} to 2013. Each variable was tested against its own alternative hypothesis (Ha). To be considered a promoter of intensification, the slaughterhouse would need to demonstrably influence the ranchers to increase their stocking rate and use calorie and protein supplementation. By the same token, to be considered a promoter of sustainability, the slaughterhouse would influence ranchers to reduce vegetation suppression and GHG emissions. For the two variables related to environmental impacts, we tested whether the slaughterhouses' start of operation is associated with decreased ΔLU (Ha : $\Delta LU_L^S < \Delta LU_F^S$) and GE (Ha: $GE_L^S < GE_F^S$). For the three variables related to intensification, we tested whether the slaughterhouses' start of operation is associated with regionally increasing the feed supply's PC (Ha: $PC_L^S > PC_F^S$) and CC (Ha: $CC_L^S > CC_F^S$) and the stocking rate SR (Ha: $SR_L^S > SR_F^S$). We tested these hypotheses for all influence zone sizes (transportation radius up to 2 h, 4 h, 6 h and 8 h). In the absence of a significant response ($p > 0.05$) in T1, no significant change could be reported in that variable (null hypothesis: Ho), and we would therefore conclude that the slaughterhouse operation had no impact on that variable.

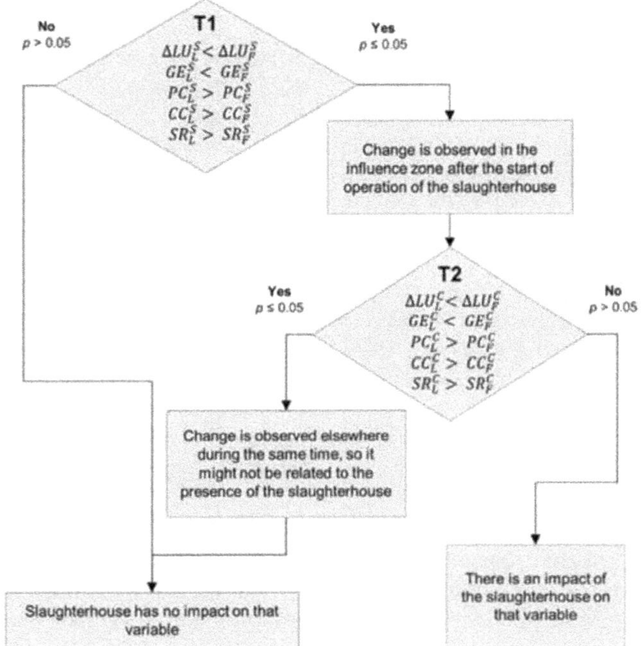

Figure 3. Flow diagram illustrating the analysis.

In the case of a significant response in any of the influence zones in T1, we used a second test (T2) to determine whether this response occurred only in the influence zones in this period (and not in the control zones). In T2, we performed a Wilcoxon paired test with the same hypotheses in the control zones (denoted by superscript C). That is, we tested whether there was a decrease in ΔLU (Ha : $\Delta LU_L^C < \Delta LU_F^C$) and GE (Ha: $GE_L^C < GE_F^C$) and an increase in the PC (Ha: $PC_L^C > PC_F^C$), CC (Ha: $CC_L^C > CC_F^C$) and SR (Ha: $SR_L^C > SR_F^C$) observed within the control zones between these time periods. A significant response ($p \leq 0.05$) in T2 means that the change in this variable was also observed elsewhere in the biome, outside of the influence zones, so it might not be directly related to the slaughterhouse. An opposite or neutral response ($p > 0.05$) means that the change observed in T1 occurred only in the slaughterhouse influence zone, and, in these cases, we would conclude that the slaughterhouse had an impact on the variable.

3. Results

3.1. Influence and Control Zones

Figure 4 shows the 12 influence zones obtained. The average sizes of the influence zones for travel times up to 2 h, 4 h, 6 h and 8 h are 0.43 Mha, 1.7 Mha, 4.1 Mha and 7.3 Mha, respectively. As the delimited extents were based on the travel time of a truck, the sizes of the influence zones vary according to the road network present near each slaughterhouse.

Due to the proximity between the slaughterhouse units, there are overlaps in some influence zones. However, just two zones (4267 and 791) have more than 50% of the 8 h zone shared by both (Figure 4). As the overlap starts at the 4 h travel time, we decided to keep the units separated instead of joining them so that the analysis has the same number of units per size of influence zone. In addition, the zones under the influence of slaughterhouses identified by SIF codes 1940, 3348 and 4333 extend over both biomes. However, just the 1940 SIF code unit was considered in both biome analyses, as a large percentage of its 8 h area is in the Cerrado biome (60% of the 8 h zone). Thus, five slaughterhouses were evaluated for the Cerrado, and eight for the Amazon.

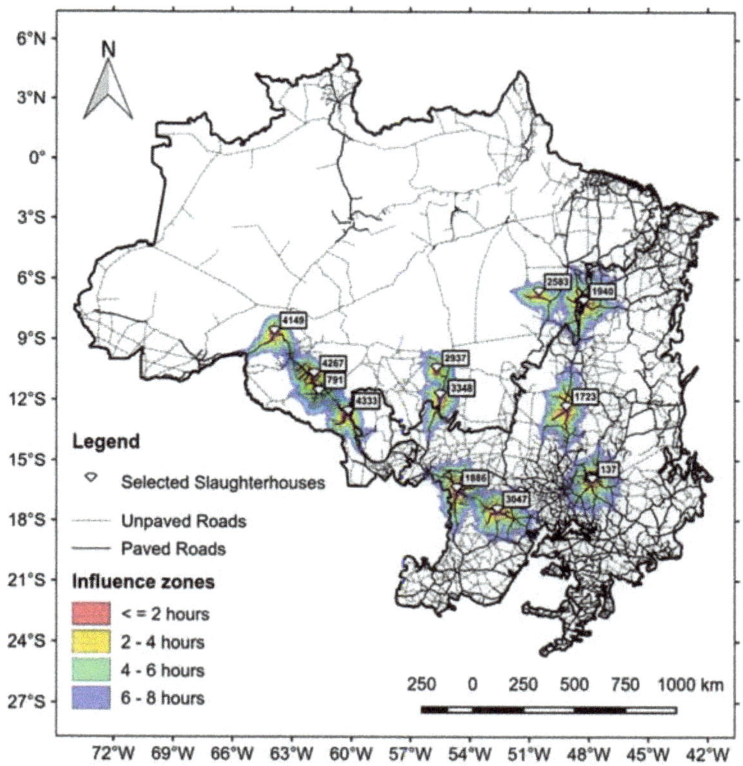

Figure 4. Locations of selected large slaughterhouses and their influence zones. Solid and dashed lines represent the paved and unpaved roads, respectively.

When choosing the control zones, first we excluded 340 Mha in both biomes, 70% in conservation units and indigenous lands and 30% in areas under the influence of selected slaughterhouses and slaughterhouses with y_{os} before 2000. We chose eight control areas in the Amazon and five control areas in the Cerrado (Figure 5). The selected zones have an average size of 7.3 Mha, the same as the average size of the 8 h influence zones.

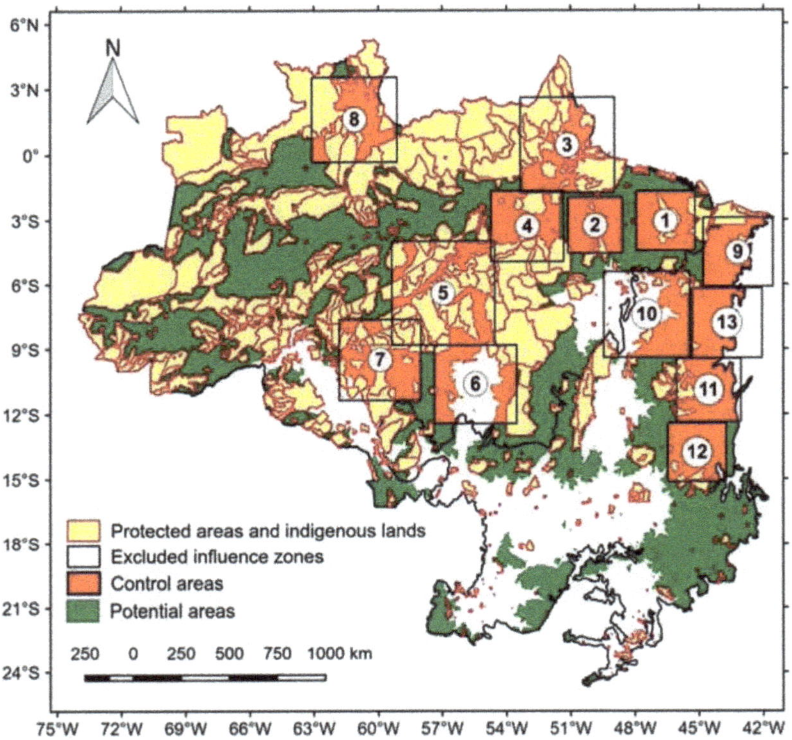

Figure 5. Locations of the control zones. The black squares indicate the zone limits. Green areas indicate areas with the potential to be control zones. White areas indicate influence zones of selected slaughterhouses and slaughterhouses with y_{os} before 2000.

3.2. Statistical Analysis

In the following sections, we show the results for each influence zone and control zone, separated by variable. Negative differences indicate a decrease in the variable analyzed with time.

3.2.1. Environmental Impact Variable: Land Use Change Rate (ΔLU)

Table 3 shows ΔLU_F and ΔLU_L results for the influence zones and the control zones. The first test (T1), a Wilcoxon paired test, determines whether there is a decrease in ΔLU inside the influence zones after the slaughterhouse start of operation. In Amazonia, there is a decrease in ΔLU in all influence zone sizes (travel times up to 2 h, 4 h, 6 h and 8 h), with similar values of probability ($p = 0.004$, Table 4). These results across the various sizes of the influence zones indicate that the distance from the slaughterhouse unit does not influence ΔLU. Results from T2 show that the decrease of ΔLU also occurs inside the control zones ($p = 0.008$, Table 4). The similar responses in both the slaughterhouse influence zones and the control zones during the same period indicate that the decrease of ΔLU might be not related to the slaughterhouse presence.

Table 3. Former and latter period values for land use change rate (ΔLU, in ha/year) for each influence zone and control zone.

	2 h		4 h		6 h		8 h		Control		
SIF Code	ΔLU_F^S (ha/year)	ΔLU_L^S (ha/year)	ΔLU_F^S (ha/year)	ΔLU_L^S (ha/year)	ΔLU_F^S (ha/year)	ΔLU_L^S (ha/year)	ΔLU_F^S (ha/year)	ΔLU_L^S (ha/year)	Control Code	ΔLU_F^C (ha/year)	ΔLU_L^C (ha/year)
791	6557.552	2387.213	28,446.332	8836.650	61,395.094	19,953.500	92,905.984	30,894.844	1	80,372.484	47,051.168
3348	9407.771	1465.027	34,195.277	7676.226	82,200.266	18,956.490	137,136.984	42,684.102	2	81,378.305	53,318.988
3047	946.861	1135.288	5235.697	5093.823	14,266.354	11,852.601	25,618.500	21,087.488	3	15,819.864	16,464.074
137	250.060	735.411	1786.634	3959.375	7293.791	11,791.568	17,400.232	25,362.859	4	59,520.324	46,821.258
1723	1816.823	1779.501	9388.703	8416.078	22,971.779	19,033.506	36,794.273	31,167.559	5	59,506.941	50,930.031
1886	1868.623	1362.502	9610.937	5900.309	20,900.621	12,623.756	35,851.840	22,329.553	6	117,978.461	46,430.785
1940	1576.978	1439.517	10,806.022	8332.689	26,263.994	18,966.988	55,317.758	35,559.426	7	105,355.867	38,125.027
2583	2439.150	563.163	9376.688	2705.178	23,052.080	7771.413	42,548.234	14,280.646	8	30,946.521	19,714.889
2937	781.155	469.791	5982.479	2420.442	22,798.094	8578.547	56,155.816	18,531.934	9	35,776.426	38,022.645
4149	4107.312	2648.438	18,282.762	11,317.136	39,151.117	24,151.670	79,550.453	41,369.051	10	39,928.773	36,683.539
4267	3596.003	1793.645	24,930.428	8992.448	64,370.426	20,787.986	122,377.789	38,633.922	11	34,875.188	35,362.727
4333	7961.677	3783.414	22,355.818	10,142.759	47,414.707	18,169.438	76,847.141	27,719.873	12	29,021.994	36,196.199
									13	22,420.396	32,550.551

Table 4. Results of Wilcoxon paired tests for T1 and T2 for land use change rate (ΔLU). T1 tests whether the introduction of large slaughterhouses was associated with the reduction of ΔLU in the influence zones (Ha: $\Delta LU_L^S < \Delta LU_F^S$). T2 tests whether reduction of ΔLU also occurred in the control zones (Ha: $\Delta LU_L^C < \Delta LU_F^C$).

	Latter Values–Former Values (ΔLU)						
		T1 $\left(\text{Ha}: \Delta LU_L^S < \Delta LU_F^S\right)$				T2 $\left(\text{Ha}: \Delta LU_L^C < \Delta LU_F^C\right)$	
	SIF Code	**2 h (ha/year)**	**4 h (ha/year)**	**6 h (ha/year)**	**8 h (ha/year)**	**Control Code**	**Control (ha/year)**
Amazonia	791	−4170.339	−19,609.682	−41,441.594	−62,011.140	1	−33,321.316
	3348	−7942.744	−26,519.051	−63,243.776	−94,452.882	2	−28,059.317
	1940	−137.461	−2473.333	−7297.006	−19,758.332	3	644.210
	2583	−1875.987	−6671.510	−15,280.667	−28,267.588	4	−12,699.066
	2937	−311.364	−3562.037	−14,219.547	−37,623.882	5	−8576.910
	4149	−1458.874	−6965.626	−14,999.447	−38,181.402	6	−71,547.676
	4267	−1802.358	−15,937.980	−43,582.440	−83,743.867	7	−67,230.840
	4333	−4178.263	−12,213.059	−29,245.269	−49,127.268	8	−11,231.632
	Median	−1839.173	−9589.343	−22,262.968	−43,654.335		−20,379.192
	p	0.004 *	0.004 *	0.004 *	0.004 *		0.008 *
Cerrado	3047	188.427	−141.874	−2413.753	−4531.012	9	2246.219
	137	485.351	2172.741	4497.777	7962.627	10	−3245.234
	1723	−37.322	−972.625	−3938.273	−5626.714	11	487.539
	1886	−506.121	−3710.628	−8276.865	−13,522.287	12	7174.205
	1940	−137.461	−2473.333	−7297.006	−19,758.332	13	10,130.155
	Median	−37.322	−972.625	−3938.273	−5626.714		2246.219
	p	0.500 NS	0.156 NS	0.156 NS	0.156 NS		0.906 NS

* Indicates significant at 5% level. NS Indicates not significant at 5% level. SIF code 1940 is used in the analyses of both biomes.

In the Cerrado, T1 shows no decrease in ΔLU (Table 4). This indicates that the slaughterhouses had no impact on ΔLU inside the slaughterhouse influence zones. Although a drop in ΔLU is observed in most of the influence zones, due to the small size of the sample, the response is not significant. By comparison, the T2 test shows interesting results: most of the control zones show increases in ΔLU. Of the five control zones, four show increases in ΔLU in the latter part of the study period ($p = 0.906$, Table 4).

3.2.2. Environmental Impact Variable: Total Greenhouse Gas Emissions (GE)

Table 5 shows GE_F and GE_L results for the influence zones and control zones. In the Amazon, T1 results show that there is a significant reduction of GE after the slaughterhouses' start of operation. As occurred with tests for ΔLU, all zones show the same level of significance, which demonstrates the absence of a distance influence ($p = 0.004$, Table 6). The similar responses between ΔLU and GE were already expected because of the large contribution of land use emissions to the total emissions. After finding a significant response in the 8 h influence zone for T1, WE used T2 to compare this result with the response in control areas outside the slaughterhouse influence zones. As occurred with ΔLU, T2 results confirm that the decrease of GE also occurred in the control zones ($p = 0.008$, Table 6). The T1 and T2 responses demonstrate that the change is observed both inside and outside of the influence zones during the same period, so the decrease of GE might be unrelated to the slaughterhouse presence.

Table 5. Former and latter period values for greenhouse gas emissions (GE, in Tg-CO_{2e}/year) for each influence zone and control zone.

	2 h		4 h		6 h		8 h			Control	
SIF Code	GE_F^S (Tg-CO_{2e}/year)	GE_L^S (Tg-CO_{2e}/year)	GE_F^S (Tg-CO_{2e}/year)	GE_L^S (Tg-CO_{2e}/year)	GE_F^S (Tg-CO_{2e}/year)	GE_L^S (Tg-CO_{2e}/year)	GE_F^S (Tg-CO_{2e}/year)	GE_L^S (Tg-CO_{2e}/year)	**Control Code**	GE_F^C (Tg-CO_{2e}/year)	GE_L^C (Tg-CO_{2e}/year)
791	4.8	2.5	19	8.2	38	17	57	24	1	59	36
3348	5.6	0.93	20	4.4	44	11	70	24	2	58	39
3047	0.35	0.35	1.9	1.7	5.1	4.4	10	9.3	3	6.1	7.0
137	0.14	0.19	0.86	1.1	3.2	3.9	7.3	8.8	4	42	34
1723	0.50	0.55	2.3	2.4	5.9	5.6	10	9.6	5	37	32
1886	0.96	0.84	3.3	2.7	6.0	4.7	9.6	7.6	6	64	28
1940	1.0	0.87	6.1	4.2	14	9.4	28	16	7	69	27
2583	1.9	0.58	7.6	2.9	18	7.8	33	14	8	18	11
2937	0.73	0.53	4.1	2.3	14	6.9	30	13	9	15	16
4149	2.3	1.4	9.5	5.6	22	13	46	24	10	7.7	6.8
4267	3.0	2.0	18	9.1	43	19	78	31	11	9.3	9.4
4333	4.2	2.1	12	5.7	25	11	41	18	12	7.3	8.3
									13	3.9	5.4

Table 6. Results of Wilcoxon paired tests for T1 and T2 for greenhouse gas emissions (GE). T1 tests whether the introduction of large slaughterhouses was associated with the reduction of GE in the influence zones (Ha: $GE_L^S < GE_F^S$) T2 tests whether reduction of GE also occurred in the control zones (Ha: $GE_L^C < GE_F^C$).

		Latter Values–Former Values (GE)					
		T1 (Ha : $GE_L^S < GE_F^S$)				T2 (Ha: $GE_L^C < GE_F^C$)	
	SIF Code	2 h ($Tg\text{-}CO_{2e}$/year)	4 h ($Tg\text{-}CO_{2e}$/year)	6 h ($Tg\text{-}CO_{2e}$/year)	8 h ($Tg\text{-}CO_{2e}$/year)	Control Code	Control ($Tg\text{-}CO_{2e}$/year)
Amazonia	791	−2.3	−11	−22	−33	1	−23
	3348	−4.7	−15	−33	−46	2	−19
	1940	−0.13	−1.9	−4.8	−12	3	0.94
	2583	−1.3	−4.6	−11	−19	4	−7.8
	2937	−0.20	−1.8	−7.1	−17	5	−4.4
	4149	−0.90	−3.8	−8.4	−22	6	−36
	4267	−0.95	−8.8	−24	−47	7	−42
	4333	−2.1	−5.9	−14	−23	8	−7.6
	Median	−1.1	−5.3	−12	−23		−13.6
	p	0.004 *	0.004 *	0.004 *	0.004 *		0.008 *
Cerrado	3047	0.0010	−0.16	−0.69	−1.0	9	0.86
	137	0.048	0.28	0.71	1.5	10	−0.92
	1723	0.046	0.075	−0.27	−0.70	11	0.14
	1886	−0.12	−0.65	−1.2	−2.0	12	1.0
	1940	−0.13	−1.9	−4.8	−12	13	1.6
	Median	0.0010	−0.16	−0.69	−1.0		0.86
	p	0.406 NS	0.219 NS	0.156 NS	0.156 NS		0.969 NS

* Indicates significant at 5% level. NS Indicates not significant at 5% level. SIF code 1940 is used in the analyses of both biomes.

In the Cerrado, T1 results show a nonsignificant response for the reduction of GE inside the slaughterhouse influence zones. As occurred in Amazonia, the GE results are very similar to the ΔLU results. In addition, for transportation distances up to 4 h, emissions due to enteric fermentation appear to have a greater influence on the total emitted. In comparison to what was observed for ΔLU, where two units show increases inside the influence zones up to 2 h and one up to 4 h, for GE, three units (SIF codes 3047, 137 and 1723) show increases in GE inside the zones up to 2 h, and two (SIF codes 137 and 1723) in the zones up to 4 h. According to the analysis framework, the T2 test is not necessary in the case of negative responses up to 8 h. As was the case with ΔLU analyses, T2 results looking at GE show that the increases also occur inside the control zones ($p = 0.969$, Table 6).

3.2.3. Intensification Variable: Protein from Crops (PC)

Table 7 shows PC_F and PC_L results for the influence zones and control zones. In Amazonia, T1 results show that there was a change in PC inside the influence zones ($p \leq 0.05$, Table 8). In addition, the decrease of p with the increase of influence zone sizes (up to 2 h, 4 h, 6 h, and 8 h) indicates that distance from the slaughterhouse unit had a likely influence. As T1 results show significant changes in PC in the influence zones, we use T2 to determine whether the changes occurred only inside the influence zones. According to T2 results, the increase of PC also occurred in the control zones ($p \leq 0.05$, Table 8), which implies the absence of slaughterhouse impact on this variable.

Table 7. Former and latter period values for protein from crops (PC, in Gg protein) for each influence zone and control zone.

	2 h		4 h		6 h		8 h			Control	
SIF Code	PC_F^S (Gg)	PC_L^S (Gg)	PC_F^S (Gg)	PC_L^S (Gg)	PC_F^S (Gg)	PC_L^S (Gg)	PC_F^S (Gg)	PC_L^S (Gg)	**Control Code**	PC_F^C (Gg)	PC_L^C (Gg)
791	8.7	15	21	38	36	67	63	1.2×10^2	1	17	37
3348	42	1.1×10^2	2.6×10^2	5.1×10^2	6.2×10^2	1.1×10^3	1.0×10^3	1.8×10^3	2	6.5	18
3047	89	1.2×10^2	3.7×10^2	4.9×10^2	9.3×10^2	1.2×10^3	1.5×10^3	2.0×10^3	3	0.20	1.0
137	32	49	1.5×10^2	2.5×10^2	4.2×10^2	7.2×10^2	7.5×10^2	1.3×10^3	4	14	22
1723	3.3	10	13	36	28	77	55	1.4×10^2	5	2.0	3.1
1886	1.0×10^2	1.5×10^2	3.5×10^2	4.8×10^2	7.1×10^2	9.5×10^2	1.1×10^3	1.4×10^3	6	5.6×10^2	1.3×10^3
1940	1.2	2.5	6.4	14	34	70	87	1.7×10^2	7	78	1.3×10^2
2583	0.20	0.19	1.4	1.4	4.0	4.2	8.5	10	8	4.4	4.4
2937	1.2	2.7	10	25	35	95	78	2.2×10^2	9	28	80
4149	0.17	0.21	0.36	0.57	0.68	1.3	1.6	3.0	10	2.8×10^2	5.6×10^2
4267	1.1	1.1	7.3	10	21	35	32	61	11	3.2×10^2	5.3×10^2
4333	16	37	63	1.3×10^2	1.6×10^2	2.8×10^2	3.4×10^2	5.4×10^2	12	3.2×10^2	5.1×10^2
									13	1.0×10^2	2.7×10^2

Table 8. Results of Wilcoxon paired tests for T1 and T2 for proteins from crops (PC). T1 tests whether the introduction of large slaughterhouses was associated with the increase of PC in the influence zones (Ha: $PC_L^S > PC_F^S$). T2 tests whether the increase of PC also occurred in the control zones (Ha: $PC_L^C > PC_F^C$).

	Latter Values–Former Values (PC)						
		T1 (Ha: $PC_L^S > PC_F^S$)				T2 (Ha: $PC_L^C > PC_F^C$)	
	SIF Code	2 h (Gg)	4 h (Gg)	6 h (Gg)	8 h (Gg)	Control Code	Control (Gg)
Amazonia	791	6.3	17	32	59	1	20
	3348	66	2.6×10^2	5.2×10^2	8.3×10^2	2	11
	1940	1.3	7.6	36	83	3	0.78
	2583	−0.011	−0.050	0.19	1.8	4	7.4
	2937	1.5	16	60	1.4×10^2	5	1.1
	4149	0.049	0.21	0.57	1.4	6	7.1×10^2
	4267	−0.032	2.5	14	30	7	51
	4333	21	70	1.3×10^2	1.9×10^2	8	−0.069
	Median	1.4	12	34	71		9.3
	p	0.020 *	0.008 *	0.004 *	0.004 *		0.008 *
Cerrado	3047	32	1.2×10^2	2.8×10^2	4.9×10^2	9	51
	137	17	1.0×10^2	3.0×10^2	5.1×10^2	10	2.9×10^2
	1723	7.0	23	49	88	11	2.1×10^2
	1886	43	1.4×10^2	2.4×10^2	3.5×10^2	12	1.9×10^2
	1940	1.3	7.6	36	83	13	1.7×10^2
	Median	17	1.0×10^2	2.4×10^2	3.5×10^2		1.9×10^2
	p	0.031 *	0.031 *	0.031 *	0.031 *		0.031 *

* Indicates significant at 5% level. SIF code 1940 is used in the analyses of both biomes.

In the Cerrado, based on T1, all sizes of influence zone show an increase in PC after the slaughterhouse start of operation at the same level of significance (Table 8). The T2 results indicate a similar increase of PC occurred inside the control zones ($p \leq 0.05$, Table 8). These similar responses indicate that the large slaughterhouses have no impact on the PC.

3.2.4. Intensification Variable: Calories from Crops (CC)

Table 9 shows CC_F and CC_L results for the study influence zones and control zones. In Amazonia, T1 shows that there is an increase in CC in all influence zone sizes (up to 2 h, 4 h, 6 h and 8 h). As occurred with PC, there is an influence of distance from the slaughterhouse, with p decreasing along with increase of zone size. T2 shows that the increase in CC between the two time periods also occurs inside the control zones ($p = 0.020$, Table 10). The similar responses in T1 and T2 indicate that the increase of CC might not be related to the slaughterhouse presence.

Table 9. Former and latter period values for calories from crops (CC, in Pcal) for each influence zone and control zone.

	2 h		4 h		6 h		8 h			Control	
SIF Code	CC_F^S (Pcal)	CC_L^S (Pcal)	CC_F^S (Pcal)	CC_L^S (Pcal)	CC_F^S (Pcal)	CC_L^S (Pcal)	CC_F^S (Pcal)	CC_L^S (Pcal)	**Control Code**	CC_F^C (Pcal)	CC_L^C (Pcal)
791	0.13	0.24	0.33	0.58	0.53	1.0	0.84	1.6	1	0.47	0.77
3348	0.39	1.2	2.4	5.6	5.8	12	10	20	2	0.17	0.32
3047	1.1	1.8	4.4	7.4	11	18	18	30	3	0.0067	0.015
137	0.45	0.74	2.1	3.6	5.3	9.7	10	17	4	0.29	0.33
1723	0.037	0.10	0.15	0.36	0.34	0.78	0.75	1.7	5	0.057	0.070
1886	1.1	1.7	3.7	5.7	7.3	11	11	17	6	5.4	14
1940	0.016	0.029	0.092	0.17	0.47	0.82	1.1	1.9	7	0.77	1.3
2583	0.0063	0.0058	0.043	0.039	0.11	0.11	0.22	0.22	8	0.070	0.056
2937	0.017	0.031	0.11	0.28	0.37	1.0	0.80	2.4	9	0.80	1.7
4149	0.0052	0.0055	0.011	0.015	0.021	0.030	0.048	0.067	10	3.0	6.1
4267	0.034	0.033	0.18	0.22	0.38	0.58	0.52	0.93	11	3.3	5.6
4333	0.16	0.41	0.62	1.5	1.6	3.2	3.4	6.0	12	3.8	5.8
									13	1.1	2.9

Table 10. Results of Wilcoxon paired tests for T1 and T2 for calories from crops (CC). T1 tests whether the introduction of large slaughterhouses was associated with the increase of CC in the influence zones (Ha: $CC_L^S > CC_F^S$).T2 tests whether the increase of CC also occurred in the control zones (Ha: $CC_L^C > CC_F^C$).

	Latter Values–Former Values (CC)						
	T1 (Ha: $CC_L^S > CC_F^S$)					T2 (Ha: $CC_L^C > CC_F^C$)	
	SIF Code	2 h (Pcal)	4 h (Pcal)	6 h (Pcal)	8 h (Pcal)	Control Code	Control (Pcal)
Amazonia	791	0.10	0.26	0.45	0.79	1	0.30
	3348	0.78	3.2	6.6	11	2	0.14
	1940	0.012	0.073	0.35	0.86	3	0.0085
	2583	−0.00056	−0.0043	−0.0065	−0.0022	4	0.037
	2937	0.014	0.17	0.67	1.6	5	0.013
	4149	0.00027	0.0034	0.0090	0.019	6	8.8
	4267	−0.0012	0.045	0.20	0.40	7	0.58
	4333	0.26	0.86	1.6	2.7	8	−0.014
	Median	0.013	0.12	0.40	0.82		0.090
	p	0.039 *	0.012 *	0.008 *	0.008 *		0.020 *
Cerrado	3047	0.76	3.0	7.3	13	9	0.90
	137	0.28	1.5	4.3	7.9	10	3.1
	1723	0.063	0.21	0.43	0.91	11	2.3
	1886	0.58	1.9	3.8	5.9	12	2.1
	1940	0.012	0.073	0.35	0.86	13	1.8
	Median	0.28	1.5	3.8	5.9		2.1
	p	0.031 *	0.031 *	0.031 *	0.031 *		0.031 *

* Indicates significant at 5% level. SIF code 1940 is used in the analyses of both biomes.

In the Cerrado, T1 shows that there is an increase in CC (Table 10). All influence zones show a significant response in T1, which indicates a change occurred after slaughterhouse start of operation. As the response of the 8 h influence zone is significant, we use T2 results to determine whether the observed result also occurred inside the control zones. The T2 results do indicate an increase of CC in the control zones ($p \leq 0.05$, Table 10), which means that the increase of CC might be unrelated to the slaughterhouse presence.

3.2.5. Intensification Variable: Stocking Rate (SR)

Table 11 shows SR_F and SR_L results for the study influence zones and control zones. In Amazonia, T1 results indicate that SR is not impacted by the slaughterhouse start of operation, with all sizes of influence zone showing nonsignificant responses for the change ($p > 0.05$, Table 12). As T1 is negative, T2 is not necessary to prove the impact of the slaughterhouse. However, contrary to the results for the slaughterhouse influence zones, the control zones show a significant increase in the SR between time periods ($p \leq 0.05$, Table 12).

Table 11. Former and latter period values for stocking rate (SR, in head/ha) for each influence zone and control zone.

	2 h		4 h		6 h		8 h			Control	
SIF Code	SR_F^S (head/ha)	SR_L^S (head/ha)	SR_F^S (head/ha)	SR_L^S (head/ha)	SR_F^S (head/ha)	SR_L^S (head/ha)	SR_F^S (head/ha)	SR_L^S (head/ha)	**Control Code**	SR_F^C (head/ha)	SR_L^C (head/ha)
791	2.023	1.915	1.990	1.855	1.946	1.856	1.936	1.873	1	0.981	1.073
3348	0.717	0.753	0.867	0.915	1.066	1.117	1.194	1.240	2	1.204	1.554
3047	0.875	1.002	0.953	1.055	0.968	1.059	1.033	1.124	3	0.249	0.286
137	1.011	1.257	0.866	1.143	0.821	1.119	0.886	1.184	4	1.452	1.504
1723	0.811	1.060	0.858	1.181	0.845	1.144	0.834	1.115	5	1.342	1.684
1886	1.543	2.033	1.130	1.433	0.980	1.186	0.921	1.073	6	1.458	1.588
1940	0.984	1.043	0.993	1.081	1.025	1.146	1.024	1.150	7	1.589	2.018
2583	2.532	2.672	1.817	1.774	1.628	1.522	1.509	1.423	8	0.584	1.453
2937	2.085	1.850	1.913	1.744	1.782	1.719	1.708	1.728	9	0.968	1.267
4149	1.347	1.668	1.407	1.604	1.421	1.691	1.577	1.902	10	0.656	0.905
4267	1.826	1.925	1.835	1.971	1.866	2.014	1.855	2.036	11	0.653	0.974
4333	1.237	1.075	1.756	1.721	1.869	1.815	1.794	1.821	12	0.612	0.936
									13	0.512	0.530

Table 12. Results of Wilcoxon paired test for T1 and T2 for stocking rate (SR). T1 tests whether the introduction of large slaughterhouses was associated with the increase of SR in the influence zones (Ha: $SR_L^S > SR_F^S$). T2 tests whether the increase of SR also occurred in the control zones (Ha: $SR_L^C > SR_F^C$).

Latter Values–Former Values (SR)							
		T1 (Ha: $SR_L^S > SR_F^S$)				T2 (Ha: $SR_L^C > SR_F^C$)	
	SIF Code	**2 h (Head/ha)**	**4 h (Head/ha)**	**6 h (Head/ha)**	**8 h (Head/ha)**	**Control Code**	**Control (Head/ha)**
Amazonia	791	−0.108	−0.135	−0.090	−0.063	1	0.092
	3348	0.036	0.048	0.051	0.046	2	0.350
	1940	0.059	0.088	0.121	0.126	3	0.037
	2583	0.140	−0.043	−0.106	−0.086	4	0.052
	2937	−0.235	−0.169	−0.063	0.020	5	0.342
	4149	0.321	0.197	0.270	0.325	6	0.130
	4267	0.099	0.136	0.148	0.181	7	0.429
	4333	−0.162	−0.035	−0.054	0.027	8	0.869
	Median	0.048	0.007	−0.002	0.037		0.236
	p	0.473 NS	0.371 NS	0.320 NS	0.125 NS		0.004 *
Cerrado	3047	0.127	0.102	0.091	0.091	9	0.299
	137	0.246	0.277	0.298	0.298	10	0.249
	1723	0.249	0.323	0.299	0.281	11	0.321
	1886	0.490	0.303	0.206	0.152	12	0.324
	1940	0.059	0.088	0.121	0.126	13	0.018
	Median	0.246	0.277	0.206	0.152		0.299
	p	0.031 *	0.031 *	0.031 *	0.031 *		0.031 *

* Indicates significant at 5% level. NS Indicates not significant at 5% level. SIF code 1940 is used in the analyses of both biomes.

In the Cerrado, all sizes of influence zone show an increase in SR after the start of operation of the slaughterhouses studied (p = 0.031, Table 12). According to T2, the control zones have the same results as the influence zones (p = 0.031, Table 12). These similar responses indicate that the large slaughterhouses are not directly responsible for SR increases in their influence zones in the Cerrado.

4. Discussion

Regarding the hypothesis that large slaughterhouses promote sustainable agricultural development and cattle ranching intensification, we expected to find significant reductions in variables that measured environmental impact (ΔLU and GE) and increases in variables that measured intensification (PC, CC, and SR) after the start of slaughterhouse operations. In Amazonia, the results show that there is a significant decrease in ΔLU and GE inside the slaughterhouse influence zones. However, since the same change happened in the control zones, this decrease might not be caused directly by the slaughterhouse presence, and might instead be part of the downward trend of deforestation over the period between 2004 and 2013 [31,32]. For agricultural intensification variables in Amazonia, PC and CC show a significant increase in both the influence and control zones, while SR does not show change in the areas under slaughterhouse influence. In the Cerrado, results for all variables are similar in the control and influence zones. Nonsignificant decreases in ΔLU and GE and significant increases of PC, CC, and SR are observed in the control zones as well as the influence zones.

The decrease in ΔLU observed both inside and outside the slaughterhouse influence zones in Amazonia demonstrates not slaughterhouse influence, but the power of conservation programs and other policies for forest protection [31,33–35]. In addition to the protection granted by the Brazilian Forest Code and monitoring programs such as the Program for Satellite Monitoring of the Brazilian Amazon Forest (PRODES—Projeto de Monitoramento da Floresta Amazônica Brasileira por Satélite) and the System for Detection of Deforestation in Real Time (DETER—System for Detection of Deforestation in Real Time), the private sector signed ambitious agreements—cattle agreements in 2009 and a Soy Moratorium in 2006 [36]—to further protect the native vegetation. The effective contribution of each measure is difficult to disentangle, but the combined result of these actions was a great success. According to INPE [32] the rate of forest loss in the Brazilian Amazon dropped from more than 2.7 Mha/year in 2004 to an average of 0.6 Mha/year in 2013, reaching the lowest rates since 1988.

Unfortunately, the same did not occur in the Cerrado. The decrease of ΔLU did not happen inside all influence zones. In the control zones, the ΔLU results indicate that there is increased suppression of Cerrado vegetation in areas away from large slaughterhouse influence. This may be linked with the absence of an effective vegetation suppression monitoring system in the biome, and the more permissive New Forest Code, which has allowed more legal suppression since 2012 [35]. Some studies [36,37] have also warned about a possible leakage of agriculture from Amazonia to the Cerrado due to the stricter conservation policies in Amazonia. According to the most recent official data available, 0.725 Mha was suppressed in the Cerrado between 2010 and 2011, which was 12% greater than observed in the previous period (0.647 Mha, between 2009 and 2010 [38]. In addition, a recent report released by Mighty Earth and Rainforest Foundation Norway (RFN) claimed that multinational companies are linked to massive and systematic suppression of native vegetation in areas of Cerrado in MATOPIBA (an acronym created from the first two letters of the states of Maranhão, Tocantins, Piauí and Bahia). The report found that areas operated by the investigated companies had 0.697 Mha of vegetation suppressed from 2011 to 2015 [39].

GE results reflect ΔLU results, as land use emissions dominate GE in both biomes. In Amazonia, even with the increase of cattle between 2000 and 2013 (from 29 to 56 million head), the emissions from enteric fermentation are not enough to exceed the emissions from land use; this result was expected due to the high Amazonian biomass. In the Cerrado, the emissions from enteric fermentation dominate GE in the influence zones up to 4 h. For GE, by contrast with the results observed for ΔLU, three slaughterhouse units showed an increase in the areas of influence up to a 2 h driving radius, and two, in a radius up to 4 h. This response suggests that, in the zones near the slaughterhouses, the native vegetation has already been suppressed for the most part, making the emissions contributions from enteric fermentation more prominent than those from land use change.

The PC and CC results show that there has been an increase in the production of protein and calories in both biomes. In Amazonia, the p calculated for the various influence zone sizes show that the farther the distance from the slaughterhouse, the greater the increase in both variables. The most likely reason for this is that areas closer to these slaughterhouses are dominated by pasture, which is unlikely to be converted to new cropping areas. According to Dias et al. [11], the Amazon and Cerrado experienced expansion of crop area and increase in production in recent decades, especially for soybeans. Considering both biomes, soybean production grew from 7.4 million tons in 1990 to approximately 45.2 million tons in 2010 [11]. As one could expect, our results indicate that the increases of PC and CC are not related to the slaughterhouses' presence. However, the large increases in crop production around slaughterhouses may contribute to future increases in animal feed availability in the region.

The SR results for Amazonia indicate that these pastures have a stable stocking rate probably related to stagnant cattle ranching technology. To complement the discussion about SR, we performed two additional tests. First, we performed a Mann−Whitney test to compare the SR of the control and influence zones before the year of start of operation. In this test, we aimed to verify whether the large slaughterhouses we studied were installed in areas with high values of SR. According to the result (Table 13), before the slaughterhouse start of operation in the Amazon, the SR in the influence zones was greater than the SR observed in the control zones (p = 0.031, Table 13). This is an indication that big companies prefer to install slaughterhouse units in areas with high production, to ensure supply to their large processing capacity.

Table 13. Results of Mann-Whitney test comparing SR_F^S and SR_F^C in Amazonia.

Former Period (Ha: $SR_F^S \neq SR_F^C$)			
SIF Code	SR_F^S **(Head/ha)**	**Control Code**	SR_F^C **(Head/ha)**
791	1.936	1	0.981
3348	1.194	2	1.204
1940	1.024	3	0.249
2583	1.509	4	1.451
2937	1.708	5	1.342
4149	1.577	6	1.458
4267	1.855	7	1.589
4333	1.794	8	0.584
Median	1.643		1.273
p		0.031 *	

* Indicates significant at 5% level.

In the second test (Table 14), to verify the stagnation of the SR inside the slaughterhouses influence zones, we performed a Mann–Whitney test to compare the SR of the control and influence zones after the start of slaughterhouse operations. The result shows that in the latter period, the SR values of the control zones are similar to the values in the influence zones (p = 0.328, Table 14). In other words, and considering also the results of Table 12, stocking rate is intensifying at much faster rates away from the large slaughterhouses than closer to them.

Table 14. Results of Mann-Whitney test comparing the SR_L^S and SR_L^C in Amazonia.

Latter Period (Ha: $SR_L^S \neq SR_L^C$)			
SIF Code	SR_L^S **(Head/ha)**	**Control Code**	SR_L^C **(Head/ha)**
791	1.873	1	1.073
3348	1.240	2	1.554
1940	1.150	3	0.286
2583	1.423	4	1.504
2937	1.728	5	1.684
4149	1.902	6	1.588
4267	2.036	7	2.018
4333	1.821	8	1.453
Median	1.775		1.529
p		0.328 ^{NS}	

^{NS} Indicates not significant at 5% level.

Our results also demonstrate that the relationship between SR and ΔLU is not easily defined. After the slaughterhouse start of operation in the Amazon, although ΔLU dropped everywhere, the process of intensification did not start in the influence zones. Through a historical comparison between the US and Brazil, Merry and Soares [18] suggested that Brazilian cattle ranching will intensify as a result of economic conditions and conservation investments (reductions in capital and land subsidies) rather than intensifying in order to produce conservation outputs. In addition, characteristics that facilitate extensive ranching practices need to be discouraged or removed. The relatively easy process of land acquisition—land grabbing and low land prices—accompanied by weak protection laws that facilitates forest clearing for new pasture areas are the main obstacles of intensive ranching profitability, and may continue to be so in the next years [10,40,41].

Finally, the main limitation of this work is related to three assumptions. First, as we assume the zone of slaughterhouse influence extends up to 8 h travel time from a slaughterhouse, we may have excluded pasture areas dedicated to the cow–calf segment of the market. This segment is the main challenge on the pathway to achieving sustainable cattle ranching in Brazil, because it is not monitored or tracked under the current cattle agreements [13]. In addition, nearly all cow–calf production continues to be dependent on extensive grazing systems in the country [9].

Second, we may underestimate the area influenced by slaughterhouses, and therefore the appropriate sizes of the influence and control zones. We do not consider variables such as cattle availability, market access and transportation cost in the zone size estimates. Today, about 49% of active slaughterhouses in Amazonia belong to companies that signed the TAC, corresponding to 70% of slaughter capacity in the biome [19]. Therefore, the similarities observed between the control and influence zones may indicate that small slaughterhouses, which are not considered in this analysis and may be found inside some areas designated as control zones, may affect their supply areas in the same way that large units do.

The third limitation is related to the assumption that only 12 selected slaughterhouses have influence in their respective supply area. As observed in Figure 2, many large slaughterhouses are near the selected ones and they may influence the variables analyzed along with the selected units. We assumed here that the effect of these older slaughterhouses has not changed in time, and the main effect measured is due to the slaughterhouses that started operations in the period of analyses. Only two slaughterhouses do not have other large units near them: SIF 4333 in Amazonia and SIF 3047 in the Cerrado (Figure 2). Although it is not possible to test statistically one slaughterhouse, SIF 4333 has the same direction of change of the set of Amazonia plants for ΔLU, GE, PC and CC at all influence zones (Table 4, Table 6, Table 8, and Table 10) and for SR at the 6 h and 8 h influence zone (Table 12). Similar results are found for SIF 3047 when compared to the Cerrado set, except for ΔLU at the 2 h influence zone. This is an indication of the effectiveness of this assumption.

5. Conclusions

This study investigated the influence of large slaughterhouses on five variables, two related to environment impact (land use change rate and GHG emissions), and three related to cattle-ranching intensification (protein from crops, calories from crops and stocking rate). The results indicate that the changes observed inside the zones influenced by slaughterhouses cannot be attributed to the start of slaughterhouse unit operation in either Amazonia or the Cerrado.

In the Amazon, the environmental impact variables we studied show the same pattern of responses inside and outside the slaughterhouse influence zones—both moving towards reduced environmental impact. The hypothesis that slaughterhouses are leverage points to reduce deforestation and suppression of native Cerrado vegetation is not confirmed, leading us to believe that conservation measures such as a strong monitoring system and more restrictive environmental policies are the main promoters of conservation in Amazonia. In addition, the slaughterhouses seem to have no effect on cattle-ranching intensification. The high stocking rates observed in the period before the slaughterhouses' start of operation indicate that large meatpackers prefer to set up their plants in areas already well established and developed in the biome.

In the Cerrado, the responses of the environmental impact variables both inside and outside the slaughterhouse influence zones indicate that there is considerable conservation work to be done in the biome. The success of sustainable agriculture in the Cerrado still relies on the implementation of conservation measures. In addition, the increase of PC, CC, and SR both inside and outside the influence zones demonstrates that, in the Cerrado, cattle-ranching intensification is a reality, and it is occurring independently of the presence of large slaughterhouses.

In conclusion, there is no evidence that large slaughterhouses have promoted either cattle-ranching intensification or improvements in the sustainability of cattle-ranching activity in the Amazon and Cerrado. The results of our study and the recent failures of some of the cattle agreements show that leaning on slaughterhouses should not be considered a reliable strategy to achieve sustainable beef production. According to Lambin et al. [42], zero-deforestation agreements signed by private sectors may not be sufficient to reduce environmental impacts in commodities supply chain; public and private policies need to complement and reinforce each other to disconnect the link between cattle production and deforestation. In addition, to achieve intensification, it is necessary to improve the ranchers' access to technologies and capital [43,44], as there are still too many cattle farmers in Amazonia and the Cerrado who are engaging in extensive ranching practices associated with low income and high environmental damage.

Author Contributions: A.B.S. and M.H.C. contributed equally to the design of the research, the analysis of the results and the writing of the manuscript.

Funding: This research was funded by CNPq—Conselho Nacional de Desenvolvimento Científico e Tecnológico (process 134430/2015-8).

Conflicts of Interest: The authors declare no conflict of interest.

Appendix Table with All Abbreviations Used in the Text

ABBREVIATIONS	
AGB	Aboveground biomass
BGB	Belowground biomass
CAR	Cadastro Ambiental Rural *(Rural Environmental Registry)*
CNPJ	Cadastro Nacional de Pessoa Jurídica *(National Register of Legal Entities)*
DETER	Sistema de Detecção de Desmatamento em Tempo Real *(System for the Detection of Deforestation in Real Time)*
DIPOA	Departamento de Inspeção de Produtos de Origem Animal *(Department for Inspection of Animal Products)*
GHG	Greenhouse Gases
GWP	Global Warming Potential
IBGE	Instituto Brasileiro de Geografia e Estatística *(Brazilian Institute of Geography and Statistics)*
INPE	Instituto Nacional de Pesquisas Espaciais *(National Institute for Space Research)*
IPCC	Intergovernmental Panel on Climate Change
LULUCF	Land Use, Land Use Change and Forestry
MAPA	Ministério da Agricultura, Pecuária e Abastecimento *(Brazilian Ministry of Agriculture, Livestock and Food Supply)*
MATOPIBA	Acronym created from the first two letters of the states of Maranhão, Tocantins, Piauí, and Bahia
NGOs	Non-Governmental Organizations
PNLT	Plano Nacional de Logística e Transporte *(National Logistics and Transportation Plan)*
PNMC	Política Nacional sobre Mudanças no Clima *(Brazil's National Policy on Climate Change)*
PRODES	Projeto de Monitoramento da Floresta Amazônica Brasileira por Satélite *(Program for Satellite Monitoring of the Brazilian Amazon Forest)*

Appendix Biomass Map and Emissions from Enteric Fermentation

The biomass map of the historic vegetation for Amazonia and the Cerrado is presented in Figure A1. The historic carbon content of native vegetation was 68.7 Pg-C for Amazonia and 10.1 Pg-C for the Cerrado. Estimation of the historic vegetation is a complicated process, and results can vary widely. Our estimate is comprehended in the range calculate by Leite et al. [45] for Amazonia (from 51.3 to 85.5 Pg-C); however, our estimate is about 53% less than the estimate for the Cerrado (from 13.8 to 28.8 Pg-C). The historic carbon content of native vegetation estimated in this study is different from the values reported in Leite et al. [45] because different methodologies and values of carbon stock were used to make the biomass maps. While Leite et al. [45] combined two maps of vegetation types (RadamBrasil and IBGE [25]) and used the values for carbon stock in vegetation from the Second National Communication of Brazil to the UNFCCC, we used the map from IBGE [25] and the data from the Third National Communication.

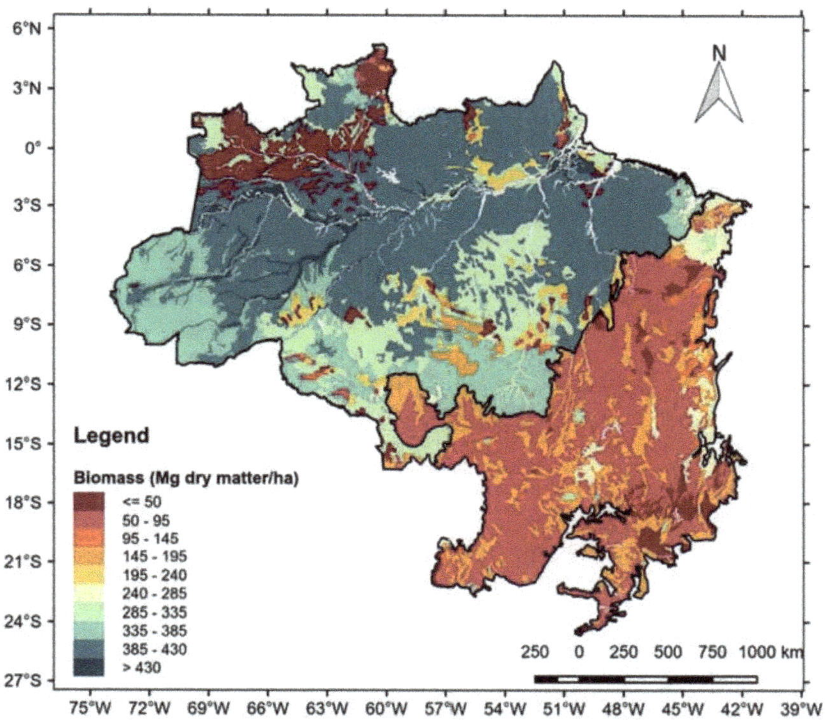

Figure A1. Biomass map for past vegetation of Amazonia and the Cerrado.

Another result is the CH_4 emissions from enteric fermentation. Between 2000 and 2013, the emissions from beef cattle increased in both biomes. Total methane emission by the two biomes in this period amounted to 2.9 Pg-CO_{2e}, about 54% of the total emitted in the country (5.3 Pg-CO_{2e} [2]). Emissions in Amazonia increased about 80% (from 41.7 Tg-CO_{2e} in 2000 to 77.5 Tg-CO_{2e} in 2013). In the Cerrado, emissions increased about 0.09% (from 82.5 Tg-CO_{2e} in 2000 to 90.5 Tg-CO_{2e} in 2013). The increase was bigger in the Amazon than in the Cerrado because of the great increase in number of cattle that occurred in this period.

Our estimates for methane emissions are very similar to other data. According to Azevedo et al. [2], for the states of the Amazon biome, the total amount of methane emitted by enteric fermentation from beef cattle was 1.0 Pg-CO_{2e} for the period, while our estimate was 0.9 Pg-CO_{2e}.

For the states of the Cerrado, Azevedo et al. [2] reported total methane emissions of 2.0 Pg-CO_2-eq for the period, about 35% greater than our estimate of about 1.3 Pg-CO_2-eq. These Cerrado estimates differ because we consider the actual geographic limits of the biome, while the Azevedo et al. [2] value includes total emissions for all Cerrado states, irrespective of how much area within the states is part of the Cerrado biome.

References

1. Ministério da Ciência, Tecnologia e Inovação. *Third National Communication of Brazil to the United Nations Framework Convention on Climate Change*; Ministério da Ciência, Tecnologia e Inovação: Brasilia, Brazil, 2016; Volume III, ISBN 9788588063198.
2. De Azevedo, T.R.; Costa Junior, C.; Brandão Junior, A.; Cremer, M.D.S.; Piatto, M.; Tsai, D.S.; Barreto, P.; Martins, H.; Sales, M.; Galuchi, T.; et al. SEEG initiative estimates of Brazilian greenhouse gas emissions from 1970 to 2015. *Sci. Data* **2018**, *5*, 1–43. [CrossRef] [PubMed]
3. Cohn, A.S.; Mosnier, A.; Havlík, P.; Valin, H.; Herrero, M.; Schmid, E.; O'Hare, M.; Obersteiner, M. Cattle ranching intensification in Brazil can reduce global greenhouse gas emissions by sparing land from deforestation. *Proc. Natl. Acad. Sci. USA* **2014**, *111*, 7236–7241. [CrossRef] [PubMed]
4. Strassburg, B.B.N.; Latawiec, A.E.; Barioni, L.G.; Nobre, C.A.; da Silva, V.P.; Valentim, J.F.; Vianna, M.; Assad, E.D. When enough should be enough: Improving the use of current agricultural lands could meet production demands and spare natural habitats in Brazil. *Glob. Environ. Chang.* **2014**, *28*, 84–97. [CrossRef]
5. Crosson, P.; Shalloo, L.; O'Brien, D.; Lanigan, G.J.; Foley, P.A.; Boland, T.M.; Kenny, D.A. A review of whole farm systems models of greenhouse gas emissions from beef and dairy cattle production systems. *Anim. Feed Sci. Technol.* **2011**, *166–167*, 29–45. [CrossRef]
6. De Oliveira Silva, R.; Barioni, L.G.; Hall, J.A.J.; Matsuura, M.F.; Zanett Albertini, T.; Fernandes, F.A.; Moran, D. Increasing beef production could lower greenhouse gas emissions in Brazil if decoupled from deforestation. *Nat. Clim. Chang.* **2016**, *6*, 493–497. [CrossRef]
7. Mazzetto, A.M.; Feigl, B.J.; Schils, R.L.M.; Cerri, C.E.P.; Cerri, C.C. Improved pasture and herd management to reduce greenhouse gas emissions from a Brazilian beef production system. *Livest. Sci.* **2015**, *175*, 101–112. [CrossRef]
8. Palermo, G.C.; D'Avignon, A.L.D.A.; Freitas, M.A.V. Reduction of emissions from Brazilian cattle raising and the generation of energy: Intensification and confinement potentials. *Energy Policy* **2014**, *68*, 28–38. [CrossRef]
9. Ferraz, J.B.S.; de Felício, P.E. Production systems—An example from Brazil. *Meat Sci.* **2010**, *84*, 238–243. [CrossRef] [PubMed]
10. Bowman, M.S.; Soares-Filho, B.S.; Merry, F.D.; Nepstad, D.C.; Rodrigues, H.; Almeida, O.T. Persistence of cattle ranching in the Brazilian Amazon: A spatial analysis of the rationale for beef production. *Land Use Policy* **2012**, *29*, 558–568. [CrossRef]
11. Dias, L.C.P.; Pimenta, F.M.; Santos, A.B.; Costa, M.H.; Ladle, R.J. Patterns of land use, extensification, and intensification of Brazilian agriculture. *Glob. Chang. Biol.* **2016**, *22*, 2887–2903. [CrossRef] [PubMed]
12. Barreto, P.; Pereira, R.; Arima, E. *A Pecuária e o Desmatamento na Amazônia na era das Mudanças Climáticas*; Instituto do Homem e Meio Ambiente da Amazônia: Belém, Brazil, 2008; ISBN 9788586212260.
13. Gibbs, H.K.; Munger, J.; L'Roe, J.; Barreto, P.; Pereira, R.; Christie, M.; Amaral, T.; Walker, N.F. Did ranchers and slaughterhouses respond to zero-deforestation agreements in the Brazilian Amazon? *Conserv. Lett.* **2016**, *9*, 32–42. [CrossRef]
14. Greenpeace. Eating up the Amazon. Available online: http://www.greenpeace.org/usa/research/eating-up-the-amazon/ (accessed on 16 May 2016).
15. Greenpeace. Slaughtering the Amazon. Available online: http://www.greenpeace.org/international/en/publications/reports/slaughtering-the-amazon/ (accessed on 16 May 2016).
16. Walker, N.; Patel, S.; Kalif, K. From Amazon pasture to the high street: Deforestation and the Brazilian cattle product supply chain. *Trop. Conserv. Sci.* **2013**, *6*, 446–467. [CrossRef]
17. Garcia, E.; Filho, F.S.V.R.; Mallmann, G.M.; Fonseca, F. Costs, benefits and challenges of sustainable livestock intensification in a major deforestation frontier in the Brazilian Amazon. *Sustainability* **2017**, *9*, 158. [CrossRef]

18. Merry, F.; Soares-Filho, B. Will intensification of beef production deliver conservation outcomes in the Brazilian Amazon? *Elem. Sci. Anth.* **2017**, *5*, 1–12. [CrossRef]
19. Barreto, P.; Pereira, R.; Brandão, A., Jr.; Baima, S. *Os Frigoríficos vão Ajudar a Zerar o Desmatamento da Amazônia?* Imazon: Belém, Brazil, 2017; ISBN 9788586212949.
20. Alves-Pinto, H.N.; Newton, P.; Pinto, L.F.G. *Certifying Sustainability: Opportunities and Challenges for the Cattle Supply Chain in Brazil*; Research Program on Climate Change; Agriculture and Food Security (CCAFS): Copenhagen, Denmark, 2010; pp. 1–8.
21. Zu Ermgassen, E.K.H.J.; de Alcântara, M.P.; Balmford, A.; Barioni, L.; Neto, F.B.; Bettarello, M.M.F.; de Brito, G.; Carrero, G.C.; Florence, E.D.A.S.; Garcia, E.; et al. Results from on-the-ground efforts to promote sustainable cattle ranching in the Brazilian Amazon. *Sustainability* **2018**, *10*, 1301. [CrossRef]
22. Barreto, P.; Silva, D. Will cattle ranching continue to drive deforestation in the Brazilian Amazon? In *Environment and Natural Resources Management in Developing and Transition Economies*; Centre of Studies and Research on International Development from the University of Auvergne: Clermont Ferrand, France, 2010; pp. 1–23.
23. Hansen, M.C.; Potapov, P.V.; Moore, R.; Hancher, M.; Turubanova, S.A.; Tyukavina, A.; Thau, D.; Stehman, S.V.; Goetz, S.J.; Loveland, T.R.; et al. High-Resolution Global Maps of 21st-Century Forest Cover Change. *Science* **2013**, *342*, 850–853. [CrossRef] [PubMed]
24. Bustamante, M.; Santos, M.M.D.O.; Shimbo, J.Z.; Cantinho, R.Z.; Mello, T.D.R.B.D.; Oliveira, P.V.D.C.; Cunha, P.W.P.D.M.; Martins, F.D.S.R.V.; Aguiar, A.P.D.; Ometto, J. *Emissões do Setor de uso da Terra, Mudança do uso da Terra e Florestas: Relatório de Referência*; Ministério da Ciência, Tecnologia e Inovação: Brasíla, Brazil, 2015; p. 343.
25. IBGE. Mapa de Vegetação. 2004. Available online: https://mapas.ibge.gov.br/tematicos/vegetacao (accessed on 13 February 2016).
26. Berndt, A.; Lemes, A.P.; Romero, L.A.; Sakamoto, L.S.; Lima, M.A. *De Emissões de Metano por Fermentação Entérica e Manejo de Dejetos de Animais: Relatório de Referência*; EMBRAPA; Ministério da Ciência, Tecnologia e Inovação: Brasília, Brazil, 2015; p. 150.
27. IPCC. IPCC Guidelines for National Greenhouse Gas Inventories. 2006. Available online: https://www.ipcc-nggip.iges.or.jp/public/2006gl/ (accessed on 1 April 2016).
28. Cardoso, E.G. Engorda de Bovinos em Confinamento (Aspectos Gerais). Embrapa, 1996. Available online: https://www.embrapa.br/busca-de-publicacoes/-/publicacao/317195/engorda-de-bovinos-em-confinamento-aspectos-gerais (accessed on 15 April 2018).
29. Da Costa, M.J.R.P.; Quintiliano, M.H.; Tseimazides, S.P. *Boas Praticas de Manejo—Transporte*; MAPA/ACS: Brasília, Brazil, 2013; ISBN 9788579910715.
30. Confederação Nacional do Transporte. Anuário CNT do Transporte—Estatísticas Consolidadas. Available online: http://anuariodotransporte.cnt.org.br/2017/ (accessed on 20 April 2018).
31. Boucher, D. How Brazil has dramatically reduced tropical deforestation. *Solutions* **2014**, *5*, 65–75.
32. Kastens, J.H.; Brown, J.C.; Coutinho, A.C.; Bishop, C.R.; Esquerdo, J.C.D.M. Soy moratorium impacts on soybean and deforestation dynamics in Mato Grosso, Brazil. *PLoS ONE* **2017**, *12*, 0176168. [CrossRef] [PubMed]
33. Nepstad, D.; McGrath, D.; Stickler, C.; Alencar, A.; Azevedo, A.; Swette, B.; Bezerra, T.; DiGiano, M.; Shimada, J.; da Motta, R.S.; et al. Slowing Amazon deforestation through public policy and interventions in beef and soy supply chains. *Science* **2014**, *344*, 1118–1123. [CrossRef] [PubMed]
34. Soares-Filho, B.; Rajao, R.; Macedo, M.; Carneiro, A.; Costa, W.; Coe, M.; Rodrigues, H.; Alencar, A. Cracking Brazil's forest code. *Science* **2014**, *344*, 363–364. [CrossRef] [PubMed]
35. Gibbs, H.K.; Rausch, L.; Munger, J.; Schelly, I.; Morton, D.C.; Noojipady, P.; Soares-Filho, B.; Barreto, P.; Micol, L.; Walker, N.F. Brazil's soy moratorium. *Science* **2015**, *347*, 377–378. [CrossRef] [PubMed]
36. INPE. Instituto Nacional de Pesquisas Espaciais 2013. Available online: http://www.inpe.br/ (accessed on 20 December 2017).
37. Macedo, M.N.; DeFries, R.S.; Morton, D.C.; Stickler, C.M.; Galford, G.L.; Shimabukuro, Y.E. Decoupling of deforestation and soy production in the southern Amazon during the late 2000s. *Proc. Natl. Acad. Sci. USA* **2012**, *109*, 1341–1346. [CrossRef] [PubMed]
38. IBAMA Projeto de Monitoramento dos Biomas Brasileiros via Satélite (PMDBBS). Available online: http://siscom.ibama.gov.br/monitora_biomas/ (accessed on 20 December 2017).

39. Bellantonio, M.; Hurowitz, G.; GrØnlud, L.; Yousefi, A. Ultimate Mystery Meat. Available online: http://www.mightyearth.org/mysterymeat/ (accessed on 16 October 2017).
40. Cardoso, D.; Brito, B.; Baima, S.; Siqueira, V. *O Estado da Amazônia Potencial de Arrecadação Financeira com a Regularização Fundiária no Pará*; Imazon: Belém, Brazil, 2018; p. 14.
41. Azevedo-Ramos, C.; Silva, J.N.M.; Merry, F. The evolution of Brazilian forest concessions. *Elem. Sci. Anth.* **2015**, *3*, 000048. [CrossRef]
42. Lambin, E.F.; Gibbs, H.K.; Heilmayr, R.; Carlson, K.M.; Fleck, L.C.; Garrett, R.D.; de Waroux, Y.L.P.; McDermott, C.L.; McLaughlin, D.; Newton, P.; et al. The role of supply-chain initiatives in reducing deforestation. *Nat. Clim. Chang.* **2018**, *8*, 109–116. [CrossRef]
43. Monzoni, M.; Vendramini, A.; Rocha, F.C.; Peirão, P.; Campanili, M.; Sobreiro, A.; Monteiro, A. *Análise dos Recursos do Programa ABC Safra 2016/17*; Centro de Estudos em Sustentabilidade da Fundação Getúlio Vargas (GVces): São Paulo, Brazil, 2017; pp. 1–36.
44. Latawiec, A.E.; Strassburg, B.B.N.; Valentim, J.F.; Ramos, F.; Alves-Pinto, H.N. Intensification of cattle ranching production systems: Socioeconomic and environmental synergies and risks in Brazil. *Animal* **2014**, *8*, 1255–1263. [CrossRef] [PubMed]
45. Leite, C.C.; Costa, M.H.; Soares-Filho, B.S.; Hissa, L.D.B.V. Historical land use change and associated carbon emissions in Brazil from 1940 to 1995. *Glob. Biogeochem. Cycles* **2012**, *26*, 1–13. [CrossRef]

sustainability

MDPI

Article

Greenhouse Gas Emissions from Beef Grazing Systems in Semi-Arid Rangelands of Central Argentina

María I. Nieto [1], Olivia Barrantes [2,3,*], Liliana Privitello [4] and Ramón Reiné [5]

[1] Instituto Nacional de Tecnología Agropecuaria (INTA), Estación Experimental Agropecuaria Catamarca, Ruta 33 Km 4 (4705), Sumalao, 4705 Catamarca, Argentina; nieto.maria@inta.gob.ar

[2] Departamento de Ciencias Agrarias y del Medio Natural, Facultad de Veterinaria, Universidad de Zaragoza, Miguel Servet 177, 50013 Zaragoza, Spain

[3] Instituto Agroalimentario de Aragón -IA2- (CITA-Universidad de Zaragoza), 50013 Zaragoza, Spain

[4] Departamento de Ciencias Agropecuarias, Facultad de Ingeniería y Ciencias Agropecuarias, Universidad Nacional de San Luis, Av. 25 de Mayo 384, Villa Mercedes, 5730 San Luis, Argentina; lilianaprivitello@hotmail.com

[5] Departamento de Ciencias Agrarias y del Medio Natural, Escuela Politécnica Superior, Universidad de Zaragoza, Carretera de Cuarte s/n, 22071 Huesca, Spain; rreine@unizar.es

* Correspondence: olivia.barrantes@unizar.es; Tel.: +34-976-761-596

Received: 30 September 2018; Accepted: 12 November 2018; Published: 16 November 2018

Abstract: The livestock sector can be a major contributor to the mitigation of greenhouse gas (GHG) emissions. Within the sector, beef production produces the largest proportion of the livestock sector's direct emissions. The objective of this study was to assess the on-farm GHG emissions in semi-arid rangelands in Argentina and to identify the relationship between emissions and current farm management practices. A survey recorded detailed information on farm management and characteristics. Assessments of GHG emissions were based on the Intergovernmental Panel on Climate Change (IPCC) Tier 2 protocols. The relationship between farm management and GHG emissions were identified using general linear models. Cluster analysis was used to identify groups of farms that differed from others in emissions and farm characteristics. Emissions per product sold were low on farms that had improved livestock care management, rotational grazing, received technical advice, and had high animal and land productivities. Emissions per hectare of farmland were low on farms that had low stocking rates, a low number of grazing paddocks, little or no land dedicated to improved pastures and forage crops, and low land productivity. Our results suggest that the implementation of realistic, relatively easy-to-adopt farming management practices has considerable potential for mitigating the GHG emissions in the semi-arid rangelands of central Argentina.

Keywords: livestock care management; rotational/continuous grazing; technical advice; stocking rate; functional units

---

1. Introduction

Livestock production is an important source of greenhouse gas (GHG) emissions worldwide. The livestock sector contributes 14.5% of global GHG emissions [1]. Since the human population is expected to increase from 7.2 to 9.6 billion by 2050 [2], together with the improvement of standard of living, there is an increasing demand for livestock products [3], which are expected to double by the mid-21st century [4]. The livestock sector will have to be a major contributor in the mitigation of GHG emissions and in the improvement of global food security [5]. Within the sector, beef production contributes the majority of emissions, producing 41% of the livestock sector's direct emissions [6]. Methane (CH_4) and nitrous oxide (N_2O) are the largest contributors to global livestock emissions in

CO_2 equivalent (CO_2e) per year [1]. In countries where pastoral agriculture is the dominant sector, CH_4 and N_2O emissions contribute up to 50% of the total GHG emissions [7]. Due to the negative relation between the efficiency of production and GHG emissions per output produced, the greatest mitigation potential lies in ruminant systems that operate with a low productivity (e.g., in South Asia, Latin America, the Caribbean, and Africa) [1].

Argentina is a major world beef producer. It is the sixth largest beef producer in the world [8] and the ninth largest beef exporter [9]. The meat chain generates around 4% of the total jobs in the country [9]. Argentina and four other Latin American main beef exporters plan to increase meat production in response to the forecasted growth in international markets [10]. Of the 156.94 Mt CO_2e emitted by the Argentinean sector relating to 'Agriculture, Forestry and other land uses', livestock directly accounts for 54 Mt CO_2e, produced by enteric fermentation (of which beef accounts for 83%); 20.26 MtCO_2e are produced by manure deposition on pastures (76% from beef); and 2.14 MtCO_2e are produced by manure management (83% from beef) [11]. Thus, methane emitted by enteric fermentation is especially important. In 2008, methane emissions from domestic ruminants in Argentina was the sixth highest in the world, and the per capita rate was one of the highest [12]. Rearte and Pordomingo [10] indicated there are ample opportunities to reduce methane emissions per unit of product in Argentina and other temperate regions of Latin America, such as Uruguay.

The GHG emissions of a product can be expressed as kg of CO_2e per kg of product, or it can be expressed as kg of CO_2e per unit of area (ha) of the production system, depending on the perspective (that of the consumer vs. that of the producer) [13–15] and product perspective vs. an IPCC inventory perspective [16,17]. From a 'product perspective', Casey and Holden [16] suggested that it is necessary to choose a functional unit (FU, kg of product vs. land area) of the function that the system delivers. However, for studies that intend to inform national GHG inventory reports, an FU coupled with land area is necessary [16,17]. Finally, environmental impacts per unit of product are more closely linked with the assessment of global issues, such as climate change [15,18], while assessing an environmental impact per hectare of land is considered a more local or regional issue, such as soil erosion, water eutrophication, or acidification [18]. Some studies have shown that the use of different FUs can produce contradictory results in assessing GHG emissions [13,14,16], illustrating the potential trade-off between carbon efficiencies per unit of product and per unit of land. Nevertheless, several studies suggest that the mitigation of emissions per unit of product and per unit of land area can be reconciled [14,16]. Casey and Holden [16], Halberg et al. [18] and Veysset et al. [19] suggested that product-based and land area-based indicators should be used to characterize the environmental impacts caused by food production.

Many studies have assessed mitigation strategies for reducing GHG emission intensity in terms of emissions per unit of animal product in several ruminant livestock farming systems worldwide, which have been reviewed extensively [1,5,20–27]. The mitigation strategies that reduce emission intensity by increasing herd productivity through improved animal husbandry practices (e.g., animal feeding, genetics, health, fertility, and the overall management of the animal operations) can be important in low–input ruminant systems [12,22] and have a greater mitigation potential in development than they do in developed economies [22].

In Argentina, >70% of the beef is produced in pasture-based grazing systems [10], mostly in extensive conditions [12]. As the opportunity for soybeans and cereal grains became structural to Argentinean agriculture, livestock businesses were displaced towards less productive, marginal lands [10,28]. Eight percent of beef production is in the semi-arid Central Region (18% of the country) [29]. Rangeland native grasslands are the main source of feed for cow-calf livestock systems, which constitute an important economic sector in the region [30]. Rangelands are the world's most common land type [31], and they provide the livelihoods for many vulnerable communities throughout the world [32]. Their relevance is linked to their multifunctional nature and provision of ecosystem services [33]. Extensively managed livestock production is the most sustainable and common form of agriculture on rangelands [33]. Global demand for livestock products will increase the pressure on rangelands, which are experiencing high degradation and losses in biodiversity worldwide [34], especially in arid and semi-arid grasslands in developing countries [33].

Identifying the relationship between GHG emissions, farm management and system productivity can help livestock producers improve operations, where productivity can be improved considerably by implementing simple management practices [35]. Furthermore, in the transition to sustainable livestock production, assessments of mitigation measures that have been tailored to the location and livestock production system in use are needed [5,14]. Our study assessed commercial farms that were representative of the extensive beef systems, based on natural rangelands in the San Luis Province, which is typical of the semi-arid Central Region of Argentina [28]. Beef cattle stocks have increased in San Luis because of the displacement of livestock into semi-arid marginal areas [36] and incipient intensification has been reported [28]. The main objective was to assess GHG emissions from representative farms in order to identify realistic farming practices that will favor low GHG emissions. Specifically, we (i) estimated the CH_4 and N_2O on-farm livestock emission intensity based on two functional units: product-based (kg CO_2e per total live weight sold) and area-based (kg CO_2e per land area used); (ii) identified farm attributes and management practices that were associated with low emission intensities; and (iii) assessed the implications of using each of the two functional units in identifying the farming practices that minimize GHG emissions.

2. Materials and Methods

The study was conducted in a 4160-km^2 area in the San Luis province, Central Region, Argentina (center of the study area: 34°17′22.46″ S; 66°25′40.89″ W), where an extensive cow-calf system, based on year-round, open-air grazing, is the main land use. The climate is semi-arid, and the annual precipitation ranges from 350 to 500 mm. The average daily temperature ranges from 8.5 °C (coldest month) to 23 °C (warmest month) [37]. The soils are shallow, poor in organic matter, have low water retention capacity, and low-medium productivity [38]. The climate and soil conditions are unsuited to croplands, and rangeland native grasslands are the main source of feed for livestock [30]. Natural vegetation consists of two main types [39]: (i) a woodland-shrubland mixture dominated by legume trees (*Prosopis flexuosa* and *Geoffroea decorticans*) and shrubs (*Larrea divaricata*); and (ii) grass-dominated steppes of *Nassella tenuis*, *Piptochaetium napostaense*, *Poa ligularis* and *Poa lanuginose*, and small scattered woodlands of legume trees (mainly *G. decorticans*). Most of the cattle are Hereford, Aberdeen Angus, or crossbreeds of the two, although some farms also have Creole [40]. A low productivity and potential improvement of the farm system have previously been reported in the region [10].

In 2014, 30 of the 67 beef cattle farms in the study area were surveyed. The farms were representative of the region based on earlier studies [30]. The survey, recorded in a structured questionnaire, collected detailed information about the size and structure of the farm, livestock management, infrastructures, productivity, as well as the ages and levels of education of the producer and the labor, referred to a one-year production cycle. With that information, a characterization of the farms was obtained, and variables were calculated, which were used in the analysis.

The general characteristics of the farms are detailed in Table 1. Seven percent of the farmers did not have any type of education, 61% had a primary or secondary education, and 32% had higher education. Half of the farms surveyed had salaried employees. In addition to natural grassland areas, 23% of the farms improved grasslands by introducing grasses, such as *Eragrostis curvula*, *Digitaria eriantha*, and *Panicum coloratm cv. verde*, and 17% had annual forage crops, such as maize (*Zea mays*), sorghum (*Sorghum vulgare*), rye (*Secale cereale*) with melilotus (*Melilotus albus*), and oats (*Avena sativa*), although in both cases, the areas were much smaller than the natural pastures (4% and 2% of the total land on average, respectively, Table 1). The annual forage crops are seeded if there have been enough rains; they are 'low-input' crops, which are not mowed but only grazed, and farmers generally do not use pesticides or chemical fertilizers. Half of the farmers purchased small amounts of maize, alfalfa or mineral complements, only in winter in critical years and not always for feeding all animal categories.

Three types of production systems were observed: (i) cow-calf (CC) systems (60% of the farms), where calves are sold at weaning; (ii) backgrounding (BG) systems (10%) (i.e., farmers purchase weaned calves that are sold once they are fattened); and (iii) cow-calf + backgrounding (CCBG) systems (30%). The calves are weaned at 6 months at 130–150 kg of live weight (LW) and sold fattened at 15 months (280–300 kg LW). On the CC and CCBG farms, the reproductive system was either (i) year-round mating (44% of the farms), or (ii) seasonal. Only 7% of farms employed artificial insemination. On the farms, water was collected in artificial dams by drilling, extraction from wells and, to a lesser extent, using natural streams.

Table 1. General characteristics of the beef cattle farms in San Luis province, central Argentina.

Variable	Mean	s.d.	Min	Max
Socio-economic data				
Age (years)	56	11.3	34	75
+ Hired labor (WU/LU) ($\times 10^{-3}$)	0.7	0.8	0	2.5
Land use				
+ Total land area (ha)	3598	4706	67	23,400
+ Land area used for native pastures (%)	95	14	33	100
+ Land area used for improved pastures (%)	4	13	0	67
Land area for annual forage crops (%)	2	6	0	29
Beef cattle				
+ Total Livestock Units [1] (LU)	337	399	17	1856
Mortality rate (%)	6.7	11.1	0.6	50.0
Stocking rate (LU/ha)	0.13	0.09	0.02	0.46
Grazing infrastructures				
Water reservoirs per total land/ha ($\times 10^{-3}$)	2.8	3.2	0.3	14.9
Water reservoirs/LU ($\times 10^{-3}$)	25	21	4	89
+ Grazing paddocks/ha ($\times 10^{-3}$)	3.7	4.5	0.6	19.6
System productivity				
Average LW of livestock (kg)	283	50	195	399
Weaning rate [2] (%)	65	17	26	95
+ Land productivity (kg LW sold/ha)	18.3	20.4	1.1	93.8
+ Animal productivity (kg LW sold/LU)	138	91	53	337

s.d., standard deviation; Min and Max, minimum and maximum values (n = 30 farms); WU, work units; LU, livestock units; LW, live weight. +Variables used for the typification of the beef cattle farms. [1] Livestock units were calculated based on Cocimano et al. [41]. [2] n = 27 (remaining three farms are backgrounding systems; they do not have breeding).

GHG emissions were estimated on-farm, based on the CH_4 emissions from cattle enteric fermentation and N_2O emissions from the managed soils used by grazing animals. Animals graze year-round, and manure is not managed, which is consistent with the Intergovernmental Panel on Climate Change (IPCC) [42]. No animal housing was involved, and crops and imported feeds were not relevant in the study area. CO_2 emissions from infrastructure, energy used for crops, and off-farm GHG emissions were not included in this study. Therefore, we focused on the relevant on-farm GHG emissions, as affected by the farm management practices. GHG emissions were expressed as CO_2 equivalents (CO_2e) for a time horizon of 100 years: CH_4 kg $\times$ 25 and N_2O kg $\times$ 298 [43]. Emissions were expressed as kgCO_2e per kg LW sold (sum of weaned calves and culled cows), and per hectare (ha) of farmland.

Assessments of GHG emissions were based on the IPCC Tier 2 protocols [42]. Appendix A shows the IPCC (2006) equations used in the calculations. Further updates of IPCC (2006) protocols did not affect those equations. Enteric CH_4 was estimated for each category of cattle on the farm: cow, weaned calf, replacement heifer, bull and steer.

Estimates of the gross energy (GE) intake of the animals were calculated based on the net energy (NE) requirements for maintenance, activity, growth, pregnancy and lactation. Enteric emissions were estimated based on GE intake and using methane conversion factors (*Ym*). We refined *Ym* calculations using the Cambra-López et al. (2008) [44] equation: $Ym = -0.0038{*}DE^2 + 0.351{*}DE\text{-}0.8111$, where DE is feed digestibility, expressed as a percentage of the GE of the feed. DE was estimated based on earlier studies on the quality of the pastures in the study area [45], percentage of land with annual forage crops, and the opinions of local experts of the 'Estación Experimental Agropecuaria San Luis' del 'Instituto Nacional de Tecnología Agropecuaria' (INTA). The average DE was 58% (range = 52–60; SD = 1.3), and the average *Ym* was 6.7% (range = 6.5–7.1; SD = 0.12).

N_2O emissions from managed soils were calculated based on the N deposited on pastures by grazing animals (urine and dung). The amount of N deposited on the pasture by each cattle type while grazing was estimated based on the number of animals, feed intake, pasture N content, and N retention of the animals, following IPCC [42] and the National Research Council [46].

Two analyses were conducted: statistical models that described GHG emissions and cluster analysis, identifying homogeneous groups of farms that differed in emissions and management practices.

The relationship between farm management practices and GHG emission was investigated using generalized linear models (GLM) [47], with the assumption that the data followed a Tweedie distribution, and a logarithmic link function. The dependent variable was CO_2e emissions, which was expressed as either (a) per kg of LW sold or (b) per hectare of land.

A set of explanatory variables was used for the models of each of the two dependent variables. The values of all explanatory variables (nominal, ordinal, or continuous) were transformed to 0–1 values and included as 'factors' (categorical predictors with values 0–1) in the models. Nominal variables included feed purchase (0, no: 1, yes); the reproductive management of the livestock (0, year-round mating; 1, seasonal mating); technical advice (0, no; 1, yes); type of production system [0, cow-calf (CC); 1, cow-calf + backgrounding (CCBG)]; and grazing system (0, continuous; 1, rotational). For the ordinal and continuous variables, the scoring criteria were based on the median (values $\leq$ median = 0; values > median = 1), except for the land area used for introduced pastures or annual forage crops. Ordinal variables included water reservoirs per total land (0, low; 1, moderate; median = 16.7×10^{-4} water reservoirs/ha); grazing paddocks per total land (0, low; 1, moderate; median = 16×10^{-4} grazing paddocks/ha); and livestock care controls (0, poor = three or fewer types of controls; 1, good = four to six types of controls; median = 3 controls). The types of livestock care controls were body condition, teeth examination, rectal palpation/ecography, parasite control, reproductive vaccine, and bull review control. Continuous variables included land area used for introduced pastures or annual forage crops (0, null/very low if area $\leq$ 4%; 1, low/moderate if area > 4%), average live weight of livestock (0, low; 1, moderate; median = 292 kg), cows-to-total animals rate (0, low; 1, moderate; median = 55%), average weight of sold calf (0, low; 1, moderate; median = 204 kg), mortality rate (0, low; 1, moderate; median = 2.6%), stocking rate (0, low; 1, moderate; median = 0.10 LU/ha), weaning rate (0, low; 1, moderate; median = 66.5%), land productivity (0, low; 1, moderate; median = 9.2 kg LW sold/ha), and animal productivity (0, low; 1, moderate; median = 100 kg LW sold/LU).

Prior to the GLM analysis, an exploratory analysis was conducted, based on the Mann–Whitney test, to identify the independent effects of variables on GHG emissions, and a Spearman's non-parametric test was used to identify co-linearity among variables. Only non-correlated variables ($r_s < 0.38$, $p > 0.05$) were included in a given GLM. Backgrounding farms were excluded from the analysis because they do not have a breeding herd.

Several analyses were performed based on all possible combinations of non-correlated variables and removing the non-significant explanatory variables one at a time (variables that did not reach $p < 0.05$ in a Wald's Chi-square test) until the final models only contained significant explanatory variables.

Only models that were significant ($p < 0.05$) based on an omnibus test were included in the analyses. The resultant models were defined as:

$$LnE = \alpha + \beta_1\, var_1 + \beta_2\, var_2 + \ldots + \beta_i\, var_i\,,$$

where E = CO_2e emission, the first term 'α' contains the regression intercept, and the remaining terms include the variables used in the model. The model indicates the partial regression β coefficients, which indicate the weights of the variables $1, 2, \ldots, i$ in the model when the variable is '0'. Thus, if β is > 0, E and the variable (level '0') are positively correlated, and if β is < 0, E and the variable (level '0') are negatively correlated. If the variable is '1', the model takes the reference value ($\beta = 0$ and, hence, $LnE = \alpha$). Emissions are calculated as:

$$E = e^{\alpha} \cdot e^{\beta_1 var_1} \cdot e^{\beta_2 var_2} \cdot \ldots \cdot e^{\beta_i var_i}$$

The statistical significance of the coefficients of individual variables in the models was tested using Wald's Chi-square test. Significant interaction effects were not detected. To express the main effects in each model, the estimated marginal means were calculated. For all possible combinations of non-correlated variables in the models, a model fit was evaluated based on Akaike's information criterion for finite samples (AICc) [48]. ΔAICc was calculated as: $\Delta AICc = AICc_i - AICc_{min}$, for $i = 1, 2, \ldots, R$, where $AICc_{min}$ denotes the minimum of the AICc values for the R models [49]. Models that had the lowest AICc were selected as the best models within a set of models that included the same set of variables [48]. Models with ΔAICc < 7 were considered plausible, and models with ΔAICc > 11 were discarded [49]. The explained deviance reflected the contributions of significant individual explanatory variables to the model as follows: $D^2 = (D_0 - D_{model})/D_0$, where D_0 is the deviance of the null model (intercept only), and D_{model} is the deviance of the analyzed model [50]. The contribution of each explanatory variable was estimated based on the change in D^2 after the variable was deleted from the model divided by the total explained deviance [51], which is expressed here as 'D^2 change on deletion' (%DCD). As the values of the variables were 0 or 1, standardization of the explanatory variables was not conducted. The statistical significance of the independent effects of each management variable on GHG emissions was assessed based on Spearman's correlation non-parametric tests.

For the typification of the farms, 7 continuous and 4 discrete variables were selected. To identify the main factors (eigenvalues > 1) that characterized the changes observed, 11 variables were subjected to principal component analysis (PCA), with varimax rotation. The Bartlett sphericity test and a Kaiser–Meyer–Olkin (KMO) test for sampling adequacy were used to validate the sampling. To identify a typology of the farms, we subjected the main factors of the PCA to a hierarchical cluster analysis (CA), with a squared Euclidean distance and Ward's aggregation method. In that way, five groups of farms were identified. To validate the results, we used a non-parametric Kruskal–Wallis test, known as 'analysis of variance by ranges' [52], which verifies which continuous variables, either those used in the PCA-CA (7 variables) or not (10 variables), are significant in explaining the differences between the groups. To identify which groups differ according to each continuous variable, we used the non-parametric Dunn–Bonferroni post-hoc test. To identify differences between groups for the discrete variables, we used the Pearson's chi-square test. The testing of variables not included in the CA is known as 'criterion validity' [53] and has been used to characterize livestock farms [54].

The statistical analyses were performed using IBM SPSS Advanced Statistics software ver. 22 [55].

3. Results

3.1. Farm Greenhouse Gas (GHG) Emissions

Among the beef farms in the San Luis province, central Argentina, the mean GHG emission intensity was 19.6 kg CO_2e/kg LW sold, but varied widely (range = 6.2–39.7). Backgrounding (BG) farms produced fewer emissions than did cow-calf (CC) farms, and mixed CCBG farms produced average emissions. On a farm-area basis, the average emission rate was 261 kg CO_2e/ha (range = 26 to 1042), which did not differ significantly among types of production systems (Table 2).

Table 2. Farm greenhouse gas emissions of the beef cattle farms in San Luis province, central Argentina.

Farm Greenhouse Gases Emission Intensity	Production System	Mean [1]	s.d.	Min	Max	*n*
kg CO_2e/kg LW sold	Cow-calf	23.6 ^{b}	7.3	12.4	39.7	18
	Backgrounding	6.9 ^{a}	1.1	6.2	8.1	3
	Cow-calf + Backgrounding	15.7 ^{ab}	6.3	7.0	22.6	9
	Overall	*19.6*	*8.6*	*6.2*	*39.7*	*30*
kg CO_2e/ha	Cow-calf	243	225	26	1042	18
	Backgrounding	345	70	270	409	3
	Cow-calf + Backgrounding	269	200	83	671	9
	Overall	*261*	*205*	*26*	*1042*	*30*

s.d., standard deviation; Min and Max, minimum and maximum values; LW, live weight. [1] Different letters in the same column indicate significant differences between production system groups (p = 0.002). Kruskal–Wallis test.

3.2. *Effects of Farm System and Management on GHG Emissions*

Considered independently, six variables had a significant effect on emission intensity per kg of LW sold (Table 3). Emissions were significantly lower under good than under poor livestock care management controls, if technical advice was sought, if rotational grazing was used, and in CCBG rather than in CC systems. Land and animal productivity affected the emissions, with lower emission intensities under higher land and animal productivity. Furthermore, emission intensity was negatively correlated with land and animal productivities ($r = -0.46$, $p < 0.05$; $r = -0.87$, $p < 0.001$, respectively). Weaning rate and emission intensity were negatively correlated ($r = -0.39$, $p < 0.05$); however, a Mann-Whitney test did not indicate a significant effect of weaning rate on emission intensity ($p < 0.10$, Table 3).

The set of variables that, considered independently, had a significant effect on emissions per hectare of farmland differed from those that affected emission intensity per kg of LW sold, with the exception of land productivity, which significantly affected both types of emissions, but in opposite directions (Table 3). Emissions per hectare were significantly lower if little or no land had been dedicated to improved pastures or annual forage crops, if mortality rate was low, if stocking rate was low, and if the number of grazing paddocks per total land was low. Emissions were higher under moderate than under low land productivity. Furthermore, land productivity and emissions per hectare were positively correlated ($r = 0.66$, $p < 0.001$).

Eleven models for emissions per kg of LW sold and eight models for emissions per hectare of farmland had significant ($p < 0.05$) values for the intercept and explanatory variables (Table 4). All of the variables that individually had a significant effect on emission intensity (Table 3) yielded significant models.

In the best model for explaining emissions per kg LW sold (Model 1, lowest AICc and highest D^2), animal and land productivities were significant explanatory variables (Table 4 and Figure 1). Systems that had higher animal and higher land productivity emitted less than those systems with a lower productivity. The calculated square deviance (D^2) indicated that the model explained 51.2% of the variation in the response variable. Models 2–3 performed worse in terms of AICc and D^2 but were plausible in terms of ΔAICc (ΔAICc < 7) and should rarely be dismissed [49]. Models 2–3 included management care controls of livestock as an explanatory variable: systems that had good management controls emitted less than those that did not. Models 4–9 performed worse than 1–3 but were not necessarily dismissed (ΔAICc < 11) [49], and they included the type of production system, grazing system and technical advice as significant explanatory variables: CCBG systems emitted less than CC systems, rotational emitted less than did continuous grazing, and systems that received technical advice emitted less than did those that did not. Models 10–11 had relatively little empirical support

(ΔAICc > 11) [49] and were dismissed. Within each model, the partial regression coefficients and the 'D2 change on deletion' (%DCD, results not shown) indicated that animal productivity (Model 1), livestock care management (Model 3), type of production system (Model 4), and grazing system (Model 5) had more weight in influencing emission intensity than land productivity.

Table 3. Individual effects of farm characteristics and management on greenhouse gas emission intensity.

Variable	Level/Type	Farm Greenhouse Gases Emissions Intensity					
		kg CO_2e/kg LW Sold			kg CO_2e/ha		
		Mean ± s.d.	*n*	Sig.[a]	Mean ± s.d.	*n*	Sig.[a]
Land use							
Land area used for improved pastures or annual forage crops (%)	0/very low Low/moderate	21.29 ± 7.76 18.28 ± 9.72	24 3	n.s.	221 ± 199 503 ± 184	24 3	**
Feed purchase	No Yes	22.28 ± 9.19 19.54 ± 6.12	14 13	n.s.	232 ± 263 273 ± 153	14 13	n.s.
Beef cattle							
Average live weight of livestock (kg)	Low Moderate	21.94 ± 7.39 19.91 ± 8.47	14 13	n.s.	257 ± 251 246 ± 175	14 13	n.s.
Cows to total animals rate (%)	Low Moderate	19.82 ± 6.25 22.18 ± 9.37	14 13	n.s.	237 ± 171 268 ± 258	14 13	n.s.
Average weight of sold calf (kg)	Low Moderate	22.00 ± 6.94 19.84 ± 8.85	14 13	n.s.	258 ± 258 246 ± 164	14 13	n.s.
Mortality rate (%)	Low Moderate	21.31 ± 8.70 20.58 ± 7.13	14 13	n.s.	170 ± 75 341 ± 277	14 13	*
Stocking rate (LU/ha)	Low Moderate	20.83 ± 9.65 21.10 ± 5.67	14 13	n.s.	145 ± 73 367 ± 256	14 13	***
Grazing infrastructures							
Water reservoirs/ha^{-1} ($\times 10^{-3}$)	Low Moderate	20.30 ± 8.21 21.67 ± 7.68	14 13	n.s.	196 ± 114 312 ± 268	14 13	n.s.
Grazing paddocks/ha ($\times 10^{-3}$)	Low Moderate	20.09 ± 8.56 21.90 ± 7.21	14 13	n.s.	157 ± 71 354 ± 268	14 13	*
Technical management of the farm							
Livestock care management controls	Poor Good	24.23 ± 7.36 16.87 ± 6.59	15 12	*	262 ± 245 239 ± 177	15 12	n.s.
Reproductive management of the livestock (mating)	Year-round Seasonal	23.72 ± 9.51 18.75 ± 5.60	12 15	n.s.	227 ± 269 272 ± 165	12 15	n.s.
Technical advice	No Yes	23.96 ± 7.40 17.21 ± 6.92	15 12	*	254 ± 245 249 ± 177	15 12	n.s.
Grazing system	Continuous Rotational	27.65 ± 7.74 18.62 ± 6.55	7 20	**	323 ± 350 227 ± 146	7 20	n.s.
Type of production system	CC CCBG	23.60 ± 7.29 15.68 ± 6.31	18 9	*	243 ± 225 269 ± 200	18 9	n.s.
Reproductive efficiency							
Weaning rate (%)	Low Moderate	23.67 ± 7.53 18.04 ± 7.34	14 13	(*)	262 ± 252 241 ± 173	14 13	n.s.
System productivity							
Land productivity (kg LW sold/ha)	Low Moderate	24.28 ± 7.78 17.38 ± 6.40	14 13	*	148 ± 67 364 ± 261	14 13	***
Animal productivity (kg LW sold/LU)	Low Moderate	26.00 ± 6.27 15.53 ± 5.39	14 13	***	262 ± 253 241 ± 172	14 13	n.s.

LU, Livestock Units. LW, Live Weight. CC, Cow-calf. CCBG, Cow-calf + Backgrounding. [a] Sig. = significance based on Mann-Whitney test. (*) = $p < 0.10$, * = $p < 0.05$, ** = $p < 0.01$, *** = $p < 0.001$.

Table 4. Generalized linear models (GLM) for the response in emission intensity (kg CO_2e/kg LW sold, and as kg CO_2e/ha) of beef cattle farms to farm management and characteristics.

Emission	Model	Livestock Care	Technical Advice	Grazing System	Production System	Animal Productivity	Land Productivity	Land Use	Mortality Rate	Stocking Rate	Grazing Paddocks	Sig.	AICc	ΔAICc	D^2 (%)
	1					+0.459 ***	+0.218 *					***	176.67	0.0	51.23
	2					+0.515 ***						***	177.78	1.1	43.81
	3	+0.360 **					+0.332 **					**	182.43	5.8	39.83
	4				+0.342 **		+0.261 **					**	184.88	8.2	34.22
kgCO_2e kg^{-1} LW sold	5			+0.330 **			+0.264 *					**	185.23	8.6	33.39
	6		+0.282 *				+0.287 **					**	186.07	9.4	31.34
	7				+0.409 **							**	186.29	9.6	23.45
	8			+0.396 **								**	186.69	10.0	22.34
	9	+0.362 **										*	187.00	10.3	21.44
	10						+0.334 **					*	187.94	11.3	18.74
	11		+0.331*									*	188.20	11.5	17.97
	1′						−0.784 ***				−0.675 ***	***	334.64	0.0	57.39
	2′						−0.748 ***		−0.442 *			***	341.82	8.2	45.25
	3′									−0.929 ***		***	342.02	8.4	39.32
kg CO_2e ha^{-1}	4′						−0.903 ***					***	342.98	9.3	37.29
	5′										−0.811 ***	**	346.01	12.4	30.45
	6′							−0.699 **	−0.620 *			**	346.83	13.2	34.93
	7′								−0.698 **			*	349.08	15.4	22.81
	8′							−0.823 **				*	351.084	17.4	17.45

Partial regression ß coefficients, with their statistical significance when the variable is '0', statistical significance of the model (Sig.) based on an omnibus test, Akaike's information criteria (AICc), ΔAICc and square deviance (D^2) are given. If β is > 0, emissions and the variable are positively correlated and if β is < 0, emissions and the variable are negatively correlated. Only statistically significant variables (based on Wald's chi-square test) are shown. Empty cells indicate variables not included in a given model. * = $p < 0.05$, ** = $p < 0.01$, *** = $p < 0.001$. ΔAICc calculated as: $\Delta AICc = AICc_i - AICc_{min}$, for $i = 1, 2, \ldots, R$, where $AICc_{min}$ denotes the minimum of the AICc values for the R models. D^2 calculated as: $D^2 = (D_0 - D_{model})/D_0$, where D_0 is the deviance of the null model (with intercept, only), and D_{model} is the deviance of the analyzed model.

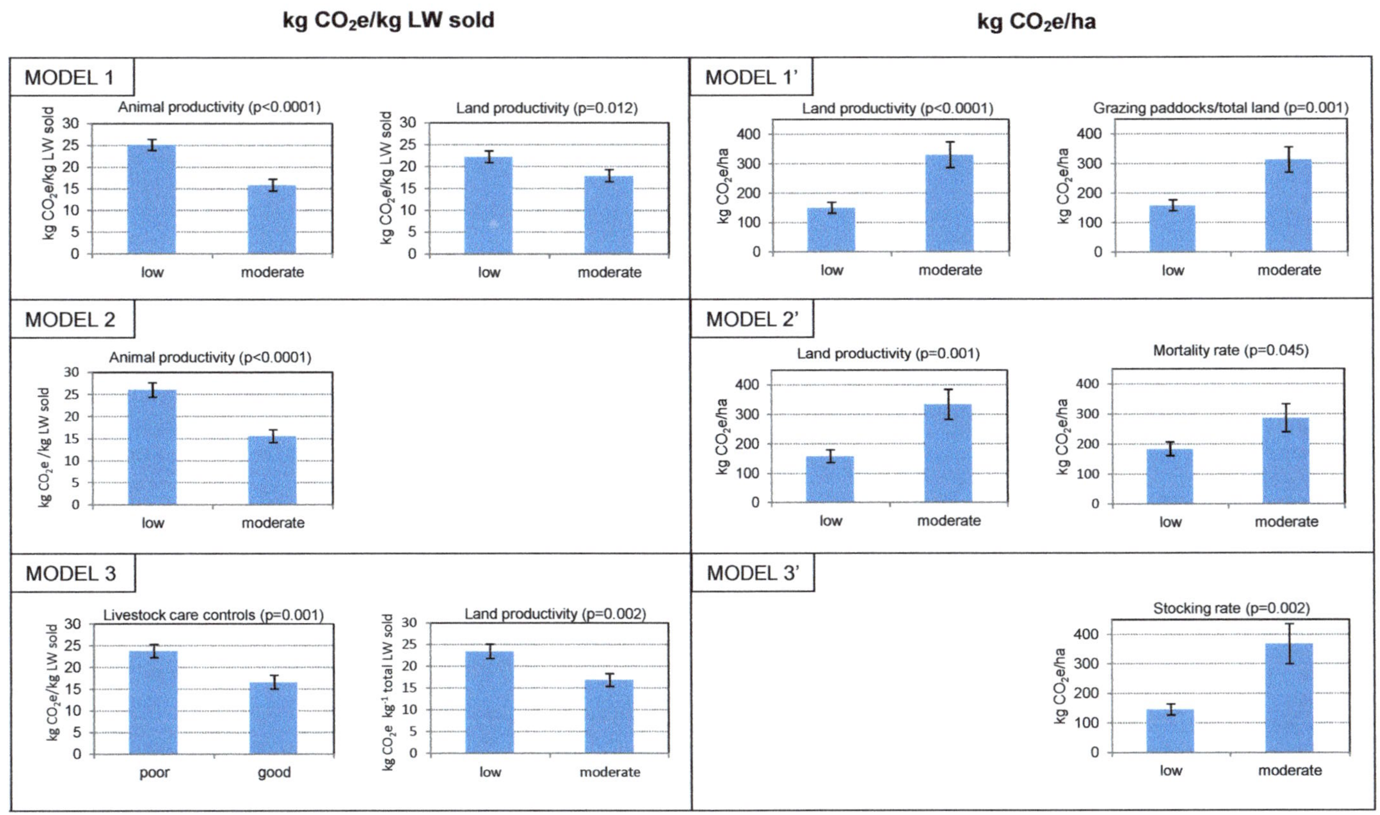

Figure 1. Marginal means and standard error of the most significant GLM models for the response in emission intensity (kg CO_2e/kg LW sold, and kg CO_2e/ha) of beef cattle farms in San Luis province, central Argentina. Differences were tested using Wald's chi-square test.

In the best model for explaining emissions per hectare of farmland (Model 1′, lowest AICc and highest D^2), land productivity and the number of grazing paddocks per total land were significant explanatory variables (Table 4 and Figure 1). Systems that had lower land productivity emitted less than did those systems with a higher land productivity. In addition, systems that had less grazing paddocks emitted less. In this model, the partial regression coefficients indicated that land productivity had more weight than the number of grazing paddocks. Calculated square deviance (D^2) indicated that the model explained 57.4% of the variation in the response variable. Models 2′-4′ performed worse than model 1′ in terms of AICc and D^2 but were not necessarily dismissed (ΔAICc < 11) [49], and they included the mortality rate and stocking rate as significant explanatory variables: systems that had a lower mortality rate emitted less than did those that had a higher mortality rate, and systems that had a lower stocking rate emitted less than did those that had a higher stocking rate. Models 5′–8′ had relatively little empirical support (ΔAICc > 11) [49] and were dismissed. Within each model, the partial regression coefficients and the 'D2 change on deletion' (%DCD, results not shown) indicated that the number of grazing paddocks (Model 1′) and mortality rate (Model 2′) had less weight in influencing emissions than land productivity.

3.3. Farm Typification

The PCA identified the following five groups of farms (Tables 5–7):

- Group I (23% of farms): 'High emitters per LW sold' (higher emitters than group V) and 'low emitters per hectare' (lower emitters than group II). 'Worse management' and 'low stocking rates' (lower stocking rates than group II). Only cow-calf systems. Highest proportion of farms with continuous grazing. On those farms, all of the land area consists of natural grasslands, and off-farm feeds are not used. Lowest percentage of farms that have three or more types of livestock care management controls. Highest percentage of farms with year-round mating. Highest proportion of farms without any technical advice. Low weaning rates (lower weaning rates than group IV). 'Low land and animal productivities' (lower land productivity than groups II and V, and lower animal productivity than group V).
- Group II (30% of farms): 'Intermediate emitters per LW sold' and 'high emitters per hectare' (higher emitters per hectare than group I). 'Medium level of management' and 'high stocking rates' (higher stocking rates than group I). Cow-calf systems and rotational grazing predominate. Almost all of the land area consists of natural grasslands, and most of the farms use off-farm feeds (higher percentage than group I and III). Most of the farms have three or more livestock management controls, 50% of farms have year-round mating, and 78% of farms receive no technical advice (less technical advice than groups III and IV). Intermediate weaning rates. 'High land productivity' (higher than group I) and 'intermediate animal productivity'.
- Group III (17% of farms): 'Intermediate emitters per LW sold and per hectare'. 'Suitable farm management' and 'intermediate stocking rates'. Cow-calf or mixed CCBG systems, and all the farms use rotational grazing. The entire land area consists of natural grasslands, and off-farm feeds are not used. All of the farms implemented at least 3 types of livestock management controls, 80% of the farms had seasonal mating of the herd, and all farms have technical advice. Intermediate weaning rates. 'Intermediate land and animal productivity'.
- Group IV (20% of farms): 'Intermediate emitters per LW sold and per hectare'. 'Suitable farm management', and 'intermediate stocking rates'. Cow-calf, BG or mixed CCBG systems, and all the farms have rotational grazing. Almost all of the land area consists of natural grasslands, and most farms use off-farm feeds. All of the farms implement at least 3 types of livestock management controls and have seasonal mating of the herd, and 83% of farms have technical advice. High weaning rates (higher weaning rates than group I). 'Intermediate land and animal productivities'.

- Group V (10% of the farms): 'Low emitters per LW sold' (lower emitters than group I), and 'intermediate emitters per hectare'. 'Good farm management' and 'intermediate stocking rates'. None of the farms were exclusively cow-calf systems. All of the farms had rotational grazing. Relatively high proportion of land used for introduced pastures and annual forage crops, and 67% of farms used off-farm feeds (higher percentage of farms than in groups I and III). All of the farms implemented at least 3 types of livestock management controls, had seasonal mating of the herd, and 67% of farms had technical advice. Intermediate weaning rates. 'High land and animal productivities' (higher than group I).

Table 5. Mean values for continuous variables by cluster group.

Variable	Cluster Group					Sig.[a]
	I	II	III	IV	V	
	$n = 7$	$n = 9$	$n = 5$	$n = 6$	$n = 3$	
Socio-economic data						
Age (years)	61	58	48	49	55	n.s.
+ Hired labor (WU/LU) ($\times 10^{-3}$)	1.0 a	0.2 a	4.6 b	2.9 ab	1.3 ab	***
Land use						
+ Total land area (ha)	1077 a	1673 a	7010 b	3284 ab	10,200 ab	**
+ Land area used for native pastures (%)	100 b	98 b	100 b	94 ab	64 a	**
+ Land area used for improved pastures (%)	0 b	0 b	0 b	6 ab	26 a	***
Land area used for forage crops (%)	0 b	2 ab	0 b	0 b	10 a	*
Beef cattle						
+ Total livestock units	51.6 a	194.4 ab	482.1 b	328.0 ab	1207.0 b	***
Mortality rate (%)	11.9	10.0	2.3	2.2	2.2	n.s.
Stocking rate (LU/ha)	0.07 a	0.19 b	0.08 ab	0.10 ab	0.19 ab	*
Grazing infrastructures						
Water reservoirs per ha ($\times 10^{-3}$)	3.0 ab	5.2 a	0.9 b	1.4 ab	1.4 ab	*
Water reservoirs/LU ($\times 10^{-3}$)	46 b	27 ab	12 a	17 ab	8 a	**
+ Grazing paddocks/ha ($\times 10^{-3}$)	2.6	7.2	1.1	2.4	2.2	n.s.
System productivity						
Average live weight of livestock (kg)	272	284	302	283	271	n.s.
Weaning rate (%)	49 a	63 ab	69 ab	82 b	73 ab	*
+ Land productivity (kg LW sold/ha)	4.7 a	20.3 b	11.3 ab	19.6 ab	52.9 b	***
Animal productivity (kg LW sold/LU)	74 a	123 ab	160 ab	144 ab	283 b	*
Farm greenhouse gases emission intensity						
kg CO_2e/kg LW sold	27 b	20 ab	15 ab	19 ab	8 a	*
kg CO_2e/ha	121 a	372 b	166 ab	266 ab	403 ab	**

+ Variables used in the principal component analysis and in the cluster analysis. [a] Sig. = significance based on the Kruskal–Wallis test. * = $p < 0.05$, ** = $p < 0.01$, *** = $p < 0.001$. Different letters in the same row indicate significant differences between groups (Dunn–Bonferroni post-hoc test).

Table 6. Frequency (% of farms) of discrete variables by cluster group.

Variable	Cluster Group					Sig.[a]
	I	II	III	IV	V	
	$n = 7$	$n = 9$	$n = 5$	$n = 6$	$n = 3$	
Socio-economic data						
+ Level of education of farmer [1]						
None	0	22	20	0	0	n.s.
Primary or secondary school	100 b	78 b	20 a	0 a	100 b	***
Higher education	0 a	0 a	60 b	83 b	0 a	***

Table 6. *Cont.*

Variable	Cluster Group					Sig. [a]
	I	II	III	IV	V	
	$n = 7$	$n = 9$	$n = 5$	$n = 6$	$n = 3$	
Type of production system						
Cow-calf	100^{b}	78^{b}	60^{ab}	50^{a}	0^{a}	*
Backgrounding	0	11	0	17	33	n.s.
Cow-calf + backgrounding	0	11	40	33	67	n.s.
Grazing system						
Continuous	71^{b}	22^{a}	0^{a}	0^{a}	0^{a}	**
Rotational	22^{a}	78^{ab}	100^{b}	100^{b}	100^{b}	**
+ Feed purchase						
Yes	0^{a}	89^{b}	0^{a}	83^{b}	67^{b}	**
No	100^{b}	11^{a}	100^{b}	17^{a}	33^{a}	*
Technical management of the farm						
+ Livestock care controls [2,3]						
None	43	13	0	0	0	n.s.
One or two control types	43	25	0	0	0	n.s.
Three or more control types	14^{a}	62^{b}	100^{b}	100^{b}	100^{b}	***
Reproductive management of the livestock [3]						
Year-round mating	100^{b}	50^{a}	20^{a}	0^{a}	0^{a}	**
Seasonal mating	0^{a}	50^{b}	80^{b}	100^{b}	100^{b}	**
+ Technical advice						
None	100^{c}	78^{b}	0^{a}	17^{a}	33^{ab}	**
Veterinarian and/or agronomist	0^{a}	22^{ab}	100^{c}	83^{c}	67^{bc}	**

+ Variables used in the principal component analysis and in the cluster analysis. [1] Remaining farmers, Do not know/No answer. [2] Types of livestock care controls: body condition, teeth examination, rectal palpation/ecography, parasite control, reproductive vaccine, bull review control. [3] $n = 27$ (backgrounding farms excluded). [a] Sig. = significance based on Pearson's chi-squared test. * = $p < 0.05$, ** = $p < 0.01$, *** = $p < 0.001$. Different letters in the same row indicate significant differences between groups.

Table 7. Main characteristics of each cluster group: socio-economic data, land use, beef cattle, feed purchase, technical management of the farm, system productivity, greenhouse gases emission, grazing system and type of production system. Based on Tables 5 and 6.

	Cluster Group				
	I	II	III	IV	V
Main education level	PS	PS	HE	HE	PS
Hired labor	a	a	b	ab	ab
Total land area	a	a	b	ab	ab
Land area used for native pastures	b	b	b	ab	a
Land area used for improved pastures	a	a	a	ab	b
Land area used for annual forage crops	a	ab	a	a	b
Total livestock units	a	ab	b	ab	b
Stocking rate	a	b	ab	ab	ab
Dependence on off-farm feeds	a	b	a	b	b
Livestock care controls: ≥3 control types	a	b	b	b	b
Technical advice: Veterinarian and/or Agronomist	a	ab	c	c	bc
Weaning rate	a	ab	ab	b	ab
Land productivity	a	b	ab	ab	b
Animal productivity	a	ab	ab	ab	b
Emission intensity per LW sold	b	ab	ab	ab	a
Emission intensity per hectare	a	b	ab	ab	ab
Main grazing system	CON	ROT	ORO	ORO	ORO
Main system	OCC	CC	CC	CC	CCBG
Main reproductive management	OYR	YRS	S	S	S

PS = Primary or secondary school. HE = Higher education. CON = Continuous, ROT = Rotational, ORO = Only rotational, OCC = Only cow-calf, CC = Cow-calf, CCBG = Cow-calf and backgrounding, OYR = Only year-round, YR = Year-round, YRS = Year-round/seasonal, S = Seasonal. Different letters in the same row indicate significant differences between groups (based on Tables 5 and 6).

4. Discussion

GHG farm emissions varied widely among the 30 farms surveyed in the semiarid rangelands of central Argentina, which reflected the high diversity in the types of production systems [26,56]. Variability is especially high in studies that have been based on actual farm survey data [14,26].

In our study, on a product sold basis, cow-calf systems emitted more GHG than backgrounding systems. Similar results have been reported in grassland-based beef systems in Uruguay [14,15,57] and Argentina [58]. In our study, the GHG emissions of cow-calf systems were similar to those of 295 cow-calf farms in Canada [26] and the cow-calf systems based on native and improved grasslands in Uruguay [14]. Emissions from backgrounding systems were similar to those from background-finishing systems that had seeded pastures and feedlots in Uruguay [57].

On a farm-area basis, in our study, GHG emissions did not differ significantly among types of systems. The average was much lower than previously reported values, which ranged between 265 and 9782 [26], and between 2334 and 3037 [59] in Canadian beef cattle production systems, between 1490 and 2827 in Uruguayan beef systems [14], and between 7902 and 10,913 in New Zealand pasture-based dairy systems [60]. The higher stocking rates in those studies (0.31, 0.77 and 2.3–3.0 LU/ha in Canadian, Uruguayan and New Zealand systems, respectively, versus 0.13 LU/ha in our study) were mainly responsible for the differences in emissions between those studies and ours. In our study, the emissions per hectare and stocking rate were positively correlated ($r = 0.900$, $p < 0.001$). In beef systems in the Brazilian Amazon [61] and in dairy systems in Ireland [62], emissions per hectare and stocking rates were positively correlated. Livestock density on extensively managed grazing lands are relatively low; therefore, CH_4 emissions per unit area from these grazing lands is much lower than those from intensively managed grazing lands [33,63]. The contribution of extensively managed grasslands to GHG emissions is expected to be low per unit area because of low livestock densities and agronomic inputs, although the absolute global contribution might be high because of their large land area [63].

In our study, on a product sold basis, animal productivity was the variable that best explained the largest amount of variance in emission intensity, which was negatively correlated with productivity. To a lesser degree, land productivity and emission intensity were negatively correlated. Improving production efficiency has been recommended as a strategy to mitigate GHG emissions in beef systems [14,15,26,57,64–66]. For instance, Alemu et al. [26] found that low-emitting farms had higher animal and land productivities than high-emitting farms in Canadian cow-calf systems. In French suckler-beef production farms, animal productivity was the main factor influencing GHG emissions [64], which suggested that technical efficiency was a factor. Becoña et al. [14] found that beef farm productivity was one of the main determinants of GHG emissions in Uruguayan cow-calf systems. The same negative correlation was found in dairy systems [67,68], mainly because emissions are spread over more units of output per cow, which dilutes emission intensity. Productivity gains are generally achieved through improved husbandry practices and technologies that increase the proportion of resources used for production purposes rather than for the maintenance of the animals, which contribute to emission reductions [1]. Improved farm productivity can result from a combination of several types of strategies.

On the beef farms in our study, continuous stocking practices emitted significantly more GHG per product sold than rotational stocking. Beef cattle in rotational stocking systems emitted less methane than cattle in continuous grazing [35]. Furthermore, good grazing management can have a positive impact on soil carbon sequestration [1].

Improved livestock care management was associated with reduced GHG emission intensity per kg LW sold in our study. Improved animal health can increase herd productivity and reduce GHG emission intensity [24]. Along with improved reproduction management, improved animal health helps to reduce the unproductive portion of the herd and associated emissions, and concomitantly, these measures increase productivity [1]. Preventive health measures can play a role in increasing growth and fertility rates, which improve animal and herd performance [1]. Llonch et al. [27] reported a reduction in rumen methanogenesis in response to an increase in production efficiency caused by improvements in the health status of the herd, which is a win-win strategy, because it increases environmental sustainability and animal welfare.

In our study, farms that had received technical advice had lower emissions per unit of product sold, which reflected the importance of technical advice in grazing management planning, feeding, health care, the reproductive management of the herd, and overall farm system management [69,70].

Land-related variables can affect GHG emissions from animals through diet quality [26]. Diet digestibility directly reduces CH_4 emission intensity [64,71], which was apparent on farms that had an increased area of improved pastures, including seeded pastures, oversowing with legumes, and annual winter crops for grazing [14]. In our study, such an effect was not apparent, probably because of the small proportion of the farmland that had been used for improved pastures or annual forage crops (mean = 6%, vs. 20.5% in the study by Becoña et al. [14]).

Many of those husbandry practices are associated with increases in productivity, which suggests that an economic benefit can be realized with a concurrent reduction in GHG emissions [27]. Strategies that both improve production efficiency and reduce GHG emissions are those most attractive to and most likely to be adopted by farmers [26]. Further studies should compare the economic impact of several measures to mitigate GHG emissions and willingness to adopt them in our study area.

In our study, emission intensity per hectare of farmland was positively correlated with stocking rate and land productivity. Similar results were reported by Becoña et al. [14] in beef cow-calf systems. In Irish dairy farms, Casey and Holden [16] found a significant positive correlation between stocking rate and the amount of GHG emissions per hectare. Bava et al. [68] found a strong positive correlation between emissions per land area and stocking rate in dairy systems. Stocking rate and total dry matter intake are the main factors driving production per hectare and GHG emissions from grazed pastoral systems [7]. The number of grazing paddocks per hectare and the proportion of land used for improved pastures and annual forage crops were positively correlated with GHG emissions per hectare of land area in our study. Higher stocking rates and land productivity, coupled with higher density of grazing paddocks and land use for improved pastures and forage crops, reflect a certain degree of intensification of the farming system, i.e., intensification implied higher emissions per hectare. Bava et al. [68] concluded that intensification, defined as the increase in output per hectare, invariably led to higher emissions on a per-area basis. Nevertheless, the emissions per unit of product and land productivity were negatively correlated, which illustrates the potential trade-off between carbon efficiencies per unit of product and per unit of land, i.e., is it possible to reduce emissions per unit of land and per unit of product at the same time?

The CA indicated that, if GHG emissions are evaluated on a land-unit basis, farms of group 1 had low emissions and were very extensive in terms of land use. They had low stocking rates, a low dependence on off-farm feeds, land productivity, and low proportion of land used for improved pastures or annual crops. Farms in that group, however, had the lowest level of husbandry practices in terms of livestock care controls and reproductive management, technical advice, and grazing system, low weaning rate and animal productivity, and concomitantly, they had high emissions per product sold. From that 'base-line' traditional farming system, strategies can differ considerably in practice and results, in terms of farm productivity and emissions. Farms in group II intensified the system by increasing the stocking rates and dependence on off-farm feeds, and they improved some husbandry practices, maintained emissions per product sold, but increased emissions per hectare. Farms in group V had a higher proportion of land as improved pastures and annual forage crops, medium stocking rates, improved livestock husbandry practices, and had intermediate levels of emissions per hectare. This group had lower emissions per product sold than group I because of those improvements, but also because of the high proportion of backgrounding on the farms in this group. Nevertheless, it has to be taken into account that only three farms belonged to group V. Groups III and IV had higher levels of husbandry practices than groups I and II, but they did not have stocking rates that were as high as those in group II. Thus, those groups (III and IV) had intermediate levels of farm productivity and emissions per product sold and per hectare.

The CA suggested that farms that had a high level of husbandry 'intensification' through livestock care and reproductive management achieved high animal productivity and, therefore, low GHG emissions per product sold compared to 'base-line farms' (group I). Thus, if land productivity is increased by using that high-output animal strategy, emissions per hectare can be limited to intermediate levels. However, if land productivity is maximized through high stocking rates, emissions per hectare is the highest, as in the case of group II. Becoña et al. [14] stated that both emissions per unit of land and per unit of product can be reduced concurrently and suggested that the key factor is reducing stocking rate (or increasing forage allowance) in grazing beef cow-calf systems. GHG emission intensity can be reduced through changes in animal husbandry practices that increase animal outputs [14,64]. Casey and Holden [16] suggested that it is physically and biologically possible to achieve low emissions, both per unit of land and per product, by using high-output cows at low stocking rates in dairy systems. A move toward fewer cows producing more milk at lower stocking rates is required, representing an extensification in terms of area, but an intensification in terms of animal husbandry practices. In a simulation experiment on pasture-based dairy farms in New Zealand, Beukes et al. [60] maintained production but reduced GHG emissions per unit of land and per unit of product by increasing efficiency (e.g., reducing the number of non-productive animals in the herd, among other mitigation strategies), which allowed stocking rates to be reduced. The mitigation of GHG emissions per unit of product should be based on the intensification of husbandry systems rather than on land intensification, which might lead to potential losses in ecosystem service provisioning, increases in GHG emissions per unit of area and other environmental impacts, such as eutrophication and acidification [15].

Among the beef cattle farms in our study, those in groups III and IV could further reduce emission intensity by adopting practices, such as improving feed quality [22,26,64], using superior animal genetics [72], or increasing the proportion of backgrounding vs. cow-calf in the farm system. Feed quality can be increased by applying seeding grasses to improve native pastures, annual forage cropping, and by purchasing high-quality off-farm feeds. However, introduced grasses can increase the impact on native grasslands, with potential biodiversity, wildlife habitat and landscape losses [15,57]. The mitigation of climate change should not be associated with directly reducing biodiversity [15]. In several regions of the world, pasture intensification has been used to increase productivity, incomes, and mitigate GHG, but has increased rangeland degradation [32]. Annual cropping systems have relatively high levels of agronomic inputs and nutrient leakage, frequent and significant disturbances of the soil surface, and net losses of soil organic content [33]. In addition, CO_2 emissions derived from fertilizers and machinery operations for annual forage crops are high [26]. Feed quality can be improved by purchasing high-quality feeds, but the embedded emissions associated with feed production should not be ignored. Alemu et al. [26] found that minimizing purchased cereal grain and forage per unit cow reduced the emissions associated with the production and transportation of farm inputs. In strategies, such as improving genetic merit, the animals have to be selected not only for their high efficiency in transforming feeds, but also for their ability to adapt to rough environments and low-quality feeds [73], which are characteristic of the semi-arid rangelands of central Argentina. In addition, to reduce emissions per unit of product, farmers can increase the proportion of backgrounding versus cow-calf in their system. However, this strategy can transfer the negative environmental impacts of the cow-calf phase to other areas, i.e., the emissions of the replacement stock, if purchased, have occurred elsewhere on other farms [74].

Our results from actual semi-arid rangeland beef systems in central Argentina suggest that the implementation of relatively easy-to-adopt farming management practices has considerable potential for reducing GHG emissions per unit of product and per unit of land area. At the same time, the preservation of rangeland ecosystem services should be a target.

The expansion of agriculture and an increase in the intensification of livestock systems have challenged the integrity of rangelands in Argentina and worldwide. Future research should assess the ecosystem services provided by the beef production systems in the semi-arid rangelands of Argentina, e.g., wildlife biodiversity and landscape preservation, animal welfare, nutrient cycling, hydrologic conditions, control of invasive plant species, and carbon sequestration. Grazing lands have a high potential for carbon sequestration [23,75,76], which can, at least partially, mitigate the GHG emissions from ruminant production systems [77]. Extensive livestock grazing systems had a lower GHG emission intensity if soil carbon uptake had been included in the emission inventory [15,65,78–81], particularly for low-input or small-scale grazing systems [20,81]. Therefore, land-use decisions should be informed by all environmental factors, negative impacts—not only GHG emissions—and ecosystem services. In order to increase the sustainability and efficiency of beef livestock systems in the Argentinean semi-arid rangelands, future studies should use an integrated, holistic approach.

5. Conclusions

This study assessed the relationships between GHG emissions and characteristics and the management practices of commercial farms in extensive beef systems that are based on the natural rangelands in the semi-arid Central Region of Argentina. The results suggest that the implementation of realistic, relatively easy-to-adopt farming management practices has a considerable potential to mitigate GHG emissions. Emissions per product sold were low on farms that had improved livestock care management, had rotational grazing, received technical advice, and had high animal and land productivities. The emissions per hectare of farmland were low on farms that had low stocking rates, a low number of grazing paddocks, little or no land dedicated to improved pastures and annual forage crops, and low land productivity.

Therefore, in our study, the set of variables that influenced the emissions per hectare of farmland differed from those that affected the emissions per unit of product, and land productivity affected the two types of emission expressions in opposite directions, which suggests a potential trade-off between the mitigation of GHG emissions per unit of product and per unit of land. Given that GHG emissions per product and per hectare of farmland differ in their implications for the assessment of the environmental impacts of food production (e.g., global vs. local scales, intensification processes), both measures should be taken into account and reconciled as much as possible.

To identify ways to increase the sustainability and efficiency of the management of beef livestock systems in the Argentinean semi-arid rangelands, future studies should use an integrated, holistic approach in which all negative environmental impacts and ecosystem service provisioning, e.g., diversity preservation and carbon sequestration, should be assessed.

Author Contributions: Conceptualization, M.I.N., O.B., L.P. and R.R.; Data curation, M.I.N.; Formal analysis, M.I.N., O.B. and R.R.; Funding acquisition, M.I.N.; Investigation, M.I.N. and L.P.; Methodology, M.I.N., O.B. and R.R.; Resources, M.I.N. and L.P.; Supervision, O.B. and R.R.; Validation, O.B. and R.R.; Writing—original draft, O.B. and M.I.N.; Writing—review and editing, O.B., R.R. and M.I.N.

Funding: This research was funded by the Instituto Nacional de Tecnología Agropecuaria (INTA), Argentina.

Acknowledgments: The authors are grateful to C.A. Frasinelli for his contribution and technical advice, and to the farmers that participated in the study.

Conflicts of Interest: The authors declare no conflict of interest. The funders had no role in the design of the study; in the collection, analyses, or interpretation of data; in the writing of the manuscript; or in the decision to publish the results.

Appendix A

The IPCC (2006) [42] equations used in the calculations of the on-farm CH_4 and N_2O gases emissions were: 10.3, 10.4, 10.6, 10.8, 10.13, 10.14, 10.15, 10.16, 10.17, 10.18, 10.19, 10.20, 10.21, 10.31, 10.32, 10.33, 11.1, 11.5, and 11.11.

References

1. Gerber, P.J.; Steinfeld, H.; Henderson, B.; Mottet, A.; Opio, C.; Dijkman, J.; Falcucci, A.; Tempio, G. *Tackling Climate Change through Livestock e a Global Assessment of Emissions and Mitigation Opportunities*; Food and Agriculture Organization of the United Nations (FAO): Rome, Italy, 2013; ISBN 978-92-5-107921-8.
2. United Nations Department of Economic and Social Affairs. Available online: http://www.un.org/en/development/desa/news/population/un-report-world-population-projected-to-reach-9-6-billion-by-2050.html (accessed on 26 September 2018).
3. Delgado, C.L. Rising demand for meat and milk in developing countries/implications for grasslands-based livestock production. In *Grassland: A Global Resource*; McGilloway, D.A., Ed.; Wageningen Academic Publishers: Wageningen, The Netherlands, 2005; pp. 29–40, ISBN 978-90-76998-71-8.
4. Garnett, T. Livestock-related greenhouse gas emissions: Impacts and options for policy makers. *Environ. Sci. Policy* **2009**, *12*, 491–503. [CrossRef]
5. Rojas-Downing, M.M.; Nejadhashemi, A.P.; Harrigan, T.; Woznicki, S.A. Climate change and livestock: Impacts, adaptation, and mitigation. *Clim. Risk Manag.* **2017**, *16*, 145–163. [CrossRef]
6. Opio, C.; Gerber, P.; Mottet, A.; Falcucci, A.; Tempio, G.; Macleod, M.; Vellinga, T.; Henderson, B.; Steinfeld, H. *Greenhouse Gas Emissions from Ruminant Supply Chains—A Global Life Cycle Assessment*; Food and Agriculture Organization of the United Nations (FAO): Rome, Italy, 2013; p. 191, ISBN 978-92-5-107945-4.
7. De Klein, C.A.M.; Pinares-Patiño, C.; Waghorn, G.C. Greenhouse gas emissions. In *Environmental Impacts of Pasture-Based Farming*; McDowell, R.W., Ed.; CABI: Wallingford, UK, 2008; pp. 1–32, ISBN 978-18-4-593411-8.
8. Ministerio de Economía y Finanzas Públicas. Complejo Ganadero Vacuno. Serie "Complejos Productivos". Secretaría de Política Económica y Planificación del Desarrollo. Gobierno Argentino. Available online: https://www.economia.gob.ar/peconomica/docs/Complejo_Ganadero_vacuno.pdf (accessed on 26 September 2018).
9. Miazzo, D.; Pisani Claro, N. *Carnes Argentinas. Actualidad, Propuestas y Futuro*; Fundación Agropecuaria para el desarrollo de Argentina (FADA), Sitio Argentino de Producción Animal: Río Cuarto, Cordoba, Argentina, 2015; pp. 1–37. Available online: http://www.produccion-animal.com.ar/informacion_tecnica/origenes_evolucion_y_estadisticas_de_la_ganaderia/00-origenes_evolucion_y_estadisticas_de_la_ganaderia.htm (accessed on 28 September 2018).
10. Rearte, D.H.; Pordomingo, A.J. The relevance of methane emissions from beef production and the challenges of the Argentinean beef production platform. *Meat Sci.* **2014**, *98*, 355–360. [CrossRef] [PubMed]
11. Segundo Informe Bienal de Actualización de la República Argentina a la Convención Marco de las Naciones Unidas sobre el Cambio Climático. Ministerio de Ambiente y Desarrollo Sustentable. República Argentina. 2017, pp. 37–69. Available online: http://euroclimaplus.org/intranet/_documentos/repositorio/02%20Bienal%20Convenci%C3%B3n%20ONU%20cambio%20clim%C3%A1tico_2017.pdf (accessed on 28 September 2018).
12. Bárbaro, N.; Gere, J.; Gratton, R.; Rubio, R.; Williams, K. First measurements of methane emitted by grazing cattle of the Argentinean beef system. *N. Z. J. Agric. Res.* **2008**, *51*, 209–219. [CrossRef]
13. Reisinger, A.; Ledgard, S. Impact of greenhouse gas metrics on the quantification of agricultural emissions and farm-scale mitigation strategies: A New Zealand case study. *Environ. Res. Lett.* **2013**, *8*, 025019. [CrossRef]
14. Becoña, G.; Astigarraga, L.; Picasso, V.D. Greenhouse gas emissions of beef cow–calf grazing systems in Uruguay. *Sustain. Agric. Res.* **2014**, *3*, 89–105. [CrossRef]
15. Picasso, V.D.; Modernel, P.D.; Becoña, G.; Salvo, L.; Gutiérrez, L.; Astigarraga, L. Sustainability of meat production beyond carbon footprint: A synthesis of case studies from grazing systems in Uruguay. *Meat Sci.* **2014**, *98*, 346–354. [CrossRef] [PubMed]
16. Casey, J.W.; Holden, N.M. The relationship between greenhouse gas emissions and the intensity of milk production in Ireland. *J. Environ. Qual.* **2005**, *34*, 429–436. [CrossRef] [PubMed]
17. Ross, S.A.; Topp, C.F.E.; Ennos, R.A.; Chagunda, M.G.G. Relative emissions intensity of dairy production systems: Employing different functional units in life-cycle assessment. *Animal* **2017**, *11*, 1381–1388. [CrossRef] [PubMed]

18. Halberg, N.; van der Werf, H.M.G.; Basset-Mens, C.; Dalgaard, R.; de Boer, I.J.M. Environmental assessment tools for the evaluation and improvement of European livestock production systems. *Livest. Prod. Sci.* **2005**, *96*, 33–50. [CrossRef]
19. Veysset, P.; Lherm, M.; Bébin, D. Energy consumption, greenhouse gas emissions and economic performance assessments in French Charolais suckler cattle farms: Model-based analysis and forecasts. *Agric. Syst.* **2010**, *103*, 41–50. [CrossRef]
20. Cottle, D.J.; Nolan, J.V.; Wiedemann, S.G. Ruminant enteric methane mitigation: A review. *Anim. Prod. Sci.* **2011**, *51*, 491–514. [CrossRef]
21. Gerber, P.J.; Hristov, A.N.; Henderson, B.; Makkar, H.; Oh, J.; Lee, C.; Meinen, R.; Montes, F.; Ott, T.; Firkins, J.; et al. Technical options for the mitigation of direct methane and nitrous oxide emissions from livestock: A review. *Animal* **2013**, *7*, 220–234. [CrossRef] [PubMed]
22. Hristov, A.N.; Oh, J.; Lee, C.; Meinen, R.; Montes, F.; Ott, T.; Firkins, J.; Rotz, A.; Dell, C.; Adesogan, C.; Yang, W.; et al. *Mitigation of Greenhouse Gas Emissions in Livestock Production—A Review of Technical Options for Non-CO2 Emissions*; Food and Agriculture Organization of the United Nations (FAO): Rome, Italy, 2013; pp. 1–206, ISBN 978-92-5-107659-0.
23. Hristov, A.N.; Oh, J.; Firkins, J.L.; Dijkstra, J.; Kebreab, E.; Waghorn, G.; Makkar, H.P.S.; Adesogan, A.T.; Yang, W.; Lee, C.; et al. Special topics-Mitigation of methane and nitrous oxide emissions from animal operations: I. A review of enteric methane mitigation options. *J. Anim. Sci.* **2013**, *91*, 5045–5069. [CrossRef] [PubMed]
24. Hristov, A.N.; Ott, T.; Tricarico, J.; Rotz, A.; Waghorn, G.; Adesogan, A.; Dijkstra, J.; Montes, F.; Oh, J.; Kebreab, E.; et al. Special topics-Mitigation of methane and nitrous oxide emissions from animal operations: III. A review of animal management mitigation options. *J. Anim. Sci.* **2013**, *91*, 5095–5113. [CrossRef] [PubMed]
25. Gerber, P.J.; Mottet, A.; Opio, C.I.; Falcucci, A.; Teillard, F. Environmental impacts of beef production: Review of challenges and perspectives for durability. *Meat Sci.* **2015**, *109*, 2–12. [CrossRef] [PubMed]
26. Alemu, A.W.; Amiro, B.D.; Bittman, S.; MacDonald, D.; Ominski, K.H. Greenhouse gas emission of Canadian cow-calf operations: A whole-farm assessment of 295 farms. *Agric. Syst.* **2017**, *151*, 73–83. [CrossRef]
27. Llonch, P.; Haskell, M.J.; Dewhurst, R.J.; Turner, S.P. Current available strategies to mitigate greenhouse gas emissions in livestock systems: An animal welfare perspective. *Animal* **2017**, *11*, 274–284. [CrossRef] [PubMed]
28. Gil, S.B.; Herrero, M.A.; Flores, M.C.; Pachoud, M.L.; Hellmers, M.M. Intensificación agropecuaria evaluada por indicadores de sustentabilidad ambiental. *Arch. Zootec.* **2009**, *58*, 413–423. [CrossRef]
29. Rearte, D.H. *La Producción de Carne en Argentina*; INTA. Sitio Argentino de Producción Animal: Córdoba, Argentina, 2007; pp. 1–25. Available online: http://www.produccion-animal.com.ar/informacion_tecnica/origenes_evolucion_y_estadisticas_de_la_ganaderia/00-origenes_evolucion_y_estadisticas_de_la_ganaderia.htm (accessed on 28 September 2018).
30. Frasinelli, C.A.; Veneciano, J.H. Introducción. In *Sistemas Bovinos Sobre Gramíneas Megatérmicas Perennes en San Luis*; Frasinelli, C.A., Veneciano, J.H., Eds.; Ediciones INTA: Buenos Aires, Argentina, 2014; pp. 13–14, ISBN 978-987-521-472-9.
31. Morris, C.D. Rangeland management for sustainable conservation of natural resources. In *Grassland Productivity and Ecosystem Services*; Lemaire, G., Hodgson, J., Chabbi, A., Eds.; CABI: Wallingford, UK; Cambridge, MA, USA, 2011; pp. 198–207, ISBN 978-1-84593-809-3.
32. Herrero, M.; Havlík, P.; Valin, H.; Notenbaert, A.; Rufino, M.C.; Thornton, P.K.; Blümmel, M.; Weiss, F.; Grace, D.; Obersteiner, M. Biomass use, production, feed efficiencies, and greenhouse gas emissions from global livestock systems. *Proc. Natl. Acad. Sci. USA* **2013**, *110*, 20888–20893. [CrossRef] [PubMed]
33. Schacht, W.H.; Reece, P.E. Impacts of livestock grazing on extensively managed grazing lands. In *Environmental Impacts of Pasture-Based Farming*; McDowell, R.W., Ed.; CABI: Cambridge, MA, USA, 2008; pp. 122–143, ISBN 978-1845934118.
34. Pereira, H.M.; Leadley, P.W.; Proença, V.; Alkemade, R. Scharlemann, J.P.W.; Fernandez-Manjarrés, J.F.; Araújo, M.B.; Balvanera, P.; Biggs, R.; Cheung, W.W.L.; et al. Scenarios for Global Biodiversity in the 21st Century. *Science* **2010**, *330*, 1496–1501. [CrossRef] [PubMed]

35. DeRamus, H.A.; Clement, T.C.; Giampola, D.D. Dickison, P.C. Methane missions of beef cattle on forages: Efficiency of grazing management systems. *J. Environ Qual.* **2003**, *32*, 269–277. Available online: https://pdfs.semanticscholar.org/1d58/15b6d22d81070c173f6281434dd213a557c7.pdf (accessed on 28 September 2018). [CrossRef] [PubMed]
36. Guzmán, M.L.; Sager, R. Inventario de metano entérico de los sistemas de producción de carne para San Luis en el año 2009. *Rev. Investig. Agropecu.* **2013**, *39*, 88–94. Available online: http://www.scielo.org.ar/pdf/ria/v39n1/v39n1a13.pdf (accessed on 28 September 2018).
37. Echeverría, J.C.; Giulietti, J. Precipitación media anual y temperatura del aire. In *Aptitud Forestal de la Provincia de San Luis: II Cartografía de Variables Ambientales*; Echeverría, J.C., Jobbagy, E., Collado, A., Eds.; INTA: San Luis, Argentina, 2006; pp. 11–13.
38. Peña Zubiate, C.; Anderson, D.L.; Demmi, M.A.; Saenz, J.L.; D'Hiriart, A. *Carta de Suelos y Vegetación de la Provincia de San Luis*; Ediciones INTA: Buenos Aires, Argentina, 1998; pp. 1–116. Available online: https://inta.gob.ar/sites/default/files/script-tmp-carta_de_suelos_hoja_san_luis.pdf (accessed on 28 September 2018).
39. Anderson, D.L.; De Aguilera, J.A.; Bernardon, A.E. Las formaciones vegetales de la provincia de San Luis. *Rev. Investig. Agropecu.* **1970**, *7*, 153–183.
40. Riedel, J.L.; Frasinelli, C.A. Los sistemas de producción bovina de la provincia de San Luis, Argentina. Oportunidades y desafíos. In Proceedings of the 3rd Simposio Internacional Sobre Producción Animal. Utilización de forrajes en la Nutrición de Rumiantes, Temascaltepec de Gonzáles, México, 6–8 May 2013.
41. Cocimano, M.; Lange, A.; Menvielle, E. Equivalencias ganaderas para vacunos de carne y ovinos (escalas simplificadas). In *Colección Estudios y Métodos*, 4th ed.; AACREA: Buenos Aires, Argentina, 1983; pp. 1–32.
42. IPCC. *IPCC Guidelines for National Greenhouse Gas Inventories. Agriculture, Forestry and other Land Use*; Eggleston, H.S., Buendia, L., Miwa, K., Ngara, T., Tanabe, K., Eds.; Prepared by the National Greenhouse Gas Inventories Programme; IGES: Kanagawa, Japan, 2006; Volume 4, Chapter 10; pp. 10.1–10.87, ISBN 4-88788-032-4.
43. Forster, P.; Ramaswamy, V.; Artaxo, P.; Berntsen, T.; Betts, R.; Fahey, D.W.; Haywood, J.; Lean, J.; Lowe, D.C.; Myhre, G.; et al. Changes in Atmospheric constituents and in radiative forcing. In *Climate Change 2007: The Physical Science Basis. Contribution of Working Group I to the Fourth Assessment Report of the IPCC*; Solomon, S., Qin, D., Manning, M., Chen, Z., Marquis, M., Averyt, K.B., Tignor, M., Miller, H.L., Eds.; Cambridge University Press: Cambridge, UK, 2008; Chapter 2; p. 234, ISBN 978-0521705967.
44. Cambra-López, M.; García Rebollar, P.; Estellés, F.; Torres, A. Estimación de las emisiones de los rumiantes en España: El factor de conversión de metano. *Arch. Zootec.* **2008**, *57*, 89–101.
45. Aguilera, M. Uso ganadero de los pastizales naturales de San Luis. In *Con las metas claras. La Estación Agropecuaria San Luis. 40 años en favor del Desarrollo Sustentable*; Aguilera, M.O., Panigatti, J.L., Eds.; Ediciones INTA: Buenos Aires, Argentina, 2003; pp. 89–124, ISBN 978-9875210745.
46. National Research Council. *Nutrient Requirements of Beef Cattle*; Seventh Revised Edition: Update 2000; The National Academies Press: Washington, DC, USA, 1996.
47. McCullagh, P.Y.; Nelder, J. *Generalized Linear Models. Monographs on Statistics and Applied Probability 37*, 2nd ed.; Chapman and Hall: London, UK, 1989; pp. 1–512, ISBN 978-0412317606.
48. Burnham, K.P.; Anderson, D.R. *Model Selection and Multimodel Inference: A Practical Information-Theoretic Approach*, 2nd ed.; Springer: New York, NY, USA, 2002; pp. 1–488, ISBN 978-0387953649.
49. Burnham, K.P.; Anderson, D.R.; Huyvaert, K.P. AIC model selection and multimodel inference in behavioral ecology: Some background, observations, and comparisons. *Behav. Ecol. Sociobiol.* **2011**, *65*, 23–35. [CrossRef]
50. Tsegaye, D.; Haile, M.; Moe, S.R. The effect of land use on the recruitment and population structure of the important food and fodder plant, *Dobera glabra* (Forssk.) Poir., in northern Afar, Ethiopia. *J. Arid Environ.* **2010**, *74*, 1074–1082. [CrossRef]
51. Manning, P.; Putwain, P.D.; Webb, N.R. Identifying and modelling the determinants of woody plant invasion of lowland heath. *J. Ecol.* **2004**, *92*, 868–881. [CrossRef]
52. Zar, J.H. *Biostatistical Analysis*, 4th ed.; Pearson Prentice Hall: Upper Saddle River, NJ, USA, 1999; ISBN 0130823902.
53. Hair, J.F., Jr.; Black, W.C.; Babin, B.J.; Anderson, R.E. *Multivariate Data Analysis. A Global Perspective*, 7th ed.; Pearson: London, UK, 2006; pp. 1–785, ISBN 978-1292021904.

54. Garcia-Martinez, A.; Olaizola, A.; Bernues, A. Trajectories of evolution and drivers of change in European mountain cattle farming systems. *Animal* **2009**, *3*, 152–165. [CrossRef] [PubMed]
55. IBM. *IBM SPSS Advanced Statistics 20*; IBM: Armonk, NY, USA, 2011.
56. Sheppard, S.C.; Bittman, S.; Donohoe, G.; Flaten, D.; Wittenberg, K.M.; Small, J.A.; Berthiaume, R.; McAllister, T.A.; Beauchemin, K.A.; McKinnon, J.; et al. Beef cattle husbandry practices across Ecoregions of Canada in 2011. *Can. J. Anim. Sci.* **2015**, *95*, 305–321. [CrossRef]
57. Modernel, P.; Astigarraga, L.; Picasso, V. Global versus local environmental impacts of grazing and confined beef production systems. *Environ. Res. Lett.* **2013**, *8*, 035052. [CrossRef]
58. Nieto, M.I.; Guzmán, M.L.; Steinaker, D. Emisiones de gases de efecto invernadero: Simulación de un sistema ganadero de carne típico de la región central Argentina. *Rev. Investig. Agropecu.* **2014**, *40*, 92–101.
59. Beauchemin, K.A.; Henry Janzen, H.; Little, S.M.; McAllister, T.A.; McGinn, S.M. Life cycle assessment of greenhouse gas emissions from beef production in western Canada: A case study. *Agric. Syst.* **2010**, *103*, 371–379. [CrossRef]
60. Beukes, P.C.; Gregorini, P.; Romera, A.J.; Levy, G.; Waghorn, G.C. Improving production efficiency as a strategy to mitigate greenhouse gas emissions on pastoral dairy farms in New Zealand. *Agric. Ecosyst. Environ.* **2010**, *136*, 358–365. [CrossRef]
61. Bogaerts, M.; Cirhigiri, L.; Robinson, I.; Rodkin, M.; Hajjar, R.; Costa Junior, C.; Newton, P. Climate change mitigation through intensified pasture management: Estimating greenhouse gas emissions on cattle farms in the Brazilian Amazon. *J. Clean. Prod.* **2016**, *162*, 1539–1550. [CrossRef]
62. Casey, J.W.; Holden, N.M. Quantification of GHG emissions from sucker-beef production in Ireland. *Agric. Syst.* **2006**, *90*, 79–98. [CrossRef]
63. Clark, H.; Pinares-Patiño, C.; De Klein, C. Methane and nitrous oxide emissions from grazed grasslands. In *Grassland: A Global Resource*; McGilloway, D.A., Ed.; Wageningen Academic Publishers: Wageningen, The Netherlands, 2005; pp. 279–293, ISBN 978-90-76998-71-8.
64. Beauchemin, K.A.; Janzen, H.H.; Little, S.M.; McAllister, T.A.; McGinn, S.M. Mitigation of greenhouse gas emissions from beef production in western Canada—Evaluation using farm-based life cycle assessment. *Anim. Feed Sci. Technol.* **2011**, *166–167*, 663–677. [CrossRef]
65. Veysset, P.; Lherm, M.; Bébin, D.; Roulenc, M.; Benoit, M. Variability in greenhouse gas emissions, fossil energy consumption and farm economics in suckler beef production in 59 French farms. *Agric. Ecosyst. Environ.* **2014**, *188*, 180–191. [CrossRef]
66. Legesse, G.; Beauchemin, K.A.; Ominski, K.H.; McGeough, E.J.; Kroebel, R.; MacDonald, D.; Little, S.M.; McAllister, T.A. Greenhouse gas emissions of Canadian beef production in 1981 as compared with 2011. *Anim. Prod. Sci.* **2016**, *56*, 153–168. [CrossRef]
67. Gerber, P.; Vellinga, T.; Opio, C.; Steinfeld, H. Productivity gains and greenhouse gas emissions intensity in dairy systems. *Livest. Sci.* **2011**, *139*, 100–108. [CrossRef]
68. Bava, L.; Sandrucci, A.; Zucali, M.; Guerci, M.; Tamburini, A. How can farming intensification affect the environmental impact of milk production? *J. Dairy Sci.* **2014**, *97*, 4579–4593. [CrossRef] [PubMed]
69. Moreau, J.; Delaby, L.; Duru, M.; Guerin, G. Advices given about the forage system: Evolutions and conceptions concerning the steps to be taken and the tools to be used. *Fourrages* **2009**, *200*, 565–586.
70. Mugnier, S.; Magne, M.A.; Pailleux, J.Y.; Poupart, S.; Ingrand, S. Management priorities of livestock farmers: A ranking system to support advice. *Livest. Sci.* **2012**, *144*, 181–189. [CrossRef]
71. Pinares-Patiño, C.S.; Waghorn, G.C.; Hegarty, R.S.; Hoskin, S.O. Effects of intensification of pastoral farming on greenhouse gas emissions in New Zealand. *N. Z. Vet. J.* **2009**, *57*, 252–261. [CrossRef] [PubMed]
72. Pickering, N.K.; Oddy, V.H.; Basarab, J.; Cammack, K.; Hayes, B.; Hegarty, R.S.; Lassen, J.; McEwan, J.C.; Miller, S.; Pinares-Patino, C.S.; et al. Animal board invited review: Genetic possibilities to reduce enteric methane emissions from ruminants. *Animal* **2015**, *9*, 1431–1440. [CrossRef] [PubMed]
73. Berndt, A.; Tomkins, N.W. Measurement and mitigation of methane emissions from beef cattle in tropical grazing systems: A perspective from Australia and Brazil. *Animal* **2013**, *7*, 363–372. [CrossRef] [PubMed]
74. Young, J.; Kingwell, R.; Bathgate, A.; Thomoson, A. Agricultural greenhouse gas emission reporting: The implications for farm management. *Agroecol. Sustain. Food Syst.* **2016**, *40*, 261–276. [CrossRef]
75. Lal, R. Soil carbon sequestration impacts on global climate change and food security. *Science* **2004**, *304*, 1623–1627. [CrossRef] [PubMed]

76. Follet, R.F.; Schuman, G.E. Grazing land contribution to carbon sequestration. In *Grassland: A Global Resource*; McGilloway, D.A., Ed.; Wageningen Academic Publishers: Wageningen, The Netherlands, 2005; pp. 265–277, ISBN 978-90-76998-71-8.
77. Soussana, J.F.; Tallec, T.; Blanfort, V. Mitigation the greenhouse gas balance of ruminant production through carbon sequestration in grasslands. *Animal* **2010**, *4*, 334–350. [CrossRef] [PubMed]
78. Schils, R.L.M.; Verhagen, A.; Aarts, H.F.M.; Sebek, L.B.J. A farm level approach to define successful mitigation strategies for GHG emissions from ruminant livestock systems. *Nutr. Cycl. Agroecosyst.* **2005**, *71*, 163–175. [CrossRef]
79. Pelletier, N.; Pirog, R.; Rasmussen, R. Comparative life cycle environmental impacts of three beef production strategies in the upper midwestern United States. *Agric. Syst.* **2010**, *103*, 380–389. [CrossRef]
80. Zhuang, M.H.; Gongbuzeren; Li, W.J. Greenhouse gas emission of pastoralism is lower than combined extensive/intensive livestock husbandry: A case study on the Qinghai-Tibet Plateau of China. *J. Clean. Prod.* **2017**, *147*, 514–522. [CrossRef]
81. Salvador, S.; Corazzin, M.; Romanzin, A.; Bovolenta, S. Greenhouse gas balance of mountain dairy farms as affected by grassland carbon sequestration. *J. Environ. Manag.* **2017**, *196*, 644–650. [CrossRef] [PubMed]

sustainability

MDPI

Article

Effect of Three Husbandry Systems on Environmental Impact of Organic Pigs

Gwendolyn Rudolph [1], Stefan Hörtenhuber [1,*], Davide Bochicchio [2], Gillian Butler [3], Roland Brandhofer [1], Sabine Dippel [4], Jean Yves Dourmad [5], Sandra Edwards [3], Barbara Früh [6], Matthias Meier [7], Armelle Prunier [5], Christoph Winckler [1], Werner Zollitsch [1] and Christine Leeb [1]

1 Department for Sustainable Agricultural Systems, University of Natural Resources and Life Sciences Vienna, 1180 Vienna, Austria; gwendolyn.rudolph@fibl.org (G.R.); r.brandhofer@gmx.at (R.B.); christoph.winckler@boku.ac.at (C.W.); werner.zollitsch@boku.ac.at (W.Z.); christine.leeb@boku.ac.at (C.L.)
2 Research Centre for Animal Production and Aquaculture, Council for Agricultural Research and Economics, 41100 Modena, Italy; davboch@hotmail.com
3 School of Natural and Environmental Sciences, Newcastle University, NE1 7RU Newcastle Upon Tyne, UK; gillian.butler@newcastle.ac.uk (G.B.); sandra.edwards@newcastle.ac.uk (S.E.)
4 Institute of Animal Welfare and Animal Husbandry, Federal Research Institute for Animal Health, 29223 Celle, Germany; sabine.dippel@fli.de
5 Physiology of Adaptation, Animal Nutrition and Health, INRA, 35590 Saint-Gilles, France; jean-yves.dourmad@inra.fr (J.Y.D.); armelle.prunier@inra.fr (A.P.)
6 Department of Extension, Training and Communication, FiBL, 5070 Frick, Switzerland; barbara.frueh@fibl.org
7 Department of Socio-Economic Sciences, FiBL, 5070 Frick, Switzerland; matthias.meier@fibl.org
* Correspondence: stefan.hoertenhuber@boku.ac.at; Tel.: +43-1-47654-93229

Received: 29 September 2018; Accepted: 17 October 2018; Published: 20 October 2018

Abstract: This study examined the environmental impact of the three common organic pig husbandry systems, indoor (n = 24), partly outdoor (n = 30), and outdoor (n = 10), in eight European countries. Global warming (GWP), acidification (AP), and eutrophication potential (EP) was assessed per 1000 kg pig live weight on 64 farrow-to-finish pig production chains (cradle to farm gate). GWP, AP, and EP varied greatly, and the most important source was feed production, followed by housing. GWP did not differ between systems ($p = 0.934$), but AP in indoor systems and EP in outdoor systems were higher than in partly outdoor systems ($p = 0.006$ and $p = 0.010$, respectively). The higher AP in indoor systems can mainly be explained by NH_3 arising from manure spreading, while PO_4-eq arising from feed consumption and emissions on pasture accounted for the higher EP in outdoor systems. Associations of farm characteristics with (reduced) environmental impacts were mainly found for AP and EP, and included: (Increasing) farm size, numbers of piglets born and weaned per litter, (bought-in) mineral feed, and high-protein by-products, the latter probably connected to beneficial effects of appropriate dietary digestible lysine levels and feed conversion ratio. Increasing carcass weights and dietary cereal proportions were associated with higher environmental impacts. Overall, variation was mostly higher within than between systems, and measures to mitigate environmental impact were identified.

Keywords: life cycle assessment; global warming potential; acidification potential; eutrophication potential; cradle to farm gate; indoor; outdoor

1. Introduction

Together with the general growth of organic farming in many European countries, organic pig farming has expanded in recent decades [1]. However, the pig meat sector still ranks relatively low in organic product sales, particularly in comparison to the sheep and bovine sectors [2].

Climate change as well as animal health and welfare are among the most current issues in the public and scientific debate concerning livestock [3–6]. Likewise, sustainability and its assessment, covering environmental, economic, and social aspects, have gained importance in recent years. For instance, the Sustainability Assessment of Food and Agriculture systems (SAFA) Guidelines [7] have been developed as an international reference document to allow for consistent analyses and assessments despite the complexity of sustainability. According to the SAFA guidelines, sustainability consists of four dimensions: Good governance, environmental integrity, economic resilience, and social well-being. In the dimension of environmental integrity, the following themes are addressed: Atmosphere, water, land, materials and energy, biodiversity, and animal welfare.

Livestock production exerts severe impacts on soil, water, and air quality due to the related emissions [4]. The world's livestock sector contributes 14.5% of global greenhouse gas emissions, with pigs accounting for 9% of overall livestock-related emissions [5].

The life cycle assessment (LCA) provides a valuable and consistent methodological framework to quantify the environmental impact within the life cycle of a product [8]. Hence, several LCAs have been conducted in recent years to quantify the environmental impact, mainly greenhouse gas emissions (global warming potential, GWP), acidification potential (AP), and eutrophication potential (EP), of animal husbandry systems [9–11]. Until now, due to high CH_4 emissions from enteric fermentation, ruminants have mainly been in the focus, but in light of the high consumption of pork and pork products in the European Union, pork production must also be considered.

Since pig production is dominated by non-organic production systems, most studies cover conventional systems [4]. As for most LCA studies dealing with pigs, for organic pig production, which is still comparatively small-scale, but nevertheless rapidly developing, only a few modelling studies are available [12–14]. Only a few studies have conducted an LCA using individual farm data [15], and even fewer have been based on a considerable number of organic farrow-to-finish pig farms.

As organic pigs are produced according to the general principles of organic farming [16], national, and international regulations (e.g., EC Nos. 2018/848 and 889/2008, [17,18]) as well as private standards [19], organic pig farms have been treated as a uniform system in most studies. However, it has recently been shown that organic pig farms in Europe can clearly be distinguished into three main "husbandry systems": Pigs may be kept completely outdoors in paddocks on pasture, as in most UK and Italian farms, or indoors, with access to a limited concrete outside run, as in most farms in German speaking countries. Furthermore, both systems, indoor and outdoor, may be combined on one farm for different production stages or during different seasons, as is common, for example, in Denmark or France [1,20].

Keeping pigs on pasture has a potentially higher risk regarding nutrient losses [14,21] compared to pigs kept indoors, where manure is collected, stored, and spread in a controlled way on fields. Furthermore, due to a more controlled (thermal) environment, pigs kept indoors might have better feed conversion ratios and a higher number of piglets weaned, both reducing the environmental impact [22,23].

Besides the effects on environmental impact caused by the husbandry system, as described above, other important influencing factors contribute to variation. As LCA data from 27 conventional farms show, other factors, such as diet characteristics (e.g., level of by-products), influence the environmental impacts [9]. Therefore, also for organic pig production, individual farm data are needed to describe and quantify the potential influence of the three husbandry systems as well as those individual farm characteristics with potential impact on the environment.

Consequently, the present study aims to deliver first indications to support strategic decisions in the organic sector (policy makers, extension services) or at the farm level (farmer). The analysis focuses on the environmental impact of three common organic pig husbandry systems, indoor (IN), partly outdoor (POUT), and outdoor (OUT), regarding their GWP, AP, and EP. Furthermore, farm characteristics influencing these impacts are assessed by cluster and correlation analysis. Two hypotheses are tested: (1) The null hypothesis that there are no differences between husbandry systems regarding the environmental impact (GWP, AP, EP), and (2) that there are specific farm characteristics, which explain variation independent from the husbandry systems.

2. Materials and Methods

2.1. Overall Study Design and Participating Farms

The present assessments are based on data collected during visits to 74 organic pig farms, representing three different husbandry systems, IN (34 farms), POUT (28 farms), and OUT (12 farms), in eight European countries (Austria, Czech Republic, Denmark, France, Germany, Italy, Switzerland, and the United Kingdom) during 2012 and 2013 (see Supplementary Materials). In each country, all assessments were carried out by one trained observer. All observers attended a common training course to standardize assessments. The different husbandry systems were defined as described in Table 1.

Table 1. Definition of the three organic pig husbandry systems: IN, POUT, and OUT.

System	Abbreviation	Definition [1,2]
indoor	IN	Pigs live in buildings with access to a concrete outside run, or to an outside run on soil, which is a small sacrifice area for permanent pig use and not integrated into crop rotation.
partly outdoor	POUT	Pigs spend part of the production cycle indoors and part outdoors. There can be at least one production stage (dry sows, lactating sows, group suckling, weaned piglets, or finishing pigs) outside while the rest is housed, or pigs spending part of the year outside and the rest indoors (seasonal housing).
outdoor	OUT	Pigs live permanently outdoors in paddocks with shelter (temporary hut or permanent building), but unrestricted access to the soil. The paddock is usually integrated as pasture in a crop rotation.

[1] Weaned piglets kept in an enclosure directly on soil in fields are considered OUT if only the lying area is roofed, and considered IN if the entire enclosure is roofed; [2] if only a small percentage of the animals (<10% in herds of up to 300 pigs in total, or <5% in larger herds) are kept in a different husbandry system, the farm is classified according to the dominant system.

Organic pig farms were recruited through farm advisors, producer associations, agricultural journals and their websites, or personal contacts. Farms had to be certified organic for at least two years and preferably combined farrow-finish farms with more than 20 sows and 100 finishing places. Recruitment was also based on the type of husbandry system, as the objective was to compare the three different organic pig systems.

As this study investigated the environmental impact of pork production from piglet production until slaughter, production chains from farrowing to finishing (PC) were the statistical unit. This comprised 64 PCs (24 IN, 30 POUT, 10 OUT), which were either farrow-to-finish farms (15 IN, 24 POUT, 9 OUT) or which were formed of co-operating farrowing-only and fattening-only farms (6, 6, and 1 IN, POUT and OUT PCs, respectively). Additionally, for three other PCs (IN), data from the co-operating farm was not available during the study, and the average of the other farms was then used for the missing part of the PC. For each individual PC, environmental impact categories were calculated as outlined below, with pairs of farrowing-only and fattening-only farms being treated as farrow-to-finish farms.

Based on the literature [10,15,24] and expert knowledge, a standardized on-farm assessment protocol for use on a tablet computer was developed and a supplementary dictionary was translated to the languages of involved countries. The final protocol included an interview with the farm manager, the analysis of farm records, and direct observations. It was structured in thematic sections, including the following parameters:

- Farm and manure management, description of husbandry systems;
- productivity data and medicinal treatments (records);
- land use (crop production);
- diet composition and dietary nutrients contents; and
- provision of resources (husbandry system) for weaners (defined as pigs from post weaning with at least six weeks until transfer to a fattening unit with approximately 12 weeks), fatteners (includes the growing and finishing phase from approximately 30 kg to slaughter), pregnant, and lactating sows (direct observations)

The PC-specific data used in the present study are based on inventory data either collected on the day of visit or covering a period of 12 months prior to the farm visits.

2.2. *Scope of the Study and Life Cycle Assessment Methodology*

2.2.1. System Boundaries, Functional Unit, and Allocation Approach

As the study focused on pig farming systems, transport, slaughter, and processing of the carcasses were not included. The system boundaries were defined as cradle-to-farm gate, in this case from piglet birth to pigs ready for slaughter, and included relevant farm inputs needed for this (e.g., electricity, transport-related (fossil) energy for feedstuffs, etc.). While the energy-gain from co-digestion of manure in biogas plants was not considered (as not directly related to the pig unit), digestion-related emission factors were used to derive results from co-digestion of manure. Following LCA-related guidelines, such as PAS2050 [25], for the assessment of greenhouse gas emissions, emissions from the construction of farm infrastructure (e.g., livestock barns, machinery, farm buildings) were excluded from the LCA. Furthermore, veterinary treatments and inputs, such as cleaning agents and disinfectants, were not considered within the system boundaries as a lower use is assumed in organic farming [26]. The functional unit was 1000 kg of live weight of fattening pigs leaving the farm (live weight at slaughter), including culled sows. The live weight at slaughter was calculated based on carcass weight records using an equation used for organic pigs from different countries [27]. An energy allocation approach was used throughout the study, for instance, for feedstuffs' emission factors.

2.2.2. Environmental Impacts Considered, Data Sources, and Methodological Details

GWP (100 year-horizon) [28], AP, and EP were calculated using a modified version of an Excel tool developed by Dourmad et al. [15]. Emissions were calculated from measured as well as modelled data per individual PC, using typical emission factors for livestock-related emissions [29]. For EP and AP from the pig unit and feed production, characterization factors were used as described in [8] and [30]. Gaseous emissions from animal husbandry and manure storage were calculated for NH_3, N_2O, and NO_X as well as CH_4 based on [28,29,31] and other sources as described in [15].

Calculations for nitrogen (N) and phosphorous (P) retention and excretion have been described in detail in [29,32]. Results from [33], adapted according to [34], were used to estimate losses from outdoor paddocks into water, assuming a constant rate between P losses and the body mass (live-weight) of pigs on pasture (see Equation (S1) in the Supplementary Materials). P losses from manure storage, treatment, and spreading were assumed to be negligible and ignored [29]. All emission factors used are given in the Supplementary Material. All traits used for deriving the indicators of the pig unit's environmental impact are provided in Table 2.

Table 2. Input data for the characterization of the environmental impact of organic pig systems.

Category	Parameter
sow performance	number of weaned piglets per sow and year, replacement rate (%), live weight at slaughter (kg), feed intake during gestation and lactation (kg/period), duration of lactation (days)
weaner performance	weight at weaning (kg), piglet mortality (%), daily feed intake (kg), feed conversion rate (kg feed/kg live weight gain), duration of weaning period (days)
fattener performance	weight at beginning of fattening phase (kg), mortality (%), daily feed intake (kg), feed conversion rate (kg feed/kg live weight gain), daily weight gain (kg/day), live weight at slaughter (kg), age at slaughter (days), duration of fattening period (days)
diets	diet composition (% of individual feed ingredients), diet nutrient, i.e., crude protein (CP, g/kg) and phosphorus (P, g/kg), and metabolizable energy content (MJ/kg)
land use	on-farm crop production
animal husbandry	type of system, i.e., outdoor (OUT), partly outdoor (POUT), indoor (IN) with outside run, type of floor (solid floor, slats/partly slatted, deep litter)
manure	manure type (liquid, solid), manure handling (cleaning frequency), manure storage (type and duration), manure treatment (composting, anaerobic/aerobic digestion), type and distance of spreading (wide spreading, injection), mean distance of manure transport to place of spreading, crop rotation, and stocking rate (animals/ha)
bedding quality	very good: 100% of litter is clean, dry, and not mouldy good: >50% of litter is clean, dry, and not mouldy poor: >50% of litter is dirty, wet, or mouldy very poor: 100% of litter is dirty, wet, or mouldy

Parameters to calculate the environmental impact from the production of all feedstuffs, except fishmeal and fish oil, were adopted from previous studies [35,36]. The nitrogen (N)-related part of feeds' EP was calculated by multiplying typical quantities of N applied in manure and commercial fertilizers on organic farms in Austria with typical NO_3-N leaching factors identified in [37]. For P-losses from feed production, it was assumed that, on average, a surplus of 5% over plant requirements (according to yield) was applied and lost. The GWP of the rarely used components, fishmeal and fish oil, originate from [38]. The GWP, AP, and EP for monocalcium phosphate and mineral premixes are based on the Danish LCA food database [39].

The total feed consumption per animal, including bought–in-feedstuffs, was calculated from farm-specific data (based on daily consumption per animal, feed conversion rate, duration of the periods, average daily gains, etc.). For individual missing values, results from [40] and [41] were used in the fattening and the weaner stages, respectively. For farms lacking full records on the feed intake and feed conversion ratio, these values were estimated according to farm and animal category-specific dietary contents for MJ ME, CP, and P, and recommended nutrient requirements at the respective physiological stage [42]. Furthermore, the relative amount of digestible lysine per unit of energy (g/MJ NE) was calculated based on the online tool, Evapig [43], and compared to requirements at different stages [44]; a 10% tolerance was accepted due to the uncertainty of NE and lysine content. Diets for growing and finishing pigs were classified as sufficient, deficient, or exceeding requirements (excess) using the content of digestible lysine relative to the NE content as an indicator. Feed components and feedstuffs were categorized as described in Table 3.

2.2.3. Characteristics of the Husbandry Systems

Animal Performance, Production Data, and Feeding Characteristics

Animal performance and production data, as well as feed characteristics, varied substantially within and between systems, see Supplementary Materials (Tables S4 and S5). In OUT-PCs, the number of piglets born and weaned per sow was lower than in the other systems, but the median age at culling was higher, and hence the replacement rate was lower in OUT. OUT-PCs showed lower mortality and heavier pigs at weaning. Some of the diverging characteristics, such as numerically smaller sows in OUT, are probably related to the use of local, traditional breeds (e.g., Cinta Senese, Tamworth). Median live weight (kg) of slaughter pigs was 131.0 kg in IN, 117.0 kg in POUT, and 124.0 kg in OUT. Diet characteristics and amounts of feed used differed in many parameters between the husbandry systems. However, average metabolizable energy and crude protein contents were similar between systems for sow diets, and similar average metabolizable energy were recorded for weaners and finishing farms.

Table 3. Categorization of feed components and feedstuffs.

Feed Stuff Category	Component
animal or microbial origin	brewer's yeast, fish meal, fish oil, whey powder (sweet), whey concentrate
compound feed	different compound feeds for growing and/or finishing pigs
grains (cereals)	barley, maize, rye, oat, triticale, wheat
high-protein by-products	false flax seed cake, potato protein, rapeseed cake, rapeseed meal, sunflower seed cake
leguminous grains	fava beans, peas, soybean, soybean cake
minerals	clay, mineral premix, monocalcium phosphate
others (based on processing of plant raw materials)	alfalfa (lucerne) green meal or similar roughage, brewer's grains (dried), grass cobs, rapeseed oil, spelt husks, sugar beet molasses, sunflower seed oil, wheat bran, wheat starch
supplementary compound feed	high-protein supplements

Housing (Floor Type) and Manure Management

The housing, and consequently the manure management, varied only between the individual PCs in IN and POUT, while in OUT, all animals were kept outdoors. Detailed information on housing (floor type) is provided for all animal categories in IN and POUT-PCs in the Supplementary Materials (Table S6). All animal categories in IN were kept on (partly slatted) concrete floors, with bedded lying areas. For POUT, 83% of PCs kept lactating sows outdoors throughout the year, and 45% kept dry sows outside, with a much lower proportion keeping weaners and fatteners outdoors.

The floor types for weaners and fatteners housed on POUT farms varied largely. Depending on the floor type, IN and POUT PCs produced either solid manure only or solid manure and slurry. In IN, 95.5% of the PCs generated a combination of solid manure and slurry, while only 4.5% had only solid manure. In POUT, the proportion of PCs with both solid manure and slurry was 69%. In OUT, no manure was stored and handled.

The type and frequency of manure treatments by husbandry system are reported in the Supplementary Material (Table S7). In IN, 16.6% of PCs used aerobic digestion. Thirty percent of the POUT-PCs applied a treatment (composting, aerobic, or anaerobic digestion) to the slurry. Across all systems, most PCs did not treat the solid manure before application; composting was the only treatment found on 26.3% of IN- and 15.3% of POUT-PCs.

2.3. Statistical Data Analysis

All statistical analyses were performed with each individual PC as the statistical unit in SAS 9.2 and 9.3 (SAS-Institute, Cary, North Carolina, USA, 2008). Non-parametric Kruskal-Wallis tests were used for comparisons between husbandry systems. When significant effects ($p < 0.05$) were revealed in global tests, pairwise comparisons were performed using the Wilcoxon Two-Sample (Rank sum) test. p-Values were adjusted for multiple comparisons using Bonferroni correction. Additionally, to investigate associations between the environmental impact categories of AP, EP, and GWP and farm characteristics, Spearman rank correlations were calculated. Only those characteristics were included which were not directly considered in the LCA calculation, except for piglets weaned per sow per year [n, 1 yr mean] and carcass weight [kg, 1 yr mean].

Since AP, EP, and GWP did not significantly correlate with each other, they were subjected to a hierarchical cluster analysis using the average linkage method. Values were standardized using the procedure, STDIZE, in SAS using the mean as a location measure and the sample standard deviation as a scale measure. For this analysis, five outliers (2 IN, 2 POUT, 1 OUT) were excluded, as identified from boxplots, resulting in 59 PCs being included in the final cluster analysis. The number of clusters was based on R-Squared (SAS-Institute, 2008), Pseudo F, and Pseudo t2 statistics. Additionally, the average distance between the clusters was graphically checked in a dendrogram.

3. Results

3.1. GWP, AP, and EP of the Three Husbandry Systems

The total environmental impacts (AP, EP, and GWP) of the 64 PCs in the different husbandry systems are presented in Table 4. Statistical comparison of the three pig husbandry systems with respect to GWP, AP, and EP revealed inconsistent results.

Table 4. Environmental impact (global warming, GWP; acidification potential, AP; eutrophication potential, EP) of organic pig production in three husbandry systems, IN, POUT, and OUT.

Parameter		Husbandry System	n (PC)	Min.	Q25%	Median	Q75%	Max.
GWP	kg CO_2-eq/1000 kg live weight at slaughter	IN	24	1605	1860	2204	2347	2962
		POUT	30	1663	1997	2213	2407	3393
		OUT	10	1470	1593	2210	2705	3480
AP	kg SO_2-eq/1000 kg live weight at slaughter	IN	24	38.0	55.2	61.9 ^{a}	78.4	114.4
		POUT	30	37.8	47.0	51.9 ^{b}	61.0	88.4
		OUT	10	34.8	38.4	55.4 ^{ab}	72.3	91.0
EP	kg PO_4-eq/1000 kg live weight at slaughter	IN	24	13.6	18.2	21.6 ^{ab}	25.7	48.7
		POUT	30	13.3	17.8	20.1 ^{b}	25.1	43.2
		OUT	10	17.8	19.9	28.7 ^{a}	36.8	46.2

^{a,b} Different superscript letters indicate differences between groups for which $p < 0.05$ (p-values adjusted according to Bonferroni correction for triple testing).

No significant differences were found regarding the GWP between systems. With 2204, 2213, and 2209 kg CO_2-eq per 1000 kg live weight at slaughter, the median estimate for the GWP was similar for the three systems, IN, POUT, and OUT ($p = 0.934$). Across systems, the PCs with the lowest and highest GWP were both found in OUT, and the variation in the GWP was numerically smaller in IN and POUT than in OUT.

In all systems, feed production most strongly contributed to the GWP, followed by animal housing (direct emissions originating from the animal and excreta inside houses), and, in IN and POUT, manure storage (Figure 1). Manure treatment and manure spreading contributed only a small percentage to the GWP in IN and POUT, while in OUT, manure is directly excreted onto the field by the animal and therefore is neither stored, treated, nor spread. Relative contributions of housing emissions tended to be lower in IN, whereas the relative contribution of manure storage was highest for this system. Consistently, the highest relative contribution of housing (including field deposition of manure) was found for OUT.

The median AP was significantly higher in IN (61.9 kg SO_2-eq per 1000 kg live weight at slaughter) than in POUT (51.9 kg SO_2-eq per 1000 kg live weight at slaughter; $p = 0.006$), mainly due to more NH_3 arising from manure spreading in IN. In OUT, AP was numerically slightly higher than in POUT. Across systems, the individual PC with the lowest AP was found in OUT and the one with the highest AP was in IN. The variation (interquartile range) was smaller in POUT compared to IN and OUT.

Similar to GWP, feed production and animal housing contributed most to AP (Figure 1). The relative contribution of manure spreading in IN and POUT to both AP and EP was higher than the corresponding contributions to GWP. Regarding AP, IN showed higher relative amounts of SO_2-eq originating from feed, housing of the animals (direct emissions), and, especially, manure storage and spreading. In POUT, some manure remains directly on the paddock and is not stored or spread on the field, thus leading to lower AP due to lower NH_3 emissions. Manure spreading was the main explanation for differences in the AP between IN and POUT. An even higher difference was found between OUT and IN, again due to the lack of manure spreading in OUT.

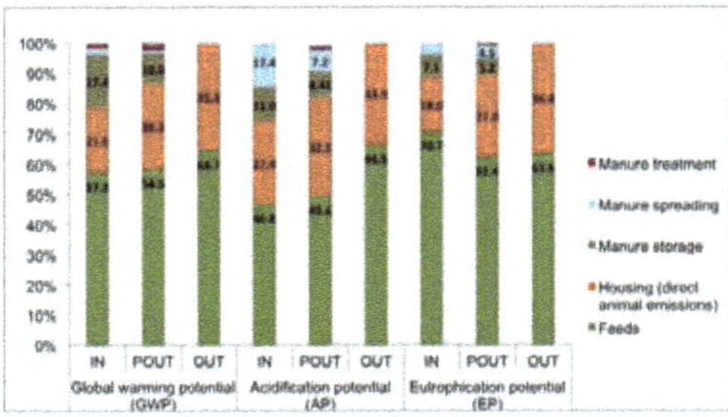

Figure 1. Mean relative contribution of the different sources to GWP, AP, and EP based on data from IN (n = 24), POUT (n = 30), and OUT (n = 10) production chains.

The median EP, expressed in kg PO_4-eq per 1000 kg of live weight at slaughter, was significantly higher in OUT than in POUT ($p = 0.010$), mainly due to more PO_4-eq resulting from feed consumption and housing. Total EP of IN was similar to the EP of POUT, but did not differ significantly from OUT. Variation (interquartile range) in OUT was larger than in the other systems. Across all systems, the most important source of EP was feed production, followed by animal housing. In IN and POUT, manure storage, treatment, and spreading also contributed to EP, but to a lesser extent. The highest contribution of feed and housing regarding EP was found in OUT, which had a higher median feed conversion rate than POUT- or IN-PCs, and consequently needed more feed to achieve 1000 kg of slaughter weight.

When considering the environmental impact indicators (GWP, AP, EP) based on animal production stages, across all systems the fattening phase had the highest influence (between 68% and 74% of the totals), with little variation between the systems and impact indicators (Figure 2).

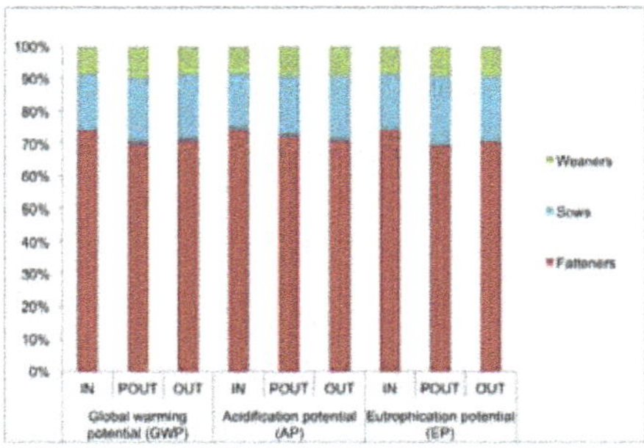

Figure 2. Mean relative contribution of the different animal categories to GWP, AP, and EP based on data from IN (n = 24), POUT (n = 30), and OUT (n = 10) production chains.

3.2. Correlations between Farm Characteristics and Environmental Impacts

Correlation coefficients of farm characteristics with environmental impact indicators are presented in Table 5. Two characteristics relating to size of the pig units (number of slaughtered pigs and livestock units per PC) negatively correlated with AP and EP, with larger PCs having lower AP and EP. The current number of sows on the farm, which also relates to farm size, was also negatively correlated with AP.

Table 5. Coefficients of correlation (Spearman's Rho's) between farm characteristics (rows) and environmental impact indicators (columns) for 61 production chains (three production chains had to be excluded due to a missing animal category at the farm).

Parameter		GWP	AP	EP
n sows [present at visit]		ns	−0.33 **	ns
slaughtered finishers [n/1 yr]		ns	−0.30 *	−0.31 *
Livestock Unit		ns	−0.37 **	−0.30 *
piglets born per litter (life born + still born) [n, 1 yr mean]		ns	ns	−0.44 ***
number of weaned piglets per sow per year		−0.35 **	−0.27 *	−0.37**
carcass weight [kg, 1 yr mean]		ns	0.30 *	0.30*
age at culling [n farrowings]		ns	ns	ns
Mastitis Metritis Agalactia (MMA) treatment of sows (%)		ns	ns	ns
percentage of bought-in feed for finishers (%)		ns	−0.40 **	ns
relative contribution of feed stuff category (%)	grains (Cereals)	ns	ns	0.29 *
	leguminous crops	ns	ns	ns
	high-protein by-products	ns	ns	−0.26 *
	others	ns	ns	ns
	components of animal or microbial origin	ns	ns	ns
	minerals	ns	ns	−0.40 **
	compound feed	ns	ns	ns
	supplementary compound feed	ns	ns	ns

ns: Not significant, * $p < 0.05$, ** $p < 0.01$, *** $p < 0.001$.

The total number of piglets born per litter negatively correlated with EP, whilst piglet numbers weaned per sow and year negatively correlated with all environmental impact indicators. The average final carcass weight correlated positively with AP and EP, while the percentage of bought-in feed in fattener diets was negatively associated with AP. Additionally, the relationship between the relative contribution of feedstuff categories to the diets and the environmental impact indicators was investigated, but associations were only found for EP. The relative contribution of the high-protein by-product feed and mineral supplements negatively correlated with EP, while the amount of grains correlated positively.

3.3. Cluster Analysis Regarding Impacts of Organic Pig Husbandry Systems

The hierarchical cluster analysis revealed two suitable classification levels, with five and four clusters, respectively. For statistical reasons (number of observations per cluster), the higher aggregation level, i.e., combining clusters 1 and 4 to cluster 1/4, was chosen. The number of PCs per husbandry system and cluster are shown in Table 6.

Table 6. Number of production chains per system (IN, POUT, OUT) in the four clusters identified.

Number of Production Chains Per System in Cluster				
System	**Cluster 1/4 ***	**Cluster 2 ***	**Cluster 3 ***	**Cluster 5**
Impact	**Intermediate**	**Low**	**High**	**Highest**
IN	5	8	9	0
POUT	13	10	5	0
OUT	4	3	0	2
total	22	21	14	2

* Clusters subjected to further statistical analysis.

In Table 7 all clusters are presented: Cluster 2, representing 35.6% of the total, on average, showed numerically the lowest environmental impacts (referred to as "low impact cluster" below). Cluster 1/4 can be considered as the "intermediate impact cluster", with values of AP, EP, and GWP between those of Cluster 2 and Cluster 3. Cluster 3 had higher median values for AP, EP, and GWP as compared to Cluster 1/4 and Cluster 2. Consequently, Cluster 3 can be described as the "high impact cluster". Cluster 5 resulted in overall highest environmental impacts, but due to the very small number of farms, included data are only presented descriptively, but not further considered in the analysis.

Table 7. Environmental impacts (GWP, AP, and EP) per 1000 kg of live weight at slaughter by cluster (total N = 59 production chains) (cluster 5 not subjected to further statistical tests is highlighted in grey).

Parameter	Cluster Number	n (PC)	Min.	Q25%	Median	Q75%	Max.
GWP [kg CO_2-eq per 1000 kg live weight at slaughter]	1/4	22	2008	2153	2222	2405	2962
	2	21	1470	1668	1776	1906	2102
	3	14	1977	2316	2348	2416	2695
	5	2	2705	2705	2908	3111	3111
AP [kg SO_2-eq per 1000 kg live weight at slaughter]	1/4	22	46.0	51.9	56.0	59.0	63.3
	2	21	34.7	40.0	47.0	50.9	59.1
	3	14	69.5	72.1	77.3	81.2	88.2
	5	2	72.3	72.3	74.1	76.0	76.0
EP [kg PO_4-eq per 1000 kg live weight at slaughter]	1/4	22	15.8	19.7	21.1	24.6	29.1
	2	21	13.3	17.0	17.6	18.6	20.5
	3	14	22.7	24.8	25.9	28.0	30.5
	5	2	36.8	36.8	37.5	38.1	38.1

Generally, environmental impact indicators were predominantly influenced by emissions from feed and housing (direct emissions during animal keeping). Numerically, the low impact cluster had the lowest contributions from these sources and Cluster 5 the highest.

Regarding GWP, the intermediate and high impact clusters had relatively comparable values, which were slightly higher in the high impact cluster, but the contribution of the different sources to total GWP differed numerically; the intermediate impact cluster was characterized by higher amounts from feed and housing, while the contribution of manure storage and spreading was higher in the high impact cluster. Considering AP and EP, the main difference leading to numerically higher amounts in the high impact cluster compared to the intermediate impact cluster, were emissions from manure spreading (Figure 3).

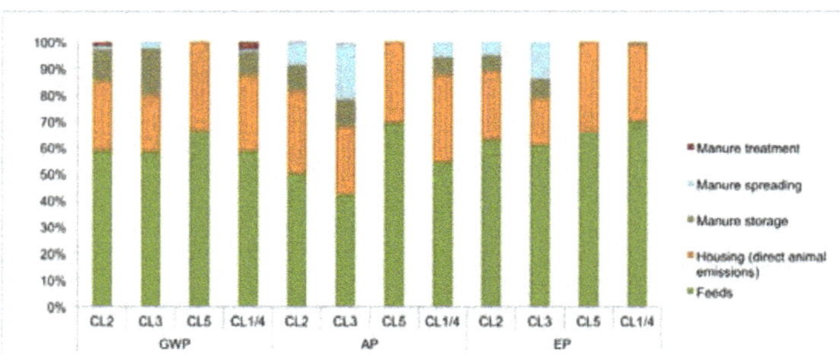

Figure 3. Mean relative contribution of the different sources to GWP, AP, and EP by cluster (total n = 59). CL2 = Cluster 2 (low environmental impact cluster, n = 21), CL3 = Cluster 3 (high environmental impact cluster, n = 14), CL5 = Cluster 5 (n = 2), CL1/4 = Cluster 1/4 (intermediate environmental impact cluster, n = 22).

Median numbers of sows, livestock units, and finishing pigs slaughtered per year per cluster ranged between 28 and 110, 38 to 121, and 324 to 859, respectively. These characteristics relating to farm size were numerically lowest for the high impact cluster, while the highest values were found in the intermediate impact cluster. The values for the low impact cluster were between those for the intermediate and high impact cluster. There were, however, no statistical differences between impact clusters in these parameters (see Supplementary Materials, Table S8).

The average weight of pigs at the end of the post-weaning period differed between clusters as shown in Table 8. In the intermediate impact cluster, pigs entering the finishing phase were heavier than in the high impact cluster ($p = 0.012$). Lower weights at the start of the fattening phase had a negative effect on environmental impact indicators. Feed conversion for fattening pigs in the low impact cluster was lower than in the intermediate and high impact clusters ($p < 0.001$). In line with this, the total amount of feed per finished pig was lower in the low impact cluster in comparison to the high impact cluster ($p < 0.001$) and to the intermediate impact cluster ($p = 0.006$; Table 9). Improved feed conversion and lower feed consumption in the low impact cluster had a beneficial effect on environmental impacts. However, mortality of finishers was higher in the low impact cluster than in the high impact cluster ($p < 0.001$; Table 8). The percentage of bought-in feed in finishing diets (Table 9) was lower for the high impact cluster in comparison to the low impact cluster ($p = 0.000$) and the intermediate impact cluster ($p < 0.001$). Additionally, in the low and high impact cluster, more leguminous grains were fed than in the intermediate impact cluster (Table 10).

Table 8. Characteristics of the animal categories by cluster.

Animal Category/Parameter [3]	Cluster impact Category [Cluster Number]	n (PC)	Min.	Q25%	Mean	Median	Q75%	Max.
sows								
piglets born per litter (live born + still born) [n, 1 yr mean] [1]	low [2]	19	12.0	12.1	13.2	13.3	14.0	14.5
	intermediate [1/4]	20	6.0	10.8	12.1	12.3	14.0	16.5
	high [3]	13	8.0	12.0	12.8	13.3	14.0	14.5
piglets weaned per sow per year [n, 1 yr mean] [2]	low [2]	20	10.0	18.2	19.2	19.9	21.1	25.0
	intermediate [1/4]	22	10.0	16.0	18.3	19.0	21.0	23.8
	high [3]	14	14.0	16.0	18.3	19.1	19.7	25.0
sow replacement rate [%, 1 yr mean] [2]	low [2]	20	12.0	26.0	31.8	30.5	35.0	50.0
	intermediate [1/4]	22	8.0	20.0	32.5	28.0	45.0	87.0
	high [3]	14	20.0	20.0	29.0	26.5	33.0	53.0
live weight at culling [kg at culling] [2]	low [2]	20	187	200	237	240	257	310
	intermediate [1/4]	22	180	220	249	245	275	325
	high [3]	14	197	200	234	233	253	277
Weaners								
weight at weaning [kg, 1 yr mean] [2]	low [2]	20	0.0	10.0	11.3	10.7	13.5	22.0
	intermediate [1/4]	22	5.5	10.0	13.1	12.5	15.0	28.0
	high [3]	14	8.0	10.0	11.5	10.9	12.5	22.0
weight at end of post-weaning [kg, 1 yr mean] [2]	**low [2]**	**20**	**24.0**	**25.8**	**29.4**	**30.0 ^{ab}**	**30.0**	**42.0**
	intermediate [1/4]	**22**	**25.0**	**30.0**	**31.6**	**31.5 ^{b}**	**35.0**	**40.0**
	high [3]	**14**	**23.0**	**25.0**	**28.3**	**29.0 ^{a}**	**30.0**	**34.0**
mortality rate weaners [%, 1 yr mean] [2]	low [2]	20	0.0	3.5	5.7	5.0	5.0	20.0
	intermediate [1/4]	22	0.0	1.0	3.2	3.0	5.0	10.0
	high [3]	14	0.0	1.0	3.1	3.0	5.0	5.0
fattening pigs								
live weight at slaughter [kg, 1 yr mean] [2]	low [2]	20	104	112	124	117	129	165
	intermediate [1/4]	21	86	112	128	120	136	200
	high [3]	14	116	121	131	129	140	150
mortality rate fattening pigs [%, 1 yr mean] [2]	**low [2]**	**20**	**1.0**	**1.5**	**2.9**	**3.0 ^{a}**	**4.0**	**5.0**
	intermediate [1/4]	**21**	**0.0**	**1.0**	**1.9**	**2.0 ^{ab}**	**2.0**	**6.0**
	high [3]	**14**	**0.0**	**0.0**	**1.0**	**1.0 ^{b}**	**1.0**	**4.0**

[1] Number of observations differs from number of production chains, as each parameter was not always assessable for all farms; [2] number of observations differ from number of production chains, as the environmental impact was calculated using mean values for missing animal categories for three farms (piglet production for two farms; fattening pigs for one farm); [3] parameters that showed significant differences between clusters are printed in bold; [a,b] different superscript letters indicate differences between groups ($p < 0.05$, p-values adjusted according to Bonferroni correction for triple testing).

Table 9. Dietary characteristics by cluster.

Animal Category/ Parameter [1]	Cluster Impact Category [cluster Number]	n (PC)	Min.	Q25%	Mean	Median	Q75%	Max.
sows								
feed per sow [kg/year]	low [2]	20	675	1028	1415	1458	1595	2053
	intermediate [1/4]	22	827	1332	1603	1702	1834	2236
	high [3]	14	1060	1203	1383	1385	1473	1946
average dietary content of								
ME [MJ//kg]	low [2]	20	9.6	11.9	12.1	12.5	12.9	13.1
	intermediate [1/4]	22	12.0	12.3	12.7	12.7	12.9	13.7
	high [3]	14	11.8	12.4	12.6	12.6	13.0	13.3
CP [g/kg]	low [2]	20	124	148	155	155	164	185
	intermediate [1/4]	22	112	137	149	152	162	180
	high [3]	14	133	146	150	151	157	165
total P [g/kg]	low [2]	20	3.0	4.7	5.2	5.5	5.9	7.8
	intermediate [1/4]	22	3.1	4.5	5.4	5.8	6.1	6.5
	high [3]	14	3.3	4.4	5.0	5.3	5.5	6.0
Weaners								
feed per weaner produced [kg/weaner]	low [2]	20	18	30	38	39	45	56
	intermediate [1/4]	22	12	33	43	38	50	111
	high [3]	14	20	30	37	36	45	59
average dietary content of								
ME [MJ//kg]	low [2]	20	9.5	12.5	12.4	12.8	13.1	13.4
	intermediate [1/4]	22	12.0	12.6	12.9	12.9	13.2	13.6
	high [3]	14	12.0	12.9	13.0	13.1	13.4	13.5
CP [g/kg]	low [2]	20	138	168	177	179	189	203
	intermediate [1/4]	22	112	170	176	181	195	208
	high [3]	14	147	172	180	182	188	208
total P [g/kg]	low [2]	20	3.5	3.8	5.2	5.7	6.0	6.8
	intermediate [1/4]	22	3.1	5.5	5.6	6.0	6.2	6.9
	high [3]	14	3.5	4.7	5.5	5.5	5.8	8.9
Fattening pigs								
feed per fattening pig [kg]	**low [2]**	**20**	**209**	**222**	**269**	**242** ^{a}	**322**	**378**
	intermediate [1/4]	**21**	**159**	**266**	**381**	**337** ^{ab}	**385**	**981**
	high [3]	**14**	**291**	**323**	**384**	**390** ^{b}	**443**	**494**
percentage of bought-in feed stuff in fattener diets	**low [2]**	**18**	**20**	**58**	**79**	**100** ^{a}	**100**	**100**
	intermediate [1/4]	**19**	**30**	**100**	**87**	**100** ^{a}	**100**	**100**
	high [3]	**14**	**0**	**7**	**35**	**25** ^{b}	**69**	**100**
fattening pig FCR [kg feed/kg pig]	**low [2]**	**20**	**2.4**	**2.7**	**2.9**	**2.8** ^{a}	**3.0**	**3.3**
	intermediate [1/4]	**21**	**3.0**	**3.2**	**3.8**	**3.6** ^{b}	**4.2**	**6.0**
	high [3]	**14**	**3.1**	**3.3**	**3.7**	**3.8** ^{b}	**4.0**	**4.6**
average dietary content of								
ME [MJ//kg]	low [2]	20	9.5	12.5	12.3	12.8	12.9	13.2
	intermediate [1/4]	21	12.0	12.6	13.0	12.9	13.2	14.5
	high [3]	14	12.0	12.7	13.0	12.9	13.0	14.8
CP [g/kg]	low [2]	20	139	153	163	165	172	190
	intermediate [1/4]	21	112	144	162	171	183	202
	high [3]	14	118	153	157	158	164	177
total P [g/kg]	low [2]	20	2.4	3.6	4.5	4.9	5.1	6.0
	intermediate [1/4]	21	3.1	4.0	5.0	5.5	5.8	6.4
	high [3]	14	3.3	4.0	4.6	4.8	5.1	6.0

[1] Parameters that showed significant differences between clusters are highlighted in bold; ^{a,b} different superscript letters indicate differences between groups ($p < 0.05$, p-values adjusted according to Bonferroni correction for triple testing).

Table 10. Dietary proportion of feedstuff categories by cluster.

Feed Category [1]	Cluster Impact Category [Cluster Number]	n (PC)	Min.	Q25%	Mean	Median	Q75%	Max.
cereal grains	low [2]	19	0.0	50.0	53.2	61.8	67.8	74.5
	intermediate [1/4]	21	0.0	50.0	53.8	65.0	71.4	90.0
	high [3]	14	41.4	65.0	67.2	68.1	71.8	85.5
leguminous grains	**low [2]**	**19**	**0.0**	**16.4**	**20.9**	**25.0 [ab]**	**27.5**	**31.6**
	intermediate [1/4]	**21**	**0.0**	**7.5**	**16.3**	**17.4 [b]**	**21.9**	**44.4**
	high [3]	**14**	**14.5**	**20.5**	**26.0**	**23.1 [a]**	**31.0**	**44.5**
high-protein by-products	low [2]	19	0.0	2.0	5.3	4.7	8.4	13.1
	intermediate [1/4]	21	0.0	0.0	3.3	1.1	4.2	13.8
	high [3]	14	0.0	0.0	2.7	1.6	6.4	6.9
others	low [2]	19	0.0	0.0	7.2	3.0	6.8	80.2
	intermediate [1/4]	20	0.0	0.0	3.9	0.1	5.1	17.1
	high [3]	14	0.0	0.0	1.0	0.0	1.1	5.1
components of animal or microbial origin	low [2]	19	0.0	0.0	0.4	0.0	1.0	3.2
	intermediate [1/4]	21	0.0	0.0	2.3	0.0	0.0	42.3
	high [3]	14	0.0	0.0	0.0	0.0	0.0	0.0
minerals	low [2]	19	0.0	2.0	2.6	2.7	3.1	5.4
	intermediate [1/4]	21	0.0	0.0	1.6	1.6	2.8	5.6
	high [3]	14	0.0	2.0	2.2	2.4	2.9	4.0
compound feed	low [2]	19	0.0	0.0	10.5	0.0	0.0	100
	intermediate [1/4]	21	0.0	0.0	19.1	0.0	0.0	100
	high [3]	14	0.0	0.0	0.0	0.0	0.0	0.0
supplementary compound feed	low [2]	19	0.0	0.0	0.0	0.0	0.0	0.0
	intermediate [1/4]	21	0.0	0.0	0.0	0.0	0.0	0.0
	high [3]	14	0.0	0.0	1.0	0.0	0.0	11.4

[1] Parameters that showed significant differences between clusters are printed in bold; [a,b] different superscript letters within columns indicate differences between groups ($p < 0.05$, p-values adjusted according to Bonferroni correction for triple testing).

The high impact cluster showed a higher proportion of PCs feeding sufficient digestible lysine to finishing pigs, compared to the intermediate impact cluster (64.3% vs. 6.3%, respectively; $p = 0.001$; Table 11). There were no differences in the proportion of deficient or excess diets for growing and finishing pigs, but the low impact cluster still showed a high proportion of diets with sufficient digestible lysine content (41.2%). Across the growing and finishing diets, the low impact cluster showed the highest proportion of diets without deficient digestible lysine (82.4%).

Table 11. Classification of animal categories by cluster according to deficient, excess, or sufficient proportions of digestible lysine in the grower and finisher diets; N = total number of production chains per cluster.

Animal Category [1]	Status of Diet	Cluster According to Impact [Cluster Number]	n [2]	Frequency (n)	Percentage (%)
growers	dLys deficient [<0.72 g dLys/MJ NE]	low [2]	17	2	11.8
		intermediate [1/4]	15	6	40.0
		high [3]	14	3	21.4
	dLys excess [>0.88 g dLys/MJ NE]	low [2]	17	4	25.5
		intermediate [1/4]	15	2	13.3
		high [3]	14	4	28.6
	dLys sufficient [0.72–0.88 g dLys/MJ NE]	low [2]	17	11	61.7
		intermediate [1/4]	15	7	64.7
		high [3]	14	7	50.0

Table 11. *Cont.*

Animal Category [1]	Status of Diet	Cluster According to Impact [Cluster Number]	n [2]	Frequency (n)	Percentage (%)
finishers	dLys deficient [<0.63 g dLys/MJ NE]	low [2]	17	2	11.8
		intermediate [1/4]	16	7	43.8
		high [3]	14	3	21.4
	dLys excess [>0.77 g dLys/MJ NE]	low [2]	17	8	47.1
		intermediate [1/4]	16	8	50.0
		high [3]	14	2	14.3
	dLys sufficient [0.63–0.77 g dLys/MJ NE]	**low [2]**	**17**	**7**	**41.2** ^{ab}
		intermediate [1/4]	**16**	**1**	**6.3** ^{b}
		high [3]	**14**	**9**	**64.3** ^{a}

[1] Parameters that showed significant differences between clusters are printed in bold; [2] classification of diets was not possible for all PCs; ^{a,b} different superscript letters indicate differences between groups ($p < 0.05$, p-values adjusted according to Bonferroni correction for triple testing).

4. Discussion

4.1. General Discussion and Comparison with Other Studies

Both hypotheses were confirmed: For all environmental impact categories, variation was greater within than between systems, with no distinct differences between husbandry systems regarding their environmental impact. Furthermore, farm characteristics contributing to environmental impact were identified, and they were mainly related to performance (e.g., piglets weaned per sow and year, fattening pig feed conversion ratio) and feed characteristics (e.g., percentage of bought-in feed stuff in fattener diets; see Section 4.2).

The selected classification was chosen according to [1], who found that specific characteristics (e.g., farm size, breeds, live-weights at slaughtering, etc.) of organic farms suggest this differentiation into the three husbandry systems. It may be used to inform the organic sector regarding the strengths and weaknesses of these three widely used systems' variants, and the results may serve as a basis for system-specific mitigation strategies. The large range of results indicates potential for improvement within all three systems. Cluster analysis was used to investigate further classification. It was interesting to find high, medium, and low impact clusters independently from husbandry systems, but with characteristics regarding feeding and performance, which are mainly affected by management decisions.

Comparing different LCA studies should always be done with caution due to differences in underlying methodology, assumptions, and chosen system boundaries [45]. The present study addressed, for the first time, the environmental impact of different organic pig husbandry systems based on data obtained from 64 individual farrow-to-finish production chains across several European countries. In accordance with [9], there was a high variation in environmental impact among farms in the present study, indicating that individual farm characteristics highly influenced the outcome.

Other authors report comparable GWP, but rather different AP and EP (Table 12). A possible reason is that the GWP originating from the main emission source feedstuffs does not differ widely, except for those loaded with high emissions from land use change. However, in organic systems, comparably low emissions occur from land use change. Contrarily, emission factors regarding AP and EP show a higher variation between animals kept indoors or outdoors. Additionally, methodological differences between studies (e.g., system boundaries, emission factors, type of allocation) could have contributed to the great variation.

In general, most of our results are within the ranges reported in the literature, e.g., as provided in a review of six comparable LCA studies of pork products [4] (Table 12). The authors also found a large variation, especially regarding acidification and eutrophication potential.

Our results are particularly comparable to results reported by [15], who used a similar methodology and system boundaries for an evaluation of 15 pig farming systems from five European countries. For each pig farming system, data from five to 10 farms were obtained from surveys, and systems were categorized into conventional, adapted conventional, traditional, and organic. Feed production contributed less to EP in organic systems than in the others [15]. In the present study, the results regarding EP (medians ranging from 20.1 to 28.7 g PO_4-eq per kg live weight) exceed the values for organic farms found by [15]; 16 g PO_4-eq), and are in the range reported for conventional systems. The higher EP in the present study may, on the one hand, be due to a poorer median feed conversion ratio in IN, POUT, and OUT than found by [15]. On the other hand, due to the assumption of a 5% PO_4-surplus over yield-related plant requirements, our methodological approach might have resulted in higher PO_4-emission estimates from feed production.

Regarding AP, [15] found higher medians for organic systems than for POUT and OUT in the present study. The even higher median AP found in IN compared to [15] may be attributed to higher emissions from manure management (storage and spreading) as well as from feed, which resulted mainly from a poorer feed conversion ratio in fatteners in the present study compared to [15].

4.2. Factors Influencing the Environmental Impacts

In terms of the relative impact of feed on GWP, AP, and EP, results presented herein agree with those of other LCA studies [10,22,33,46,47]. Dolman et al. [9] calculated correlations between farm characteristics and environmental impact: All environmental indicators correlated positively with feed intake and dietary composition. Likewise, and in accordance with [22], who identified the feed conversion ratio in fatteners as the parameter with the greatest influence on LCA outcomes, in the present study, the feed conversion ratio was significantly lower for fatteners in the low impact cluster. The percentage of bought-in feed for fattener diets (%) was weakly negatively correlated with AP. This might indicate that these feeds either were loaded with relatively low AP or the feeds contributed to a better balanced diet and hence an improved feed conversion.

Regarding the relative contribution (%) of feedstuff categories to the environmental impact indicators, significant correlations were only found for EP. The proportion of high-protein by-products and minerals negatively correlated with EP, while the proportion of cereal grains was positively correlated. This might indicate that high-protein by-products and minerals contribute to a better balanced diet. Furthermore, for fed by-products, a relatively low proportion of impacts from crop production and processing is allocated to the co-product feed (e.g., molasses or oil cakes), and a rather high share to the main product, e.g., sugar or oil.

The average fattener carcass weight (kg) was positively correlated with AP and EP, as during the fattening stage a large proportion of emissions arise. Towards the end of the fattening period the amount of feed needed for 1 kg of weight gain increases in comparison to earlier growing stages [42].

Furthermore, reproductive performance, especially the number of piglets born alive per litter, has been described as an important influencing factor [23,48]. However, as for all associations found in the present study, only weak correlations regarding sow fertility and the environmental impacts were found. The number of total born (still and live born) piglets per litter negatively correlated with EP. However, the correlation coefficient again indicated only a weak relationship, therefore, the results do not allow for strong conclusions. Emissions originating from the sows are allocated to their offspring, and, therefore, higher litter sizes may be regarded as beneficial if all piglets survive and grow adequately. However, in light of the effect of litter size on piglet mortality [20], the focus should be on an adequate litter size, with robust and viable piglets. This suggestion fits to the number of piglets weaned per sow and year being negatively associated with all environmental impact indicators, which may indicate that farms with good management and productivity are at an advantage.

Table 12. Greenhouse gas emissions (GWP, kg CO_2-eq/FU), acidification potential (AP, g SO_2-eq/FU), and eutrophication potential (EP, g PO_4-eq/FU) from selected LCA studies on pig production (FU = Functional unit).

Source	System/Study Case	FU	GWP	AP	EP
present results	IN (indoor)	kg live weight	2.2	61.9	21.6
present results	POUT (partly outdoor)	kg live weight	2.2	51.9	20.1
present results	OUT (outdoor)	kg live weight	2.2	55.4	28.7
[4]	literature review: range across different conventional systems	kg pork	3.9–10	43–741	up to 20
[49]	organic (outdoors)	kg live weight	4.0	92.8	4.1
[15]	range across conventional, adapted conventional, traditional, organic	kg live weight	2.2–3.4	44-57	16–34
[15]	organic	kg live weight	2.4	57	16
[24] [1,2]	conventional, base scenario	kg live weight	2.4	44.8	16.1
[10]	good agricultural practice	kg live weight	2.3	21	44
[10]	'Label Rouge'	kg live weight	3.5	23	17

[1] Results were recalculated to the functional unit of 1 kg of live weight by using the carcass yield mentioned in the studies; [2] recalculated excluding the slaughter process.

Two farm characteristics describing the size of the pig units (number of slaughtered pigs, livestock units per PC) negatively correlated with AP and EP, with larger PCs having lower AP and EP. The current number of sows per PC was also weakly negatively correlated with AP, in line with the results for other farm size-related characteristics. These results may indicate that larger farms who are more efficient in managing their pigs, eventually connected with better trained farm staff. Similarly, negative correlations were found between the average number of fattening pigs and environmental impact indicators in [9], but this relationship is not considered as causal and may have been related to potentially better management in larger farms.

4.3. Limitations of the Method and Uncertainties

The present study used primary data from a non-representative sample of farms instead of modelled data based on assumptions. Therefore, sensitivity analyses, which are used for specific scenarios, were not applied here. Besides the influence of the PC-specific characteristics, LCA results depend on underlying methodological assumptions (e.g., emission factors for different floor types, allocation approach [49]). The degree of uncertainty for estimates revealed from these assumptions might vary between PCs, and estimates may therefore not be fully appropriate for the diverse husbandry systems in organic pig farming. A Monte Carlo analysis for the calculation of uncertainties of primary data as well as emission factors, describing the uncertainty areas of GWP, AP, and EP results, was not applied in the present study. In future studies on organic pig production, an analysis of the uncertainties with data from a representative sample of farms should be implemented. In addition, expanded system boundaries (including, for instance, the impacts of infrastructure) should be considered in future studies.

In case of emissions from feed production, uncertainty also arises from the fact that all countries' estimates were based on data for organic feed production in Austria. This approach was considered as the best option, as comprehensive region-specific data for the production of organic feedstuffs is not available and calculations of these would have been beyond the project resources.

Furthermore, farm-specific primary data, such as the feed conversion ratio, may be a source of uncertainty. The calculation of the feed conversion ratio is based on feed use (kg feed fed per day) as reported by the farmers; losses of feed, occurring especially outdoors, were probably not taken into consideration by some farmers. Therefore, the calculated feed conversion ratio might be higher (i.e., poorer) than the true feed conversion ratio. However, in terms of environmental impact, the amount of feed should include feed losses.

These limitations offer possible improvements for future studies. As the limitations concern all husbandry systems, they are not expected to affect the analysis of differences between husbandry systems. Furthermore, LCA takes numerous aspects into account; if weaknesses concern single aspects, this does not reduce the meaningfulness of the calculation. Finally, it must be mentioned that, despite these limitations, the data available for the present PC-specific cradle-to-farm gate LCA with a high number of organic pig farms included and taking different husbandry systems into account are unique.

5. Conclusions

Regarding environmental impact, a substantial variation was found between individual PCs. The ranking of the husbandry systems was not consistent regarding environmental impact; whereas the (median) GWP was similar in all systems; POUT had less AP than IN, and less EP than OUT.

The huge variation among PCs indicates that LCAs based on mean values of model scenarios will not necessarily be representative for individual farms, which may be subsumed in the modelled scenario. This reflects the importance of farm specific cradle-to-farm-gate assessments, rather than generalized scenarios for identifying the extent and the main sources of environmental impact in the different husbandry systems (IN, POUT, OUT).

In all husbandry systems, PCs with low environmental impacts were found, indicating that IN and POUT as well as OUT may be managed in an environmentally friendly way. However, a lack of consistent differences between husbandry systems, as well as results from cluster analysis, indicate that factors other than the husbandry systems affect the environmental impact of organic pig production.

Feeds generally constitute an important source for environmental impact, with the feed conversion ratio of fattening pigs being particularly important. Based on cluster analysis, an appropriate dietary digestible lysine content as well as buying in feedstuffs for supplementation of protein or minerals appears to be beneficial for a low environmental impact.

Furthermore, manure management (storage and spreading) was identified as a main source of emissions. Measures, such as covering slurry tanks and direct application of manure to the soil, offer mitigation options.

The results indicate that an overall good farm management and adequate productivity reduces the environmental impacts, and hence optimization of management should be the focus irrespective of the production system.

Supplementary Materials: The following are available online at http://www.mdpi.com/2071-1050/10/10/3796/s1, Equation S1: P-losses into water bodies, Table S1: Emission factors for NH_3, N_2O, N_2 and NO_3 in outdoor paddocks, Table S2: Emission factors for housing and manure management systems depending on floor type and manure type, litter quality, storage period, spreading according to Rigolot et al. [29], Table S3: Characteristics of the animal production stages by system, Table S4: Characteristics of dietary nutrient content and feed consumption by system. The values reflect an average dietary content over all diets fed to the animal group. FCR = Feed conversion rate, n = Number of production chains, Table S5: Percentages of floor type for lactating and pregnant sows, weaners and fattening pigs kept in the systems indoor (IN) and partly outdoor (POUT), Table S6: Type and frequency of manure treatment by system, Table S7: Cluster characteristics regarding number of sows on PC at farm visit, average number of slaughtered fattening pigs/year and number of livestock units (LSU).

Author Contributions: G.R. provided a draft manuscript, which was revised and edited by S.H., S.E., C.W. and C.L., G.B., J.Y.D., W.Z. additionally reviewed advanced versions of the manuscript. All authors approved the final version of the manuscript. G.R., D.B., R.B., B.F. and A.P. collected data. G.R., S.H., G.B., R.B., J.Y.D., M.M. and W.Z. developed the methodological approach and conducted the analysis; J.Y.D. and S.H. provided the software. Conceptualization and funding acquisition of the project included D.B., S.D., J.Y.D., S.E., B.F., M.M., A.P. and C.L.

Funding: This research was funded by the FP7 ERA-Net, CORE Organic II project "ProPIG", i.e. the national CORE Organic II funding agencies of Austria, Switzerland, United Kingdom, Italy, France, Denmark, Germany, Czech Republic.

Acknowledgments: The authors would like to thank all involved farmers and the ProPIG partners Tine Rousing, University of Aarhus in Denmark and Gudrun Illmann, Czech University of Life Sciences Prague.

Conflicts of Interest: The authors declare no conflict of interest.

References

1. Früh, B.; Bochicchio, D.; Edwards, S.; Hegelund, L.; Leeb, C.; Sundrum, A.; Werne, S.; Wiberg, S.; Prunier, A. Description of organic pig production in Europe. *Org. Agric.* **2014**, *4*, 83–92. [CrossRef]
2. Lernoud, J.; Willer, H. Current statistics on organic agriculture worldwide: Organic area, producers, markets and selected crops. In *The World of Organic Agriculture. Statistics and Emerging Trends 2015*; Willer, H., Lernoud, J., Eds.; IFOAM: Bonn, Germany; FiBL: Frick, Switzerland, 2015; pp. 32–116. ISBN 978-3-03736-271-6.
3. Goodland, R. *Livestock and Climate Change: What if the Key Actors in Climate Change Are... Cows, Pigs and Chicken?* Worldwatch Institute: Washington, DC, USA, 2009.
4. De Vries, M.; de Boer, I.J.M. Comparing environmental impacts for livestock products: A review of life cycle assessments. *Livest. Sci.* **2010**, *128*, 1–11. [CrossRef]
5. Gerber, P.J.; Steinfeld, H.; Henderson, B.; Mottet, A.; Opio, C.; Dijkman, J.; Falcucci, A.; Tempio, G. *Tackling Climate Change through Livestock—A Global Assessment of Emissions and Mitigation Opportunities*; Food and Agriculture Organization of the United Nations: Rome, Italy, 2013.
6. Jacques, S. Science and animal welfare in France and European Union: Rules, constraints, achievements. *Meat Sci.* **2014**, *98*, 484–489. [CrossRef] [PubMed]
7. FAO. *SAFA Sustainability Assessment of Food and Agriculture Systems Guidelines Version 3.0*; Food and Agriculture Organization of the United Nations: Rome, Italy, 2014.
8. Guinée, J.B.; Gorrée, M.; Heijungs, R.; Huppes, G.; Kleijn, R.; Koning, A.; de Oers, L.; van Wegener Sleeswijk, A.; Suh, S.; Udo de Haes, H.A.; et al. *Handbook on Life Cycle Assessment. Operational Guide to the ISO Standards*; Kluwer Academic Publishers: Dordrecht, The Netherlands, 2002.
9. Dolman, M.A.; Vrolijk, H.C.J.; de Boer, I.J.M. Exploring variation in economic, environmental and societal performance among Dutch fattening pig farms. *Livest. Sci.* **2012**, *149*, 143–154. [CrossRef]
10. Basset-Mens, C.; van der Werf, H.M.G. Scenario-based environmental assessment of farming systems: The case of pig production in France. *Agric. Ecosyst. Environ.* **2005**, *105*, 127–144. [CrossRef]
11. Halberg, N.; Hermansen, J.E.; Kristensen, I.S.; Eriksen, J.; Tvedegaard, N.; Petersen, B.M. Impact of organic pig production systems on CO2 emission, C sequestration and nitrate pollution. *Agron. Sustain. Dev.* **2010**, *30*, 721–731. [CrossRef]
12. Olea, R.; Guy, J.; Edge, H.; Stockdale, E.A.; Edwards, S.A. Pigmeat supply chain: Life Cycle Analysis of contrasting pig farming scenarios. *Asp. Appl. Biol.* **2009**, *95*, 91–96.
13. Williams, A.G.; Audsley, E.; Sandars, D.L. *Determining the Environmental Burdens and Resource Use in the Production of Agricultural and Horticultural Commodities*; Main Report. Defra Research Project IS0205; Cranfield University and Defra: Bedford, UK, 2006.
14. Jakobsen, M.; Preda, T.; Kongsted, A.; Hermansen, J. Increased Foraging in Outdoor Organic Pig Production—Modeling Environmental Consequences. *Foods* **2015**, *4*, 622–644. [CrossRef] [PubMed]
15. Dourmad, J.Y.; Ryschawy, J.; Trousson, T.; Bonneau, M.; Gonzàlez, J.; Houwers, H.W.J.; Hviid, M.; Zimmer, C.; Nguyen, T.L.T.; Morgensen, L. Evaluating environmental impacts of contrasting pig farming systems with life cycle assessment. *Animal* **2014**, *8*, 2027–2037. [CrossRef] [PubMed]
16. IFOAM. *The IFOAM Norms for Organic Production and Processing Version 2014*; IFOAM: Bonn, Germany, 2014; ISBN 978-3-944372-10-5.
17. Algers, B.; Blokhuis, H.J.; Bøtner, A.; Broom, D.M.; Costa, P.; Domingo, M.; Greiner, M.; Hartung, J.; Koenen, F.; Müller-Graf, C.; et al. Scientific opinion of the Panel on Animal Health and Welfare on a request from the European Commission on species-specific welfare aspects of the main systems of stunning and killing of farmed carp. *EFSA J.* **2009**, *1013*. [CrossRef]
18. The European Parliament and the Council of the European Union. *Regulation (EU) 2018/848 of the European Parliament and of the Council of 30 May 2018 on Organic Production and Labelling of Organic Products*; The European Parliament and the Council of the European Union: Brussels, Belgium, 2018.
19. Edwards, S.A.; Prunier, A.; Bonde, M.; Stockdale, E.A. Special issue—Organic pig production in Europe—Animal health, welfare and production challenges. *Org. Agric.* **2014**, *4*, 79–81. [CrossRef]
20. Prunier, A.; Dippel, S.; Bochicchio, D.; Edwards, S.; Leeb, C.; Lindgren, K.; Sundrum, A.; Dietze, K.; Bonde, M. Characteristics of organic pig farms in selected European countries and their possible influence on litter size and piglet mortality. *Org. Agric.* **2014**, *4*, 163–173. [CrossRef]

21. Eriksen, J.; Hermansen, J.E.; Strudsholm, K.; Kristensen, K. Potential loss of nutrients from different rearing strategies for fattening pigs on pasture. *Soil Use Manag.* **2006**, *22*, 256–266. [CrossRef]
22. Reckmann, K.; Krieter, J. Environmental impacts of the pork supply chain with regard to farm performance. *J. Agric. Sci.* **2015**, *153*, 411–421. [CrossRef]
23. Garcia-Launay, F.; van der Werf, H.M.G.; Nguyen, T.T.H.; Le Tutour, L.; Dourmad, J.Y. Evaluation of the environmental implications of the incorporation of feed-use amino acids in pig production using Life Cycle Assessment. *Livest. Sci.* **2014**, *161*, 158–175. [CrossRef]
24. Nguyen, T.L.T.; Hermansen, J.E.; Mogensen, L. Fossil energy and GHG saving potentials of pig farming in the EU. *Energy Policy* **2010**, *38*, 2561–2571. [CrossRef]
25. The British Standards Institution. *Publicly Available Specification. PAS 2050: 2011 Specification for the Assessment of the Life Cycle Greenhouse Gas Emissions of Goods and Services*; British Standards Institution: London, UK, 2011.
26. Jespersen, L.M.; Baggesen, D.L.; Fog, E.; Halsnæs, K.; Hermansen, J.E.; Andreasen, L.; Strandberg, B.; Sørensen, J.T.; Halberg, N. Contribution of organic farming to public goods in Denmark. *Org. Agric.* **2017**, *7*, 243–266. [CrossRef]
27. Kool, A.; Blonk, H.; Ponsioen, T.; Sukkel, W.; Vermeer, H.M.; De Vries, J.W.; Hoste, R. *Carbon Footprints of Conventional and Organic Pork: Assessment of Typical Production Systems in the Netherlands, Denmark, England and Germany*; Wageningen UR: Wageningen, The Netherlands, 2009.
28. IPCC. *Guidelines for National Greenhouse Gas Inventories*; Eggleston Buendia, L., Miwa, K., Ngara, T., Tanabe, K.S., Eds.; IPCC: Geneva, Switzerland, 2006.
29. Rigolot, C.; Espagnol, S.; Pomar, C.; Dourmad, J.Y. Modelling of manure production by pigs and NH_3, N_2O and CH_4 emissions. Part I: Animal excretion and enteric CH_4, effect of feeding and performance. *Animal* **2010**, *4*, 1401–1412. [CrossRef] [PubMed]
30. Wenzel, H.; Hausschild, M.; Alting, L. *Environmental Assessment of Products: Methodology, Tools and Case Studies in Product Development*; Chapman & Hall: London, UK, 1997; Volume 1.
31. Nemecek, T.; Kägi, T. *Life Cyce Inventories of Swiss and European Agricultural Production Systems*; Final Report Ecoinvent Report v2.0., No. 15; Agroscope Reckenholz-Taenikon: Zürich and Dübendorf, Switzerland, 2007.
32. Brandhofer, R. Umweltwirkungen Biologischer Schweinehaltung: Vergleich Zweier Haltungssysteme auf Basis des Treibhausgas-Potenzials Sowie Einer Stickstoff und Phosphorbilanz. Master's Thesis, University of Natural Resources and Life Sciences, Vienna, Austria, 2014.
33. Basset-Mens, C.; van der Werf, H.M.G.; Robin, P.; Morvan, T.; Hassouna, M.; Paillat, J.M.; Vertès, F. Methods and data for the environmental inventory of contrasting pig production systems. *J. Clean. Prod.* **2007**, *15*, 1395–1405. [CrossRef]
34. Dourmad, J.Y.; Physiology of Adaptation, Animal Nutrition and Health, INRA, Saint-Gilles, France. Personal communication, 2013.
35. Hörtenhuber, S.; Lindenthal, T.; Amon, B.; Markut, T.; Kirner, L.; Zollitsch, W. Greenhouse gas emissions from selected Austrian dairy production systems—Model calculations considering the effects of land use change. *Renew. Agric. Food Syst.* **2010**, *25*, 316–329. [CrossRef]
36. Hörtenhuber, S.J.; Lindenthal, T.; Zollitsch, W. Reduction of greenhouse gas emissions from feed supply chains by utilizing regionally produced protein sources: The case of Austrian dairy production. *J. Sci. Food Agric.* **2011**, *91*, 1118–1127. [CrossRef] [PubMed]
37. Kolbe, H. Wasserbelastung in Abhängigkeit von der Landnutzung. Available online: http://orgprints.org/801/ (accessed on 19 October 2018).
38. Samuel-Fitwi, B.; Meyer, S.; Reckmann, K.; Schroeder, J.P.; Schulz, C. Aspiring for environmentally conscious aquafeed: Comparative LCA of aquafeed manufacturing using different protein sources. *J. Clean. Prod.* **2013**, *52*, 225–233. [CrossRef]
39. LCA Food database. Aalborg University and 2.-0 LCAconsultants. 2007. Available online: http://www.lcafood.dk/ (accessed on 19 Oct 2018).
40. Leeb, C.; Bernardi, F.; Winckler, C. *Einführung und Monitoring von BetriebsEntwicklungsPlänen (BEP) Tiergesundheit und Wohlbefinden in Österreichischen Bioschweinebetrieben*; Institut für Nutztierwissenschaften: Vienna, Austria, 2010.
41. Bayern, L. *Schweinefütterung am Ökobetrieb II—Fütterungsversuche, Fütterungsempfehlungen*; LfL Bayern: Poing, Germany, 2011.

42. Kirchgessner, M.; Roth, F.X.; Schwarz, F.J.; Stangl, G. *Tierernährung. Leitfaden für Studium, Beratung und Praxis*, 13th ed.; DLG Verlag: Frankfurt, Germany, 2011.
43. EvaPig®. A Calculator of Energy, Amino Acid and Phosphorus Values of Ingredients and Diets for Growing and Adult Pigs. Available online: http://www.evapig.com (accessed on 19 October 2018).
44. Alibert, L. Besoins Alimentaires des Animaux en Fonction du Stade Physiologique. Available online: http://www.itab.asso.fr/downloads/porc-bio/cahier_porc_0.pdf (accessed on 10 Dec 2014).
45. Lammers, P.J. Life-cycle assessment of farrow-to-finish pig production systems: A review. *CAB Rev. Perspect. Agric. Vet. Sci. Nutr. Nat. Resour.* **2011**, *6*, 1–8. [CrossRef]
46. Pelletier, N.; Lammers, P.; Stender, D.; Pirog, R. Life cycle assessment of high- and low-profitability commodity and deep-bedded niche swine production systems in the Upper Midwestern United States. *Agric. Syst.* **2010**, *103*, 599–608. [CrossRef]
47. Alig, M.; Grandl, F.; Mieleitner, J.; Nemecek, T.; Gaillard, G. *Life Cycle Assessment of Beef, Pork and Poultry*; Executive Summary; Reckenholz-Tänikon: Zürich, Switzerland, 2012.
48. Dalgaard, R.; Halberg, N.; Hermansen, J.E. *Danish Pork Production—An Environmental Assessment*; University of Aarhus: Tjele, Denmark, 2007; Volume 82.
49. Mackenzie, S.G.; Leinonen, I.; Ferguson, N.; Kyriazakis, I. Accounting for uncertainty in the quantification of the environmental impacts of Canadian pig farming systems Supplementary Material. *J. Anim. Sci.* **2015**, *93*, 3130–3143. [CrossRef] [PubMed]

sustainability

MDPI

Article

Effects of Alternative Uses of Distillery By-Products on the Greenhouse Gas Emissions of Scottish Malt Whisky Production: A System Expansion Approach

Ilkka Leinonen [1,*], Michael MacLeod [1] and Julian Bell [2]

1 Land Economy, Environment and Society Research Group, Scotland's Rural College (SRUC), Peter Wilson Building, Kings Buildings, West Mains Road, Edinburgh EH9 3JG, UK; michael.macleod@sruc.ac.uk

2 Rural Business Unit, SAC Consulting, 2 Technopole, Bush Estate, Penicuik, Midlothian EH26 0PJ, UK; Julian.Bell@sac.co.uk

* Correspondence: ilkka.leinonen@sruc.ac.uk; Tel.: +44-131-5354044

Received: 6 April 2018; Accepted: 4 May 2018; Published: 8 May 2018

Abstract: Agricultural by-products are an important component of livestock feed. In Scotland, distillery by-products are protein rich and traditionally cost competitive feed ingredients in cattle production. However, during recent years, distilleries in the UK (including Scotch whisky producers) have started to use the by-products also as a source of renewable energy, in order to reduce the carbon footprint of alcohol production. In this study, a systems-based material and energy flow analysis was performed to calculate the life-cycle greenhouse gas (GHG) emissions of whisky production for two scenarios where distillery by-products were used either (1) as beef cattle feed to replace other protein sources (namely soya bean meal and rapeseed meal); or (2) as anaerobic digester (AD) feedstock in order to generate renewable energy (heat and electricity). System expansion was used to quantitatively handle the by-products in the analysis. The results show that considerable reductions in GHG emissions could be achieved by either replacing feed crops with by-products or by using the by-products in AD plants to generate bio-energy. The biggest reductions in the GHG emissions were achieved when by-products were used to replace soya meal in animal feed. However, the results are highly sensitive to methodological choices, including the accounting method of the land use change emissions arising from soya production.

Keywords: agricultural by-products; whisky production; cattle; beef; livestock feed; renewable energy; greenhouse gas emissions

1. Introduction

Various agricultural by-products (i.e., products originating from food and drink production but not normally used for human consumption) are widely used as part of livestock feed. They can form an important source of protein and metabolizable energy needed for animal production [1–4] and are currently used in production of a wide variety of livestock species [5]. For example, brewery, spirit distillery and bio-ethanol production by-products, such as brewers' grains, usually made from barley, wheat, maize or rice, can form a significant part of the animal diet especially in ruminant production. These products are generally favoured due to their high protein and fibre content [6–10] and inclusion of them in ruminant feed has been found to have positive effects on the animal performance [11,12]. Brewers' grains, or further processed distillery by-product "distillers dark grains with solubles" or "dried distillers grains with solubles" (DDGS), can be also used in non-ruminant production, for example, as part of pig feed [13–17]. According to recent studies, relative high proportions of DDGS can be applied in pig diets while still maintaining acceptable level of growth performance [15,16,18].

Further application of brewery and distillery by-products such as DDGS and yeast products can also be found in chicken industry [19–23], or even in fish production [24,25]. However, high variability in the nutritional value of these by-products, depending for example on the variety of the cereal, harvest time and the malting and mashing processes [10,26,27] may limit their use in non-ruminant livestock production where the content of certain nutrients, especially the balance of essential amino acids, can have major impact on animal performance [28]. Despite such limitations, brewery and distillery by-products can be potentially seen as a partial replacement of soya as a protein source in livestock production [20,29,30]. Reducing the dependency on soya is an increasing trend in livestock systems due to environmental concerns related to its production. The cultivation of this crop is generally associated with recent land use changes, causing greenhouse gas emissions from deforestation and conversion of other land uses to arable production, resulting in loss of carbon previously stored in soil and biomass [31]. Other environmental issues associated with soya production include the loss of biodiversity and freshwater and groundwater contamination [32,33].

The availability and usability of agricultural by-products or co-products can vary strongly depending on the location of production and this can also affect the distribution and profitability of livestock industries. In agricultural production in Scotland, barley is by far the most important arable crop species, in terms of area of cultivation, the total mass of grain production and the amount of crop protein produced [34]. The barley grown in Scotland is used either directly for animal feed, or as a raw material of alcohol (including whisky) production. The whisky distillation process utilizes only the carbohydrates of the barley grains, so therefore the use of the remaining compounds (including protein) in livestock production has been seen as a useful way to fully utilize this important crop. As a result, the main by-products of Scottish malt whisky production, namely "draff" (unprocessed by-product of the mashing process), "pot ale" (liquid by-product of distillation) and "distillers dark grains with solubles" (DDGS, dried and pelletized product made of draff and pot ale) are widely used as protein rich and low-cost ingredients of livestock feed, particularly in beef and dairy cattle production. In 2012, it was estimated that a total of 346,000 t (on a dried product basis) of distillery by-products from the whisky industry in Scotland were potentially available for use in animal feed [35]. About 60% of this amount was consumed in Scotland. Therefore, in this specific region, whisky industry and the feed that it provides for livestock can be considered as a significant part of the beef and dairy production chains.

Distillery by-products used as feed have been considered to have low environmental impacts and especially low embedded greenhouse gas (GHG) emissions, compared to alternative protein sources such as rapeseed meal and imported soya [35,36]. One of the reasons for this finding is that in agricultural life cycle assessment (LCA) studies, economic allocation of the environmental burdens is generally applied to distribute the burdens between various co-products originating from the same production process [37–39]. In this specific method, the burdens are allocated to different co-products proportionally to their economic value [3,36]. Therefore, due to a relatively low price of the distillery by-products such as draff, the produced alcohol is considered to be the main product with the highest share of the environmental burdens and only a small proportion of the overall environmental impacts associated with whisky production is allocated to by-products [35].

In addition to the traditional use of distillery by-products as livestock feed, distilleries in Scotland (and other alcohol producers in the UK) have during recent years started to use these products as a source of renewable energy. This development is partly motivated by the government subsidies for such energy sources due to their low embedded GHG emissions. Furthermore, the reduction of the use of fossil fuels in the production process can also provide a greener image for the whisky industry through reducing its carbon footprint [40]. A recent trend in this area has been to increase the use of by-products as a feedstock of anaerobic digestion and utilize the produced biogas to generate electricity and heat that can be directly used in the distillation process. This strategy is expected to have multiple benefits in terms of reduction of GHG emission. For example, in addition to the renewable energy generated in digestion, the digestate (material remaining after the anaerobic digestion process) contains nutrients that can be utilized by crops and therefore can replace the use of synthetic fertilisers and reduce the greenhouse gas emissions related to production of the fertilisers [41,42].

The proposed alternative use of distillery by-products has raised some concerns amongst livestock farmers in Scotland and elsewhere in the UK [43]. As the use as renewable energy increases, the farmers are getting worried about the future availability and cost of by-products as animal feed. However, it should be noted that also the feed use of these by-products can be expected to have beneficial effects on greenhouse gas emissions due to reduction of the use of carbon-intensive feed ingredients. For this reason, a systematic comparison of GHG emissions associated with alternative uses of by-products is needed to quantify these possible effects.

The aim of this study was to carry out a systems-based analysis of the material and energy flows within the whisky production chain, in order to quantitatively analyse the processes associated with distillery by-products and assess the life-cycle greenhouse gas emissions arising from the entire whisky production chain, including the end use of by-products. Instead of applying an economic allocation approach for the outputs of the distillery processes, we selected to use a system expansion method for the by-products, specified as "expanding the product system to include the additional functions related to the co-products" in the ISO [44] LCA standards. Those standards also prefer the system expansion approach over any allocation methods and suggest that allocation between co-products should be avoided when possible. The system expansion approach makes it possible to directly compare the changes in greenhouse gas emissions when using the by-products for alternative purposes (livestock feed, bioenergy generation) and replacing alternative commodities (protein crops grown for cattle feed, fossil fuels used in distillation process, synthetic fertilisers used in crop production). As an outcome, the greenhouse gas emissions can be calculated for a unit of produced alcohol while taking into account both the burdens and benefits achieved through the end use of the by-products.

2. Materials and Methods

A systems-based material flow analysis linking the Scottish malt whisky production and the end use of its by-products either in a cattle production system or in energy generation processes was carried out (Figure 1) to quantify the life-cycle GHG emissions arising from whisky production. The analysis applied a system expansion approach, that is, whisky was considered as the main product of distilleries, the by-products of the process were used to replace alternative commodities and the changes in greenhouse gas emissions as a result of this replacement were accounted for when quantifying the overall emissions of whisky production. In order to achieve this, the whisky production system was linked to other systems affected by the by-product use, namely beef cattle production system, crop production system and energy generation system (see detailed system diagram in Supplementary Materials). The changes in the material and energy flows within these connected systems were analysed and the associated GHG emissions quantified as detailed below. This approach was used to calculate the greenhouse gas emissions for two baseline scenarios: Scenario 1: distillery by-products were used as cattle feed to replace either soya bean meal or rapeseed meal as a protein source (while maintaining the nutritional quality of the feed) and Scenario 2: distillery by-products were used as anaerobic digester (AD) feedstock in order to generate renewable energy (heat and electricity). For both of these scenarios, the functional unit was 1 litre of pure alcohol (LPA) and the system boundary was from cradle to the end of the distillation process. However, instead of the whisky production chain itself (which remained the same in both baseline scenarios), the main focus of this study was in the effects of alternative uses of the by-products on GHG emissions in other, connected systems, analysed through the system expansion approach.

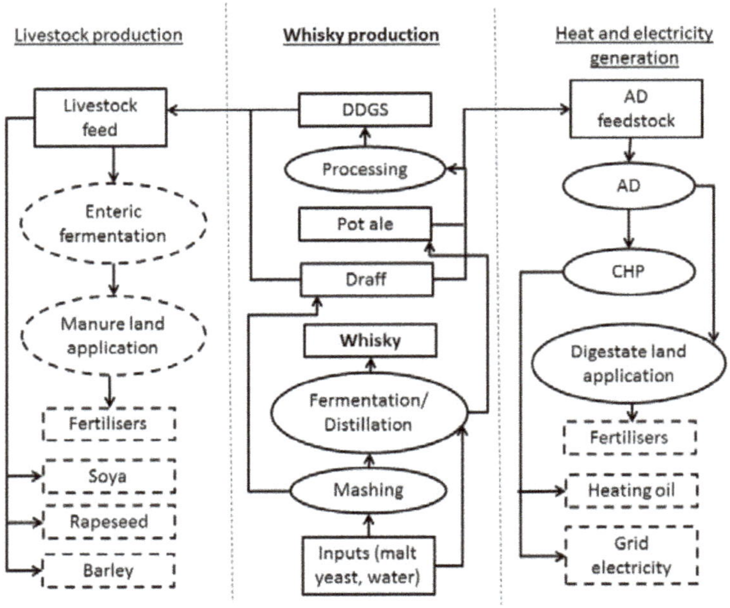

Figure 1. A simplified diagram of the production systems analysed in this study and processes affecting the greenhouse gas (GHG) emissions. The rectangles indicate inputs or outputs and the ovals indicate processes. The objects with broken lines indicate replaced products or processes partly affected by the replacement.

The greenhouse gas emissions associated with whisky production (including production, processing and transport the raw materials but excluding the by-product processing and their end use) were quantified based on data from an earlier study [35] and the changes of these emissions, as affected by the alternative uses of the by-products, were included in the calculations. Details of the data and the calculated GHG emissions associated with different processes within the whisky production chain can be found in the Supplementary Materials. Following this analysis, the processing and end use of the by-products was analysed in different scenarios, the changes in the material and energy flows within connected systems were quantified and the resulting increases or reductions of GHG emissions were included in the total emissions of the whisky production chain. For this analysis, the quantities of the primary by-products produced per one litre of pure alcohol were obtained from the calculations by Bell et al. [35] and were estimated to be 0.56 kg DM of draff and 0.36 kg DM of pot ale. In all the systems included in the calculations, the changes in the emissions of carbon dioxide (CO_2), methane (CH_4) and nitrous oxide (N_2O) were quantified and the overall greenhouse gas emissions were expressed in terms of CO_2 equivalent: with a 100-year timescale, where 1 kg CH_4 and N_2O are equivalent to 25 and 298 kg CO_2 respectively [45].

2.1. Scenario 1: Distillery By-Products as Livestock Feed

In this scenario, the amounts of replaced feed ingredients in livestock production were quantified, together with the resulting changes in GHG emissions associated with feed production, enteric fermentation, manure management and fertiliser use in crop production. Two alternative by-products of malt whisky distilleries were used as a protein source in cattle feed, namely draff and DDGS. Draff was used fresh, without further processing, while DDGS was produced as a combination of draff and pot ale and additional processing (drying and pelletizing) was needed in its production. The use of distillery by-products was assumed to replace two alternative protein sources widely used in Scottish cattle industry, namely soya bean meal and rapeseed meal. As a result, four different options in the feed use of the by-products were considered separately:

- Scenario 1a: draff is used to replace soya bean meal (and pot ale is disposed)
- Scenario 1b: draff is used to replace rapeseed meal (and pot ale is disposed)
- Scenario 1c: draff and pot ale are used to produce DDGS, which is used to replace soya bean meal, and
- Scenario 1d: draff and pot ale are used to produce DDGS, which is used to replace rapeseed meal

The amounts of the replaced alternative ingredients in animal feed were quantified based on a feeding strategy according to which the digestible protein and metabolizable energy contents of the feed were kept unchanged for all scenarios, although the intakes of the total combustible "gross" energy and the total (digestible + non-digestible) protein (and therefore also the total nitrogen intake) could vary between feeding options. Using this approach, also the animal performance was assumed to remain unchanged (for example, the same amount of meat was always produced per unit of energy or protein fed into the beef production system) and would therefore have no effect on the greenhouse gas emissions when compared between the scenarios. Because the protein content of the distillery by-products is lower than that of the replaced soya meal or rapeseed meal, on dry matter (DM) basis higher quantities of the by-products were needed, compared to the replaced ingredients. The higher DM quantity of the by-products brought also additional energy to the feed. Therefore, other changes in the feed were also needed to achieve equal energy content. This was done in this study by reducing the inclusion of barley grains (which are a typical source of energy in cattle feed in Scotland) when the distillery by-products were used in the feed. Linear optimization was used (with the constraints of equal protein and energy content) to determine the quantities of the replaced ingredients to equal 1 kg DM of the by-product for both draff and DDGS feeding scenarios.

The nutrient contents of different feed ingredients were based on the Feedipedia [46] database. Assuming unchanged digestible protein and energy intake, the nutritional data were used to quantify the differences in the animal intake of dry matter and nutrients when either distillery by-products or alternative ingredients were used in the feed and this information was used to quantify the changes of the greenhouse gas emissions arising from the livestock production system (see below).

The greenhouse gas emissions of the production of the replaced feed ingredients (soya bean meal, rapeseed meal, barley) were calculated using the Scottish Agricultural Emission Model (SAEM) [47], which is based on the FAO GLEAM livestock model [48,49]. Details of SAEM can be found in the Supplementary Materials. The emissions related to land use changes (LUC) arising from South American soya production were also included in the calculations following the FAO [48] approach and emission factors [47]. However, since there are plenty of uncertainties in the LUC emissions and differences in the methods for accounting for those in the context of agricultural production chains, alternative estimates of those emissions were evaluated in the sensitivity analysis (see below). The other replaced feed crops apart from soya, that is, rapeseed and barley, were assumed to be originated from "mature" agricultural land in the UK and therefore no LUC emissions were associated with those feed ingredients [50,51].

The changes in the greenhouse gas emissions arising from enteric fermentation and manure management, resulting from the changes in the feed composition, were also quantified and taken into account in the analysis. The calculation of these emissions was based on the intake of nitrogen and gross energy by the animal, applying the IPCC [45] Tier 2 approach and SAEM [47] emission factors (Table 1, Supplementary Materials). In this approach, the emissions of CH_4 are proportional to the feed gross energy intake of the animal and the N_2O and NH_3 emissions are proportional to the amount of nitrogen excretion (affected by the protein content of the animal feed intake).

Table 1. Emission factors and other constants used in the life cycle assessment (LCA) model.

Factor	Value	Source
GHG emissions from rapeseed production, kg CO_2e per kg DM	0.86	[47]
GHG from barley production, kg CO_2e per kg DM	0.325	[47]
LUC emissions from soya meal production, kg CO_2e per kg DM	3.25	[47,48]
Other GHG emissions from soya meal production, kg CO_2e per kg DM	0.83	[47]
GHG emissions from N fertiliser production, kg CO_2e per kg N	6.8	[53]
GHG emissions from P fertiliser production, kg CO_2e per kg P	3.3	[53]
GHG emissions from K fertiliser production, kg CO_2e per kg K	1.0	[53]
GHG emissions from UK grid electricity, kg CO_2e per kWh	0.465	[54]
GHG emissions from heating oil, kg CO_2e per kWh	0.267	[54]
Volatile solids of DM of distillery by-products	91%	[55]
CH_4 yield, m^3 per kg VS	0.385	[55]
Biogas CO_2 content	40%	[56]
AD/digestate CH_4 leakage	3%	[56]
Other digestate C losses	10%	[57]
Feed/AD feedstock DM carbon content	47.5%	[58]
Long-term (20 years) soil C storage of all soil C input	15%	[59]
CHP electricity yield of total CH_4 energy	36%	[56]
CHP heat yield of total CH_4 energy	40%	[56]
Electricity fed back to AD	12%	[56]
Heat fed back to AD	9%	[56]
Proportion of net heat utilized	60%	[56]
Digestate/manure N fertiliser replacement value	65%	[52]
Digestate/manure P fertiliser replacement value	100%	[52]
Digestate/manure K fertiliser replacement value	100%	[52]
CH_4 from enteric fermentation, kg CH_4 per MJ GE	0.0012	[45,47]
Direct manure management N_2O emissions, % N_2O-N of total N	0.2%	[45,47]
Indirect manure management N_2O emissions, % N_2O-N of total N	0.13%	[45,47]
Total manure management N losses: volatilized	12.5%	[47]
Total manure management N losses: leached	1.7%	[47]

The differences in the contents of main available plant nutrients, namely nitrogen (N), phosphorus (P) and potassium (K) in the manure, as affected by alternative scenarios, were calculated on the basis of nutrient intake and subsequent losses (assuming that the nutrient retention in animal body remained unchanged regardless of the feed ingredients used) and their benefits in reducing the greenhouse gas emissions by replacing synthetic fertilisers were quantified. The replacement rates of synthetic fertilisers applied in this study were based on the study by DeVries et al. [52]. The greenhouse gas emissions from the replaced synthetic fertiliser production were quantified by using the emission factors of the SAEM beef model [47,53].

Changes in the soil carbon (C) content (carbon sequestration) as a result of manure fertiliser use were not considered in the baseline calculations. However, the possible effects of changes of the carbon input to soil are evaluated in the sensitivity analysis (see below).

2.2. Scenario 2: Distillery By-Products as a Source of Renewable Energy

In this scenario, the amounts of replaced commodities were quantified when the by-products were used as an energy source. These commodities include grid electricity, fossil fuels used for heating and fertilisers used in crop production. The changes in GHG emissions associated with energy generation, management of the digestate and production of the fertiliser were then quantified as detailed below. Two by-products of whisky production, namely draff only (Scenario 2a) and a combination of draff and pot ale (Scenario 2b), were used as feedstock of anaerobic digestion (AD) to produce biogas. These scenarios were selected so that in terms of the input material, Scenario 2a was comparable to the feed use scenarios 1a and 1b (using draff directly as a feed ingredient) and Scenario 2b comparable to Scenarios 1c and 1d (using draff and pot ale to produce DDGS for feed). Unlike in Scenarios 1c and

1d, no further processing of by-products (needed to produce DDGS) was applied in Scenario 2b, as such processing would have no value in biogas production.

The produced biogas was assumed to be subsequently used to generate heat needed in mashing, fermentation and distillation processes and electricity to be fed to electrical grid, using the combined heat and power (CHP) technology.

The yield of CH_4 per kg of DM of distillery by-products was obtained from Luna-del Risco et al. [55] and the same value was used for both Scenario 2a and 2b (Table 1). The yields of electricity and heat generated in the CHP process, the proportions of the generated electricity and heat needed in the AD process and efficiency of the utilization of the net heat were obtained from an earlier study carried out at SRUC [56]. The heat obtained from the use of the biogas was assumed to replace part of the use of fuel oil, which is a commonly used fuel in Scottish malt whisky production as many distilleries are remote and not on the natural gas grid. Finally, the replacement of the grid electricity (i.e., the electricity fed to the grid minus the electricity used in the AD process) and the utilizable heat (net heat production minus the heat used in the AD process) were quantified, the associated greenhouse gas emissions were calculated based on the UK Department for Business, Energy & Industrial Strategy emission factors [54] and these avoided emissions were subtracted from the total greenhouse gas emissions of whisky production.

The direct and indirect greenhouse gas emissions arising from the AD and CHP processes and from the management of the digestate were also quantified and added to the emissions of whisky production. The methane leakage was based on values from an earlier study carried out at SRUC [56] and the N_2O and NH_3 emissions where based on the nutrient content of the feedstock, obtained from the Feedipedia [46] database and calculated in a same way as the emissions from manure management.

Similarly, as in the use of the by-products as feed, the fertiliser value of the nutrients obtained as an output of the AD process was quantified. All phosphorus and potassium of the feedstock were assumed to remain in the digestate and the nitrogen emissions described above were subtracted from the total nitrogen content of the digestate. The remaining nutrients could then be expected to replace synthetic fertilisers. It should be noted that part of the grid electricity is also generated from biomass (in 2016, this was estimated to be 7.6%, according to Ofgem [60] data) and was therefore assumed to replace a similar amount of fertiliser per kWh of generated electricity as the AD system considered here. To avoid double counting, a proportion equal to this share was subtracted from the applicable organic fertilisers produced in the AD. Similarly, as in Scenario 1, the changes in soil carbon content were not included in this baseline but they are evaluated in the sensitivity analysis.

A summary of the emission factors and other constants used in the calculations for different scenarios can be found in Table 1.

2.3. Sensitivity Analysis

2.3.1. LUC Emissions

In the baseline analyses, the emissions arising from the land use changes associated with the soya production were based on the FAO [48] default calculations methods, resulting in an emission factor of 3.25 kg CO_2 equivalent per kg DM soya bean meal [47]. However, alternative estimates of this value exist, so two alternatives were used here for comparison. First, recent calculations by Williams et al. [59] estimated the direct LUC emissions to be 1.46 kg CO_2e per kg DM soya bean meal (based on weighted average of the origins of soya used in the UK). Second, assuming that the soya would originate from sustainable sources, (i.e., no LUC associated with its production) and following the methods of PAS 2050 carbon footprinting guidelines [50,51], the LUC emissions would be zero.

2.3.2. Carbon Sequestration

Although soil carbon content is not normally considered in agricultural LCA, changes in the ecosystem carbon stock are included in the IPCC [45] guidelines for calculating GHG emissions.

For this reason, alternative calculations were carried out where the changes of soil carbon content in different scenarios were estimated. First, the amount of carbon in the digestate was determined based on the carbon content of the AD feedstock [58] and the carbon losses as CO_2 and CH_4 during the digestion process. For the animal feed, it was assumed that the dry matter excretion rate is inversely proportional to the organic matter digestibility, obtained from the Feedipedia [46] database and a constant carbon content of the excreted organic matter was used in the calculations [58]. Then, the changes of the carbon input entering soil (in the fertiliser use of digestate or manure) were quantified in each scenario. Finally, following the method introduced by Williams et al. [59] and assuming the same decomposition rate of organic matter as in their study, the proportion of the carbon remaining in the soil within a 20-year time scale (consistent with the IPCC [45] guidelines) was determined to be 15% of the total addition of carbon to soil. This was then converted to avoided CO_2 emissions and subtracted from the greenhouse gas emissions of whisky.

3. Results and Discussion

The potential of the distillery by-products to replace different feed ingredients in either soya meal based (Scenarios 1a and 1c) or rapeseed meal based (Scenarios 1b and 1c) feed is presented in Table 2. The results show that using an equal amount of the by-product, generally a bigger amount of rapeseed meal could be replaced than soya bean meal, due to the lower protein content of the former. It can be also seen that on the dry matter basis, DDGS is more efficient than draff in replacing alternative protein sources.

Table 2. The amounts of different feed ingredients in cattle feed replaced by the distillery by-products (Draff or dried distillers grains with solubles (DDGS)) in different scenarios. The quantities are based on equal content of metabolizable energy and digestible protein.

By-Product (1 kg DM)	Replaced Soya Meal, kg DM	Replaced Rapeseed Meal, kg DM	Replaced Barley, kg DM
Draff: Scenario 1a	0.28		0.41
Draff: Scenario 1b		0.44	0.32
DDGS: Scenario 1c	0.39		0.54
DDGS: Scenario 1d		0.60	0.42

Replacement of alternative feed ingredients resulted in considerable reductions in greenhouse gas emissions arising from feed production (Table 3). When comparing different alternative protein sources (soya and rapeseed), higher reductions could be achieved when soya bean meal was removed from the feed, due to avoided emissions that would arise from land use changes associated with soya production.

Table 3. Effects of replacing soya bean meal or rapeseed meal in cattle feed by distillery by-products (Draff or DDGS) on GHG emissions arising from different sources (kg CO_2 equivalent per 1 kg DM of by-product). Negative signs indicate reduction of emissions and positive signs increase of emissions.

Source of GHG Emission	Draff Replacing Soya (Scenario 1a)	Draff Replacing Rapeseed (Scenario 1b)	DDGS Replacing Soya (Scenario 1c)	DDGS Replacing Rapeseed (Scenario 1d)
Feed production	−1.288	−0.480	−1.773	−0.655
Manure N_2O	0.002	−0.001	0.003	−0.001
CH_4 enteric fermentation	0.063	0.049	0.14	0.066
N fertiliser replacement	−0.005	0.001	−0.007	0.001
P fertiliser replacement	0.001	−0.005	−0.016	−0.024
K fertiliser replacement	0.009	0.002	0.002	−0.007
Processing (DDGS)	0	0	0.518	0.518
Total	−1.219	−0.433	−1.133	−0.101

There were only minor changes in emissions directly related to livestock production, other than those arising from feed production, when distillery by-products were replaced with alternative feed ingredients (Table 3). The reason for that is that the inputs (gross energy intake, nutrient intake) affecting those emissions in the calculations based on IPCC Tier 2 method remained very similar in all feeding scenarios. This is expected, as the quantities of the feed ingredients consumed by the animals were specified so that the metabolizable energy intake and the digestible protein intake (affecting N emissions) remained constant in all scenarios. The differences in those emissions were thus a result of small differences in the contents of nutrients in the feed and different digestibility of energy and protein in different feed ingredients. These factors affected the avoided emissions related to the production of replaced fertilisers but the differences in those emissions were also rather small between the feeding scenarios with different feed ingredients.

The outputs of the scenario where the distillery by-products were used to generate renewable energy are presented in Table 4. In addition to the generated electricity and heat (replacing grid electricity and heating oil used in whisky production), considerable amounts of synthetic fertilisers could also be replaced when the digestate obtained from the anaerobic digestion would be used as a fertiliser in crop production.

Table 4. Outputs of use of distillery by-products as a source of renewable energy by applying a combination of anaerobic digester (AD) and combined heat and power (CHP) (per 1 kg of dry matter (DM) of the by-product).

Outputs per 1 kg DM by-Product	Draff (Scenario 2a)	Draff + Pot Ale (Scenario 2b)
Methane, m^3	0.351	0.351
Electricity to grid, kWh	1.146	1.146
Utilized heat, kWh (replacing oil)	0.790	0.790
Replaced N as fertiliser, kg	0.018	0.024
Replaced P as fertiliser, kg	0.003	0.009
Replaced K as fertiliser, kg	0.0003	0.009

Since the same conversion factor for methane yield was used for different types of distillery by-products when used as AD feedstock [55], there were no differences in the amount of energy generated per unit of DM between the options where draff only or a combination of draff and pot ale were used in AD. However, there were differences in the amount of fertiliser that could be replaced by the digestate, depending on the feedstock used. This was due to the high nutrient content of pot ale and for this reason, the digestate from the combination of draff and pot ale had a higher capacity to replace fertilisers than the digestate originating from draff only.

The biggest part of the reductions in the greenhouse gas emissions, when distillery by-products were used as AD feedstock, was associated with the replaced grid electricity (Table 5). However, considerable reductions were achieved also by using the digestate to replace synthetic fertilisers, most notably nitrogen fertilisers. Unlike in the Scenario 1, most of the nutrients entering to the AD process as part of the feedstock could actually be credited to whisky production. In contrast, in the feed use, the difference in the nutrient output between the by-products and the alternative feed ingredients was rather small. The reduction of the GHG emissions was also affected by the emissions of methane and nitrous oxide arising from the digestion process and storage of the digestate (Table 5).

Table 5. Effects of using distillery by-products as a source of renewable energy (by applying a combination of AD and CHP) on GHG emissions arising from different sources (kg CO_2 equivalent per 1 kg DM of by-product). Negative signs indicate reduction of emissions and positive signs increase of emissions.

Source of GHG Emission	Draff (Scenario 2a)	Draff + Pot Ale (Scenario 2b)
Electricity replacement	−0.549	−0.549
Oil replacement	−0.25	−0.25
N fertiliser replacement	−0.12	−0.159
P fertiliser replacement	−0.011	−0.03
K fertiliser replacement	−0.000	−0.009
N_2O emissions	0.051	0.063
CH_4 emissions	0.176	0.176
Total	−0.703	−0.759

Total GHG emissions per volume unit of produced alcohol were 2.6 kg CO_2 when the end use of the by-products was excluded from the calculations [35]. The changes in these emissions when the draff obtained from the mashing process was used either as animal feed or as AD feedstock are presented in Table 6 and Figure 2. It can be seen that the highest overall reductions were achieved when draff was used to replace soya in animal feed. However, nearly as high reduction could be achieved when draff was used to generate renewable energy.

Table 6. Effects of the alternative uses of draff on the GHG emissions of Scottish malt whisky production. The following scenarios are compared: Scenario 1a: draff replacing soya bean meal and barley, Scenario 1b: draff replacing rapeseed meal and barley and Scenario 2a: draff replacing grid electricity and oil.

	Amount of Draff Produced, kg DM per Litre of Alcohol	Effect on GHG Emissions, kg CO_2e per Litre of Alcohol (Relative Change in Parentheses)	Final GHG Emissions per Litre of Alcohol, kg CO_2e
Scenario 1a: Replacing soya	0.56	−0.68 (−26%)	1.92
Scenario 1b: Replacing rapeseed	0.56	−0.24 (−9%)	2.36
Scenario 2a: Renewable energy	0.56	−0.39 (−15%)	2.21

When comparing the alternative uses of DDGS (or a combination of draff and pot ale), the patterns were somewhat different (Table 7, Figure 2). Compared to draff, the combination of draff and pot ale was able to produce higher reductions in GHG emissions when used either to replace soya in animal feed, or to generate renewable energy. The reason for this is partly in the larger amount of the by-product utilised per unit of produced alcohol and also in the high nutrient content of pot ale. This enabled the combination of draff and pot ale to replace higher amount of alternative protein sources in animal production, or replace more fertilisers when the digestate is used in crop production. In contrast, only minimal reductions in the GHG emissions were achieved when DDGS was used to replace rapeseed meal in animal feed. This is due to the fact that the benefits achieved by reducing the amount of rapeseed in feed are counteracted by the high energy use of producing DDGS from the primary by-products draff and pot ale.

Table 7. Effects of the alternative uses of combination of draff and pot ale on the GHG emissions of Scottish malt whisky production. The following scenarios are compared: Scenario 1c: DDGS replacing soya bean meal and barley, Scenario 1d: DDGS replacing rapeseed meal and barley and Scenario 2b: draff + pot ale replacing grid electricity and oil.

	Amount of Draff + Pot Ale Produced, kg DM per Litre of alcohol	Effect on GHG Emissions, kg CO_2e per Litre of Alcohol (Relative Change in Parentheses)	Final GHG Emissions per Litre of Alcohol, kg CO_2e
Scenario 1c: Replacing soya	0.92	−1.04 (−40%)	1.57
Scenario 1d: Replacing rapeseed	0.92	−0.09 (−3%)	2.51
Scenario 2b: Renewable energy	0.92	−0.70 (−27%)	1.91

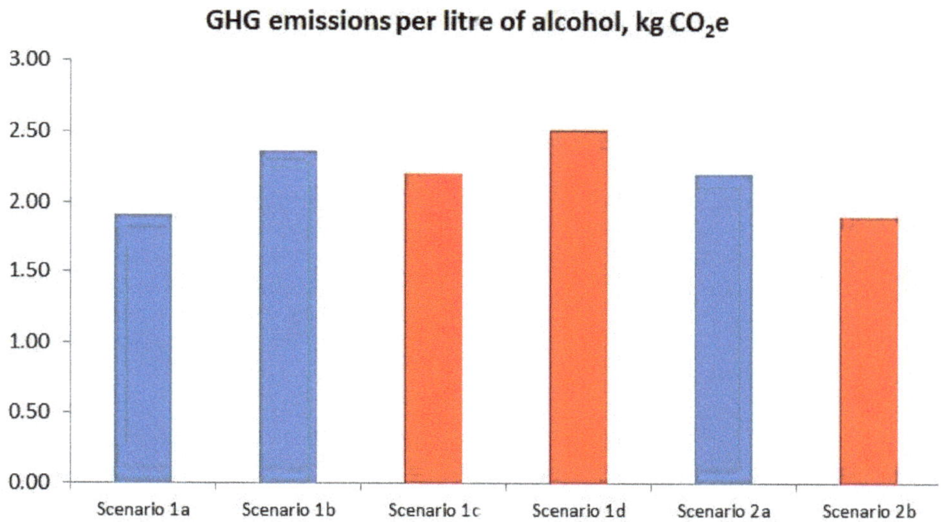

Figure 2. GHG emissions of Scottish malt whisky production with different end uses of by-products. Scenarios applying draff only are presented in blue colour and scenarios applying both draff and pot ale in red colour. Scenario 1a: draff replacing soya bean meal and barley, Scenario 1b: draff replacing rapeseed meal and barley, Scenario 1c: DDGS replacing soya bean meal and barley, Scenario 1d: DDGS replacing rapeseed meal and barley, Scenario 2a: draff replacing grid electricity and oil and Scenario 2b: draff + pot ale replacing grid electricity and oil.

The sensitivity analysis shows that the method used for quantifying the GHG emissions associated with the land use changes has a major effect on the comparison between different uses of the by-products and on conclusions drawn from this comparison. It can be seen in Table 8 that when lower LUC emissions are applied for soya, the benefits of by-product feed use in reduction of the GHG emissions are largely lost and the use of the by-products for heat and electricity generation becomes the most efficient option. In fact, in the option where zero LUC emissions were assigned to soya production, the use of DDGS to replace soya became the least efficient way to reduce the GHG emissions of whisky production, while in the baseline scenario it resulted in the highest reductions.

Including the carbon sequestration (i.e., changes in soil carbon content as a result of the alternative uses of the by-products) also affected the results. In general, when such changes were accounted for, the efficiency of energy generation option in reducing the GHG emissions increased. This is because the net input of C to soil increases when digestate is used as a fertiliser. However, an increase in soil carbon was also observed when the by-products were used as animal feed. This increase was a result of the lower digestibility (i.e., higher proportion of non-digestible carbon compounds per unit of digestible energy) of the by-products and hence higher dry matter excretion when the by-products were used as feed, compared to the alternative feed ingredients barley and soya/rapeseed.

Table 8. The results of the sensitivity analysis showing the GHG emissions (kg CO_2e per litre of alcohol) with different assumptions of land use changes (LUC)-related emissions and carbon sequestration. Scenario 1a: draff is used to replace soya; Scenario 1b: draff is used to replace rapeseed; Scenario 1c: DDGS is used to replace soya; Scenario 1d: DDGS is used to replace rapeseed meal; Scenario 2a: draff is used to replace grid electricity and oil; Scenario 2b: draff and pot ale are used to replace grid electricity and oil.

	Scenario 1a	Scenario 1b	Scenario 1c	Scenario 1d	Scenario 2a	Scenario 2b
Baseline	1.92	2.36	1.57	2.51	2.21	1.91
Alternative LUC emissions (Williams et al. 2016)	2.21	2.36	2.21	2.51	2.21	1.91
No LUC emissions	2.44	2.36	2.73	2.51	2.21	1.91
Including carbon sequestration	1.88	2.33	1.53	2.50	2.16	1.83

General Discussion

Overall, the results confirm the expectations of the whisky industry that the carbon footprint of whisky production can be reduced considerably when the by-products are used to generate heat and electricity and therefore the use of fossil fuels in the distillation process can be reduced. However, the by-products can also have other, indirect effects outside the distillery system, affecting greenhouse gas emissions. To understand and quantify such effects, other systems, in addition to whisky production itself, need to be included in the analysis and this can be done by applying a systems-based modelling approach.

In general, when the greenhouse gas emissions related to agricultural by-products (or co-products) are considered using LCA or other environmental assessment methods, the methodological choices related to allocation of the environmental burdens between those products often become a central issue of the analysis. Unfortunately, such choices can be very much subjective and they can also strongly affect the outcomes of the study in question. In studies related to livestock production and especially animal feeding, the use of various by-products as feed ingredients has often been found as an environmentally friendly option [3,36]. The reason for this that amongst different potential approaches to handling the co-products in environmental impact assessment, economic allocation has been widely used in agricultural LCA studies [38,39,61]. As an outcome of such analyses, smaller proportions of the environmental impacts are allocated to low value by-products, compared to the "main" product. In many cases, such an approach is justified and preferred, keeping in mind that the main product, not the by-products, is actually the driving force of the production.

Despite the general applicability of the economic allocation, alternative allocation methods have also been suggested for agricultural LCA. The idea behind this is that the ISO [44] standards prefer the use of a "causal" or "physical" allocation method, in cases where allocation cannot be avoided altogether for example through system separation. Another reason for avoiding economic allocation is the potentially varying prices of the co-products, which has sometimes been considered problematic and causing inconsistencies in LCA studies. However, alternative methods, such as system separation, or so called "biophysical allocation", where the aim is to link the inputs to outputs through actual physical flows or causalities [61–65] are problematic due to the unique nature of agricultural LCA. The fact is that agricultural products are always an outcome of complicated biological processes with various interactions. Therefore, attempts to physically separate the processes behind each co-product are not meaningful as such an approach would necessarily be based on arbitrary, subjective decisions [66]. In addition to attempts to model the physical flows, other "physical" allocation methods, used to avoid economic allocation, can be based for example on mass (fresh or dry matter) or protein or energy content of the products. Again, using such methods in agricultural LCA in a systematic way can be very difficult, due to potentially varying end uses of the by-products and possible difficulties in distinguishing between actual products and waste materials, for example, in case of manure [66].

To avoid allocation, in cases where system separation is not possible, the ISO [44] standards recommend using a system expansion approach. System expansion is also preferred over co-product allocation in various carbon footprinting guidelines such as PAS 2050 [50]. Over recent years, this approach has been widely applied for example in LCA studies on livestock production [67,68]. However, as pointed out by Mackenzie et al. [66], the use of this approach cannot be considered as becoming a general practice in agricultural LCA, for example due to possible difficulties in identifying the main product and by-products. Another difficulty with this approach is that it could require large amount of additional data from other sub-processes or systems [66,69,70]. Despite possible shortcomings, the use this method can be considered to be justified in many cases and especially when 1) the process in question is clearly targeted to produce one specific main product and 2) the focus of the analysis is in alternative uses of the emerging by-products, not in the production process itself. This was the case in this study and in fact, it would be difficult to systematically compare the very different uses of whisky by-products by applying any other of those alternative methods mentioned above.

The system expansion (usually as part of so called "consequential LCA" approach [71]) has been earlier used in other studies evaluating the environmental consequences of alternative uses of agricultural by-products, including the use of those products in generating renewable energy. For example, Styles et al. [57] used a consequential approach to quantify the reduction of the GHG emissions when manure and food waste are used in AD in a dairy farm. Their conclusion was that the achieved benefits were dependent on how much crops were used as co-digestate in addition to the manure, as this would determine the need for animal feed imported to the farm, which had its own effect on the feed-related GHG emissions. In another study, van Zanten et al. [72] applied consequential LCA to compare uses of two by-products, namely wheat middlings and beet tails. In the case of the beet tails, the alternative use also in their study was bioenergy generation using AD, as opposed to using them as cattle feed.

Williams et al. [59] assessed the effects of alternative end uses of turkey litter on the environmental impacts of turkey meat production. Also in that study, a system expansion approach was used in order to compare the use of litter either as a biofuel (used for electricity generation in a large-scale power plant) or directly as a fertiliser through land spreading. Similarly, as in the current study, the availability of nutrients for crop production, obtained from the litter, was quantified for both scenarios and the displacement of the synthetic fertilisers (and the grid energy in the case of bioenergy use) was taken into account when calculating the total emissions related to turkey production. That study also demonstrated the potentially significant effect of carbon sequestration on the overall GHG emissions arising from agricultural production.

4. Conclusions

The most notable reductions of greenhouse gas emissions of whisky production were achieved when the distillery by-products replaced soya meal, the production of which is associated with land use changes. When using the by-products to replace alternative feed ingredients, there were changes also in emissions from enteric fermentation, manure management and the end use of manure and its potential to replace synthetic fertilisers but these had only minor effect on the overall greenhouse gas emissions, compared to effect achieved by changes in the production of the feed and related LUC emissions. Different by-products had different environmental effects when used as livestock feed. Compared to draff, DDGS could reduce more greenhouse gas emissions related to production of the replaced feed ingredients but on the other hand, it had higher emissions arising from processing, especially from drying and pelletizing.

The use of by-products as a source of renewable energy, that is, production and combustion of biogas, reduced the greenhouse gas emissions by replacing grid electricity and fossil fuels used for heating. Additional benefits were achieved by using the digestate as a fertiliser and thus replacing the production of synthetic nitrogen fertilisers. Overall, the energy use of by-products could produce reduction of greenhouse gas emissions with similar magnitude as the use of the by-products as feed.

It should be also noted that in the case of this study, similarly as in earlier studies on agricultural by-products discussed above, the outcome is largely dependent on methodological choices. The sensitivity analysis demonstrates that the calculated reductions in the GHG emissions are strongly affected by the method how the emissions associated with land use changes and land management are accounted for in the calculations. Currently, there is no generally accepted approach to this process and this is causing difficulties in comparing the results of agricultural LCAs, especially in cases where the products are strongly associated with LUC. In assessment of greenhouse gas emissions in agricultural production, probably the most important single product where direct LUC is involved is soya bean meal imported from South America. This has been demonstrated for example by Leinonen et al. [51], who applied LCA modelling to quantify the environmental impacts of UK poultry production systems with alternative scenarios using different protein crops as a basis of diet formulations for broilers and laying hens. The general conclusion of their study was that inclusion of alternative protein sources (e.g., beans or peas) to replace soya bean meal could slightly reduce the Global Warming Potential of broiler and egg production but also in that study, this observed reduction was highly dependent on the LUC emission accounting method.

A high sensitivity to the LUC emissions was also found by van Zanten et al. [72]. According to their findings, higher reductions in GHG emissions could be achieved with the feed use of beet tails, through displaced emissions related to barley production, when compared to the bioenergy use. However, this was only the case when indirect LUC emissions were accounted for in barley productions, which is not a common practice in agricultural LCA studies [73]. If these emissions were excluded, higher reductions in the emissions could be achieved through the bioenergy option [72]. To avoid these methodological problems, some attempts have been made to harmonize the LUC accounting methods in agricultural LCA [73,74]. Despite this, current inconsistencies in the methods will necessarily remain a challenge in interpretation of the results of studies on environmental sustainability of agricultural products, especially in the context of livestock systems.

In addition to carbon losses due to land use changes, agricultural systems can also contribute to carbon sequestration. Although a large proportion of the carbon in the UK soils has been lost during recent decades, partly as a result of agricultural practices [75,76], this trend can be partially reversed for example, by the use of organic fertilisers. Although the changes in soil carbon (other than those related to land use changes) are not usually considered in agricultural LCA, we explored this process in the sensitive analysis of this study, using a similar approach as Williams et al. [59]. In general, soil carbon balance is potentially an important component of the overall effect of agricultural and bioenergy systems on global GHG emissions and therefore should not be ignored in such studies.

Supplementary Materials: The following supplementary materials are available online at http://www.mdpi.com/2071-1050/10/5/1473/s1, 1. Diagram of the production systems and material flows, 2. Quantifying GHG emissions from malt whisky production and 3. Outline of the Scottish Agricultural Emission Model (SAEM).

Author Contributions: All authors contributed to the design of the study, analysis of the data and writing the paper. I.L. led the writing process and data analysis, M.M. led the development of the livestock modelling tool and J.B. provided unpublished data for the analysis.

Acknowledgments: This research was undertaken within the Scottish Government Rural Affairs and the Environment Portfolio Strategic Research Programme 2016–2021, WP 1.4 "Integrated and Sustainable Management of Natural Assets" and 2.4 "Rural Industries."

Conflicts of Interest: The authors declare no conflict of interest.

References

1. Woyengo, T.A.; Beltranena, E.; Zijlstra, R.T. Nonruminant Nutrition Symposium: Controlling feed cost by including alternative ingredients into pig diets: A review. *J. Anim. Sci.* **2014**, *92*, 1293–1305. [CrossRef] [PubMed]
2. Zijlstra, R.T.; Beltranena, E. Swine convert co-products from food and biofuel industries into animal protein for food. *Anim. Front.* **2013**, *3*, 48–53. [CrossRef]
3. Mackenzie, S.G.; Leinonen, I.; Ferguson, N.; Kyriazakis, I. Can the environmental impact of pig systems be reduced by utilising co-products as feed? *J. Clean. Prod.* **2016**, *115*, 172–181. [CrossRef]
4. Nonhebel, S. Energy from agricultural residues and consequences for land requirements for food production. *Agric. Syst.* **2007**, *94*, 586–592. [CrossRef]
5. Ajila, C.M.; Brar, S.K.; Verma, M.; Tyagi, R.D.; Godbout, S.; Valéro, J.R. Bio-processing of agro-byproducts to animal feed. *Crit. Rev. Biotechnol.* **2012**, *32*, 382–400. [CrossRef] [PubMed]
6. Dhiman, T.R.; Bingham, H.R.; Radloff, H.D. Production response of lactating cows fed dried versus wet brewers' grain in diets with similar dry matter content. *J. Dairy Sci.* **2003**, *86*, 2914–2921. [CrossRef]
7. Miyazawa, K.; Sultana, H.; Hirata, T.; Kanda, S.; Itabashi, H. Effect of brewer's grain on rumen fermentation, milk production and milk composition in lactating dairy cows. *Anim. Sci. J.* **2007**, *78*, 519–526. [CrossRef]
8. Chiou, P.W.S.; Chen, C.R.; Chen, K.J.; Yu, B. Wet brewers' grains or bean curd pomance as partial replacement of soybean meal for lactating cows. *Anim. Feed Sci. Technol.* **1998**, *74*, 123–134. [CrossRef]
9. Gallo, M.; Sommer, A.; Mlynar, R.; Rajcakova, L. Effect of dietary supplementation with brewery draff on rumen fermentation and milk production in grazing dairy cows. *J. Farm. Anim. Sci.* **2001**, *34*, 107–113.
10. Mussatto, S.I.; Dragone, G.; Roberto, I.C. Brewers' spent grain: Generation, characteristics and potential applications. *J. Cereal Sci.* **2006**, *43*, 1–14. [CrossRef]
11. Belibasakis, N.G.; Tsirgogianni, D. Effects of wet brewers grains on milk yield, milk composition and blood components of dairy cows in hot weather. *Anim. Feed Sci. Technol.* **1996**, *57*, 175–181. [CrossRef]
12. Sawadogo, L.; Sepehri, H.; Houdebine, L.M. Presence of a factor stimulating prolactin and growth hormone secretion in brewers' spent grains. *Reprod. Nutr. Dev.* **1989**, *29*, 139–146. [CrossRef] [PubMed]
13. Yaakugh, I.D.I.; Tegbe, T.S.B.; Olorunju, S.A.S.; Aduku, A.O. Replacement value of brewers' dried grain for maize on performance of pigs. *J. Sci. Food Agric.* **1994**, *66*, 465–471. [CrossRef]
14. Dung, N.N.X.; Manh, L.H.; Udén, P. Tropical [illegible] sources for pigs—Digestibility, digesta retention and estimation of fibre digestibility in vitro. *Anim. [illegible] Sci. Technol.* **2002**, *102*, 109–124. [CrossRef]
15. Gutierrez, N.A.; Kil, D.Y.; Liu, Y.; Pettigrew, J.E.; Stein, H.H. Effects of co-products from the corn-ethanol industry on body composition, retention of protein, lipids and energy and on the net energy of diets fed to growing or finishing pigs. *J. Sci. Food Agric.* **2014**, *94*, 3008–3016. [CrossRef] [PubMed]
16. Stein, H.H.; Shurson, G.C. Board-invited review: The use and application of distillers dried grains with solubles in swine diets. *J. Anim. Sci.* **2009**, *87*, 1292–1303. [CrossRef] [PubMed]
17. Duttlinger, A.J.; Derouchey, J.M.; Tokach, M.D.; Dritz, S.S.; Goodband, R.D.; Nelssen, J.L.; Houser, T.A.; Sulabo, R.C. Effects of increasing crude glycerol and dried distillers grains with solubles on growth performance, carcass characteristics and carcass fat quality of finishing pigs. *J. Anim. Sci.* **2012**, *90*, 840–852. [CrossRef] [PubMed]
18. Graham, A.B.; Goodband, R.D.; Tokach, M.D.; Dritz, S.S.; Derouchey, J.M.; Nitikanchana, S.; Updike, J.J. The effects of low-, medium- and high-oil distillers dried grains with solubles on growth performance, nutrient digestibility and fat quality in finishing pigs. *J. Anim. Sci.* **2014**, *92*, 3610–3623. [CrossRef] [PubMed]
19. Olumu, J.M. Industrial by-products as alternative to maize in ration. *Anim. Sci. J.* **1988**, *15*, 61–63.
20. Carias, D.; Millan, N. Brewery waste as a substitute for soy protein in soy-brewer's yeast mixtures to feed broiler chickens. *Arch. Latinoam. Nutr.* **1996**, *46*, 67–70. [PubMed]
21. Onifade, A.A.; Babatunde, G.M. Comparison of the utilization of palm kernel meal, brewer's dried grains and maize offal by broiler chicks. *Br. Poult. Sci.* **1998**, *39*, 245–250. [CrossRef] [PubMed]
22. Kratzer, F.H.; Earl, L. The feeding value of the protein of brewers' dried grains for chicks. *Poult. Sci.* **1980**, *59*, 2361–2364. [CrossRef]
23. Swain, B.K.; Naik, P.K.; Chakurkar, E.B.; Singh, N.P. Effect of feeding brewers' dried grain on the performance and carcass characteristics of Vanaraja chicks. *J. Appl. Anim. Res.* **2012**, *40*, 163–166. [CrossRef]

24. Kaur, V.I.; Saxena, P.K. Incorporation of brewery waste in supplementary feed and its impact on growth in some carps. *Bioresour. Technol.* **2004**, *91*, 101–104. [CrossRef]
25. Muzinic, L.A.; Thompson, K.R.; Morris, A.; Webster, C.D.; Rouse, D.B.; Manomaitis, L. Partial and total replacement of fish meal with soybean meal and brewers' grains with yeast in practical diets for Australian red claw crayfish *Cherax quadricarinatus*. *Aquaculture* **2004**, *230*, 359–376. [CrossRef]
26. Huige, N.J. Brewery by-products and effluents. In *Handbook of Brewing*; Hardwick, W.A., Ed.; Marcel Dekker: New York, NY, USA, 1994; pp. 501–550. ISBN 9780585139173.
27. Santos, M.; Jiménez, J.J.; Bartolomé, B.; Gómez-Cordovés, C.; del Nozal, M.J. Variability of brewers' spent grain within a brewery. *Food Chem.* **2003**, *80*, 17–21. [CrossRef]
28. National Research Council. *Nutrient Requirements of Swine*, 11th ed.; The National Academies Press: Washington, DC, USA, 2012; ISBN 978-0-309-22423-9.
29. Ghazalah, A.A.; Abd-Elsamee, M.O.; Moustafa, E.S. Use of distillers dried grains with solubles (DDGS) as replacement for soybean meal in laying hen diets. *Int. J. Poult. Sci.* **2011**, *10*, 505–513. [CrossRef]
30. Hoffman, L.A.; Baker, A. Estimating the Substitution of Distillers' Grains for Corn and Soybean Meal in the U.S. Feed Complex. Economic Research Service/USDA. Available online: https://www.ers.usda.gov/publications/pub-details/?pubid=36472 (accessed on 27 April 2018).
31. Guo, L.B.; Gifford, R.M. Soil carbon stocks and land use change: A meta analysis. *Glob. Chang. Biol.* **2002**, *8*, 345–360. [CrossRef]
32. Soya and the Cerrado: Brazil's Forgotten Jewel. WWF Report UK 2012. Available online: http://assets.wwf.org.uk/downloads/soya_and_the_cerrado.pdf (accessed on 27 April 2018).
33. Clay, J. *World Agriculture and the Environment*, 4th ed.; Island Press: Washington, DC, USA, 2003; ISBN 9781559633703.
34. Economic Report on Scottish Agriculture. Available online: http://www.gov.scot/Topics/Statistics/Browse/Agriculture-Fisheries/PubEconomicReport (accessed on 26 March 2018).
35. Bell, J.; Morgan, C.; Dick, G.; Reid, G. Distillery Feed By-Products Briefing, An AA211 Special Economic Study for the Scottish Government. 2012. Available online: http://www.sruc.ac.uk/download/downloads/id/1057/distillery_feed_by-products_briefing (accessed on 26 March 2018).
36. Meul, M.; Ginneberge, C.; Van Middelaar, C.E.; de Boer, I.J.M.; Fremaut, D.; Haesaert, G. Carbon footprint of five pig diets using three land use change accounting methods. *Livest. Sci.* **2012**, *149*, 215–223. [CrossRef]
37. De Vries, M.; de Boer, I.J.M. Comparing environmental impacts for livestock products: A review of life cycle assessments. *Livest. Sci.* **2010**, *128*, 1–11. [CrossRef]
38. Ardente, F.; Cellura, M. Economic allocation in life cycle assessment. *J. Ind. Ecol.* **2012**, *16*, 387–398. [CrossRef]
39. Brankatschk, G.; Finkbeiner, M. Application of the Cereal Unit in a new allocation procedure for agricultural life cycle assessments. *J. Clean. Prod.* **2014**, *73*, 72–79. [CrossRef]
40. Scots Whisky Association. Environmental Strategy. Available online: http://www.scotch-whisky.org.uk/what-we-do/environmental-strategy/ (accessed on 26 March 2018).
41. Risberg, K.; Cederlund, H.; Pell, M.; Arthurson, V.; Schnürer, A. Comparative characterization of digestate versus pig slurry and cow manure—Chemical composition and effects on soil microbial activity. *Waste Manag.* **2017**, *61*, 529–538. [CrossRef] [PubMed]
42. Ohdoi, K.; Miyahara, S.; Iwashita, K.; Umeda, M.; Shimizu, H.; Nakashima, H.; Miyasaka, J. Optimization of fertiliser application schedule: Utilization of digestate after anaerobic digestion as liquid fertiliser. *IFAC Proc. Vol.* **2013**, *46*, 317–322. [CrossRef]
43. Scottish Tenant Farmers Association News Release, 21st February 2017. Available online: http://www.tfascotland.org.uk/stfa-urges-government-to-rethink-policy-of-turning-valuable-livestock-protein-feed-into-renewable-energy/ (accessed on 4 April 2018).
44. BS EN ISO 14044:2006. Environmental Management—Life Cycle Assessment—Requirements and Guidelines. Available online: https://www.iso.org/standard/38498.html (accessed on 26 March 2018).
45. Eggleston, H.S.; Buendia, L.; Miwa, K.; Ngara, T.; Tanabe, K. (Eds.) *2006 IPCC Guidelines for National Greenhouse Gas Inventories, Prepared by the National Greenhouse Gas Inventories Programme, Volume 4: Agriculture, Forestry and Other Land Use*; Institute for Global Environmental Strategies (IGES): Hayama, Japan, 2006; Available online: http://www.ipcc-nggip.iges.or.jp/public/2006gl/index.html (accessed on 26 March 2018).
46. Feedipedia: An On-Line Encyclopedia of Animal Feeds. Available online: https://www.feedipedia.org (accessed on 26 March 2018).

47. MacLeod, M.; Sykes, A.; Leinonen, I.; Eory, V. *Quantifying the Greenhouse Gas Emission Intensity of Scottish Agricultural Commodities: CXC Project*; Technical Report; SRUC: Edinburgh, UK, 2017.
48. Food and Agriculture Organization of the United Nations. *Global Livestock Environmental Assessment Model Version 2.0 Model Description Revision 4, June 2017*; Food and Agriculture Organization of the United Nations (FAO): Rome, Italy, 2017.
49. Global Livestock Environmental Assessment Model (GLEAM). Food and Agriculture Organization of the United Nations (FAO), Rome. Available online: http://www.fao.org/gleam/en/ (accessed on 26 March 2018).
50. Specification for the Assessment of the Life Cycle Greenhouse Gas Emissions of Goods and Services. PAS 2050:2011. Available online: http://www.bsigroup.com/Standards-and-Publications/How-we-can-help-you/Professional-Standards-Service/PAS-2050/ (accessed on 26 March 2018).
51. Leinonen, I.; Williams, A.G.; Waller, A.H.; Kyriazakis, I. Comparing the environmental impacts of alternative protein crops in poultry diets: The consequences of uncertainty. *Agric. Syst.* **2013**, *121*, 33–42. [CrossRef]
52. De Vries, J.W.; Vinken, T.M.W.J.; Hamelin, L.; de Boer, I.J.M. Comparing environmental consequences of anaerobic mono- and co-digestion of pig manure to produce bio-energy—A life cycle perspective. *Bioresour. Technol.* **2012**, *125*, 239–248. [CrossRef] [PubMed]
53. Jenssen, T.K.; Kongshaug, G. *Energy Consumption and Greenhouse Gas Emissions in Fertiliser Production*; Proceedings No. 509; The International Fertilizer Society: York, UK, 1998.
54. Greenhouse Gas Reporting—Conversion Factors 2016. Department for Business, Energy & Industrial Strategy, UK. Available online: https://www.gov.uk/government/publications/greenhouse-gas-reporting-conversion-factors-2016 (accessed on 26 March 2018).
55. Luna-delRisco, M.; Normak, A.; Orupõld, K. Biochemical methane potential of different organic wastes and energy crops from Estonia. *Agron. Res.* **2011**, *9*, 331–342.
56. Eory, V.; MacLeod, M.; Topp, C.F.E.; Rees, R.M.; Webb, J.; McVittie, A.; Wall, E.; Borthwick, F.; Watson, C.; Waterhouse, A.; et al. *Review and Update the UK Agriculture Marginal Abatement Cost Curve to Assess the Greenhouse Gas Abatement Potential for the 5th Carbon Budget Period and to 2050*; Final Report Submitted for the Project Contract "Provision of Services to Review and Update the UK Agriculture MACC and to Assess Abatement Potential for the 5th Carbon Budget Period and to 2050"; Committee on Climate Change: London, UK, 2015.
57. Styles, D.; Gibbons, J.; Williams, A.P.; Stichnothe, H.; Chadwick, D.R.; Healey, J.R. Cattle feed or bioenergy? Consequential life cycle assessment of biogas feedstock options on dairy farms. *GCB Bioenergy* **2015**, *7*, 1034–1049. [CrossRef]
58. Knowledge Reference for National Forest Assessments—Modeling for Estimation and Monitoring. Food and Agriculture Organization of the United Nations (FAO), Rome. Available online: http://www.fao.org/forestry/8758/en/ (accessed on 26 March 2018).
59. Williams, A.G.; Leinonen, I.; Kyriazakis, I. Environmental benefits of using turkey litter as a fuel instead of a fertiliser. *J. Clean. Prod.* **2016**, *113*, 167–175. [CrossRef]
60. Electricity Generation Mix by Quarter and Fuel Source (GB). Office of Gas and Electricity Markets (Ofgem). Available online: https://www.ofgem.gov.uk/data-portal/electricity-generation-mix-quarter-and-fuel-source-gb (accessed on 26 March 2018).
61. Van der Werf, H.M.G.; Nguyen, T.T.H. Construction cost of plant compounds provides a physical relationship for co-product allocation in life cycle assessment. *Int. J. Life Cycle Assess.* **2015**, *20*, 777–784. [CrossRef]
62. Eady, S.; Carre, A.; Grant, T. Life cycle assessment modelling of complex agricultural systems with multiple food and fibre co-products. *J. Clean. Prod.* **2012**, *28*, 143–149. [CrossRef]
63. Thoma, G.; Jolliet, O.; Wang, Y. A biophysical approach to allocation of life cycle environmental burdens for fluid milk supply chain analysis. *Int. Dairy J.* **2013**, *31*, S41–S49. [CrossRef]
64. Gac, I.A.; Salou, T.; Espagnol, S.; Ponchant, P.; Dollé, J.-B.; van der Werf, H.M.G. An original way of handling coproducts with a biophysical approach in LCAs of livestock systems. In Proceedings of the 9th International Conference on Life Cycle Assessment in the Agri-Food Sector (LCA Food 2014), San Francisco, CA, USA, 8–10 October 2014; Schenck, R., Huizenga, D., Eds.; ACLCA: Vashon, WA, USA, 2014; pp. 443–449.
65. Wiedemann, S.G.; Ledgard, S.F.; Henry, B.K.; Yan, M.-J.; Mao, N.; Russell, S.J. Application of life cycle assessment to sheep production systems: Investigating co-production of wool and meat using case studies from major global producers. *Int. J. Life Cycle Assess.* **2015**, *20*, 463–476. [CrossRef]

66. Mackenzie, S.G.; Leinonen, I.; Kyriazakis, I. The need for co-product allocation in the Life Cycle Assessment of agricultural systems—Is "biophysical" allocation progress? *Int. J. Life Cycle Assess.* **2017**, *22*, 128–137. [CrossRef]
67. Cederberg, C.; Stadig, M. System Expansion and Allocation in Life Cycle Assessment of Milk and Beef Production. *Int. J. LCA* **2003**, *8*, 350–356. [CrossRef]
68. Wiedemann, S.; Yan, M. Livestock meat processing: Inventory data and methods for handling co-production for major livestock species and meat products. In Proceedings of the 9th International Conference on Life Cycle Assessment in the Agri-Food Sector (LCA Food 2014), San Francisco, CA, USA, 8–10 October 2014; Schenck, R., Huizenga, D., Eds.; ACLCA: Vashon, WA, USA, 2014; pp. 1512–1520.
69. Parker, G. Measuring the environmental performance of food packaging: Life cycle assessment. In *Environmentally Compatible Food Packaging*; Chiellini, E., Ed.; Woodhead Publishing Ltd.: Cambridge, UK, 2008; pp. 211–237, ISBN 9781845691943.
70. Curran, M.A. Nanomaterials life cycle assessment: Framing the opportunities and challenges. In *Life Cycle Analysis of Nanoparticles*; Vaseashta, A., Ed.; DEStech Publications Inc.: Lancaster, UK, 2015; pp. 24–55, ISBN 978-1-60595-023-5.
71. Ekvall, T.; Weidema, B.P. System boundaries and input data in consequential life cycle inventory analysis. *Int. J. LCA* **2004**, *9*, 161–171. [CrossRef]
72. Van Zanten, H.H.E.; Mollenhorst, H.; de Vries, J.W.; van Middelaar, C.E.; van Kernebeek, H.R.J.; de Boer, I.J.M. Assessing environmental consequences of using co-products in animal feed. *Int. J. Life Cycle Assess.* **2014**, *19*, 79–88. [CrossRef]
73. Leinonen, I.; Williams, A.G.; Kyriazakis, I. Evaluating methods to account for the greenhouse gas emissions from Land Use Changes in agricultural LCA. In Proceedings of the 9th International Conference on Life Cycle Assessment in the Agri-Food Sector (LCA Food 2014), San Francisco, CA, USA, 8–10 October 2014; Schenck, R., Huizenga, D., Eds.; ACLCA: Vashon, WA, USA, 2014; pp. 711–717.
74. Williams, A.G.; Dominguez, H.; Leinonen, I. A simple approach to land use change emissions for global crop commodities reflecting demand. In Proceedings of the 9th International Conference on Life Cycle Assessment in the Agri-Food Sector (LCA Food 2014), San Francisco, CA, USA, 8–10 October 2014; Schenck, R., Huizenga, D., Eds.; ACLCA: Vashon, WA, USA, 2014; pp. 1527–1534.
75. Bellamy, P.H.; Loveland, P.J.; Bradley, R.I.; Lark, R.M.; Kirk, G.J.D. Carbon losses from all soils across England and Wales 1978–2003. *Nature* **2005**, *437*, 245–248. [CrossRef] [PubMed]
76. Smith, P.; Chapman, S.J.; Scott, W.A.; Black, H.I.; Wattenbach, M.; Milne, R.; Campbell, C.D.; Lilly, A.; Ostle, N.; Levy, P.E.; et al. Climate change cannot be entirely responsible for soil carbon loss observed in England and Wales, 1978–2003. *Glob. Chang. Biol.* **2007**, *13*, 2605–2609. [CrossRef]

sustainability

MDPI

Article

Influence of Reduced Protein Content in Complete Diets with a Consistent Arginine–Lysine Ratio on Performance and Nitrogen Excretion in Broilers

Cristina Ullrich [1], Marion Langeheine [2], Ralph Brehm [2], Venja Taube [3], Diana Siebert [4] and Christian Visscher [1,*]

[1] Institute for Animal Nutrition, University of Veterinary Medicine Hannover, Foundation, Bischofsholer Damm 15, D-30173 Hanover, Germany; Cristina.ullrich@tiho-hannover.de

[2] Institute for Anatomy, University of Veterinary Medicine Hannover, Foundation, Bischofsholer Damm 15, D-30173 Hanover, Germany; Marion.langeheine@tiho-hannover.de (M.L.); Ralph.brehm@tiho-hannover.de (R.B.)

[3] BEST 3 Gefluegelernaehrung GmbH, Ringstrasse 16, D-27239 Twistringen, Germany; v.taube@best-3.de

[4] CJ Europe GmbH, Ober der Röth 4, D-65824 Schwalbach, Germany; d.siebert@cj.net

* Correspondence: Christian.visscher@tiho-hannover.de; Tel.: +49-511-856-7415

Received: 30 September 2018; Accepted: 19 October 2018; Published: 23 October 2018

Abstract: The current discussion concerning resource-efficient broiler production inevitably leads to diets with lowered crude protein (CP) levels. Therefore, the hypothesis was formed that crude protein reduction far below the recommended levels can significantly lower the nitrogen (N) content in litter, if essential amino acids are added and a constant lysine-arginine ratio is guaranteed. In a five-week feeding trial, 360 ROSS 308 broilers of both sexes were randomly assigned to four feeding groups with six replicates each with a standard three-phase feeding program (d 1–7, d 8–14, d 15–35). The control group was offered a complete diet with a common protein content found in practice (CP-% as fed; starter: 21.5, grower: 20.5, finisher: 20.0; lysine/arginine: 100/115). In the experimental diets the lysine/arginine ratio was constant, whereas the protein content was lowered in steps of 1.00 percent each with simultaneous supplementation of growth limiting amino acids. Feeding a diet with a 2.00 percent reduced protein content led to higher body weights after 34 days compared to the control (2329 g vs. 2192 g). The N content in the total litter decreased significantly with a 2.00 and 3.00 percent reduction in the CP content (51.2 vs. 46.2 or rather 36.2 g/kg dry matter (DM)). Meticulous balanced protein-reduced diets therefore allow a significant environmental relief.

Keywords: arginine; broiler; crude protein; lysine; nitrogen; resource efficiency

1. Introduction

The impact of livestock production on the environment is significant [1–4]. Due to the intensification of pig and poultry production, owing to efficiency, environmental problems occurred in some areas of the world [5,6]. Differences in environmental impact among different production systems (e.g., pork, chicken and beef) can be explained by the following three factors: feed efficiency, differences in enteric CH_4 emission between monogastric animals and ruminants, and differences in reproduction rates [1,7]. The production of 1 kg of beef protein also has the highest impact, followed by pork protein, whereas chicken protein has the lowest impact when only meat production is considered [1,8]. Although feed is the primary input source of nutrients, the amount of nutrients ultimately emitted into the environment is dependent on the efficiency of nutrient utilization of the animal [2,5].

Low protein diets with the correct amino acid (AA) supplementation promote a reduction in N excretion and ammonia emission from the litter of broiler chickens [9]. Reduction in dietary crude protein (CP) content resulted in a 10–27% reduction in the total amount of N excreted during a six-week broiler rearing period [10]. Low CP diets, namely to supply sufficient amounts of essential AA to meet the requirements only, can in turn help to decrease the amounts of excess dietary non-essential AA [11]. Although reducing dietary CP reduces N content and, therefore, pollution potential of the resulting litter, adverse effects can occur in live performance [11]. In some studies, low protein diets failed to support equal growth performance of that of high protein control diets [9,12–14]. Feeding broiler chickens diets containing a high proportion of crystalline amino acids with low intact CP can cause retarded growth when using diets below 19% CP [15]. In a recent study, crystalline amino acid supplementation based on a similar amino acid profile could reduce N excretion and foot pad dermatitis without having any negative effects on growth performance by reducing dietary CP levels from 19% to 17% of free-range yellow broilers [16].

In particular, arginine is important in broiler nutrition because in uricotelic animals exogenous arginine is needed for the urea cycle [17,18]. The National Research Council (NRC) recommends an arginine–lysine ratio of 1.14 (lysine content in diet: 1.10% at 90% dry matter (DM)) for broilers in the first three weeks of fattening and a ratio of 1.10 in the fourth to sixth week (lysine content in diet: 1.00% at 90% DM [19]). The information in the literature, however, is not without contradiction. In younger broilers males optimized body weight gain was found for 1.15% dietary arginine (1.21% lysine; 20.6% CP; [20]). In older birds, the final body weight together with body weight gain and feed conversion throughout the 42-to-56-d experimental period were optimized at 0.98% arginine and 0.85% lysine (ratio = 1.15; [21]). In a preliminary study a constant arginine–lysine ratio of 1.15 in protein was found to be the optimum for broiler nutrition under European conditions during the starter period [22].

Therefore, the objective of the present study was to evaluate the effects of reduced protein content in complete diets with a consistent arginine–lysine ratio on performance and N excretion in broilers during a 34-day rearing period.

2. Materials and Methods

Animal experiments were carried out in accordance with German regulations. These animal experiments require no notification or approval in accordance with the Animal Protection Act (§ 7, paragraph 2, sentence 3). Interventions before dissection were not carried out. The animals were killed in accordance with § 4, paragraph 3 of the Animal Protection Act, exclusively to use their organs or tissues for scientific purposes. The experiments were approved by the Animal Welfare Officer of the University of Veterinary Medicine Hannover, Germany (reference: TVT-2018-V-102).

2.1. *Animals, Housing and Experimental Design*

Experiments were performed with a common line of broilers kept for fattening purposes (as hatched; ROSS 308; BWE-Brüterei Weser-Ems GmbH & Co. KG; Visbek-Rechterfeld, Germany). In total 360 one-day-old broiler chickens were divided into four different groups depending on the CP content or rather reduction in CP content (in steps of 1%) in the diet (n = 4: CP-C, CP-1; CP-2; CP-3; with six replicates each; 15 animals per replicate/subgroup). If an animal died or showed weakness during the first two weeks of the trial, it was replaced by a broiler from a reserve group. After that time, no replacement was made.

From day 1 onwards the animals were kept in 24 boxes (1.20 × 0.80 m; AviMax, Big Dutchman International GmbH, Vechta-Calveslage, Germany). One subgroup of each group was placed in a block (=4 boxes) in a randomised sequence, with six blocks in the same stable. A vacuum air ventilation system was installed in the ceiling in two rows above the pens. The boxes were littered with approximately 1 cm (1 kg per square metre) of wood shavings (GOLDSPAN®, Goldspan GmbH and Co. KG, Goldenstedt, Germany). By the end of the trial stocking densities had reached a maximum of 35 kg per square metre. On the left-hand side of each pen, there was a scratching area, and on the right-hand side there was a feeding area equipped with one hanging-type feeder (Klaus Gritsteinwerk

GmbH & Co. KG, Bünde, Germany). Drinking lines with nipple drinkers (two nipples per box) for broilers (Big Dutchman International GmbH, Vechta, Germany) were used. The environmental temperature was gradually reduced from about 33 °C for the one-day-old birds to about 20 °C by day 34. Lights were continuously on at days one, two, and three, and the photoperiod from day four onwards amounted to 16 h of light and 8 h of darkness with dimmed night lighting.

2.2. Diets and Feeding Concept

All birds were fed ad libitum with specifically prepared pelleted diets. For the trial, 12 different diets (four different starter diets, four grower diets, and four finisher diets) were produced at the Institute for Animal Nutrition, University of Veterinary Medicine Hannover, Foundation, Hannover, Germany. One component (80% of diet; Best 3 Geflügelernährung GmbH, Twistringen, Germany) of each of the 12 diets was the same supplementary feed with vitamins, minerals and coccidiostats (narasin/nicarbacin). This maximized the comparability of the final diets with each other.

The remaining 20% of the specific compound feedingstuffs consisted of corn, soybean meal, amino acids and some minerals (limestone, monocalciumphosphate, salt, sodium bicarbonate) to make up the 12 different experimental diets (Table 1).

Table 1. Ingredient composition of the diets during the whole experimental period. CP: crude protein.

Parameter [%]	Starter				Grower				Finisher			
	CP-C	CP-1	CP-2	CP-3	CP-C	CP-1	CP-2	CP-3	CP-C	CP-1	CP-2	CP-3
Supplementary Feedingstuff *	80.0	80.0	80.0	80.0	80.0	80.0	80.0	80.0	80.0	80.0	80.0	80.0
Corn	0.30	4.51	8.83	13.6	4.31	9.44	13.1	9.95	7.15	10.8	15.3	7.77
Soybean meal	14.8	10.8	6.83	2.40	10.8	6.00	2.40	0.00	8.50	5.25	1.00	0.00
Plant oil	3.40	2.68	1.85	1.00	4.15	3.30	2.60	4.35	4.10	3.30	2.50	5.60
Limestone	0.75	0.74	0.73	0.72	0.30	0.29	0.30	0.28	0.05	0.04	0.02	0.00
Mono-Ca-phosphate	0.50	0.57	0.63	0.70	0.25	0.33	0.39	0.52	0.00	0.06	0.13	0.26
Salt	0.12	0.08	0.04	0.00	0.10	0.05	0.02	0.00	0.07	0.04	0.00	0.00
Sodium Bicarbonate	0.00	0.00	0.05	0.00	0.00	0.06	0.11	0.14	0.00	0.05	0.09	0.10
Arginine	0.00	0.12	0.23	0.36	0.00	0.14	0.25	0.35	0.00	0.10	0.22	0.29
Isoleucine	0.00	0.07	0.14	0.21	0.00	0.08	0.15	0.21	0.00	0.06	0.13	0.17
L-Lysine-HCl	0.04	0.17	0.30	0.44	0.04	0.20	0.32	0.42	0.04	0.15	0.29	0.35
L-Methionine	0.03	0.05	0.06	0.08	0.00	0.02	0.04	0.06	0.00	0.01	0.03	0.05
Threonine	0.07	0.13	0.19	0.26	0.05	0.01	0.18	0.24	0.08	0.13	0.19	0.24
Valine	0.00	0.07	0.13	0.20	0.00	0.08	0.14	0.20	0.00	0.05	0.12	0.17
Caolin	0.00	0.00	0.00	0.00	0.00	0.00	0.00	3.32	0.00	0.00	0.00	5.00

* Containing 51.4% wheat, 24.1% soybean meal, 15.0% corn, 5.08% plant oil, 1.30% limestone, 0.94% mono-Ca-phosphate, 0.70% formic acid, 0.40% Alimet, 0.25% enzyme mixture (phytase, xylanase), 0.25% premix, 0.22% sodium bicarbonate, 0.22% salt, 0.10% lysine, 0.06% betaine (33%); feed additives per kg: 12,500 IU vitamin A, 6250 IU vitamin D3, 43 mg Vitamin E, 18.7 mg Cu, 25 mg Fe, 87.5 mg Mn, 62.5 mg Zn, 2.5 mg I, 0.37 mg Se, acetic acid, formic acid, 625 FTU 6-phytase, 3000 U Endo 1,4-ß-xylanase, 62.5 mg narasin, 62.5 mg nicarbacin.

One group received a standard control diet (starter: d 1–7, grower: d 8–14; finisher: d 15–35), the other groups, specially prepared experimental diets. The control diet was a complete feedingstuff with a common protein content (as fed; starter: 21.5% CP; grower: 20.5% CP; finisher: 20.0% CP; each designed for an arginine–lysine ratio of 115:100; Table 2). In the experimental diets, the arginine–lysine ratio was constant; the protein was lower in steps of 1.00 per cent. The levels of essential amino acids were held at the same level due to the supplementation of single amino acids. To ensure the optimum composition and to verify the ratio between the components, these 12 different diets (in total: four starters, growers and finishers each) were analyzed (Table 2).

Table 2. Concentrations of ingredients and energy content after chemical analysis in the starter, grower and finisher diets.

Item [g/kg; 88% DM]	Starter				Grower				Finisher			
	CP-C	CP-1	CP-2	CP-3	CP-C	CP-1	CP-2	CP-3	CP-C	CP-1	CP-2	CP-3
Crude ash	58.7	56.4	54.4	51.9	50.5	48.6	47.6	73.8	45.4	46.5	43.3	84.4
Crude fat (EE)	78.8	75.3	68.7	64.4	89.8	83.3	79.3	93.7	93.3	83.5	76.4	103
Crude fibre	20.5	19.6	18.5	16.2	19.2	19.1	18.2	18.4	20.7	20.7	19.3	18.2
Crude protein	222	215	202	192	205	194	186	177	199	192	181	170
Nitrogen free extract [1]	500	514	536	556	516	535	549	517	522	537	560	504
Starch	335	356	382	410	355	384	406	380	376	385	412	371
Sugar	52.5	51.9	47.2	43.3	48.7	44.6	41.1	37.4	49.1	44.4	39.8	35.8
Calcium	9.67	9.72	9.69	9.67	8.02	7.62	7.59	7.60	6.03	6.04	6.04	6.07
Phosphorus	7.43	6.59	7.25	7.47	6.34	6.42	5.99	6.13	5.77	5.45	5.24	5.31
Potassium	9.85	8.91	8.12	7.29	8.87	7.92	7.22	7.61	8.41	7.73	7.44	8.37
Arginine	14.8	14.8	14.3	14.8	13.5	13.6	14.1	13.6	13.7	13.1	12.8	12.6
Cysteine	2.23	2.28	2.03	2.02	2.24	2.23	2.05	1.84	2.02	2.30	1.97	1.66
Histidine	5.87	5.53	4.76	4.72	5.28	4.98	4.68	4.28	5.22	5.00	4.33	4.04
Isoleucine	9.62	9.42	8.99	9.38	9.24	8.72	9.04	8.73	7.96	8.55	8.18	8.11
Leucine	16.9	16.1	14.6	14.0	16.3	14.7	14.4	12.7	15.5	15.0	13.3	12.5
Lysine	12.8	12.9	12.4	12.6	12.0	11.7	12.3	11.7	11.7	11.4	11.3	10.8
Methionine	5.18	5.73	5.43	5.59	5.04	5.45	5.01	5.00	5.18	5.10	4.70	5.16
Phenylalanine	10.8	10.2	9.04	8.74	10.4	9.30	8.99	7.96	10.1	9.41	8.24	7.66
Threonine	8.30	8.64	8.14	8.64	7.77	6.73	8.18	7.56	7.86	8.11	7.82	7.36
Valine	10.3	10.5	9.86	10.3	9.83	9.47	10.1	9.69	9.30	9.33	9.07	8.85
Metabolizable energy AME_N [2] [MJ/kg DM]	12.4	12.5	12.5	12.6	12.8	12.9	12.9	12.8	13.2	12.8	12.8	12.8

[1] Nitrogen-free extract = DM − (crude ash + crude fat + crude fibre + crude protein); [2] AME_N = nitrogen-corrected apparent metabolizable energy; AME_N (per kg) = 0.1551 × % CP + 0.3431 × % ether extracts (EE) + 0.1669 × % starch + 0.1301 × % sugar (as sucrose).

2.3. Measurements

2.3.1. Technical Performance

The individual body weight (BW) was measured weekly on the same day. One exception was the last week, in which the interval had to be shortened by one day for technical reasons (six days). Feed intakes (FI) as well as losses were determined at subgroup level. The feed conversion ratio (FCR) was estimated from feed consumed and body weight growth throughout the experimental period at box level.

2.3.2. Excreta and Litter Sampling

Excreta samples were obtained by means of a rubber mat placed on top of the litter for one hour in each box once a week. Excreta samples were collected without wood shavings contaminating the samples. Litter samples for measuring the DM content were collected weekly from three defined locations along a diagonal line through the box (at both ends and a central one). At each area, over the whole bedding height a sample was punched out from the full depth of the litter using a cup with a 5 cm diameter. At the end of the experiment, the entire litter material was removed, weighed and homogenised with a rotating machine. An aliquot was taken from this material and used for further analyses.

2.3.3. Foot Pad Dermatitis Scoring Criteria

Foot pads were examined weekly, starting at the end of week one. Dirty feet were carefully washed with a wet cloth to remove slightly adhering litter and excreta. Only the dried central plantar of foot pads were scored. The foot pads of the animals were scored with a scoring system from 0–7 designed by Mayne, et al. [23]: score 0 represents healthy skin with no swelling or redness, whereas score 7 stands for a foot pad more than 50% necrotic (Figure 1). Other measures concerning behaviour, use of space, use of the sandbox, and other welfare indicators might have been very useful, but were not part of the study. They should be considered in further studies.

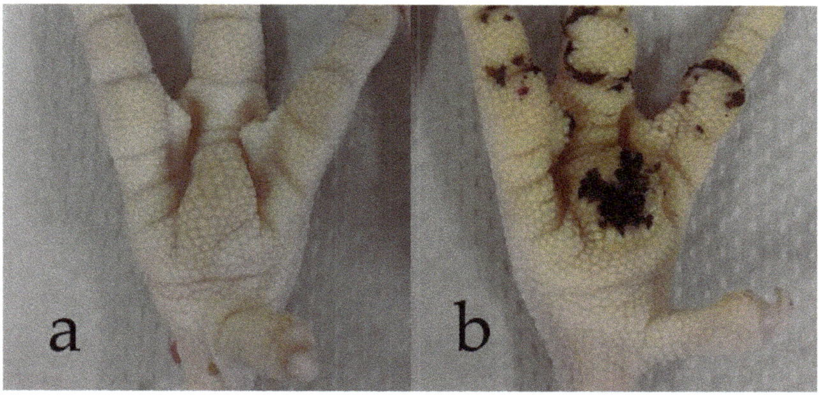

Figure 1. Foot pad lesions with different scores. (**a**) Score 0: no alterations detected; and (**b**) Score 7: more than half of the foot pad covered with necrotic cells (photo: ©University of Veterinary Medicine, Hannover).

2.3.4. Dissection

After 35 days, 120 animals (n = 30 per group) were dissected. A percussive blow to the head was used as anaesthesia or rather the stunning method in accordance with Annex I of Council Regulation (EC) No. 1099/2009, Chapter I, Methods, Table 1—Mechanical methods, No. 6. After bleeding the animals, the body cavity was opened, the sternum lifted up and samples of the ileum were taken for histological investigations. All remaining birds were sold.

2.3.5. Histological Investigations

For histological investigations, an approximately 1 cm long piece was removed from the ileum (in the middle between the diverticulum and cranial part of the caecum) and fixed in 4% formaldehyde for 48 h. After fixation, tissue samples were embedded in paraffin using standard techniques [24]. For histological evaluation, 4 µm sections of all samples were stained with haematoxylin and eosin (HE) using established protocols [24]. The villus height was measured from the tip of the villi to the villus crypt junction; villus width was measured at the base of the villus above the villus crypt junction using a Zeiss axioscope (Carl Zeiss Jena GmbH, Jena, Germany).

2.3.6. Analysis of Feed, Excreta, and Litter Samples

The official methods of the VDLUFA were the basis of the standard procedures being used to analyze all diets [25]. Determining the dry matter content and the crude ash content works through action of heat; drying the samples to the weight constancy at 103 °C delivers the results for the DM content. Through combustion of the sample at 600 °C in a muffle furnace the crude ash content was analyzed. The total N content was determined by the DUMAS combustion method (Vario Max, Elementar, Hanau, Germany). The crude fat content was determined by standard protocol (soxhlet apparatus). Crude fiber was analyzed after washing in dilute acids and alkalis. Starch determination was carried out polarimetrically (Polatronic E, Schmidt und Haensch GmbH & Co., Berlin, Germany). The Luff-Schoorl method was used to determine the sugar content. Minerals were analyzed by atomic absorption spectrometry (Unicam Solaar 116, Thermo, Dreieich, Germany). Amino acids were determined by ion-exchange chromatography (AA analyzer LC 3000, Biotronic, Maintal, Germany).

2.3.7. Estimations of Efficiencies of Nitrogen or Rather Protein Utilisation

The single box was the observation unit for assessing nutrient efficiency between groups. The basis for the considerations was the total FI in the box, the increase in body weight (final body weight minus weight at hatch; including the increase in deceased animals), and the amount of final litter material in the box. The N content in the different diets and in the litter were analyzed by standard methods described above (2.2.6). To calculate the N content in the fresh whole body of birds (including feathers), crude protein contents from various publications were used as a basis [12,26–28]. From the means (in g N/100 g fresh total body: 3.312 [27], 3.016 [26], 2.754 [28], 2.598 [12]) a factor was derived (2.920) and used as the basis for calculating parameters in accordance with Aletor, Hamid, Niess and Pfeffer [26], modified.

$$\text{N-retention efficiency (\%)} = \text{g N retained/g N consumed} \times 100$$

$$\text{N-efficiency rate} = \text{g weight gain/g N consumed}$$

$$\text{N excretion (apparent)} = \text{g N consumed} - \text{g N retained}$$

2.4. Statistical Analysis

Data analyses were performed using the SAS statistical software package version 7.1 (SAS Inst., Cary, NC, USA). Mean values, as well as the standard deviation of the mean (standard deviation (SD)), were calculated for all parameters. The mean footpad dermatitis (FPD) scores were evaluated by using the mean of both feet. For the values of body weight (BW), the foot pad dermatitis scores and the results of the histopathological investigations as well as the individuum were the basis of observation. All other parameters were analyzed at box-level. The group comparisons were performed by one-way analysis of variance (ANOVA) for independent samples. In general, the Ryan–Einot–Gabriel–Welsch multiple-range test (REGWQ) was used for multiple pairwise means comparisons between the four groups. All statements of statistical significance were based on $p < 0.05$.

3. Results

The experiments ran without complications. From the start of week 3, six out of 360 animals died (CP-C: 2; CP-1: 1; CP-2: 1; CP-3:2). There was no antibiotic treatment during the trial.

3.1. Technical Performance

Results relating to growth performance of broilers are shown in Table 3. At day 1, the average weight per chicken was 43.0 g. In broilers, after 34 days of fattening, BW exceeded the performance goals of the breeding company [29]. At the end of the first and the second week of fattening the BW already showed significant differences between groups, being lowest in group CP-3 (Table 3). In weeks three, four, and five, birds in group CP-2 were significantly heavier than in groups CP-C and CP-3.

Table 3. Average body weight and feed conversion ratio (FCR) during the trial (mean ± standard deviation (SD)).

Item	Group	Week 1	Week 2	Week 3	Week 4	Week 5 *
Body weight [in g; end of week]	CP-C	216 ^{A} ± 19.8	535 ^{A} ± 58.1	989 ^{B,C} ± 116	1589 ^{B,C} ± 189	2192 ^{B,C} ± 258
	CP-1	218 ^{A} ± 16.5	541 ^{A} ± 51.7	1004 ^{A,B} ± 116	1632 ^{A,B} ± 199	2244 ^{A,B} ± 275
	CP-2	217 ^{A} ± 15.6	552 ^{A} ± 47.1	1034 ^{A} ± 106	1678 ^{A} ± 182	2329 ^{A} ± 266
	CP-3	206 ^{B} ± 20.1	517 ^{B} ± 45.6	956 ^{C} ± 99.8	1548 ^{C} ± 175	2131 ^{C} ± 244
FCR [Ø in week]	CP-C	0.987 ± 0.073	1.169 ± 0.023	1.363 ± 0.043	1.488 ± 0.039	1.848 ± 0.066
	CP-1	0.952 ± 0.034	1.185 ± 0.025	1.342 ± 0.026	1.474 ± 0.026	1.817 ± 0.020
	CP-2	0.996 ± 0.044	1.165 ± 0.009	1.334 ± 0.047	1.494 ± 0.032	1.776 ± 0.050
	CP-3	1.025 ± 0.070	1.199 ± 0.042	1.389 ± 0.047	1.518 ± 0.026	1.827 ± 0.104

^{A,B,C} averages differ significantly within a column ($p < 0.05$); * only six days—measurement end of day 34.

No differences were observed concerning the feed conversion rate (FCR) in broilers depending on the dietary feeding concept. The FCR showed numerically lowest values for group CP-2 during the entire period (CP-C: 1.474 ± 0.036; CP-1: 1.457 ± 0.010; CP-2: 1.454 ± 0.017; CP-3: 1.489 ± 0.016, respectively).

3.2. Excreta and Litter Quality

There were no differences in DM content of excreta between groups CP-C, CP-1 and CP-2 (Table 4). Excreta of group CP-3 were the driest from week two onwards. The control group had the wettest litter in the fourth week of fattening. At the end of the experiment, groups CP-C, CP-1 and CP-2 were not significantly different concerning DM content in litter. The group CP-3 had the driest litter at week five. The litter material was finally removed from the boxes. The dry matter content was significantly higher in material from group CP-3 than in the other groups (in g DM/kg; CP-C: 584 ^{B} ± 31.3; CP-1: 602 ^{B} ± 39.7; CP-2: 603 ^{B} ± 44.4; CP-3: 698 ^{A} ± 30.7, respectively).

Table 4. Dry matter content of excreta and litter material during the five-week trial period (mean ± SD).

Item	Group	Week 1	Week 2	Week 3	Week 4	Week 5 *
Dry matter (DM) excreta † [in g/kg; end of week]	CP-C	188 ^{B} ± 8.34	200 ^{B} ± 10.8	189 ^{B} ± 7.92	188 ^{B} ± 9,31	184 ^{B} ± 9.46
	CP-1	193 ^{A,B} ± 9.27	208 ^{B} ± 9.27	188 ^{B} ± 6.62	186 ^{B} ± 11.9	184 ^{B} ± 8.18
	CP-2	199 ^{A,B} ± 14.8	207 ^{B} ± 12.1	192 ^{B} ± 16.9	184 ^{B} ± 14.2	181 ^{B} ± 8.01
	CP-3	206 ^{A} ± 9.47	228 ^{A} ± 7.25	228 ^{A} ± 20.2	232 ^{A} ± 2.76	219 ^{A} ± 9.77
Dry matter litter ‡ [in g/kg; end of week]	CP-C	815 ± 36.2	816 ^{B} ± 23.9	712 ^{A,B} ± 53.8	588 ^{B} ± 54.3	609 ^{B} ± 69.8
	CP-1	829 ± 29.3	837 ^{A,B} ± 24.6	698 ^{B} ± 42.2	596 ^{A,B} ± 67.6	610 ^{B} ± 31.3
	CP-2	817 ± 31.5	831 ^{A,B} ± 43.9	718 ^{A,B} ± 58.0	617 ^{A,B} ± 58.0	584 ^{B} ± 61.6
	CP-3	830 ± 30.7	856 ^{A} ± 24.3	763 ^{A} ± 49.7	667 ^{A} ± 66.7	709 ^{A} ± 45.1

^{A,B} averages differ significantly within a column ($p < 0.05$); * only six days—measurement end of day 34; † REGWQ-test; ‡ LSD-test.

Analyzing the N content in the excreta showed that diets with lower CP content led to significantly lower N values in the excreta (Table 5). In comparison to the control, in week 5 the N content was reduced by 36.6%. Also, the mean N content in the total litter was significantly lower in group CP-3 (36.2 ± 1.64 g/kg DM) than in all other groups. Additionally, group CP-2 (46.2 ± 1.30 g/kg DM) showed significantly lower concentrations of N in the total litter compared to the material from the groups CP-1 (49.7 ± 1.57 g/kg DM) and CP-C (51.2 ± 2.22 g/kg DM).

Table 5. Nitrogen content in excreta during the five-week trial period (mean ± SD).

Item	Group	Week 1	Week 2	Week 3	Week 4	Week 5
Nitrogen content [in g/kg DM; end of week]	CP-C	44.7 ^{A} ± 2.16	43.5 ^{A} ± 1.45	44.8 ^{A} ± 1.28	46.3 ^{A} ± 1.43	50.0 ^{A} ± 3.13
	CP-1	43.4 ^{A,B} ± 2.05	38.7 ^{B} ± 0.71	42.1 ^{B} ± 1.79	45.0 ^{A} ± 1.59	47.9 ^{A} ± 3.02
	CP-2	40.4 ^{A,B} ± 2.12	36.4 ^{C} ± 2.01	36.4 ^{C} ± 2.56	41.2 ^{B} ± 1.94	42.4 ^{B} ± 2.64
	CP-3	35.4 ^{C} ± 3.03	29.1 ^{D} ± 0.86	28.4 ^{D} ± 1.29	29.7 ^{C} ± 1.63	31.7 ^{C} ± 1.99

^{A,B,C} averages differ significantly within a column ($p < 0.05$).

The level of FI and the protein content in the feed determined the average total N intake per box. The birds in group CP-2 showed the significantly highest FI compared to groups CP-1 and CP-3 (per animal; CP-C: 3169 ^{B} g, CP-1: 3210 ^{A,B} g; CP-2: 3323 ^{A} g; CP-3 3107 ^{C} g, respectively). Total N uptake was significantly higher in the control group than in CP-2 and CP-3 (Table 6). The total amount of N in the litter material harvested at the end of the experiment was gradually reduced analogous to the protein reduction in the feed. The group CP-2 showed the highest absolute weight gain per box and accordingly the highest absolute N retention per box compared to the other groups.

Table 6. Estimations of nitrogen balance over the entire trial period (mean ± SD).

Group	N-Intake From Feed [g/box]	N-Amount in Final Litter [g/box]	Weight Gain Δ Final-Start [g/box]	N-Retained [g/box]
CP-C	1517 ^{A} ± 52.5	581 ^{A} ± 39.6	32,033 ^{B,C} ± 781	938 ^{B} ± 22.8
CP-1	1484 ^{A,B} ± 52.9	522 ^{B} ± 28.2	32,923 ^{A,B} ± 1147	938 ^{B} ± 22.9
CP-2	1451 ^{B} ± 35.8	477 ^{C} ± 25.3	34,190 ^{A} ± 987	1001 ^{A} ± 28.8
CP-3	1278 ^{C} ± 30.8	416 ^{D} ± 20.0	31,220 ^{C} ± 803	914 ^{B} ± 23.5

^{A,B,C,D} averages differ significantly within a column ($p < 0.05$).

The N retention efficiency in the groups CP-C and CP-1 was significantly worse than in the groups CP-2 and CP-3 (Table 7). Compared to the control, the efficiency could be improved step by step by 1.4, 7.1 and 9.6 percentage points from CP-1 to CP-3, respectively.

Table 7. Estimations of efficiencies of nitrogen utilization (mean ± SD).

Group	N-Retention Efficiency * [%]	N-Efficiency Ratio † [g/g]	N-Excretion (Apparent) ‡ [g/box]	N-Excretion (Apparent) ‡ [g/animal]
CP-C	61.9 ^{C} ± 1.46	21.1 ^{D} ± 0.50	579 ^{A} ± 39.8	38.6 ^{A} ± 2.66
CP-1	63.3 ^{C} ± 2.76	22.2 ^{C} ± 0.17	546 ^{A} ± 56.7	36.4 ^{A} ± 3.78
CP-2	69.0 ^{B} ± 0.81	23.6 ^{B} ± 0.28	450 ^{B} ± 15.2	30.0 ^{B} ± 1.01
CP-3	71.5 ^{A} ± 0.70	24.4 ^{A} ± 0.24	364 ^{C} ± 12.6	24.2 ^{C} ± 0.84

^{A,B,C,D} averages differ significantly within a column ($p < 0.05$). * Nitrogen retention efficiency (%) = (g N retained/g N consumed) × 100; † Nitrogen efficiency rate = g weight gain/g N consumed; ‡ Nitrogen excretion (apparent) = g N consumed − g N retained.

3.3. Footpad Dermatitis

In this trial there were no clinical issues concerning foot pad health. The foot pad health was very good overall (Table 8). A comparison between groups showed significant differences at the end of the third week. In group CP-2, the scores were significantly higher.

Table 8. Foot pad (FPD) scores in broilers during the five-week trial period (mean ± SD).

Item	Group	Week 1	Week 2	Week 3	Week 4	Week 5
FPD Scores [averages of both feet; end of week]	CP-C	0.04 ± 0.13	0.46 ± 0.38	0.94 ^{B} ± 0.20	0.97 ± 0.15	0.68 ± 0.39
	CP-1	0.03 ± 0.12	0.43 ± 0.37	1.02 ^{A,B} ± 0.23	0.97 ± 0.12	0.61 ± 0.41
	CP-2	0.05 ± 0.17	0.38 ± 0.37	1.04 ^{A} ± 0.23	0.97 ± 0.12	0.63 ± 0.40
	CP-3	0.03 ± 0.12	0.45 ± 0.38	0.99 ^{A,B} ± 0.28	1.01 ± 0.12	0.74 ± 0.41

^{A,B} averages differ significantly within a column ($p < 0.05$).

3.4. Histological Investigations

There were no differences in the results of the histological examinations (Table 9).

Table 9. Average villus height * and villus width in the ileum of broilers (mean ± SD).

Item	Group	Dissection
Villus height * [in μm]	CP-C	454 ± 68.3
	CP-1	448 ± 58.1
	CP-2	455 ± 50.4
	CP-3	446 ± 75.9
Villus width † [in μm]	CP-C	139 ± 23.0
	CP-1	142 ± 18.1
	CP-2	140 ± 26.9
	CP-3	140 ± 25.8

* Villus height was measured from the tip of the villi to the villus crypt junction; † villus width was measured at the base of the villus above the villus crypt junction.

The intestinal wall of the ileum showed no alterations (Figure 2).

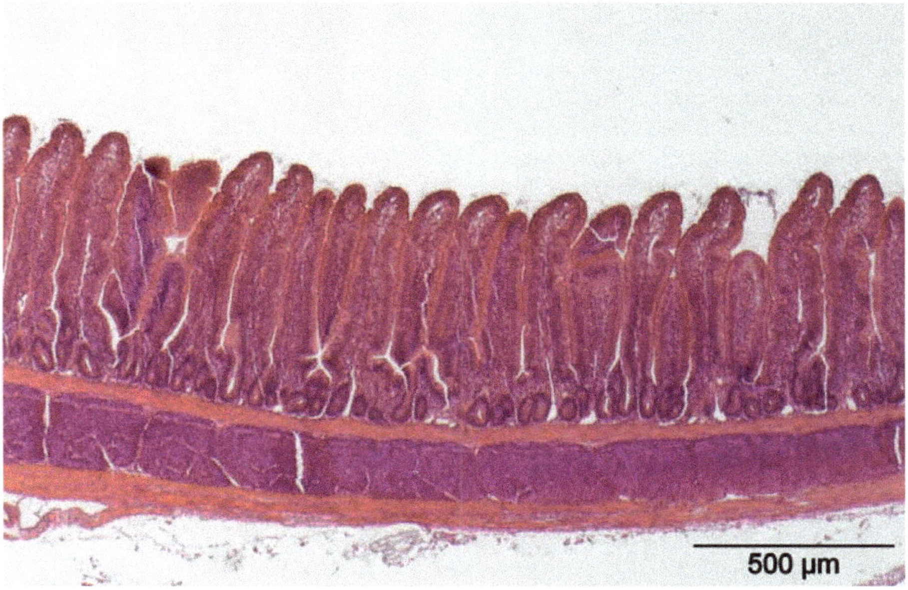

Figure 2. HE staining from ileal intestinal wall of broiler chickens with CP-1 diet. Scale bars = 500 µm.

4. Discussion

The investigations were conducted without incidents. Animal losses were very low (1.66% mortality) from day 14 onwards. Performance parameters were above performance goals of the breeding company [29]. With a final average body weight between 2131 g (CP-3) and 2329 g (CP-2), the body weight development excelled the guidelines of the breeding organization (2050 g). Thus, the performance was higher in each group (CP-C: +6.93%; CP-1: +9.46%; CP-2: +13.6%; CP-3: +3.95%, respectively). Moreover, the feed utilization per kg body weight gain was lower in all groups compared to the performance goals (cumulative feed expenditure in accordance with Aviagen [29]: 1528; reduction: CP-C: −3.53%; CP-1: −4.65%; CP-2: −4.84%; CP-3: −2.55%, respectively). Moran et al. [11] observed in early studies on CP reduction (downwards to 17.8%; weeks 4–6) an increase in feed to gain ratio from 1.72 to 1.76. However, the rations were balanced only with L-lysine HCl, DL-methionine, and L-threonine. The arginine–lysine ratio was therefore not optimal, this being 1.04 in the first three weeks (20.8% CP) and 0.99 in weeks 4–6 (17.8% CP). In a frequently cited study by Bregendahl et al. [12], the performance deteriorated when reducing the protein content to a level comparable to ours in the present study. In the study mentioned above [12], at protein levels of 185–186.5 g CP per kg diet and supplementation of identical amino acids used in our study, complemented by glutamine, the performance was lower compared to the control (23% CP). Nonetheless, the two investigations differ in terms of the absolute levels of supplemented amino acids. Especially methionine seemed to be much higher in the study of Bregendahl et al. [12]. Namroud et al. [15] were able to show a comparable body weight in broilers using crystalline amino acids in protein reduced diets (reduction of 23% to 19% CP at day 28). A further reduction to 17% CP led to a significant decrease in the final body weight at day 28 by 17.5%. In comparison to Namroud et al. [15], no comparable effect occurred in the present study, although reduction in CP content was nearly the same (a decrease to 17.7% percent by day 14 or to 17% thereafter). The greatest difference between Namroud et al. [15] and the present trial can be seen in the methionine content (approx. 1 g higher for Namroud et al. [15]), and leucine (up to 5 g surplus in the study of Namroud et al. [15]). In studies with free-range chickens, however, observations were in line with the present study [16]. The performance was comparable when reducing from 19% to 17% CP and balancing essential amino acid content. From these observations, it can be concluded that with increasing crude protein reduction, the targeted application of certain amino acids must be given more emphasis if performance is to be maintained. Feed costs represent about 70% of the cost of poultry production [30]. Therefore, this makes a bird's ability to use feed efficiently very important [30]. Precisely following the nutritional requirements of poultry is the guarantor for optimum feed efficiency [30]. Therefore, studies like this can help to maintain profitability while at the same time taking into account other aspects of sustainability like ecological dimensions of production.

In the present investigations, the N content in the excreta could be lowered significantly using CP reduced diets. The N content in the excreta was reduced by up to 36.6% at a comparable performance by reducing the protein content from 20% to 17% (reduction in N content in excreta compared to CP-C in week 5: CP-1: −4.20%; CP-2: −15.2%; CP-3: −36.6%, respectively). The ratios were also reflected in the content of the entire litter material. The protein reduction of up to 3% starting from a diet with moderate protein content in the present study resulted in a reduction in the N content in the total litter material of up to 37.3% (N reduction in litter compared to CP-C: CP-1: −5.70%; CP-2: −22.3%; CP-3: −37.3%, respectively). The N efficiency could be further increased by almost 10 percentage points (CP-C: 61.9%; CP-3: 71.5%, respectively) due to the feeding concept, in spite of an already high efficiency. The present investigations are in line with the results of Ospina-Rojas et al. [9]. This research group was able to reduce the N content in the litter from 47.2 g N/kg DM to 31.9 g N/kg DM (−32.4%) by reducing the protein content in the diet from 19% to 16% by supplementing valine, isoleucine, arginine and glycine achieving a constant performance. When supplementation of valine, isoleucine and glycine or valine, isoleucine and arginine or of all four aforementioned amino acids was not made, a significant performance depression occurred [9]. Blair et al. [10] were able to achieve a reduction in the N content in the excreta from 52.5 g/kg DM to 47.2 g/kg DM without affecting the performance by reducing the protein contents from 21% to 18% in weeks 3 to 6 while balancing essential amino acids. The N retention could be increased from 47% to 51%. In the same study, the N retention could be increased from 61% to 63% under conditions of a total of 6 weeks' adjustment in the protein content. Namroud et al. [15] were able to reduce the N content in excreta from 50.3 g/kg DM to 36.3 g/kg DM (−27.8%) by achieving a constant amino acid content in the diet, reducing the crude protein content from 23% to 17%. Shao et al. [16] were able to reduce the N content in the excreta from 65.2 g/kg DM to 47.3 g/kg DM (−27.4%) in free-range chickens by reducing the protein content from 19% to 17%. In the present investigations, the reduction in the N content in excreta or litter material was higher in its peak than in the aforementioned publications. Overall, the protein reduction was accompanied to a certain extent by an increased performance (CP-2), which was also reflected in the high absolute N efficiency. The current global production of ammonia, CH_4, and N_2O by the poultry industry is significant [30]. Improvements in feed conversion ratio have a favorable effect on environmental emissions and decrease the environmental impact of poultry production [30]. Therefore, the present investigations thus show that the environment can be relieved to the maximum by continuously optimizing the rations.

In this study, no clinical problems concerning the foot pad health of the broilers could be determined, independent of CP level in the diets. After 34 days of fattening, nearly 100% of the broilers had an FPD score ≤1, despite the high stocking density of about 35 kg/m^2. Only one animal in group CP-C and three animals in group CP-3 had a score of 2 in one foot. Therefore, the results of this study show neither negative nor positive effects of using the different dietary concepts.

The results of the histological examination were used to check whether the reduction in the protein content also had an influence on intestinal histology. The results of the study gave no indication of this.

With maximum protein reduction (CP-3), we used a diet containing kaolin at the beginning of the third fattening week effectuating a dilution of the ration. Due to the kaolin the effect of differences in diet composition between the feeding phases and between the groups in one feeding phase (80% supplement concept) was minimized. However, it cannot be ruled out that this ingredient might have influenced the FI. The results of the investigations suggest that further analyses would be useful. In these investigations the use of kaolin should be avoided on the one hand. Furthermore, it should be examined to what extent very expensive amino acids such as isoleucine etc. have to be completely balanced in the rations. If reductions are possible, this would allow the already economical concept to be continuing optimized and at the same time the environmental burden could be further reduced.

5. Conclusions

The potential environmental impact is an important evaluation factor for the sustainability of animal production. Meticulous balanced protein-reduced diets allow a significant reduction in the use of nutrients, especially in high N containing components. Under these conditions, the adjustment of the amino acid content becomes more and more important if the performance is at least kept constant. As the degree of crude protein reduction increases, N excretion falls disproportionately. This nutritional intervention enables significant improvements in terms of the environmental impact of animal production.

Author Contributions: Conceptualization, C.V., V.T. and D.S.; Methodology, C.V., M.L. and R.B.; Validation, C.V., C.U., M.L. and R.B.; Formal Analysis, C.U. and C.V.; Investigation, C.U., M.L. and C.V.; Resources, C.V., V.T. and R.B.; Data Curation, C.U. and C.V.; Writing—Original Draft Preparation, C.U. and C.V.; Writing—Review & Editing, C.U., M.L., R.B., V.T., D.S. and C.V.; Visualization, C.U., C.V., M.L. and R.B.; Supervision, C.V. and R.B.; Project Administration, C.V. and R.B.; Funding Acquisition, C.V.

Funding: The study was financially supported by CJ Europe GmbH, Schwalbach, Germany. The role of the funding body was merely financial. The CJ Europe GmbH did not have any role in collecting, interpreting, analyzing or writing up the data. This publication was also supported by the Deutsche Forschungsgemeinschaft and University of Veterinary Medicine Hannover, Foundation, Germany within the funding program Open Access Publishing.

Acknowledgments: We would like to thank Frances Sherwood-Brock for proof-reading the manuscript to ensure correct English.

Conflicts of Interest: Venja Taube is an employee of BEST 3 Gefluegelernaehrung GmbH, Twistringen, Germany. The BEST 3 Gefluegelernaehrung GmbH had no role in the design of the study; in the collection, analyses, or interpretation of data; in the writing of the manuscript, or in the decision to publish the results. Diana Siebert is an employee of CJ Europe GmbH, Schwalbach, Germany. The CJ Europe GmbH had no role in the design of the study; in the collection, analyses, or interpretation of data; in the writing of the manuscript, or in the decision to publish the results.

References

1. De Vries, M.; de Boer, I.J.M. Comparing environmental impacts for livestock products: A review of life cycle assessments. *Livestig. Sci.* **2010**, *128*, 1–11. [CrossRef]
2. Steinfeld, H.; Gerber, P.; Wassenaar, T.; Castel, V.; Rosales, M.; Rosales, M.; de Haan, C. *Livestock's Long Shadow: Environmental Issues and Options*; Food & Agriculture Org.: Rome, Italy, 2006.
3. Leip, A.; Billen, G.; Garnier, J.; Grizzetti, B.; Lassaletta, L.; Reis, S.; Simpson, D.; Sutton, M.A.; De Vries, W.; Weiss, F. Impacts of european livestock production: Nitrogen, sulphur, phosphorus and greenhouse gas emissions, land-use, water eutrophication and biodiversity. *Environ. Res. Lett.* **2015**, *10*, 115004. [CrossRef]
4. Meier, M.S.; Stoessel, F.; Jungbluth, N.; Juraske, R.; Schader, C.; Stolze, M. Environmental impacts of organic and conventional agricultural products—Are the differences captured by life cycle assessment? *J. Environ. Manag.* **2015**, *149*, 193–208. [CrossRef] [PubMed]
5. Ferket, P.R.; van Heugten, E.; van Kempen, T.A.T.G.; Angel, R. Nutritional strategies to reduce environmental emissions from nonruminants. *J. Anim. Sci.* **2002**, *80*, E168–E182. [CrossRef]
6. Mallin, M.A.; Cahoon, L.B. Industrialized animal production—A major source of nutrient and microbial pollution to aquatic ecosystems. *Popul. Environ.* **2003**, *24*, 369–385. [CrossRef]
7. Clark, M.; Tilman, D. Comparative analysis of environmental impacts of agricultural production systems, agricultural input efficiency, and food choice. *Environ. Res. Lett.* **2017**, *12*, 064016. [CrossRef]
8. Leip, A.; Weiss, F.; Lesschen, J.; Westhoek, H. The nitrogen footprint of food products in the european union. *J. Agric. Sci.* **2014**, *152*, 20–33. [CrossRef]

9. Ospina-Rojas, I.C.; Murakami, A.E.; Duarte, C.R.A.; Eyng, C.; Oliveira, C.A.L.; Janeiro, V. Valine, isoleucine, arginine and glycine supplementation of low-protein diets for broiler chickens during the starter and grower phases. *Br. Poult. Sci.* **2014**, *55*, 766–773. [CrossRef] [PubMed]
10. Blair, R.; Jacob, J.P.; Ibrahim, S.; Wang, P. A quantitative assessment of reduced protein diets and supplements to improve nitrogen utilization. *J. Appl. Poult. Res.* **1999**, *8*, 25–47. [CrossRef]
11. Moran, E.T.; Bushong, R.D.; Bilgili, S.F. Reducing dietary crude protein for broilers while satisfying amino-acid-requirements by least-cost formulation—Live performance, litter composition, and yield of fast-food carcass cuts at 6 weeks. *Poult. Sci.* **1992**, *71*, 1687–1694. [CrossRef]
12. Bregendahl, K.; Sell, J.L.; Zimmerman, D.R. Effect of low-protein diets on growth performance and body composition of broiler chicks. *Poult. Sci.* **2002**, *81*, 1156–1167. [CrossRef] [PubMed]
13. Laudadio, V.; Passantino, L.; Perillo, A.; Lopresti, G.; Passantino, A.; Khan, R.; Tufarelli, V. Productive performance and histological features of intestinal mucosa of broiler chickens fed different dietary protein levels. *Poult. Sci.* **2012**, *91*, 265–270. [CrossRef] [PubMed]
14. Laudadio, V.; Dambrosio, A.; Normanno, G.; Khan, R.U.; Naz, S.; Rowghani, E.; Tufarelli, V. Effect of reducing dietary protein level on performance responses and some microbiological aspects of broiler chickens under summer environmental conditions. *Avian Biol. Res.* **2012**, *5*, 88–92. [CrossRef]
15. Namroud, N.F.; Shivazad, M.; Zaghari, M. Effects of fortifying low crude protein diet with crystalline amino acids on performance, blood ammonia level, and excreta characteristics of broiler chicks. *Poult. Sci.* **2008**, *87*, 2250–2258. [CrossRef] [PubMed]
16. Shao, D.; Shen, Y.R.; Zhao, X.; Wang, Q.; Hu, Y.; Shi, S.R.; Tong, H.B. Low-protein diets with balanced amino acids reduce nitrogen excretion and foot pad dermatitis without affecting the growth performance and meat quality of free-range yellow broilers. *Ital. J. Anim. Sci.* **2018**, *17*, 698–705. [CrossRef]
17. Khajali, F.; Wideman, R. Dietary arginine: Metabolic, environmental, immunological and physiological interrelationships. *World Poult. Sci. J.* **2010**, *66*, 751–766. [CrossRef]
18. Wu, G.; Bazer, F.W.; Dai, Z.; Li, D.; Wang, J.; Wu, Z. Amino acid nutrition in animals: Protein synthesis and beyond. *Annu. Rev. Anim. Biosci.* **2014**, *2*, 387–417. [CrossRef] [PubMed]
19. Council, N.R. *Nutrient Requirements of Poultry*; National Academic Press: Washington, DC, USA, 1994.
20. Corzo, A.; Kidd, M. Arginine needs of the chick and growing broiler. *Int. J. Poult. Sci.* **2003**, *2*, 379–382.
21. Corzo, A.; Moran, E., Jr.; Hoehler, D. Arginine need of heavy broiler males: Applying the ideal protein concept. *Poult. Sci.* **2003**, *82*, 402–407. [CrossRef] [PubMed]
22. Siebert, D.; Khan, D.R.; Esteve-Garcia, E. Investigations on the optimal arginine:lysine ratio in the starter phase of meat-type chickens. In *Proceedings of the 21st European Society of Veterinary and Comparative Nutrition Congress*; Harris, P., Moore-Colyer, M., Eds.; European Society of Veterinary & Comparative Nutrition: Cirencester, UK, 2017; Volume 21, p. 144.
23. Mayne, R.; Else, R.; Hocking, P. High litter moisture alone is sufficient to cause footpad dermatitis in growing turkeys. *Br. Poult. Sci.* **2007**, *48*, 538–545. [CrossRef] [PubMed]
24. Slaoui, M.; Fiette, L. Histopathology procedures: From tissue sampling to histopathological evaluation. In *Drug Safety Evaluation*; Springer: Cham, Switzerland, 2011; pp. 69–82.
25. Naumann, C.; Bassler, R. *Methoden der Landwirtschaftlichen Forschungs-und Untersuchungsanstalt, Biochemische Untersuchung von Futtermitteln*; Methodenbuch iii (einschließlich der achten Ergänzungen); VDLUFA: Darmstadt, Germany, 2012.
26. Aletor, V.A.; Hamid, I.I.; Niess, E.; Pfeffer, E. Low-protein amino acid-supplemented diets in broiler chickens: Effects on performance, carcass characteristics, whole-body composition and efficiencies of nutrient utilisation. *J. Sci. Food. Agric.* **2000**, *80*, 547–554. [CrossRef]
27. Ernährungsphysiologie, G.F. *Empfehlungen zur Energie-und Nährstoffversorgung der Legehennen und Masthühner (Broiler)*; DLG Verlag: Frankfurt am Main, Germany, 1999.
28. Kamran, Z.; Sarwar, M.; Nisa, M.; Nadeem, M.; Ahmad, S.; Mushtaq, T.; Ahmad, T.; Shahzad, M. Effect of lowering dietary protein with constant energy to protein ratio on growth, body composition and nutrient utilization of broiler chicks. *Asian Austral J. Anim.* **2008**, *21*, 1629–1634. [CrossRef]

29. Aviagen. *Ross 308 Broiler Performance Objectives*; Aviagen Group: Huntsville, AL, USA, 2014.
30. Willems, O.; Miller, S.; Wood, B. Aspects of selection for feed efficiency in meat producing poultry. *World Poult. Sci. J.* **2013**, *69*, 77–88. [CrossRef]

sustainability

MDPI

Article

Production of Pig Feed under Future Atmospheric CO_2 Concentrations: Changes in Crop Content and Chemical Composition, Land Use, Environmental Impact, and Socio-Economic Consequences

Henrik Saxe [1,*], Lorie Hamelin [2], Torben Hinrichsen [3] and Henrik Wenzel [4]

1 Mindful Food Solutions, Engbakkevej 3C, 2900 Charlottenlund, Denmark
2 Department of Engineering of Biological Systems and Processes (LISBP), National Institute of Applied Sciences (INSA), INRA UMR792 and CNRS UMR5504, Federal University of Toulouse, 135 Avenue de Rangueil, F-31077 Toulouse, France; hamelin@insa-toulouse.fr
3 DSM Nutritional Products Ltd., Kirkebjerg Alle 88, 2605 Brøndby, Denmark; Torben.hinrichsen@dsm.com
4 Institute of Chemical Engineering, Biotechnology and Environmental Technology, SDU Life Cycle Engineering, University of Southern Denmark, Campusvej 55, 5230 Odense M, Denmark; Henrik.wenzel@kbm.sdu.dk
* Correspondence: Henriksaxe@gmail.com; Tel.: +45-60742225

Received: 14 August 2018; Accepted: 2 September 2018; Published: 6 September 2018

Abstract: With the rising atmospheric CO_2, crops will assimilate more carbon. Yields will increase in terms of carbohydrates while diluting the content of protein and minerals in compound pig feed, calling for an altered formulation with more protein and less carbohydrate crops to maintain its nutritional value. Using crop response data from CO_2 exposures in a linear modeling of feed formulation, we apply a consequential life cycle assessment (cLCA) to model all of the environmental impacts and socio-economic consequences that altered crop yields and chemical composition at elevated CO_2 levels have on feed formulation, targeting altered amino acid contents rather than overall protein. An atmospheric CO_2 of 550 µmole mole^{-1} gives rise to a 6% smaller demand for land use for pig feed production. However, feed produced at this CO_2 must include 23% more soymeal and 5% less wheat than at present in order to keep its nutritional value. This counteracts the yield benefit. The monetized environmental cost of producing pig feed, where sunflower and soy contribute the most, equals the direct feed price in both scenarios. If external costs were internalized, honoring the Rio Declaration, feed prices would double. In contrast, the future composition of pig feed will increase the direct price by only 0.8%, while the external cost decreases by only 0.3%.

Keywords: amino acids; compound pig feed formulation; consequential life cycle assessment; cLCA; land-use changes (LUC); monetized environmental impact; protein crops; starch (energy) crops

1. Introduction

During this century, the rising atmospheric carbon dioxide concentration (CO_2) is expected to increase crop yields due to increased carbon assimilation, for example in wheat grains, by 10% [1]. On the other hand, the higher carbon uptake causes increased starch content, and consequently also a dilution of the relative protein and mineral content in food and feed crops [2]. Additional consequences of future atmospheric CO_2 include altered patterns of temperature and precipitation, which together with regional changes, e.g., ozone pollution and nitrogen (N) fertilization, will further modify plant growth and its chemical composition.

The animal feed market, which currently absorbs approximately 45%, 58%, and 80% of the world cereal, maize, and soy, respectively [3], is likely to be affected by such changes in crop yield and composition. In fact, since the composition of animal feed for intensive livestock production is carefully optimized in order to meet both the applying legislation and the nutritional requirements in a cost-effective manner, any change in both the chemical composition and the yield of feed crops will affect the future formulation of the feed.

Although the environmental consequences of the forecasted future increases in meat (and thus feed) demand have been overly studied (e.g., references [4–6]), there are no studies, to the authors' knowledge, that have attempted to address the environmental impacts related to the changes that a higher CO_2 would induce to the feed market. From an environmental perspective, a (CO_2-induced) higher crop yield would contribute to reduce both the land use and the land-use changes that are associated with increased demand for animal feed. On the other hand, lower protein content would trigger an increased need for protein-rich crops such as soybeans, which in turns would involve an increased land-use change, the environmental impact of which may be considerable (e.g., references [7–9]). Yet, it is not clear which effect would dominate over the other. Further, a change in protein content has to be addressed with regards to the specific amino acids affected, as a change in some non-essential amino acids (e.g., proline) is rather useless for the overall feed composition.

Focusing on the effects of elevated CO_2 alone, this study endeavors to assess the environmental consequences, the associated environmental cost, and the direct price of compound pig feed based on the yield and chemical composition of crops grown at the "present" atmospheric CO_2 (taken at 380 µmole CO_2 mole^{-1} dry air) and at a "future" CO_2 (550 µmole mole^{-1} dry air by 2050 [10]). From this point onwards, the former will be referred to as "present feed", and the latter will be referred to as "future feed", with the understanding that these are both compound feeds. Further, the socio-economic consequences are, in this study, represented by the sum of the direct price and the cost of environmental impacts (so-called shadow price) of producing compound pig feed. Compound feed means multi-component animal feed produced at a competitive price to satisfy all of the nutritional and technical needs in terms of carbohydrates, protein, oils, fiber, enzymes, vitamins, etc.

2. Materials and Methods

2.1. *Life Cycle Assessment (LCA) Model*

Consequential life cycle assessment (cLCA, e.g., references [11,12]) modeling was used to assess the environmental burdens of producing one extra tonne (the functional unit) of pig feed produced today (present CO_2 scenario) and in a future with 550 µmole CO_2 mole^{-1} (future CO_2 scenario).

The life cycle impact assessment was carried out using the stepwise method, which allows for the full monetarization of environmental impacts (expressed in €2003). This method combines characterization models from IMPACT2002 + v2.1 and EDIP2003 and is further described in reference [13]. The assessment was facilitated with the LCA software Simapro® 7.3.3. Translating all of the environmental impact into monetized costs enables the selection and presentation of the most important environmental impact categories causing the socio-economic impacts of compound pig feed (other impact categories are presented in Supplementary). Biogenic carbon (C), including soil C changes and C sequestration by feed crops, was included in all of the calculations in order to allow for a transparent carbon balance, as described by reference [14].

Background LCA data were taken from the Ecoinvent v.2.2 database [15], while foreground data were essentially related to the cultivation, yield, and chemical composition of the crops used in feed formulation. These are described in Section 2.4 and Supplementary. The geographical scope considered for the foreground system was Denmark, i.e., the pig feed was considered to be manufactured for the Danish market. The Danish example is taken as representative for most industrialized countries in a temperate climate.

2.2. LCA System Boundary

All of the processes affected by the production of one tonne of pig feed, i.e., from crop cultivation to harvest, and up to the mixing of the feed, were included within the LCA system boundary. Direct as well as indirect land-use changes were also included in the study (Section 2.4). As the focus of the study is on the production of the feed itself, processes downstream from the feed production (e.g., consumption by the animals) were excluded, as these were outside of the study's scope.

Based on the cLCA principles, system expansion was performed for all of the feed ingredients affected by the considered change of atmospheric CO_2.

2.3. Feed Formulation and Life Cycle Inventory (LCI)

2.3.1. Formulation of Present and Future Pig Feed

To apply a market-based approach, the commercial feed formulation software Bestmix® [16] was used to establish both the present and future feed formulation for piglets, sows, and slaughter pigs (a feed mix consisting of 20% piglet feed, 20% sow feed, and 60% fattening pig feed was considered, because this is what is used in a normal pig life cycle). Through optimization algorithms, Bestmix® formulated the best compromise between profitability, nutritional value, and animal health, while ensuring that the Danish and European legal requirements of the feed were met. This procedure allowed determining the exact ingredient proportions constituting the present and future pig feed.

One important input to Bestmix® is the chemical composition of crop ingredients, in terms of starch, protein, amino acids, macronutrients, and micronutrients. Another important input parameter to Bestmix® is the price of feed ingredients. However, as the aim of this study is primarily to investigate the consequences of altered crop composition, and since no reliable estimate is available on future crop prices, the prices of crop ingredients under a high CO_2 future were taken to be equal to the present prices.

The chemical composition of all of the crop ingredients, when grown under today's atmospheric CO_2, was established based on the Danish Pig Research Centre feed ingredients database [17]. Data on the yield and chemical composition of the crop ingredients grown at 550 μmole CO_2 mole^{-1} was established based on data from growth experiments with increased CO_2 exposure taken from the scientific literature. Reliable data was only found for the four main crop ingredients (Table 1): namely wheat, barley, soy, and rape, which together make up 89% of the present pig feed by dry weight (Table 2).

Since there is a large variability reported in the literature for plant responses to elevated CO_2, the following principles for selection of the best data were adopted. Results from Free Air Carbon Dioxide Exposure (FACE) systems were assumed to provide the most reliable data, since they represent the most realistic growing conditions, although FACE responses are typically smaller than for other exposure systems, see e.g., references [18,19]. When FACE data were not available, data from open-top chambers (OTC), from outdoor closed-top chambers (CTC), or from climate chambers were used, in that order of ranking. As crops respond differently to elevated atmospheric CO_2 with different climatic or other abiotic or biotic conditions, exposure experiments that were performed near to where the feed ingredients that were considered in this study were grown were preferred over other studies. Even with the same exposure technique, all of the studies give different results for each year, and with each plant species, it varies with the studied cultivar. Under these conditions, either the best available reference was applied for each species and chemical component, or data from meta studies were applied. When data for future chemical content was not found for a given component (e.g., amino acids), the future content was presumed to be the same as today. The chosen data sources are summarized in Table 1.

2.3.2. Life Cycle Inventory (LCI) Data for Crop and Non-Crop Ingredients

Crop Ingredients

The LCI of crops grown in Denmark (wheat and barley) was based on a recent Danish consequential LCI dataset [14], which comprises all of the processes involved during the cultivation stage, up to harvest. This includes the tillage activities, liming, seed propagation, plant protection, fertilization, sowing, harvest, and transport from farm to field. A sandy soil has been considered for both crops, as well as precipitations of 964 mm y^{-1} and removal (harvest) of the straw. LCI data from the Ecoinvent (v.2.2) database were used to model the cultivation of imported soybean (Brazil), rapeseed (Germany), sunflower (Spain), and palm oil (Malaysia). Other important foreground data for crop ingredients are thoroughly described in the appendices.

Industrial Amino Acids

As shown in Table 2, two types of industrial amino acids are included in the assessed pig feed: industrial amino acids produced by fermentation (lysine, threonine, and tryptophan) and industrial amino acids produced synthetically (dl-methionine). Based on reference [19], a generic recipe was considered for 1 kg of amino acids produced by fermentation; 1 kg of sugar (from sugar beet), 0.5 kg of maize starch, 0.5 kg of wheat starch, 0.3 kg of liquid ammonia, and 36 MJ of process energy at the plant. In this study, the recipe was supplemented with 0.053 kg of sulfuric acid per kg amino acid, and 4.6 g of phosphorous based on data from the Danish lysine supplier (see Supplementary). Under Danish production conditions, the energy consumption was estimated at only 18 MJ, which was supplied as electricity (50%) and natural gas (50%). On the basis of reference [20], it was assumed that the production of 1 kg of dl-methionine required 0.43 kg of propylene, 0.27 kg of hydrogen sulfide, 0.39 kg of methanol, 0.21 kg of hydrogen cyanide, and 7.4 MJ of process energy at the plant, which was supplied by electricity (50%) and natural gas (50%).

Enzymes

No specific LCI data were found for the added enzymes (phytase, xylanase). Instead, published LCA data for phytase (i.e., characterization results for global warming, acidification, nutrient enrichment, photochemical ozone formation, land use, phosphorus consumption, and energy consumption) were used for both types of enzymes [21].

Other Ingredients

In the absence of LCI data for the vitamins added to the compound pig feed, these ingredients were calculated as for enzymes. LCI data for limestone meal ($CaCO_3$), monocalcium phosphate (MCP), phosphoric acid, and sodium chloride (NaCl) were taken from the Ecoinvent (v.2.2) database. LCI data for fish meal were taken from the LCA food database ("industrial fish, ex harbor", as available in Simapro 7.3.3).

2.4. Land-Use Changes (LUC)

In order to quantify the environmental consequences of land-use changes, three main steps are required: (i) determining the total land area undergoing conversion to cropland; (ii) identifying which land types (i.e., biomes) are converted; and (iii) determining the changes in carbon flows from these conversions. Direct and indirect land-use change (LUC) have been extensively studied over the past decade for all of the major global ecosystems [8,22–25]. LUC also impacts biogeochemical cycling [26,27] and biodiversity [28], but these aspects have not yet been sufficiently quantified, and can therefore not be included in the present analyses.

2.4.1. Land Conversion: Determination of the Area and Geographical Location (Steps i and ii)

In this study, a deterministic explorative approach based on crop yields (Table 1), carbohydrate content, lysine content, and historical market trends has been used in order to determine the above-mentioned steps (i) and (ii). This is further described below. As a starting point, the LUC involved in this study may be classified in two distinct categories: (a) those triggered by the displacement of carbohydrates crops, and (b) those triggered by the displacement of oil crops.

The former represents the LUC resulting from the cultivation of crops whose increased demand (as a result of producing one extra tonne of Danish pig feed) involves the displacement of a carbohydrate crop. This can be illustrated by the case of wheat (Figure 1). As shown in Figure 1, an extra demand of one tonne of Danish pig feed induces a need for an additional 480 kg of wheat (present feed) and 460 kg of wheat (future scenario). These figures were determined from Bestmix® data, as presented in Table 2.

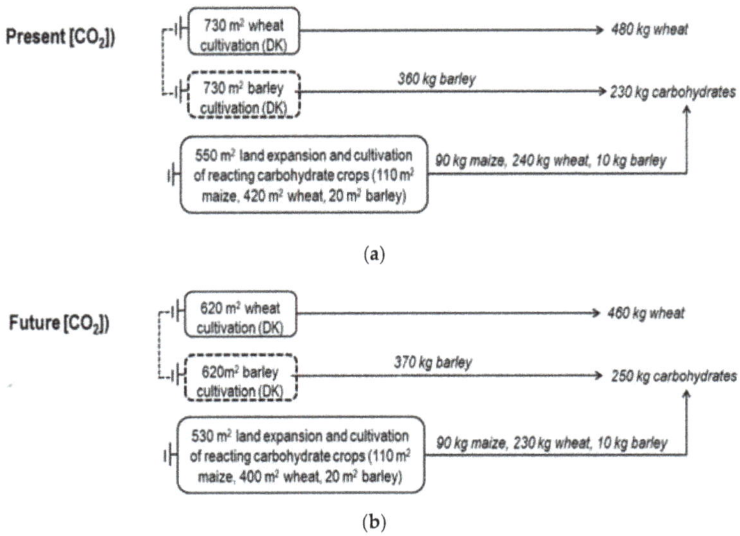

Figure 1. Consequential life cycle assessment (LCA) model for wheat grown (**a**) at the present CO_2 and (**b**) at the future CO_2. Dotted lines indicate avoided flows. All of the flows are related to the functional unit, i.e., 1 tonne of compound Danish pig feed.

In a country such as Denmark, where 65% of the total land is used for cropland, and where policies have been adopted in order to double the forested area (nowadays representing ca. 13% of the total land [29]), very limited conversion from forest or similar nature types is realistic. Most likely, the land that is needed to grow this extra wheat will be taken from actual Danish cropland, which means that one crop cultivated today will be displaced. Such a displaced crop is, in cLCA, referred to as the marginal crop. In this study, the marginal crop was assumed to be spring barley (a carbohydrate crop), based on references [30,31]. The environmental consequences of cultivating wheat instead of spring barley represent, in this case, the so-called direct land-use changes (dLUC). Indirect land-use changes (ILUC), on the other hand, represent the environmental consequences that are related to how this missing supply of Danish spring barley will be supplied on the world market. Such increased crop production may stem from increased yield, which is also referred to as intensification, or from land conversion to cropland, which is also referred to as agricultural land expansion. As illustrated in Figure 1, and as in many ILUC or LCA studies (e.g., references [32–35]), this study includes the environmental impacts of the latter only.

Similar to wheat, an increased demand for land in Europe in order to cultivate the barley, rapeseed meal, and sunflower meal (Table 2) resulting from an extra demand of Danish pig feed was considered to take place at the expense of spring barley. The process flow diagrams for these crops are illustrated in Supplementary.

An extra demand for soybean meal, rapeseed meal, sunflower meal, and palm oil also implies an interaction with the vegetable oil market (category "b" above). For the latter, this interaction is straightforward. For the three former, the interaction is indirect and happens as oil is co-produced with the desired meal. This can be illustrated with the case of soybean meal (Figure S3 in Supplementary). Increased soybean meal production involves increased soy oil co-production, causing a decrease in palm oil production (palm oil being identified as the marginal oil, on the basis of the references [30,31]), and consequently also a decrease in the co-produced palm meal (a source of both protein and carbohydrates), inducing, somewhere in the world, an increase in the production of a marginal protein and carbohydrate feed crop. Since the production of a marginal protein interacts with the oil market again, a loop system is thus created, and this loop should be stopped at the point where the consequences are so small (or the uncertainties so large) that any further expansion of the boundaries would yield no significant information for decision support [36]. Such flows are illustrated in Supplementary for all of the crops interacting with the oil market. A substitution ratio of 1:1 was assumed, i.e., that 1 t of a given vegetable oil would replace 1 t of the marginal oil (palm oil), since it is the long-term effect of the demand that should be guiding for decisions in LCA [37]. In other words, the supply of goods and services was assumed to be fully elastic, and accordingly, short-term effects are not captured.

For LCA models of soy, rape, and sunflower that all involve the displacement of palm meal as a result of the above-described oil loop, soybean meal was identified as the reacting marginal protein [30,31]. Being the market covering the greatest share of the worldwide increase in soybean production [38], soybean from South America (Argentina and Brazil) was identified as the market reacting to an extra demand for soybean meal. Based on a similar logic, palm fruit cultivation from Southeast Asia was identified as the one reacting to an extra demand.

When applying a deterministic approach to land-use changes in LCAs, it has been a common practice to determine the amount of marginal protein crop (here soybean meal) reacting to a decrease in palm meal production on the basis of the protein content of these meals, e.g., references [23,31,39]. However, it is the composition of the protein in terms of amino acids, or rather in terms of the limiting amino acids, that matters for feed. The most important limiting amino acid in pig feed is lysine. Therefore, the reacting amount of soybean meal has been identified on the basis of the lysine content of the meals (Supplementary).

For fermentation-based amino acids, which are produced from of a mix of different crops (Supplementary), the same principles as described above were applied. For each crop system, all of the details of the inventory data that were used for determining the area and geographical location of the land converted are presented in Supplementary. Further, it has been considered, on the basis of reference [34], that only 80% of the new soy and carbohydrates demand is supplied through land expansion, while this was 70% for palm oil. The remaining production must be supplied by the intensification of existing areas (the various types of intensification pathways and the extent to which each is driven by crop prices is discussed in reference [40]. As previously highlighted, the environmental impacts of such intensification were not included in this study. With input-driven intensification, the reduced environmental impact associated with increased crop yield is countered by the environmental impact of an increased use of fertilizers, pesticides, and machinery, among others.

2.4.2. Environmental Consequences of Land Conversion

The dLUC consequences, in terms of changed C and N flows, of cultivating a given crop in Europe instead of spring barley have been included and modeled on the basis of the data from reference [14].

In order to quantify the releases of carbon due to the land converted to cultivate soybean, palm fruit, or the reacting carbohydrate mix, the soil and vegetation carbon data from the Woods Hole Research Centre, as published in reference [8], was used. This allowed the calculation of the CO_2 emitted during land conversion (i.e., step (iii) referred to in the first sentence of Section 2.4), where the following has been considered:

- 25% of the carbon in the soil is released as CO_2 for all types of land-use conversion, except when forests are converted to grassland, where 0% is released;
- 100% of the carbon in vegetation is released as CO_2 for all of the forest types as well as for tropical grassland conversions, while 0% is released for the remaining biome types (e.g., shrub land, non-tropical grassland, chaparral).

The results of this calculation are shown in Supplementary; flows were annualized (distributed equally) over 20 years, which is in line with most LUC calculations used by European policy makers [41]. A similar procedure has also been applied by reference [35].

2.5. Direct Cost

Based on today's market prices for all of the ingredients (as available in Bestmix®), the price of one tonne compound pig feed for the present and future atmospheric CO_2 were calculated.

3. Results and Discussion

3.1. Changes in Crops' Chemical Composition and in the Compound Feed Formulation

Table 1 shows the established present and future chemical composition of wheat, barley, soymeal, and rape meal used as input to Bestmix® [16]. The optimization performed in Bestmix® determined that one tonne of compound pig feed produced under the future relative to the present atmospheric CO_2 will contain 5.1% less wheat, 23% more soy, unchanged amounts of barley, rape, and sunflower, 18% more beet molasses, 65% more hemoglobin meal, 16% less amino acids produced by fermentation (lysine, threonine, tryptophan), and varying amounts of other minor ingredients (Table 2). The results in Table 2 thus support the hypothesis that pig feed based on wheat, barley, and rape grown in Northern Europe under higher CO_2 will need a higher supplement of protein as well as less carbohydrate ingredients. Table 2 also highlights that both present and future feed consist of at least 46% wheat, 25% barley, 9% soy meal, 7% rape meal, and 4% sunflower seeds; these five ingredients making up approximately 93% of the feed.

3.2. Environmental Impact

Characterized LCA results are presented in Figure 2, with a focus on the four socio-economically most important impact categories, which were identified through normalization (Section 3.4); human toxicity in terms of non-carcinogenic organics and respiratory inorganics ($PM_{2.5}$); nature occupation; and global warming. The results for all of the other evaluated impact categories are given in Supplementary (Figures S8 and S9).

The environmental impact of producing one extra tonne of compound pig feed was expected to decrease in the high CO_2 future due to the higher yields (and thus lower land use) caused by the higher CO_2 uptake from the atmosphere. However, even though the yield increases were found to be large, e.g., 10% for wheat and 28% for rape, the net land use did not decrease to the same extent. The net reduction in land use per tonne of pig feed was found to be around 6%. Unexpectedly, this did not lead to a net fall in greenhouse gas emissions per tonne of feed; on the contrary, it increased by 9% for crops grown at the future CO_2 (explained in detail in Supplementary, Figure S10). This is essentially due to the change in feed composition, where considerably more soybean meal, which involves the conversion of biomes with relatively high C stock, is required under a high CO_2. This is reflected in Table 2 and Figure 2.

However, the biggest differences are found for the impact categories of respiratory inorganics (fine particles) and non-carcinogenic human toxicity (17% increase and 18% decrease, respectively). Again, these differences between the present and future CO_2 scenarios can be explained by the change in the feed's chemical composition, especially the increased need for soy meal and decreased need for wheat at elevated atmospheric CO_2.

Nearly all of the impacts are due to the crop-based ingredients, i.e., the sum of others (non-crop ingredients) is infinitesimal (reflected in Figure 3 discussed in Section 3.4). For non-carcinogenic human toxicity, rape and wheat are the biggest contributors (caused by rape and wheat cultivation, but the effect of wheat will be greatly reduced by the displaced barley; Figure 1). For respiratory inorganics, soy and wheat are the major contributors (caused by soy and wheat cultivation, but the effect of wheat will be greatly reduced by the displaced barley, and that of soy will be somewhat reduced by soy oil displacing palm oil; see Figure S3 in Supplementary), while rape contributes negatively (mainly due to displaced palm oil; Figure S4). Regarding natural occupation, all of the main crop ingredients contribute, and sunflower the most (due to the low yield of sunflower cultivation and the displaced barley; Figure S5). For global warming, soy is the biggest overall contributor (71% of its contribution is caused by indirect land-use change, 19% by soy oil displacing palm oil; Figure S3), followed by barley. These observations emphasize that each crop ingredient in compound pig feed contributes differently to the various environmental impacts of the feed based on their content, and that ILUC is important in the calculation of the overall GWP (Global Warming Potential) of feed production, second only to crop and displaced crop cultivation. Cultivation of the main crops (photosynthetic carbon uptake) as well as the reacting carbohydrate crop cultivation all contribute negatively to the GWP impact (Supplementary, Figure S10).

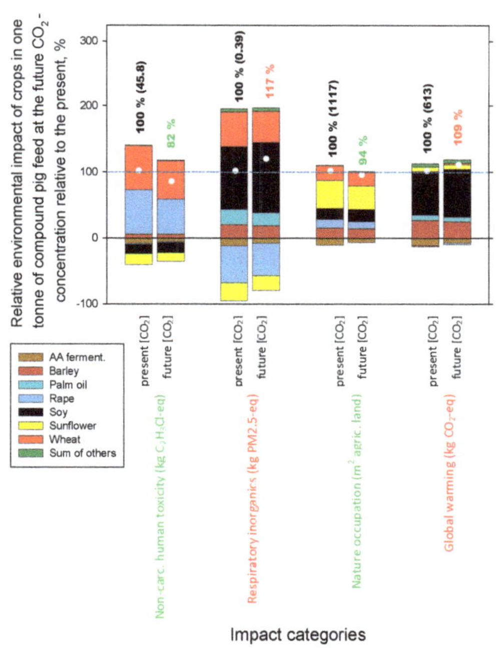

Figure 2. Relative environmental impacts of producing one tonne of compound pig feed under future and current CO_2 for the socio-economically four most important impact categories. The dots represent the net value of the environmental impact, i.e., the positive minus the negative values. Eq: equivalents; $PM_{2.5}$: ultrafine particles; AA ferment.: amino acids produced for feed by fermentation processes (lysine, threonine, tryptophan).

Table 1. Chemical composition and yield of wheat, barley, soy, and rape at the present (380 µmole mole^{-1}) and future (550 µmole mole^{-1}) atmospheric CO_2 concentrations.

Chemical Composition	Wheat, Denmark			Barley, Denmark			Soy, Argentina/Brazil			Rapeseed, Germany		
	Present [1]	Change [2] (%)	Future [7]	Present [1]	Change [3,4] (%)	Future	Present	Change [3] (%)	Future	Present [1]	Change [5] (%)	Future
Total Dry Matter, %	85	0	85	85	0	85	88	unknown	unknown	89	unknown	unknown
Starch (g/kg)	670	+7.5	720	630	+4.1 [4]	660	27	unknown	unknown	19	unknown	unknown
Raw Protein, %	9.5	−14	8.2	9.7	−15 [3]	8.2	47	−1.4	46	35	−4.6	33
Amino Acids (g/kg)												
Lysine	2.3	−10	2.0	2.6	unknown	unknown	26	unknown	unknown	15	−2.5	14
Methionine	1.3	−17	1.1	1.4	unknown	unknown	5.7	unknown	unknown	6.0	−4.3	5.7
Cystine	1.9	−17	1.6	1.8	unknown	unknown	5.8	unknown	unknown	7.0	−5.4	6.6
Threonine	2.2	−21	1.8	2.4	unknown	unknown	16	unknown	unknown	12	−1.8	11
Tryptophan	0.93	−17	0.77	0.93	unknown	unknown	5.7	unknown	unknown	3.3	−3.1	3.2
Isoleucine	2.8	−22	2.2	2.7	unknown	unknown	18	unknown	unknown	11	−3.1	10
Leucine	5.2	−19	4.2	5.1	unknown	unknown	32	unknown	unknown	20	−3.7	19
Histidine	1.9	−16	1.6	1.7	unknown	unknown	11	unknown	unknown	8.1	−3.9	7.7
Phenylalanine	3.6	−14	3.1	3.7	unknown	unknown	21	unknown	unknown	11	−1.9	11
Tyrosine	2.4	−16	2.0	2.3	unknown	unknown	16	unknown	unknown	8.2	−2.6	8.0
Valine	3.5	−8	3.2	3.6	unknown	unknown	20	unknown	unknown	14	−3.4	13
Macroelements (g/kg)					(4)							
Calcium	0.43	−15	0.36	0.43	4.8	0.45	3.5	unknown	unknown	8.5	−1.5	8.4
Phosphorus	2.6	−4	2.5	3.0	−4.6	2.8	6.8	unknown	unknown	11	1.8	11
Sodium	0.085	−6	0.080	0.17	unknown	unknown	0.18	unknown	unknown	0.36	0.1	0.36
Potassium	5.0	−1	4.9	4.8	0	4.8	22	unknown	unknown	13	0.3	13
Magnesium	1.0	−7	0.95	1.0	−3.3	0.99	3.2	unknown	unknown	4.7	0.8	4.8
Sulphur	1.1	−13	0.96	0.94	−5.0	0.89	3.6	unknown	unknown	6.5	−6.1	6.1
Microelements (mg/kg)					(4)							
Iron	27	−18	22	29	−11	26	260	unknown	unknown	230	1.0	230
Manganese	25	−3	24	12	unknown	unknown	47	unknown	unknown	68	−2.6	66
Zink	34	−13	30	24	−13	21	48	unknown	unknown	65	−5.6	61
Yield (t fm ha^{-1})	6.8 [6]	+11% [5]	7.3	5.0 [6]	+20% [8]	6.0	3.4 [9]	+15% [10]	3.9	3.8 [11]	+ 5.0% [12]	4.0

[1]: reference [17]; [2]: Based on OTC data in reference [19]; [3]: reference [42], meta-data; [4]: reference [43], FACE data; [5]: reference [44], FACE data, extrapolation assuming linear responses to elevated CO_2, [6]: reference [14]; [7]: reference [1], FACE data, extrapolation assuming linear responses to elevated [CO_2]; [8]: reference [45], OTC data; [9]: reference [31]; [10]: reference [46], FACE data; [11]: reference [38], average 2005–2010; [12]: reference [47], RERAF chamber data.

Table 2. Ingredients in Danish pig feed for piglets, sows, and slaughter pigs, and their weighted average, produced at the present (380 μmole mole^{-1}) and future (550 μmole mole^{-1}) atmospheric CO_2 concentrations. Amino acids produced by fermentation include lysine, threonine, and tryptophan. Dl-methionine is chemically synthesized.

Product	Main Function in the Feed	Sow Feed (20%)		Piglet Feed (20%)		Slaughter Pig Feed (60%)		Weighted Average Pig Feed			
		Present	Future	Present	Future	Present	Future	Present	Future	Absolute Change	Relative Change
		Percent by Mass (Dry Weight Basis)								kg	percent
Wheat	Energy	52	50	46	44	47	45	48	46	−25	−5.1%
Barley	Energy	25	25	25	25	25	25	25	25	0.0	0.0%
Soy meal	Protein	6.0	8.6	19	20	6.9	9.4	9.2	11	+22	+23%
Rape meal	Protein	4.0	4.0	0.0	0.0	10	10	6.8	6.8	0.0	0.0%
Sunflower meal	Protein	6.0	6.0	0.0	0.0	5.0	5.0	4.2	4.2	0.0	0.0%
Beet molasses	Energy	2.0	2.0	0.50	2.0	2.0	2.0	1.7	2.0	+3.0	+18%
PFAD oil, palm fatty acid distillate	Energy and technical aid	1.3	1.3	1.5	1.3	1.3	1.3	1.3	1.30	−0.39	−2.9%
Limestone meal ($CaCO_3$)	Production requirement	1.4	1.3	0.86	0.85	1.3	1.3	1.2	1.2	−0.25	−2.1%
Amino acids (fermentation)	Production requirement	0.37	0.28	0.73	0.66	0.48	0.40	0.51	0.43	−0.81	−16%
Salt, sodium chloride:	Production requirement	0.43	0.42	0.46	0.50	0.50	0.44	0.48	0.45	+0.02	+6.3%
Monocalcium phosphate:	Production requirement	0.64	0.66	0.88	0.94	0.22	0.23	0.43	0.47	+0.34	+5.5%
Protein (from fish):	Protein	0.00	0.00	2.00	2.00	0.00	0.00	0.40	0.40	0.00	0.0%
Vitamins	Production requirement	0.59	0.59	0.20	0.20	0.20	0.20	0.32	0.32	0.00	0.0%
Phytase and xylanase	Health, economy, environment	0.04	0.04	0.14	0.14	0.08	0.08	0.08	0.08	0.00	0.0%
Dl-methionine (synthetic)	Production requirement	0.00	0.00	0.10	0.11	0.00	0.01	0.02	0.03	+0.03	+14%
Hemoglobin meal	Protein	0.00	0.00	1.00	1.7	0.00	0.00	0.20	0.34	+1.4	+65%
Formic acid, calcium salt	PH-adjustment	0.00	0.00	0.80	0.80	0.00	0.00	0.16	0.16	0.00	0.0%
Sum		100	100	100	100	100	100	100	100		

Note: PFAD—Palm Fatty Acid Distillates.

Amino acids produced by fermentation contribute negatively to each of the main impact categories. This is because sugar production (one of the substrates in the fermentation process producing amino acids) gives rise to by-products (molasses and pulp), which can substitute the use of marginal carbohydrates for animal feed. The saved carbohydrates production (and the land-use changes it generated) had a greater negative impact than the positive impact from the consumed sugar substrate (see Figure S7a in Supplementary). Yet, these effects are of course highly dependent upon the data quality that is used to model them. As three crop ingredients are used to produce these amino acids (Figure S7a–c in Supplementary), and as each of these crops involve at least three co-products, a considerable degree of uncertainty is introduced in the model, as a result of the numerous assumptions involved regarding the displacement effects (i.e., system expansion).

3.3. Direct Cost

Data for direct market prices of all of the ingredients were added up for each pig growth stage (Table 3). Based on the present prices of commodities and the altered content of future feed, an increase in the direct costs for future pig feed was estimated at €1.92 per tonne of pig feed, which represents an increase of only 0.8% on top of the current price of €250.39 per tonne.

Table 3. Differences in direct costs for present and future compound pig feed, on the basis of today's prices.

	Present CO_2	Future CO_2	Difference
Sows, 200 kg feed	€47.69	€48.14	€0.45
Piglets, 200 kg feed	€60.92	€61.05	€0.13
Slaughter pigs, 600 kg feed	€141.78	€143.12	€1.34
Total pig feed, 1000 kg	€250.39	€252.31	€1.92
Per cent	100%	100.77%	0.77%

These results indicate that the direct price of future pig feed will not increase significantly due to elevated atmospheric CO_2. However, it may, very well increase due to other circumstances, e.g., increasing prices on land, energy, fertilizers and pesticides, transport, manufacture, and increased demand. If the demand for protein in animal feed continues to rise, the price of soy is likely to go up, and soy in European pig feed may be replaced with other protein sources combined with an increased application of industrial amino acids. However, all of these circumstances are impossible to foresee in any detail, and are not directly related to the rising atmospheric CO_2.

3.4. External Environmental Cost

In the future, the external cost that is associated with the environmental impact of growing crops may be included in the actual cost of food and feed according to "the polluter pays principle" [40], significantly increasing the current price for compound pig feed. This in itself makes calculations of external cost of pig feed production relevant.

Figure 3 shows the monetized environmental impact using the stepwise normalization methodology [13] for the four most important impact categories, and the sum of 12 other categories. As indicated in Figure 3 (and further detailed in Figure S11 and S12 in Supplementary), the cost of the overall environmental externalities of one tonne of pig feed is estimated at €236.05 for the present CO_2, and at €235.26 for the future CO_2, i.e., €0.79 or 0.3% less than feed produced today. These results indicate that elevated CO_2 will not increase the external environmental costs of pig feed.

As highlighted in Table 3 and Figure 3, the external socio-economic costs of producing compound pig feed are in the same order of magnitude as the direct costs of pig feed. In fact, internalizing the environmental externalities of producing pig feed nearly doubles the present (direct) cost of pig feed, which is a vast price increase if the Rio Declaration is to be taken seriously [48].

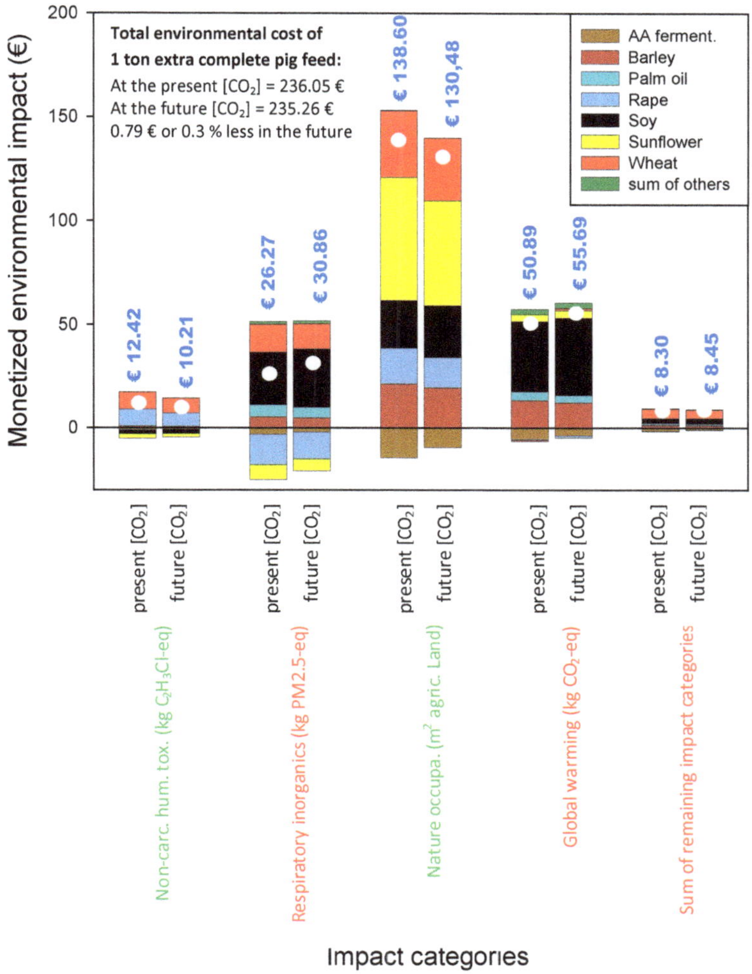

Figure 3. Breakdown of the monetized environmental impact results per ingredient. The white dots represent the net value of the environmental impact, i.e., the positive minus the negative values. Eq: equivalents; $PM_{2.5}$: ultrafine particles; AA ferment.: amino acids produced by fermentation (lysine, threonine, tryptophan).

Figure 3 (as Figure 2 and the extended Figures S9–S12 in Supplementary) highlights soy as a main contributor to most of the environmental impact categories, whereas for natural occupation, sunflower is the major contributor. Figure 3 also highlights the important contribution of the natural occupation category, when environmental impacts are translated into a common monetized environmental impact, which explains why the overall high environmental cost of sunflower is associated with a low yield. Per tonne of pig feed, soy has the largest environmental impact in monetary terms, followed by sunflower (Table 4). However, relative to their content in pig feed, sunflower has an 11-times higher (specific) impact and soy has a seven-times higher (specific) impact compared with wheat, barley, or rape (Table 4).

Table 4. The relative importance of crops in one tonne of feed at the present CO_2 in terms of monetized environmental impact. Due to the negative environmental impact that amino acids and some crops have for various impact categories, the sum of the crop impact percentages surpasses 100%.

Importance of Crop Ingredients in Terms of Monetized Impact	Wheat	Barley	Soy	Rape	Sunflower
Content, kg per tonne of feed (see Table 2)	480	250	92	68	42
Monetary weight of environmental impact, %	25	18	35	5	23
Monetized environmental impact relative to kg content, %	5.2	7.2	38.3	6.9	54.2
Factor of the above	1	1	7	1	11

In the context of a worldwide growing population with growing food and energy needs (e.g., reference [49]), alternative strategies to minimize the amount of land that is used for food, feed, or bioenergy are urgent. The main ingredients in compound pig feed could just as well be consumed directly by humans. A pig consumes 2.5 times its weight gain in terms of pig feed before only part of the pig ends up as food for humans (bones, head, and entrails are wasted). In a world with a growing population and with about 800 million people starving, a more plant-based diet would make sense, in terms of increasing food production, increasing the availability of bioenergy, and reducing the environmental impact of global agriculture.

According to Table 4, future pig feed recipes ought to consider a reduced use of sunflower and soy as protein crops. Whether rape or non-crop ingredients (e.g., industrial amino acids, insect-based protein) would be a better environmental choice could be the object of future investigations. However, since the value of protein in animal feed relies on the amino acid composition, it is important to consider balancing the overall amino acid composition of the pig feed to suit the needs of pigs. In this sense, industrial amino acids have a clear advantage, although economically it may not be the best solution measured in the present commodity prices (as in this study). On the other hand, if and when the cost of environmental impacts is included in the price of pig feed, it may likely be an advantage to include more industrial amino acids in the feed.

An alternative or supplement to the above-mentioned solutions could be to include genetically modified grain crops. These crops would need not only to contain more protein—which may in itself be useless—but rather, they should contain more of the limiting amino acids for the better digestibility of the protein in the feed.

3.5. Limits and Uncertainties

One limit of this study is that it focused only on the effect of elevated CO_2 to represent the future conditions under which pig feed will be produced. In fact, higher atmospheric CO_2 is not the only global change to determine the yield and the chemical composition of crops in the future. With elevated CO_2 follows elevated temperature and altered patterns of precipitation (10). Typically, crop yields respond positively to both elevated CO_2 and moderate (1–3 °C) temperature increases, but negatively to high (i.e., above 3 °C) temperature increases [47,50]. Moreover, the response depends on the plant species, water availability, humidity, wind, soil type, etc.

Furthermore, it has been highlighted that the current and anticipated elevated concentration of ground-level ozone may likely decrease plant productivity [51]; and contrary to elevated CO_2, ozone increases grain protein concentration [51].

Thus, the results of this study are not to be seen as representative of the future state of global climate and meteorological conditions (data for this is as yet unavailable), but rather as an illustration of the cascading consequences that any change in crop yield and composition (which is in this case triggered by an increase in atmospheric CO_2) can have for pig feed formulation. As for any LCA, the results of this study are closely linked to the quality of the inventory data and assumptions taken. For example, no changes in amino acid composition were considered for barley and soy as a result of elevated CO_2, because no data or reliable estimations could be found. Further, the overall changes in yield and chemical composition considered (Table 1) are based on the best state-of-knowledge, as available FACE experiments results are still scarce. Thus, this study would considerably benefit from a greater availability of such data. Nevertheless, the study provides a solid framework for assessing the consequences of a changed crop composition due to the effect that elevated atmospheric CO_2 would have on pig feed, which has not, to authors' knowledge, been available so far.

Another limit is that the changes in manure composition resulting from a change in feed composition have not been taken into account. Feed containing less protein from cereals, which are difficult to digest, and more easily digestible protein from e.g., soy or rapeseed meal, involves a better digestion, and thus a reduction in excreted N. This could have consequences for the subsequent use of the manure as a fertilizer, as it would involve a reduced potential for the emission of N flows (e.g., ammonia, nitrous oxide, nitrate losses). Based on Table 2, these induced changes in manure composition would likely have induced additional benefits for the future pig feed, which comprises significantly more soymeal and less wheat.

One of the most important sources of uncertainty probably lies in the estimation of the environmental consequences generated by land-use changes. As clearly emphasized in several publications, e.g., references [52–54], the estimation of land-use changes, particularly ILUC, involves multiple sources of uncertainty. In the explorative deterministic approach used in this study, one main source of uncertainty lies in the choice of the market that would react to a changed demand for the different crop ingredients presented in Table 2, i.e., the choice of the marginal crops and their geographical location. Further, as expressed by e.g., reference [52] there is also uncertainty related to the actual C stock in the biomes converted. On the other hand, the strength of the approach used in this study lies in its transparency. In fact, the assumptions used in the deterministic model can be easily changed to reflect different possible futures with regard to how the land market could develop. However, such assessment is beyond the scope of this study. Nevertheless, it should be highlighted that although the actual magnitude of environmental impacts related to land-use changes is uncertain, the potentiality of adverse effects arising from it is hardly subject to dispute [25,53].

Finally, we draw the attention to the work by Dijkman et al. [55], where the environmental impact of barley cultivation under current and future climatic conditions is analyzed using aLCA. Dijkman et al. [55] found that a predicted decrease in barley yields under future climate conditions is the main driver for increased impacts. An increased impact was not substantiated by the present study based on cLCA and including all of the ingredients of compound pig feed.

4. Conclusions

The main findings and highlights of this study can be summarized as follows:

- A methodological framework was developed in order to assess the cascading environmental consequences that a change in crop yield and chemical composition (triggered by an increase in atmospheric CO_2) can have for pig feed formulation, including land-use change consequences.
- The positive environmental effect of elevated CO_2 on crop yield (carbohydrates) was counterbalanced by a need for increased soy content in pig feed, and the land-use that consequences this generated. Therefore, the net effects are close to zero.
- The four most important environmental impact categories in pig feed production under current and future atmospheric CO_2, as determined by the stepwise normalization methodology, were human toxicity in terms of non-carcinogenic toxicity and respiratory inorganics, natural occupation, and global warming.
- The monetized environmental impact (shadow price) of compound pig feed produced today (€236.05 per tonne) was found to be of the same order of magnitude as the direct price of compound pig feed (€250.39 per tonne). Internalizing the cost of environmental impacts would nearly double the price of pig feed if the Rio Declaration was to be honored.
- Since the protein crops (soy, rape, and sunflower) account for about 60% of the overall environmental impact of pig feed, it is important to optimize their content in a future with

expected growing demands for food and bioenergy (and thus for land). In this context, it is important to optimize the protein content in the feed based on the limiting amino acids in each crop rather than on total protein.

Supplementary Materials: The following are available online at http://www.mdpi.com/2071-1050/10/9/3184/s1. References [56–64] are cited in the supplementary materials.

Author Contributions: Conceptualization: H.S.; Methodology, software, formal analyses and investigation: H.S., L.H., T.H.; Original Draft Preparation and visualization: H.S. and L.H.; Project Administration: H.S. and H.W.; Funding Acquisition, H.W.

Funding: This research was funded by the Danish Food Research Programme 2009 for supporting this project through grant number 3304-FVFP-09-B-004: "Development of genetically modified cereals adapted to the increased CO_2 levels of the future".

Acknowledgments: The authors thank DLG for participating with Bestmix® analyses performed by Torben Hinrichsen.

Conflicts of Interest: The authors declare no conflict of interest. The funders had no role in the design of the study; in the collection, analyses, or interpretation of data; in the writing of the manuscript, and in the decision to publish the results.

References

1. Högy, P.; Wieser, H.; Köhler, P.; Schwadorf, K.; Breuer, J.; Franzaring, J.; Muntefering, R.; Fangmeier, A. Effects of elevated CO_2 on grain yield and quality of wheat: Results from a 3-year free-air CO_2 enrichment experiment. *Plant Biol.* **2009**, *11*, 60–69. [CrossRef]
2. Smith, M.R.; Myers, S.S. Impact of anthropogenic CO_2 emissions on global human nutrition. *Nat. Clim. Chang.* **2018**, *8*, 834–839. [CrossRef]
3. International Grain Council. Grain Marked Report. 2018; pp. 1–8. Available online: http://www.igc.int/downloads/gmrsummary/gmrsumme.pdf (accessed on 14 August 2018).
4. Steinfeld, H.; Gerber, P.; Wasssenaar, T.; Castel, V.; Rosales, M.; de Haan, C. *Livestock's Long Shadow*; Environmental Issues and Options; FAO: Rome, Italy, 2006; pp. 1–390.
5. Nellemann, C.; MacDevette, M.; Manders, T.; Eickhout, B.; Svihus, B.; Prins, A.G.; Kaltennorn, B.P. The Environmental Food Crisis–the Environment's Role in Averting Future Food Crisis. A UNEP Rapid Response Assessment. United Nations Environment Programme, GRID-Arendal. 2009. Available online: http://www.grida.no/publications/rr/food-crisis/ (accessed on 14 August 2018).
6. Meul, M.; Ginneberge, C.; Van Middelaar, C.E.; de Boer, I.J.M.; Fremaut, D.; Haesaert, G. Carbon footprint of five pig diets using three land use change accounting methods. *Livest. Sci.* **2012**, *149*, 215–223. [CrossRef]
7. Hamelin, L.; Naroznova, I.; Wenzel, H. Environmental consequences of different carbon alternatives for increased manure-based biogas. *Appl. Energy* **2014**, *114*, 774–782. [CrossRef]
8. Searchinger, T.; Heimlich, R.; Houghton, R.A.; Dong, F.; Elobeid, A.; Fabiosa, J.; Tokgoz, S.; Hayes, D.; Yu, T.-H. Use of U.S. Croplands for Biofuels Increases Greenhouse Gases through Emissions from Land Use Change. *Science* **2008**, *319*, 1238–1240. [CrossRef] [PubMed]
9. Gibbs, H.K.; Johnston, M.; Foley, J.A.; Holloway, T.; Monfreda, C.; Ramankutty, N.; Zaks, D. Carbon payback times for crop-based biofuel expansion in the tropics: The effects of changing yield and technology. *Environ. Res. Lett.* **2008**, *3*, 1–10. [CrossRef]
10. IPCC, Climate Change 2014. Synthesis Report. 2015. Available online: https://www.ipcc.ch/pdf/assessment-report/ar5/syr/SYR_AR5_FINAL_full_wcover.pdf (accessed on 14 August 2018).
11. Weidema, B.P.; Ekvall, T.; Heijungs, R. Guidelines for Applications of Deepened and Broadened LCA. Deliverable D18 of Work Package 5 of the CALCAC Project. 2009; pp. 1–49. Available online: https://www.leidenuniv.nl/cml/ssp/publications/calcas_report_d18.pdf (accessed on 14 August 2018).
12. Earles, J.M.; Halog, A. Consequential life cycle assessment: A review. *Int. J. LCA* **2011**, *16*, 445–453. [CrossRef]
13. Weidema, B.P. Using the budget constraint to monetarise impact assessment results. *Ecol. Econ.* **2008**, *68*, 1591–1598. [CrossRef]
14. Hamelin, L.; Jørgensen, U.; Petersen, B.M.; Olesen, J.E.; Wenzel, H. Modelling the carbon and nitrogen balances of direct land use changes from energy crops in Denmark: A consequential life cycle inventory. *GCB Bioenergy* **2012**, *4*, 889–907. [CrossRef]

15. Ecoinvent Centre. 2013. Available online: http://www.ecoinvent.org/database/ecoinvent-version-2/ (accessed on 14 August 2018).
16. Adifo. Bestmix Software. 2011. Available online: https://www.adifo.com/products/bestmix-feed-formulation (accessed on 14 August 2018).
17. VSP, Videncenter for Svineproduktion, 2012. Fodermiddeldatabase (Feed Database). Available online: https://www.lf.dk/aktuelt/nyheder/2015/april/ny-organisation-i-seges-videncenter-for-svineproduktion#.W05DYtL7SUk (accessed on 14 August 2018).
18. Ainsworth, E.A.; Leakey, A.D.B.; Ort, D.R.; Long, S.P. FACE-ing the facts: Inconsistencies and interdependence among field, chamber and modelling studies of elevated [CO_2] impacts on crop yield and food supply. *New Phytol.* **2008**, *179*, 5–9. [CrossRef] [PubMed]
19. Högy, P.; Fangmeier, A. Review: Effects of elevated atmospheric CO_2 on grain quality of wheat. *J. Cereal Sci.* **2008**, *48*, 580–591. [CrossRef]
20. Mosnier, E.; van der Werf, H.M.G.; Boissy, J.; Dourmad, J.-Y. Evaluation of the environmental implications of the incorporation of feed-use amino acids in the manufacturing of pig and broiler feeds using Life Cycle Assessment. *Animal* **2011**, *5*, 1972–1983. [CrossRef] [PubMed]
21. Nielsen, P.H.; Wenzel, H. Environmental assessment of Ronozyme® P5000 CT phytase as an alternative to inorganic phosphate supplementation to pig feed used in intensive pig production. *Int. J. LCA* **2007**, *12*, 514–520. [CrossRef]
22. Schmidt, J.H.; Weidema, B.P.; Brandão, M. A framework for modelling indirect land use changes in Life Cycle Assessment. *J. Clean. Prod.* **2015**, *99*, 230–238. [CrossRef]
23. Tonini, D.; Hamelin, L.; Astrup, T.F. Environmental implications of the use of agro-industrial residues for biorefineries: Application of a deterministic model for indirect land-use changes. *GCB Bioenergy* **2016**, *8*, 690–706. [CrossRef]
24. JRC. World Atlas of Desertification. 2018. Available online: https://wad.jrc.ec.europa.eu/ (accessed on 14 August 2018).
25. Marelli, L.; Mulligan, D.; Edwards, R.; Critical Issues in Estimating ILUC Emissions. Critical Issues in Estimating ILUC Emissions. JRC Scientific and Technical Reports. Outcomes of an Expert Consultation 9–10 November 2010, Ispra, Italy. 2011; pp. 1–64, EUR 24816EN-2011. Available online: http://publications.jrc.ec.europa.eu/repository/bitstream/111111111/22908/2/reqno_jrc64429_cirtical%20issues%20in%20estimating%20iluc%20emissions%20print%20version.pdf (accessed on 14 August 2018).
26. Verchot, L.V.; Davidson, E.A.; Cattânio, J.H.; Ackerman, I.L.; Erickson, H.E.; Keller, M. Land use change and biogeochemical controls of nitrogen oxide emissions from soils in eastern Amazonia. *Glob. Biogeochem. Cycles* **1999**, *13*, 31–46. [CrossRef]
27. Wolosin, M.; Harris, N. Ending Tropical Deforestation: A Stock-Take of Progress and Challenges. Tropical Forests and Climate Change: The Latest Science. 2018. World Resources Institute, Working Paper. Available online: https://wriorg.s3.amazonaws.com/s3fs-public/ending-tropical-deforestation-tropical-forests-climate-change.pdf (accessed on 14 August 2018).
28. De Lima, R.F.; Dallimer, M.; Atkinson, P.W.; Barlow, J. Biodiversity and land-use change: Understanding the complex responses of an endemic-rich bird assemblage. *Divers. Distrib.* **2013**, *19*, 411–422. [CrossRef]
29. Nielsen, O.-K.; Mikkelsen, M.H.; Hoffmann, L.; Gyldenkærne, S.; Winther, M.; Nielsen, M.; Fauser, P.; Thomsen, M.; Plejdrup, M.S.; Albrektsen, R.; et al. *Denmark's National Inventory Report 2011-Emission Inventories 1990–2009-Submitted under the United Nations Framework Convention on Climate Change and the Kyoto Protocol*; NERI Technical Report 827; The National Environmental Research Institute, Aarhus University: Aarhus, Denmark, 2011; pp. 1–1199. Available online: http://www2.dmu.dk/pub/fr827.pdf (accessed on 14 August 2018).
30. Schmidt, J.H. Life Cycle Assessment of Rapeseed Oil and Palm Oil. Ph.D. Thesis, Department of Development and Planning, Aalborg University, Aalborg, Denmark, 15 June 2007.
31. Dalgaard, R.; Schmidt, J.; Halberg, N.; Christensen, P.; Thrane, M.; Pengue, W.A. LCA of soybean meal. *Int. J. LCA* **2008**, *13*, 240–254. [CrossRef]

32. Edwards, R.; Mulligan, D.; Marelli, L. Indirect Land Use Change from Increased Biofuels Demand. Comparison of Models and Results for Marginal Biofuels Production from Different Feedstocks. (No. EUR 24485). European Commission Joint Research Centre. 2010. Available online: https://www.energy.eu/publications/Indirect-Land-Use-Change-from-increased-Biofuels-Demand.pdf (accessed on 14 August 2018).
33. Hertel, T.W.; Golub, A.A.; Jones, A.D.; O'Hare, M.; Plevin, R.J.; Kammen, D.M. Effects of US Maize Ethanol on Global Land Use and Greenhouse Gas Emissions: Estimating Market-mediated Responses. *Bioscience* **2010**, *60*, 223–231. [CrossRef]
34. Laborde, D. Assessing the Land Use Change Consequences of European Biofuel Policies. ATLASS Consortium, 2011. Available online: http://trade.ec.europa.eu/doclib/docs/2011/october/tradoc_148289.pdf (accessed on 14 August 2018).
35. Tonini, D.; Hamelin, L.; Wenzel, H.; Astrup, T. Bioenergy Production from Perennial Energy Crops: A Consequential LCA of 12 Bioenergy Scenarios including Land Use Changes. *Environ. Sci. Technol.* **2012**, *46*, 13521–13530. [CrossRef] [PubMed]
36. Ekvall, T.; Weidema, B. System Boundaries and Input Data in Consequential Life Cycle Inventory Analysis. *Int. J. LCA* **2004**, *9*, 161–171. [CrossRef]
37. Weidema, B.P.; Bauer, C.; Hischier, R.; Mutel, C.; Nemecek, T.; Reinhard, J.; Vadenbo, C.O.; Wernet, G. The Ecoinvent Database: Overview and Methodology, Data Quality Guideline for the Ecoinvent Database Version 3. 2017. Available online: https://lca-net.com/publications/show/overview-methodology-data-quality-guideline-ecoinvent-database-version-3/ (accessed on 14 August 2018).
38. FAOSTAT. 2011. Available online: http://www.fao.org/faostat/en/#data (accessed on 14 August 2018).
39. De Vries, J.W.; Vinken, T.M.W.J.; Hamelin, L.; De Boer, I.M.J. Comparing environmental consequences of anaerobic mono- and co-digestion alternatives for pig manure to produce bio-energy—A life cycle perspective. *Bioresour. Technol.* **2012**, *125*, 239–248. [CrossRef] [PubMed]
40. Hamelin, L. Carbon Management and Environmental Consequences of Agricultural Biomass in a Danish Renewable Energy Strategy. Ph.D. Thesis, University of Southern Denmark, Odense, Denmark, February 2013.
41. Edwards, R.; Padella, M.; Vorkapic, V.; Marelli, L. Historical Deforestation Due to Expansion of Crop Demand: Implications for Biofuels. 2014. Available online: http://publications.jrc.ec.europa.eu/repository/handle/JRC83819 (accessed on 14 August 2018).
42. Erbs, M.; Manderscheit, R.; Jansen, G.; Seddig, S.; Pacholski, A.; Weigel, H.-J. Effects of free-air CO_2 enrichment and nitrogen supply on grain quality parameters and elemental composition of wheat and barley grown in a crop rotation. *Agric. Ecosyst. Environ.* **2010**, *136*, 59–68. [CrossRef]
43. Högy, P.; Franzaring, J.; Schwadorf, K.; Breuer, J.; Schütze, W.; Fangmeier, A. Effects of free-air CO_2 enrichment on energy traits and seed quality of oilseed rape. *Agric. Ecosyst. Environ.* **2010**, *139*, 239–344. [CrossRef]
44. Taub, D.R.; Miller, B.; Allen, H. Effects of elevated CO_2 on the protein concentration of food crops: A meta-analysis. *Glob. Chang. Biol.* **2008**, *14*, 65–575. [CrossRef]
45. Fangmeier, A.; Chrost, B.; Högy, P.; Krupinska, K. CO_2 enrichment enhances flag leaf senescence in barley due to greater grain nitrogen sink capacity. *Environ. Exp. Bot.* **2000**, *44*, 151–164. [CrossRef]
46. Morgan, P.B.; Bollero, G.A.; Nelson, R.L.; Dohleman, F.G.; Long, S.P. Smaller than predicted increase in aboveground net primary production and yield of field-grown soybean under fully open-air [CO_2] elevation. *Glob. Chang. Biol.* **2005**, *11*, 1856–1865. [CrossRef]
47. Clausen, S.K.; Frenck, G.; Linden, L.G.; Mikkelsen, T.N.; Lunde, C.; Jørgensen, R.B. Effects of single and multifactor treatment with elevated temperature, CO_2 and ozone on oilseed rape and barley. *J. Agron. Crop Sci.* **2011**, *197*, 442–453. [CrossRef]
48. Rio Declaration. 1992. Rio Declaration on Environment and Development. Principle 16. Available online: https://www.jus.uio.no/lm/environmental.development.rio.declaration.1992/16.html (accessed on 14 August 2018).
49. Nonhebel, S. Global food supply and the impacts of increased biofuels. *Energy* **2012**, *37*, 115–121. [CrossRef]
50. Qaderi, M.M.; Reid, D.M. *Crop Responses to Elevate Carbon Dioxide and Temperature*; Singh, S.N., Ed.; Chapter 1 in: Climatic Change and Crops, Environmental Science and Engineering; Springer: Berlin/Heidelberg, Germany, 2009; pp. 1–17.

51. Booker, F.; Muntifering, R.; McGrath, M.; Burkey, K.; Decoteau, D.; Fiscus, E.; Manning, W.; Krupa, S.; Chappelka, A.; Grantz, D. The ozone component of global change: Potential effects on agricultural and horticultural plant yield, product quality and interactions with invasive species. *J. Integr. Plant Biol.* **2009**, *51*, 337–351. [CrossRef] [PubMed]
52. Plevin, R.J.; O'Hara, M.; Jones, A.D.; Torn, M.S.; Gibbs, H.K. Greenhouse Gas Emissions from Biofuels' Indirect Land Use Change Are Uncertain but May Be Much Greater than Previously Estimated. *Environ. Sci. Technol.* **2010**, *44*, 8015–8021. [CrossRef] [PubMed]
53. Khanna, M.; Crago, C.L. Measuring Indirect Land Use Change with Biofuels: Implications for Policy. *Annu. Rev. Resour. Econ.* **2012**, *4*, 161–184. [CrossRef]
54. Sanchez, S.T.; Woods, J.; Akhurst, M.; Brander, M.; O'Hara, M.; Dawson, T.P.; Edwards, R.; Liska, A.J.; Malpas, R. Accounting for indirect land-use change in the life cycle assessment of biofuel supply chains. *J. R. Soc. Interface* **2012**, *9*, 1105–1119. [CrossRef] [PubMed]
55. Dijkman, T.J.; Birkved, M.; Saxe, H.; Wenzel, H. Environmental impacts of barley cultivation under current and future climatic conditions. *J. Clean. Prod.* **2017**, *140*, 644–653. [CrossRef]
56. Kløverpris, J. Consequential Life Cycle Inventory Modelling of Land Use Induced by Crop Consumption. Ph.D. Thesis, DTU Management Engineering, Lyngby, Denmark, November 2008. Updated Version 2010. Available online: http://orbit.dtu.dk/en/publications/consequential-life-cycle-inventory-modelling-of-land-use-induced-by-crop-consumption%286457fc93-c53a-4beb-83ff-015c8047b682%29.html (accessed on 14 August 2018).
57. Jungbluth, N.; Dinkel, F.; Doka, G.; Chudacoff, M.; Dauriat, A.; Gnansounou, E.; Sutter, J.; Spielmann, M.; Kljun, N.; Keller, M. Life Cycle Inventory of Bioenergy. Data v2.0. Ecoinvent Report no. 17. Swiss Centre for Life Cycle Inventories. A Joint Initiative of the ETH Domain and Swiss Federal Offices, Ulster, Switzerland. Available online: https://www.researchgate.net/profile/Niels_Jungbluth/publication/230725648_Life_Cycle_Inventories_of_Bioenergy_ecoinvent_report_No_17/links/0c96051b76e2fb8dce000000/Life-Cycle-Inventories-of-Bioenergy-ecoinvent-report-No-17.pdf (accessed on 14 August 2018).
58. Schmidt, J.H.; Christensen, P.; Christensen, T.S. Assessing the land use implications of biodiesel use from an LCA perspective. *J. Land Use Sci.* **2009**, *4*, 35–52. [CrossRef]
59. Nemecek, T.; Kägi, T. Life Cycle Inventory of Agricultural Production Systems. Data ver. 2.0. Ecoinvent Report no. 15. Agroscope Reckenholz-Tänikon Research Station ART, Zürich and Dübendorf, Switzerland. Available online: https://db.ecoinvent.org/reports/15_Agriculture.pdf (accessed on 14 August 2018).
60. Miyagi, K.-M.; Kinugasa, T.; Hikosaka, K.; Hirose, T. Elevated CO_2 concentration, nitrogen use, and seed production in annual plants. *Glob. Chang. Biol.* **2007**, *13*, 2161–2170. [CrossRef]
61. Buckley, T.N. The role of stomatal acclimation in modelling tree adaptation to high CO_2. *J. Exp. Bot.* **2008**, *59*, 1951–1961. [CrossRef] [PubMed]
62. Attavanich, W.; McCarl, B.A. The effect of climate change, CO_2 fertilization, and crop production technology on crop yields and its economic implications on market outcomes and welfare distribution. In Proceedings of the Agricultural & Applied Economics Association's 2001 AAEA & NAREA Joint Annual Meeting, Pittsburgh, PA, USA, 24–26 July 2011.
63. Sprague, G.F.; Duncan, W.G.; Kommedahl, T.; BeMiller, J.N. Corn. In AccessScience, ®McGraw-Hill Companies. 2008. Available online: https://www.accessscience.com/content/corn/162600 (accessed on 14 August 2018).
64. Cederberg, C.; Flysjö, A. Environmental Assessment of Future Pig Farming Systems–Quantification of Three Scenarios from the FOOD 21 Synthesis Work. SIK Rapport nr. 723. SIK, Sweden. 2004. Available online: http://www.vaxteko.nu/html/sll/institutet_livsm_bioteknik/sik-rapport/SIK723/SIK723.PDF (accessed on 14 August 2018).

sustainability

MDPI

Article

Cattle Diets Strongly Affect Nitrous Oxide in the Rumen

Katrin Gerlach [1,*], Alexander J. Schmithausen [2], Ansgar C. H. Sommer [1], Manfred Trimborn [2], Wolfgang Büscher [2] and Karl-Heinz Südekum [1]

[1] Institute of Animal Science, University of Bonn, 53115 Bonn, Germany; ansgar.sommer@blattin.de (A.C.H.S.); ksue@itw.uni-bonn.de (K.-H.S.)

[2] Institute of Agricultural Engineering, University of Bonn, 53115 Bonn, Germany; schmithausen@uni-bonn.de (A.J.S.); m.trimborn@uni-bonn.de (M.T.); buescher@uni-bonn.de (W.B.)

* Correspondence: kger@itw.uni-bonn.de; Tel.: +49-228-732281

Received: 5 October 2018; Accepted: 12 October 2018; Published: 14 October 2018

Abstract: This study aimed at assigning climate-relevant gaseous emissions from ruminants to animal- or feed-related origin. Three adult rumen-cannulated German Holstein steers and three forage types (corn silage (CS), alfalfa silage (AS) and grass hay (GH)) were used in a 3 × 3 Latin square design. Each period consisted of 12 days (d), during which animals received 10 kg dry matter/day of one forage as sole feed. Gaseous samples from forages and the steers´ rumen were taken and analyzed for CO_2, CH_4, and N_2O using gas chromatography. There were large differences in the amounts of CO_2 and N_2O emitting from the forage types. Most N_2O came from AS and only small amounts from GH and CS. Results indicate that fermented forages rich in nitrogen can release climate-relevant N_2O. The highest CO_2 amounts were measured in CS. Methane was not detected in any forage sample. Animals consuming CS showed slightly lower CH_4 concentrations in the rumen gas sample than animals fed AS or GH. Big differences were found for ruminal N_2O with the highest concentration after AS ingestion such that the N_2O measured in the rumen seems to originate from the used feedstuff.

Keywords: cow; greenhouse gas; methane; rumen; silage

1. Introduction

Animal production significantly contributes to climate-relevant greenhouse gas (GHG) emissions but also offers considerable reduction potential such that different mitigation strategies like the use of feed additives and application of feeding strategies as well as different manure, reproduction, and animal management strategies are discussed [1,2]. Ruminants are mainly responsible for the trace gases methane (CH_4) and nitrous oxide (N_2O) with the latter having a much higher carbon dioxide (CO_2) equivalence factor (298) than CH_4 (25) [3]. Methane is a product of the anaerobic fermentation of carbohydrates in the rumen, which is a pathway for the disposal of hydrogen formed during microbial metabolism [4]. Cattle lose 2–10% of their ingested gross energy as eructated CH_4 [5], and the total amount is influenced by dry matter (DM) intake (DMI) and ration composition [6–8]. The volatile N_2O (Henry´s law constant, k°_H 0.025 mol/kg × bar) is mainly produced by the microbially facilitated denitrification in manure and to a smaller extent by nitrification in soils [9]. The contribution of GHG emissions from enteric fermentation and manure management occurs in a ratio of about 9:1 [10] such that the potential for decreasing GHG emissions is mainly seen in manipulating enteric fermentation, e.g., by adjusting composition of rations. In this regard, different studies have already been performed using in vitro and in vivo measurements (e.g., recent work by Lee et al. [11] and Macome et al. [12]) as well as rumen-cannulated cows, among others resulting in different regression equations for predicting CH_4 emissions based on intake and diet characteristics [13]. When applying different regression equations to five typical Central European dairy cow rations it was shown that the best differentiation between diets was achieved with equations containing forage proportion and DMI as factors [13]. For measurement of GHG emissions on animal level the use of respiration chambers is a proven technology [5,14]. Other techniques comprise a mobile open-circuit hood system to measure the gas exchange in small ruminants [15] and a ventilated hood system for measuring GHG from cattle [16]. Most studies focused on emissions of CH_4 and CO_2, whereas approaches investigating the effect of ration composition on enteric emissions of N_2O are rare [17,18]. Rotz and Thoma [19] reviewed that N_2O emissions are in the range of 0.3–0.5 g/cow per day (d), with higher values occurring possibly under certain dietary conditions. Authors state that mechanisms and amount produced are generally not well understood but that high dietary nitrate (NO_3^-) levels might induce increased N_2O emissions. Though, they also agree that more research is needed to better quantify that source of emission as formation in the rumen is questionable [20].

However, beside the ruminant itself, the forage used as feedstuff can also act as a source of emissions: for non-fermented forages Emery and Mosier [21] measured emissions of CO_2, CH_4, and N_2O from switchgrass and corn stover under varying storage conditions. Both CH_4 and N_2O were detected and concentrations were influenced by forage DM concentration. However, when calculating the net global warming potential for the different treatments (0–2.4 g CO_2 equivalents/kg DM) authors suggested that direct emission of CH_4 and N_2O from aerobically stored (non-fermented) feedstuffs have a minor effect on net global warming potential of cellulosic biofuels. Fermented forages as an origin of gaseous emissions measured in the environment of ruminants have rarely been studied. In early studies of Wang and Burris [22] N_2O was detected in whole-crop corn silages where the gas composition was analyzed eight times within 66 h after sealing the silo. A constant increase in N_2O concentration from 1.50% (v/v) to 4.55% after 54 h was measured which declined afterwards to 1%. The origin of the N_2O was seen in the reduction of NO_3^- [22]. The reduction of NO_3^- starts a few hours after ensiling with an enrichment of the intermediate products NO und NO_2^- which normally disappear after one or two weeks of ensiling [23]. Further reduction by *Enterobacteriaceae* results in N_2O and ammonia (NH_3) [24]. Also recent work using Fourier transform infrared (FTIR) spectroscopy verified the presence of N_2O in gases formed in the early phase of ensilage of whole-crop corn [25]. Franco [26] showed that particularly forages naturally rich in nitrogen (N), especially in the form of nitrate, had significant N_2O production during silage fermentation. Up to now, only little attention has been given to N_2O possibly emitting during the feed-out phase of silages.

Gaseous emissions occurring in the environment of ruminants are often difficult to assign to a specific source (e.g., feed, rumen, manure), especially when measurements are conducted on barn level, in respiration or environmental chambers. This impedes the explanation of their formation and strategies for mitigation.

Therefore, the objective of the present study was to determine gaseous emissions from ruminants offered different forage types (corn silage (CS), alfalfa silage (AS) and grass hay (GH)) with contrasting chemical composition and to assign the emissions to animal or feed-related sources, with special emphasis on nitrous oxide. To the best of our knowledge, this is the first study determining the concentration of CO_2, CH_4, and N_2O in the ruminal gas phase of steers after ingestion of three different forage types.

2. Materials and Methods

2.1. Animals, Diets, and Experimental Design

This study was conducted at the Educational and Research Center Frankenforst of the Faculty of Agriculture, University of Bonn (Königswinter, Germany). All experimental procedures were conducted in accordance with the German guidelines for animal welfare and were approved (file number 84-02.04.2017.A247) by the Animal Care Committee of the state of North Rhine-Westphalia. Three animals and three forage types differing in chemical composition (CS, AS, GH) were used.

Three adult rumen-cannulated German Holstein steers (born and raised on the Center, 4 years old, rumen-cannulated since 2 years, about 1300 kg body weight) were housed separately in single pens (4.4 × 4.6 m) allowing visual contact. Ambient conditions within the barn were consistent throughout the experimental period with a temperature of 18.4 ± 2.1 °C and relative air humidity of 74.7 ± 8.3%. Water was continuously available allowing ad libitum intake. The whole trial consisted of 42 d (4 June to 15 July 2016) and was divided into three periods following a 3 × 3 Latin square design. Each period started with a 2-d adaptation phase during which animals were offered a ration consisting of 50% of the previous forage and 50% of the new forage. Twelve days of experimental feeding followed during which animals received one of the three forages as sole feed. During this time, each steer was offered 10 kg DM/d of the respective forage. Measurement of gaseous emissions was carried out during the last 3 d of each period. Table 1 shows the chemical composition of the forages which had been produced at the Educational and Research Center Frankenforst. The AS was produced from a fourth cut of alfalfa (harvest date 9 September 2015) and ensiled in round bales. For CS, the whole-crop corn (harvest date 20 September 2015) was chopped (6 mm theoretical chop length) and ensiled in a bunker silo. The GH was made from the second cut (harvest date 28 June 2015), and the field-dried hay was packed in round bales. To ensure constant forage qualities during each period, silages were stored anaerobically in 120-L plastic barrels. Therefore, the CS was taken from a fresh silage face and the AS was obtained from a round bale opened just before. Silages were filled into the barrels in several layers, each layer was compacted separately such that a high density was reached, and were then stored anaerobically. Forages were offered to the steers once daily at 08.00 a.m. Before feeding in the morning, remaining feed was removed and weighed to determine DMI. During the last 3 d of each period (sampling period), the DM consumed within 180 min after offering feed in the morning was also measured. Every day (d 10–12), a representative sample (500 g) of each forage was taken and composited to one sample for each period. After sampling, forages were immediately frozen until analysis.

Table 1. Chemical composition of forages used for the gaseous measurements and as feedstuffs for the steers (expressed as g/kg dry matter (DM) unless stated; (*n* = 3)).

	Corn Silage (CS)	Alfalfa Silage (AS)	Grass Hay (GH)
DM [g/kg]	366	415	881
Ash	34.9	124	70.5
Crude protein	70.7	246	79.2
Ether extract	35.9	30.2	20.2
aNDFom [1]	314	396	599
ADFom [2]	175	300	340
Acid detergent lignin	17.4	98.3	36.8
Starch	438	n.a.	n.a.
In vitro gas production [mL/200 mg DM]	64.1	39.8	50.5
Metabolizable energy [MJ/kg DM]	11.7	8.78	9.40
pH	3.9	5.77	n.a.
Lactic acid	40.7	8.2	n.a.
Acetic acid	9.9	6.3	n.a.
Butyric acid	n.d. [3]	n.d.	n.a.
Methanol	0.3	1.5	n.a.
Ethanol	1.7	1.6	n.a.
Water-soluble carbohydrates	13.4	49.8	n.a.
NH_3-N [g/kg total N]	109	96.7	n.a.
Ethyl acetate [mg/kg DM]	54.4	19.3	n.a.
Ethyl lactate [mg/kg DM]	105	n.d.	n.a.

[1] aNDFom: neutral detergent fiber assayed with heat-stable amalyse and expressed exclusive residual ash. [2] ADFom: acid detergent fiber expressed exclusive residual ash. n.d.: not detected. n.a.: not analyzed.

2.2. Sampling and Measurements of Gaseous Emissions

During the last 3 d of each period, sampling for measurements of gaseous emissions from forages and the rumen was conducted. Concurrently, forage samples for laboratory analysis and incubation experiments for gas measurement were taken. The emission measurements aimed at the acquisition of the gases CO_2, CH_4, and N_2O from the forages and the rumen gas of the steers. Sampling of emissions from the forages was conducted simultaneously to the feeding using closed containers with a volume of 10 L. The containers were made of polyethylene (PE) and were equipped with a rubber septum for gas sampling via twin needle. For each container, average temperature and relative humidity were logged continuously using data loggers (Tinytag Plus 2—TGP-4500, Gemini Data Loggers Ltd., Chichester, West Sussex, UK). For sample collection a defined amount of each forage (1 kg each of CS and AS, and 0.5 kg GH) was put in the container and sealed gas tight. Within the next 40 min five gaseous samples were taken using evacuated headspace vials directly after closure (0 min) and 10, 20, 30, and 40 min after closure (Figure 1a). Then the containers were opened for 140 min to enable unrestricted, natural air exchange before a second sealing and gas sampling period started. The headspace vials had a vacuum range below 5 mbar. The vacuum was produced by pricking a twin needle through the container septum as described by Schmithausen et al. [27]. This procedure (sampling of emissions from the forages) was conducted on 3 consecutive d.

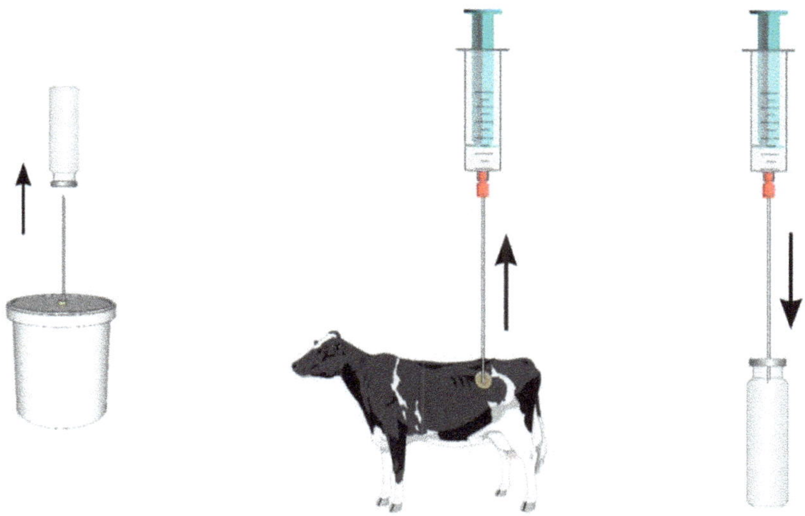

(a) (b)

Figure 1. Sampling (**a**) of gaseous emissions from the forages stored in a closed container via headspace vials and (**b**) from the gaseous phase of the rumen with a syringe through the closed lid of the rumen-cannula (**left**) and filling into a headspace vial (**right**).

Samples from the gaseous phase of the steers´ rumen were taken 180 min after offering feed in the morning. In 10-min intervals (0, 10, 20, and 30 min), one sample was obtained with a syringe (50 mL) through the closed lid of the rumen-cannula and filled into two evacuated headspace vials (20 mL each; Figure 1b). Subsequently, samples were analyzed for CO_2, CH_4, and N_2O using a gas chromatograph (GC) (8610 C, SRI Instruments, Torrance, CA, USA). The N_2O and CO_2 were determined with an electron capture detector (ECD) and CH_4 was measured with a flame ionization detector (FID) [28,29]. The detection limit of the used analytical technique for CO_2, CH_4, and N_2O is described in detail by Schmithausen et al. [30]. The emission rates of the respective gas from the incubation experiments were calculated via linear regression of the gas concentration over time, more specifically, the slope of the regression line. The detection limits of the GC result in minimally measurable increases in the concentration (slope) of the investigated gases in the incubation experiment. In the case of CH_4, for example, this minimum slope was 0.5 ppm CH_4, which corresponds to 3.3 μg CH_4/(kg of feed × h). Comparable measurements by using headspace vials and defined criteria of evaluation are described by Schmithausen et al. [27]. As a result of the ruminal gas analysis, the concentrations of CH_4 in rumen gas phase and the ratio of N_2O to the sum of CO_2 and CH_4 are shown. The amounts of N_2O formed in the rumen or the emission rates of N_2O from the rumen could not be calculated, as the total volume of air in the rumen and the total rates of formation of CO_2 and CH_4 in the rumen could not be determined in this experiment. The CO_2, N_2O, and CH_4 values are expressed as concentration in the rumen gas phase as well as ratio of CH_4 to CO_2 as an indicator of the efficiency of microbial fermentation [31].

2.3. Laboratory Analyses of the Forages

In each of the three periods, silages and hay were sampled for chemical analyses. Forages were kept at −20 °C and were then freeze-dried (Freeze-Dryer P18K-E, Piatkowski Forschungsgeräte, München, Germany) in triplicate. Afterwards, a duplicate subsample was oven-dried overnight at 105 °C for determination of the DM concentration. A correction of DM (DMcor) for the losses of volatiles during drying was done in alfalfa and corn silages with the following equations (concentrations are given as g/kg):

Alfalfa silage [32]:

$$\begin{aligned}\text{DMcor} = \text{DM} + (1.05 - 0.059 \times \text{pH}) \times \text{total volatile fatty acids (VFA, } C_2 - C_6) + \\ 0.08 \times \text{lactic acid} + 0.77 \times \text{1,2-propanediol} + 0.87 \times \text{2,3-butanediol} + 1.00 \times \text{total of other alcohols.}\end{aligned} \tag{1}$$

Corn silage [33]:

$$\begin{aligned}\text{DMcor} = \text{DM} + 0.95 \times \text{VFA } (C_2 - C_6) + 0.08 \times \text{lactic acid} + \\ 0.77 \times \text{1,2-propanediol} + 1.00 \times \text{other alcohols.}\end{aligned} \tag{2}$$

After freeze-drying, samples were ground using 3-mm and afterwards 1-mm sieves. Samples were chemically analyzed according to VDLUFA [34] and following method numbers: Analysis of ash and ether extract (EE) was done by using methods 8.1 and 5.1. Crude protein (CP) was analyzed by Dumas combustion (4.1.2, FP328, Leco 8.1, Leco Instrumente, Mönchengladbach, Germany). The concentrations of neutral detergent fiber assayed with heat-stable amylase and expressed exclusive residual ash (aNDFom; 6.5.1), acid detergent fiber expressed exclusive residual ash (ADFom; 6.5.2), and acid detergent lignin (ADL; 6.5.3) were determined with an Ankom2000 Fiber Analyzer (Ankom Technology, Macedon, NY, USA). Following point 8.8 of method 6.5.2 the analysis of ADFom was conducted sequentially for AS to avoid precipitation of pectins. In CS, the concentration of starch was determined after enzymatically hydrolyzing starch to glucose [35]. The 24 h in vitro gas production (GP [mL/200 mg DM]) of forage samples was measured with the Hohenheim gas test (method 25.1, [34]) and afterwards, the concentration of metabolizable energy (ME) was estimated as follows:

Corn silage [36]:

$$\text{ME} = 0.136 \times \text{GP} + 0.0057 \times \text{CP} + 0.000286 \times \text{EE}^2 + 2.20. \tag{3}$$

Alfalfa silage [37]:

$$\begin{aligned}\text{ME [MJ/kg organic matter]} = 11.09 - 0.01040 \times \text{ADFom} + 0.00497 \times \text{CP} + 0.00750 \times \text{EE} + \\ 0.0351 \times \text{GP; ME [MJ/kg DM]} = \text{ME (MJ/kg organic matter)} \times [1000 - \text{ash (g/kg DM)}]/1000.\end{aligned} \tag{4}$$

Grass hay [38]:

$$\text{ME} = 7.81 + 0.07559 \times \text{GP} + 0.00384 \times \text{ash} + 0.00565 \times \text{CP} + 0.01898 \times \text{EE} - 0.00831 \times \text{ADFom.} \tag{5}$$

Both silage types were analyzed for fermentation products after cold-water extraction. These analyses were conducted at the Central Analytical Laboratory of the Humboldt University, Berlin, Germany and concentrations of lactic acid, volatile fatty acids (VFA), alcohols (methanol, ethanol, propanol, 1,2-propanediol, 2,3-butanediol), acetone, ammonia, and water-soluble carbohydrates (WSC) as well as the pH were determined. Frozen forage samples (50.0 g) were blended with a mixture of 200 mL distilled water and 1 mL toluene for preparation of cold-water extracts. After keeping them overnight in a refrigerator extracts were filtered with a folded filter paper. The pH in the extract was measured potentiometrically with a calibrated pH electrode. Analysis of lactic acid was done by high performance liquid chromatography (HPLC) (RI-detector, Shimadzu Deutschland GmbH, Duisburg, Germany) [39]. Gas chromatography with FID (GC-2010; Shimadzu Deutschland, Duisburg, Germany) and a free fatty acid phase column (Permabond FFAP 0.25 Tm; Macherey-Nagel, Düren, Germany) was used for determining the VFA and alcohols. Ammonia was measured colorimetrically using a continuous flow analyzer (Skalar Analytical B.V., Breda, The Netherlands) and the concentration of WSC was analyzed using the anthrone method [40].

2.4. Statistical Analyses

All statistical analyses were performed with SAS 9.4. The following mixed model was used for the rumen samples:

$$y_{ij} = \mu + F_i + P_j + (F \times P)_{ij} + A_k + e_{ijk} \quad (6)$$

with y = observed response; μ = overall mean; F_i = fixed effect of forage type i = 1, 2, 3; P_j = fixed effect of period j = 1, 2, 3; $(B \times P)_{ij}$ = effect of interaction forage type I $\times$ period j; A = random effect of the animal k = 1, 2, 3; and e_{ijk} = residual error.

For analysis of the gas samples from forages the following mixed model was used:

$$y_{ij} = \mu + F_i + P_j + (F \times P)_{ij} + e_{ij} \quad (7)$$

y = observed response; μ = overall mean; F_i = fixed effect of forage type i = 1, 2, 3; P_j = fixed effect of period j = 1, 2, 3; $(B \times P)_{ij}$ = effect of interaction forage type I $\times$ period j; and e_{ij} = residual error.

Covariance structures were tested with the types "unstructured", "autoregressive", and "compound symmetry". "Akaike´s Information Criterion" (AIC) was used to decide which model showed the best fit and based on that, "autoregressive" was chosen for the analysis. Within the period, d was taken as a repeated measurement. Least squares means were compared using the PDIFF option in SAS. Significant treatment effects were detected by pairwise comparisons employing Tukey´s test. In all statistical analyses, differences among means with $p < 0.05$ were accepted as representing statistically significant differences.

3. Results

3.1. Gas Production from Forages

As intended, forages differed considerably in chemical composition (Table 1). The AS had high concentrations of CP (246 g/kg DM), whereas GH and CS had only low to moderate concentrations. The GH contained high concentrations of fiber fractions (e.g., aNDFom) and was low in EE. The CS was high in starch (438 g/kg DM), in vitro gas production and metabolizable energy. Both silage types were well fermented with moderate to low concentrations of acetic acid and without butyric acid. The pH value in AS, however, was higher than recommended.

The emissions from forages as influenced by forage type, period, and their interaction are shown in Table 2. There were large differences in the rates of CO_2 and N_2O emitting from the forages ($p < 0.05$). Most N_2O was released from AS (24.1 μg/(kg DM × h)) and only small amounts from GH (0.233 μg/(kg DM × h)) and CS (0.109 μg/(kg DM × h)). The CO_2 emissions were also influenced by forage type and greatest CO_2 amounts were measured in CS, followed by AS ($p < 0.01$). Both N_2O and CO_2 were influenced by forage type, but no influence ($p > 0.05$) was observed of period or the interaction between period and forage type. After 180 min, most emissions from forages were strongly reduced but 170 mg/(kg DM × h) of CO_2 were still emitting from CS. Methane was not detected in any forage sample, neither directly after silo opening nor after 180 min (detection limit for CH_4 was 3.3 μg/(kg × h)).

Table 2. Effect of forage type (F) and period (P) on emission rates of CH_2 and N_2O * of samples obtained from corn silage (CS), alfalfa silage (AF) and grass hay (GH) directly after silo opening (8 a.m.) and after 180 min of air exposure (11 a.m.).

		Least Square Means				Effect		
		CS	AS	GH	SEM	F	P	F·P
N_2O [μg/(kg dry matter × h)]	8 a.m.	0.109 ^{b}	24.1 ^{a}	0.233 ^{b}	3.81	0.02	n.s.	n.s.
	11 a.m.	0.140 ^{b}	2.46 ^{a}	0.176 ^{b}	0.172	<0.01	0.01	<0.01
CO_2 [mg/(kg dry matter × h)]	8 a.m.	391 ^{a}	141 ^{b}	8.13 ^{c}	32.0	<0.01	n.s.	n.s.
	11 a.m.	170 ^{a}	19.0 ^{b}	9.38 ^{b}	14.4	<0.01	n.s.	n.s.

* Methane was not detected in any forage sample. SEM: standard error of the mean. n.s.: not significant ($p < 0.05$).
$^{a–c}$ Values within a row with different letters are significantly ($p < 0.05$) different.

3.2. Gas Composition in the Rumen

The effect of forage type and period on DMI and composition of gaseous samples obtained from the rumen of steers is shown in Table 3. During 180 min, animals consumed 3.3 to 6.6 kg DM which was influenced by forage type. 180 min after initiation of feed ingestion, gas samples were taken from the rumen. There was a significant effect of forage type on all measured variables ($p < 0.01$). Big differences were found for N_2O with higher concentrations for AS than for CS and GH ($p < 0.01$). The N_2O concentration in the rumen atmosphere relative to the sum of CH_4 and CO_2 (%) for the different forage types obtained from the single measurements is shown in Figure 2. Elevated concentrations were only detected after ingestion of AS. Animals fed CS showed slightly lower CH_4 concentrations in the rumen gas sample than when fed AS or GH. The CH_4 to CO_2 ratio was highest for AS ($p < 0.01$) and there was no difference between CS and GH ($p > 0.05$). This ratio can be seen as an indicator of the efficiency of microbial fermentation as it directly describes the share of emitted C that has not been metabolized to CO_2 [31].

The CO_2 concentration was highest after ingestion of GH ($p < 0.01$) and did not differ between CS and AS ($p > 0.05$). The remaining gas (difference to 100%) that cannot be explained by CH_4, CO_2, and N_2O is presumably atmospheric air that may have entered the rumen or the gaseous sample via three possible ways: with the forage into the rumen during ingestion, via small leakages of the rumen cannula into the rumen or during sampling (into evacuated headspace vials). Concentrations of both O_2 and N_2 typically increase during feeding [41]. As oxygen entering during ingestion or via the cannula is depleted rapidly in the rumen atmosphere, mainly N_2 remains from the atmospheric air which could not be analyzed with the methodology applied in this study.

Table 3. Effect of forage type (F) and period (P) on dry matter intake (DMI) over 180 min and composition of gaseous samples obtained from the rumen of steers 180 min after initiation of feed intake.

	Least Square Means				Effect		
	Corn Silage (CS)	Alfalfa Silage (AS)	Grass Hay (GH)	SEM	F	P	F·P
DMI [kg/180 min]	4.60 [b]	3.22 [b]	6.64 [a]	0.473	<0.01	0.02	<0.01
N_2O [ppm]	0.246 [b]	0.857 [a]	0.171 [b]	0.068	<0.01	n.s.	0.02
CH_4 [%]	16.9 [b]	20.6 [a]	20.3 [a]	0.890	0.01	<0.01	0.03
CO_2 [%]	46.1 [b]	41.1 [b]	54.8 [a]	2.21	<0.01	n.s.	0.03
CH_4:CO_2	0.358 [b]	0.501 [a]	0.372 [b]	0.010	<0.01	0.02	<0.01

[a–c] Values within a row with different letters are significantly ($p < 0.05$) different. SEM: standard error of the mean. n.s.: not sigificant ($p > 0.05$).

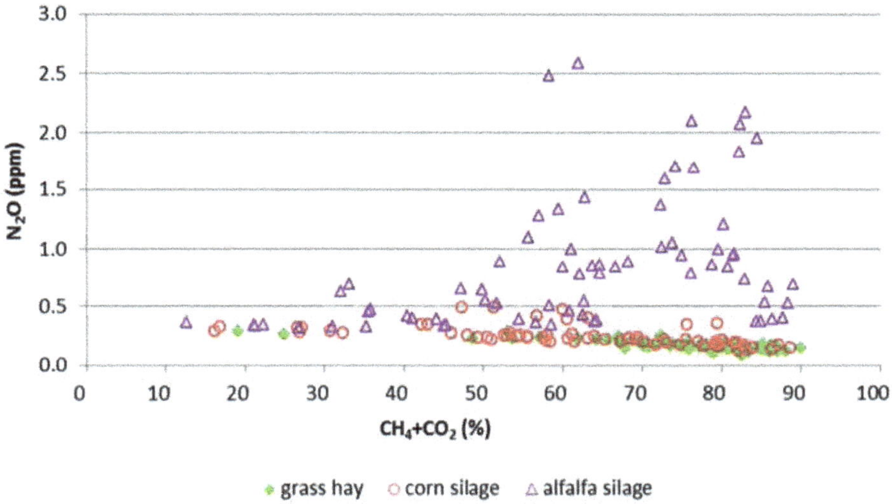

Figure 2. The N_2O concentration (ppm) in the rumen gas relative to the CH_4+CO_2 concentration (%) for alfalfa silage, corn silage, and grass hay.

4. Discussion

4.1. Emissions from Forages

Directly after silo opening, N_2O emitted from AS but there were no N_2O emissions from CS and GH. Formation of N_2O during ensiling has been described before and can be mainly ascribed to anaerobic activity of *Enterobacteriaceae* species occurring during the initial period of ensiling [23]. Plant enzymes, on the other hand were not capable of producing N_2O and NOx during ensiling such that microbial activity seems to be the main underlying process [26]. The conversion of NO_3^- during ensiling appears to be related to the duration the crop remains at a pH at which *Enterobacteriaceae* may grow and utilize NO_3^- (pH > 4.5–5.0) [24]. Due to a typically high buffering capacity (high CP, high ash concentration) and a high DM concentration of the experimental AS only a moderate drop in pH to 5.7 had been achieved. Consequently *Enterobacteriaceae* were not restricted by acidic conditions during the whole storage period. Also the increased NH_3-N concentrations in AS may reflect increased activity of *Enterobacteriaceae* [23]. In contrast to this, whole-crop corn typically has a low buffering capacity and ferments rapidly. As a result, the CS had a low pH (3.9) which inhibits *Enterobacteriaceae*. The N_2O emissions from CS and GH were very low and only slightly above detection limit (0.1 µg/(kg × DM h)).

Aerobic activity of *Enterobacteriaceae* may also occur in silages [42], but is most probably restricted to respiration. The decreased emission rates of N_2O after 180 min of air exposure indicate that N_2O emitted that had already been formed during the anaerobic fermentation process. The major part of the N_2O was released during 180 min such that an aerobic formation seems unlikely. It can be concluded that N_2O emissions from forages are possible under certain circumstances. It seems to be most pronounced from forages with high CP and NO_3^- concentrations at harvest [26] and extended and/or continuous activity of *Enterobacteriaceae* which can be caused by high silage pH [24]. It is therefore important to optimize the ensiling conditions (rapid wilting and sealing, strong compaction, use of additives in substrates that are classified as being difficult to ensile) to ensure a fast and sufficient drop in pH. More research is needed to state more precisely the conditions of formation and release of N_2O in silages. However, the total amounts of N_2O emitting from fermented forages are much lower than typical emissions from manure during storage which are in the range of 1.0 to 3.0 kg/cow per year (equaling 0.1 to 0.3 g/cow per h), mainly depending on the method of storage [19].

Besides N_2O, also CO_2 emitted from forages with an effect of forage type. As expected, only fermented forages released considerable amounts of CO_2, most likely produced at the beginning of the ensiling process. The CS emitted more CO_2 than AS. Caused by its plant structure and longer chop length in comparison to CS, alfalfa is more difficult to compact and its tubular hollow stem may even impede the removal of air during ensilage [43] or, vice versa, facilitate ingress of oxygen as soon as the silo is opened. Therefore, CO_2 might be lower in concentration and emit very quickly after silo opening or during relocation to the barrels, explaining the lower emission rates in AS. Also aerobic spoilage processes by yeasts and molds which typically take place after silo opening lead to the formation of CO_2 [44]. However, as CS still had a low pH and high concentrations of lactic acid (as an indicator of good fermentation quality) and emission rates diminished during aerobic exposure, ongoing aerobic deterioration processes can be excluded and the measured CO_2 might result from gassing out of CO_2 already being formed during ensiling. The forage gas samples were also analyzed for CH_4 but changes in concentration were below detection limit in all cases. Fermented forages seem to be an unlikely source of CH_4 emissions. To the best of our knowledge, possible CH_4 emissions from silages have also not been studied or discussed in literature. Emery and Mosier [21] measured GHG emissions from unfermented feedstuffs and detected small amounts of CH_4; however impact on the net global warming potential was assessed to be small.

4.2. Concentration Ratios in the Rumen

With the method of taking samples through the closed lid of the rumen-cannula via a syringe it was possible to obtain information on the composition of the gaseous phase in the rumen of the steers, without any interference (e.g., atmospheric air, oral contact, manure). Highest ruminal concentrations of N_2O were found for steers fed AS with values exceeding 2.5 ppm at some sampling times (Figure 2) despite the fact that DMI was lowest for AS. In contrast, the N_2O concentrations after ingestion of GH and CS were always below 0.5 ppm such that a clear effect of forage type could be shown. It is questionable whether N_2O can be formed directly in the rumen under certain conditions. Kaspar and Tiedje [45] detected traces of N_2O (up to 0.3% of added nitrogen) when investigating the dissimilatory reduction of nitrate and nitrite by the rumen microbiota of a rumen-cannulated cow. They concluded that N_2O is a by-product of dissimilatory nitrite reduction to ammonium rather than a product of denitrification which seems to be absent from the rumen habitat. However, only traces were found under those experimental feeding conditions with addition of nitrate. Also de Raphélis-Soissan et al. [46] and Lee et al. [11] fed nitrate to ruminants in an attempt to lower ruminal CH_4 production. In this regard, two main possibilities by which NO_3^- reduces enteric CH_4 production were discussed [11]: NO_3^- reduction (thermodynamically favorable in comparison to methanogenesis) as major pathway and secondly, possibly being quantitatively less important, NO_3^- and NO_2^- being toxic to methanogens in the rumen. In both cases, CH_4 production was decreased by addition of nitrate, however, de Raphélis-Soissan et al. [46] stated that, on the other hand, the N_2O emission from sheep in respiration chambers was increased which led to a reduction of the net benefit of methane mitigation on global warming potential (CO_2 equivalents/kg DMI) of 18%. This effect could be mitigated by using encapsulated NO_3^- as slow-release form, thereby lowering NO_2^- toxicity after nitrate ingestion [11]. When ruminants are fed typical rations without added nitrate, formation of N_2O under anaerobic conditions in the rumen seems unlikely such that oral ways of formation after dietary nitrate supplementation were discussed as possible mechanisms based on measurements of N_2O from dairy cows in respiration chambers [20]. A release from the rumen via eructation was excluded by the authors as there was no relationship at all between CH_4 and N_2O in ventilation air of the respiration chamber. However, the possibility of N_2O formation in the oral cavity can be excluded for the current study as the gas samples were taken directly from the rumen atmosphere without oral contact. Also, feces as a possible source of N_2O as discussed for sheep [46] can be excluded in our study due to the sampling method. As enteric formation under anaerobic conditions seems unlikely, the transfer from the forage into the rumen is the most likely way. In our study the AS emitted considerably more N_2O than the other forage types. After ingestion of AS, solved N_2O may have gassed out in the rumen, which would explain the increased concentrations in the rumen gas sample 180 min after initiation of feed intake.

Also, the CH_4 concentrations in the rumen gas sample were influenced by forage type and the lowest concentrations were detected after ingestion of CS. In contrast to N_2O methane is formed in the rumen as a product of carbohydrate fermentation, and the total amount is influenced by DMI and chemical composition of the feedstuff [8] as well as by the rumen microbial community (species, abundance, and activity of microbes) and fermentation pathways [47]. An effect of diurnal variation on rumen CH_4 concentrations as described by Bjerg et al. [48] can be excluded due to the experimental design. A decreased concentration is not necessarily connected with a decreased total CH_4 formation; however, a reduced formation of CH_4 in the rumen of cattle fed CS in comparison to other forage types has also been observed in other studies [9] and is related to the increased propionate to acetate ratio and a decreased rumen pH caused by feedstuffs with enhanced degradability (e.g., increased starch and reduced fiber concentration like CS in the present study) [4,49].

The CO_2 concentration in rumen gas samples was greatest after ingestion of GH and did not differ for CS and AS, and all concentrations were in the range of values summarized from several feeding trials [50]. The CH_4 to CO_2 ratio was lower for CS and GH than for AS. The lower ratio seems to be caused by the lower CH_4 concentration for CS as discussed before and an increased share of CO_2 for GH where DMI was highest. As the amount of consumed DM and its fermentability are the main factors influencing the CO_2 production [31] the amount of ingested fermentable substrate might explain the higher CO_2 concentration for GH. The CH_4 to CO_2 ratio can be seen as an indicator of the efficiency of microbial fermentation as it directly describes the share of emitted C that has not been metabolized to CO_2 [31]. According to this, the efficiency of microbial fermentation was lowest for AS. As the DMI was lowest for AS, a reduced passage rate of the digesta could have caused an increased methanogenesis. McAllister et al. [8] concluded from several studies that properties of forages decreasing the rate of digestion or prolonging the time of feed particles being in the rumen generally lead to a rise in the amount of CH_4 that is formed per unit of forage digested. In contrast, recent work by Dittmann et al. [51] carefully proposed the opposite way as the CH_4 production itself might influence digesta retention in the sense of a feedback mechanism to mitigate CH_4 losses by decreasing retention time at higher CH_4 production.

5. Conclusions

The experimental setup in this study with very diverging types of forages and a 3 × 3 Latin square design made it possible to assign gaseous emissions from steers to animal- or feed-related origin. Results indicate that fermented forages rich in CP or nitrate like alfalfa silage can release climate-relevant N_2O with the conditions of its formation, emitting amounts and strategies for reduction (e.g., targeted use of silage additives, feed-out management) warranting further research.

Under the aspect of mitigating GHG emissions from animal production also the feeding management of farms has to be considered. The N_2O detected in the rumen gas of the steers seems to originate from the consumed feedstuff and is probably not synthesized in the rumen. Additional studies, e.g., with high-yielding dairy cows and concurrent analyses of feedstuffs and environmental conditions are needed to make those findings applicable for ruminants in general.

Author Contributions: Conceptualization, K.G., W.B. and K.-H.S.; Data curation, A.C.H.S. and M.T.; Investigation, K.G.; Methodology, A.J.S., A.C.H.S. and M.T.; Project administration, W.B. and K.-H.S.; Supervision, K.-H.S.; Validation, M.T.; Visualization, A.J.S.; Writing—Original draft, K.G.; Writing—Review & editing, A.J.S., M.T. and K.-H.S.

Funding: This work was partially funded by the German Research Foundation (DFG, BU 1235/8-1, Germany).

Acknowledgments: The authors thank the staff of the Educational and Research Center Frankenforst for support in conducting the experiments. Furthermore, this study was partly conducted by members of the Center of Integrated Dairy Research (CIDRe), University of Bonn (Bonn, Germany).

Conflicts of Interest: The authors declare no conflict of interest.

References

1. Gerber, P.J.; Steinfeld, H.; Henderson, B.; Mottet, A.; Opio, C.; Dijkman, J.; Falcucci, A.; Tempio, G. *Tackling Climate Change through Livestock: A Global Assessment of Emissions and Mitigation Opportunities*; Food and Agriculture Organization of the United Nations (FAO): Rome, Italy, 2013.
2. Hristov, A.; Oh, J.; Lee, C.; Meinen, R.; Montes, F.; Ott, T.; Firkins, J.; Rotz, A.; Dell, C.; Adesogan, A. *Mitigation of Greenhouse gas Emissions in Livestock Production: A Review of Technical Options for Non-CO_2 Emissions*; FAO Anim. Produon and Health Paper; FAO: Rome, Italy, 2013; Volume 177, pp. 1–206.
3. IPCC. *The Physical Science Basis. Contribution of working group I to the Fourth Assessment Report of the Intergovernmental Panel on Climate Change*; Cambridge University Press: Cambridge, UK, 2007.
4. McAllister, T.; Newbold, C. Redirecting rumen fermentation to reduce methanogenesis. *Aust. J. Exp. Agric.* **2008**, *48*, 7–13. [CrossRef]
5. Johnson, K.A.; Johnson, D.E. Methane emissions from cattle. *J. Anim. Sci.* **1995**, *73*, 2483–2492. [CrossRef] [PubMed]
6. Jentsch, W.; Schweigel, M.; Weissbach, F.; Scholze, H.; Pitroff, W.; Derno, M. Methane production in cattle calculated by the nutrient composition of the diet. *Arch. Anim. Nutr.* **2007**, *61*, 10–19. [CrossRef] [PubMed]
7. Kirchgessner, M.; Windisch, W.; Müller, H.L. Nutritional Factors for the Quantification of Methane Production. In *Proceedings of the 8th Symposium on Ruminant Physiology, 1995*; Engelhart, W., Leonhard-Marek, S., Breves, G., Giesecke, D., Eds.; Ferdinand Enke: Stuttgart, Germany, 1995; pp. 333–348.
8. McAllister, T.; Cheng, K.-J.; Okine, E.; Mathison, G. Dietary, environmental and microbiological aspects of methane production in ruminants. *Can. J. Anim. Sci.* **1996**, *76*, 231–243. [CrossRef]
9. Eckard, R.J.; Grainger, C.; de Klein, C.A.M. Options for the abatement of methane and nitrous oxide from ruminant production: A review. *Livest. Sci.* **2010**, *130*, 47–56. [CrossRef]
10. Prusty, S.; Sontakke, U.; Kundu, S. Methane and nitrous oxide emission from livestock manure. *Afr. J. Biotechnol.* **2014**, *13*, 4200–4207.
11. Lee, C.; Araujo, R.C.; Koenig, K.M.; Beauchemin, K.A. In situ and in vitro evaluations of a slow release form of nitrate for ruminants: Nitrate release rate, rumen nitrate metabolism and the production of methane, hydrogen, and nitrous oxide. *Anim. Feed Sci. Technol.* **2017**, *231*, 97–106. [CrossRef]
12. Macome, F.M.; Pellikaan, W.F.; Schonewille, J.T.; Bannink, A.; van Laar, H.; Hendriks, W.H.; Warner, D.; Cone, J.W. In vitro rumen gas and methane production of grass silages differing in plant maturity and nitrogen fertilisation, compared to in vivo enteric methane production. *Anim. Feed Sci. Technol.* **2017**, *230*, 96–102. [CrossRef]
13. Hippenstiel, F.; Pries, M.; Büscher, W.; Südekum, K.-H. Comparative evaluation of equations predicting methane production of dairy cattle from feed characteristics. *Arch. Anim. Nutr.* **2013**, *67*, 279–288. [CrossRef] [PubMed]
14. Storm, I.M.L.D.; Hellwing, A.L.F.; Nielsen, N.I.; Madsen, J. Methods for measuring and estimating methane emission from ruminants. *Animals* **2012**, *2*, 160–183. [CrossRef] [PubMed]

15. Fernández, C.; López, M.; Lachica, M. Low-cost mobile open-circuit hood system for measuring gas exchange in small ruminants: From manual to automatic recording. *J. Agric. Sci.* **2015**, *153*, 1302–1309. [CrossRef]
16. Place, S.E.; Pan, Y.; Zhao, Y.; Mitloehner, F.M. Construction and operation of a ventilated hood system for measuring greenhouse gas and volatile organic compound emissions from cattle. *Animals* **2011**, *1*, 433–446. [CrossRef] [PubMed]
17. Hassanat, F.; Gervais, R.; Julien, C.; Massé, D.I.; Lettat, A.; Chouinard, P.Y.; Petit, H.V.; Benchaar, C. Replacing alfalfa silage with corn silage in dairy cow diets: Effects on enteric methane production, ruminal fermentation, digestion, N balance, and milk production. *J. Dairy Sci.* **2013**, *96*, 4553–4567. [CrossRef] [PubMed]
18. Haque, M.N.; Hansen, H.H.; Storm, I.M.; Madsen, J. Comparative methane estimation from cattle based on total CO_2 production using different techniques. *Anim. Nutr.* **2017**, *3*, 175–179. [CrossRef] [PubMed]
19. Rotz, C.A.; Thoma, G. Assessing carbon footprints of dairy production systems. In *Large Dairy Herd Management*, 3rd. ed.; Beede, D.K., Ed.; American Dairy Science Association: Champaign, IL, USA, 2017; pp. 19–31.
20. Petersen, S.O.; Hellwing, A.L.F.; Brask, M.; Højberg, O.; Poulsen, M.; Zhu, Z.; Baral, K.R.; Lund, P. Dietary nitrate for methane mitigation leads to nitrous oxide emissions from dairy cows. *J. Environ. Qual.* **2015**, *44*, 1063–1070. [CrossRef] [PubMed]
21. Emery, I.; Mosier, N. Direct emission of methane and nitrous oxide from switchgrass and corn stover: Implications for large-scale biomass storage. *GCB Bioenergy* **2015**, *7*, 865–876. [CrossRef]
22. Wang, L.C.; Burris, R. Toxic gases in silage, mass spectrometric study of nitrogenous gases produced by silage. *J. Agric. Food Chem.* **1960**, *8*, 239–242. [CrossRef]
23. Spoelstra, S.F. Nitrate in silage. *Grass Forage Sci.* **1985**, *40*, 1–11. [CrossRef]
24. Pahlow, G.; Muck, R.E.; Driehuis, F.; Oude Elferink, S.J.W.H.; Spoelstra, S.F. Microbiology of ensiling. In *Silage Science and Technology*; Buxton, D.R., Muck, R.E., Harrison, J.H., Eds.; ASA, CSSA, SSSA: Madison, WI, USA, 2003; pp. 31–93.
25. Zhao, Y.; Wexler, A.S.; Hase, F.; Pan, Y.; Mitloehner, F.M. Detecting nitrous oxide in complex mixtures using FTIR spectroscopy: Silage gas. *J. Environ. Prot.* **2016**, *7*, 1719–1729. [CrossRef]
26. Franco, R. Measuring Emissions and Developing Strategies to Mitigate Volatile Organic Compounds and Oxides of Nitrogen from Silage. Ph.D. Thesis, University of California, Davis, CA, USA, 2016.
27. Schmithausen, A.J.; Trimborn, M.; Büscher, W. Sources of nitrous oxide and other climate relevant gases on surface area in a dairy free stall barn with solid floor and outside slurry storage. *Atmos. Environ.* **2018**, *178*, 41–48. [CrossRef]
28. Wulf, S.; Maeting, M.; Clemens, J. Application technique and slurry co-fermentation effects on ammonia, nitrous oxide, and methane emissions after spreading. *J. Environ. Qual.* **2002**, *31*, 1795–1801. [CrossRef] [PubMed]
29. Clemens, J.; Trimborn, M.; Weiland, P.; Amon, B. Mitigation of greenhouse gas emissions by anaerobic digestion of cattle slurry. *Agric. Ecosyst. Environ.* **2006**, *112*, 171–177. [CrossRef]
30. Schmithausen, A.J.; Trimborn, M.; Büscher, W. Methodological comparison between a novel automatic sampling system for gas chromatography versus photoacoustic spectroscopy for measuring greenhouse gas emissions under field conditions. *Sensors* **2016**, *16*, 1638. [CrossRef] [PubMed]
31. Madsen, J.; Bjerg, B.S.; Hvelplund, T.; Weisbjerg, M.R.; Lund, P. Methane and carbon dioxide ratio in excreted air for quantification of the methane production from ruminants. *Livest. Sci.* **2010**, *129*, 223–227. [CrossRef]
32. Weißbach, F.; Strubelt, C. Correcting the dry matter content of grass silages as a substrate for biogas production. *Landtechnik* **2008**, *63*, 210–246.
33. Weißbach, F.; Strubelt, C. Correcting the dry matter content of maize silages as a substrate for biogas production. *Landtechnik* **2008**, *63*, 82–83.
34. VDLUFA. *VDLUFA-Methodenbuch, Bd. III. Die Chemische Untersuchung von Futtermitteln*; VDLUFA-Verlag: Darmstadt, Germany, 2012.
35. Brandt, M.; Schuldt, A.; Mannerkorpi, P.; Vearasilp, T. Zur enzymatischen Stärkebestimmung im Darminhalt und Kot von Kühen mit hitzestabiler Amylase. *Arch. Anim. Nutr.* **1987**, *37*, 455.
36. Menke, K.H.; Steingass, H. Schätzung des energetischen Futterwerts aus der in vitro mit Pansensaft bestimmten Gasbildung und der chemischen Analyse. II. Regressionsgleichungen. *Übers Tierernährg* **1987**, *15*, 59–94.

37. GfE. Equations for predicting metabolisable energy and digestibility of organic matter in forage legumes for ruminants. *Proc. Soc. Nutr. Physiol.* **2017**, *26*, 186–193.
38. GfE. New equations for predicting metabolisable energy of grass and maize products for ruminants. *Proc. Soc. Nutr. Physiol.* **2008**, *17*, 191–197.
39. Weiß, K.; Kaiser, E. Milchsäurebestimmung in Silageextrakten mit Hilfe der HPLC. *Wirtschaftseig. Futter* **1995**, *41*, 69–80.
40. von Lengerken, J.; Zimmermann, K. *Handbuch Futtermittelprüfung*; Deutscher Landwirtschaftsverlag: Berlin, Germany, 1991.
41. Barry, T.; Thompson, A.; Armstrong, D. Rumen fermentation studies on two contrasting diets. 1. Some characteristics of the in vivo fermentation, with special reference to the composition of the gas phase, oxidation/reduction state and volatile fatty acid proportions. *J. Agric. Sci.* **1977**, *89*, 183–195. [CrossRef]
42. Lindgren, S.; Pettersson, K.; Kaspersson, A.; Jonsson, A.; Lingvall, P. Microbial dynamics during aerobic deterioration of silages. *J. Sci. Food Agric.* **1985**, *36*, 765–774. [CrossRef]
43. McAllister, T.A.; Feniuk, R.; Mir, Z.; Mir, P.; Selinger, L.B.; Cheng, K.J. Inoculants for alfalfa silage: Effects on aerobic stability, digestibility and the growth performance of feedlot steers. *Livest. Prod. Sci.* **1998**, *53*, 171–181. [CrossRef]
44. Weinberg, Z.G.; Ashbell, G. Changes in gas composition in corn silages in bunker silos during storage and feedout. *Can. Agric. Eng.* **1994**, *36*, 155–158.
45. Kaspar, H.F.; Tiedje, J.M. Dissimilatory reduction of nitrate and nitrite in the bovine rumen: Nitrous oxide production and effect of acetylene. *Appl. Environ. Microbiol.* **1981**, *41*, 705–709. [PubMed]
46. De Raphélis-Soissan, V.; Li, L.; Godwin, I.R.; Barnett, M.C.; Perdok, H.B.; Hegarty, R.S. Use of nitrate and *Propionibacterium acidipropionici* to reduce methane emissions and increase wool growth of Merino sheep. *Anim. Prod. Sci.* **2014**, *54*, 1860–1866. [CrossRef]
47. Knapp, J.R.; Laur, G.L.; Vadas, P.A.; Weiss, W.P.; Tricarico, J.M. Invited review: Enteric methane in dairy cattle production: Quantifying the opportunities and impact of reducing emissions. *J. Dairy Sci.* **2014**, *97*, 3231–3261. [CrossRef] [PubMed]
48. Bjerg, B.; Zhang, G.; Madsen, J.; Rom, H.B. Methane emission from naturally ventilated livestock buildings can be determined from gas concentration measurements. *Environ. Monit. Assess.* **2012**, *184*, 5989–6000. [CrossRef] [PubMed]
49. Beauchemin, K.A.; Kreuzer, M.; O'Mara, F.; McAllister, T.A. Nutritional management for enteric methane abatement: A review. *Aust. J. Exp. Agric.* **2008**, *48*, 21–27. [CrossRef]
50. Hegarty, R.; Gerdes, R. Hydrogen production and transfer in the rumen. *Rec. Adv. Anim. Nutr. Aust.* **1999**, *12*, 37–44.
51. Dittmann, M.T.; Hammond, K.J.; Kirton, P.; Humphries, D.J.; Crompton, L.A.; Ortmann, S.; Misselbrook, T.H.; Südekum, K.-H.; Schwarm, A.; Kreuzer, M.; et al. Influence of ruminal methane on digesta retention and digestive physiology in non-lactating dairy cattle. *Br. J. Nutr.* **2016**, *116*, 763–773. [CrossRef] [PubMed]

sustainability

MDPI

Article

Assessing the Greenhouse Gas Mitigation Effect of Removing Bovine Trypanosomiasis in Eastern Africa

Michael MacLeod [1,*], Vera Eory [1], William Wint [2], Alexandra Shaw [3], Pierre J. Gerber [4,5], Giuliano Cecchi [6], Raffaele Mattioli [7], Alasdair Sykes [1] and Timothy Robinson [7]

1 Land Economy, Environment and Society Group, SRUC, Edinburgh EH9 3JG, UK; vera.eory@sruc.ac.uk (V.E.); Alasdair.Sykes@sruc.ac.uk (A.S.)
2 Department of Zoology, Environmental Research Group Oxford (ERGO), South Parks Road, Oxford OX1 3PS, UK; william.wint@zoo.ox.ac.uk
3 AP Consultants, 22 Walworth Enterprise Centre, Duke Close, Andover SP10 5AP, UK; alex@apconsultants.co.uk
4 The World Bank Group, Agriculture and Food Global Practice, 1818 H Street N.W., Washington, DC 20433, USA; pgerber@worldbank.org
5 Animal Production Systems Group, Wageningen University, P.O. Box 338, Wageningen 6708, The Netherlands
6 Food and Agriculture Organization of the United Nations (FAO), Sub-regional Office for Eastern Africa, CMC Road, P.O. Box 5536, Addis Ababa, Ethiopia; Giuliano.Cecchi@fao.org
7 Food and Agriculture Organization of the United Nations (FAO), Animal Production and Health Division, Viale delle Terme di Caracalla, 00153 Rome, Italy; raffaelemattioli55@libero.it (R.M.); Timothy.Robinson@fao.org (T.R.)
* Correspondence: michael.macleod@sruc.ac.uk; Tel.: +44-131-535-4124

Received: 25 April 2018; Accepted: 15 May 2018; Published: 18 May 2018

Abstract: Increasing the production of meat and milk within sub-Saharan Africa should provide significant food security benefits. However, greenhouse gas (GHG) emissions represent a challenge, as cattle production in the region typically has high emissions intensity (EI), i.e., high rates of GHG emissions per unit of output. The high EI is caused by the relatively low production efficiencies in the region, which are in turn partly due to endemic cattle diseases. In theory, improved disease control should increase the efficiency and decrease the emissions intensity of livestock production; however quantitative analysis of the potential GHG mitigation effects of improved disease control in Africa is lacking. This paper seeks to respond to this by using a hybrid modelling approach to quantify the production and emissions effects of removing trypanosomiasis from East African cattle production systems. The emissions are quantified for each cattle production system using an excel version of GLEAM, the Food and Agriculture Organization's Global Livestock Environmental Assessment Model. The results indicate that removing trypanosomiasis leads to a reduction in the emissions intensity per unit of protein produced of between 0% and 8%, driven mainly by the increases in milk yields and cow fertility rates. Despite the limitations, it is argued that the approach provides considerable scope for modelling the GHG impacts of disease interventions.

Keywords: cattle health; climate change; livestock modelling; GLEAM; sustainable intensification

1. Introduction

In developing countries growing populations, rising incomes, and urbanization are translating into increasing demand for livestock products. Livestock is now one of the fastest growing sub-sectors of agriculture: a doubling of demand for animal-source foods is expected within developing countries between 2000 and 2050, and a 70% increase for the world as a whole [1,2]. However, it has been noted that: "Meeting this demand in a way that is socially desirable and environmentally sustainable is a major challenge facing agriculture today" [3] (p. 9).

A major part of the challenge is the emission of greenhouse gases (GHGs) arising from livestock production. It has been estimated through life-cycle assessments that the global livestock sector accounts for around 14.5% of all anthropogenic GHG emissions, with cattle meat and milk accounting for 65% of the livestock emissions [4]. If increased demand is to be met without significant increases in GHG emissions, then ways of reducing the emissions intensity (EI, i.e., the amount of GHG emitted per unit of commodity produced) need to be identified and deployed.

Improving livestock health is potentially a cost-effective way of increasing production, while reducing EI. The World Organisation for Animal Health (OIE) has estimated that: "at the worldwide level, average losses due to animal diseases are more than 20%" [5]. The main direct farm level losses arise from mortality (including involuntary culling), a lowering of the efficiency of the production process, and reduction in output quantity or quality [6]. Reducing the disease burden can lead to significant reductions in EI by, for example, improving the feed conversion ratio of individual animals or changing herd structures (i.e., the proportion of each animal cohort in the herd). However, disease reduction is not yet widely recognised as a mitigation measure [7]. In fact, the Intergovernmental Panel on Climate Change Fifth Assessment Report [8] does not specifically mention livestock health, and current evidence is limited to a small number of studies of European livestock [9–13].

Understanding the impacts of improving cattle health on production and emissions is of particular relevance in sub-Saharan Africa (SSA), given that cattle are the predominant source of livestock GHGs [4], growth in demand between 2000 and 2030 is forecast to be 113% for beef and 107% for dairy products [1] and that the EIs for bovine milk and meat tend to be higher in SSA than in other regions [14]. The higher EIs are largely due to inefficiencies in converting natural resources into edible animal products, with relatively high feed conversion ratios and high energy (methane) and nitrogen (nitrous oxide) losses occurring along the process [4]. In these systems, efficiency is closely and positively related to productivity and key factors such as: milk yield, growth rates, fertility and mortality rates, and ages at slaughter. These, in turn, are a function of the genetics, feeding, management, environmental stress, and health status of the animals.

While various studies have investigated the effects of improved genetics, feeding, and management on livestock [15], evidence on the mitigation potential of improving health is scarce. It has been noted that, "Simulation results seem promising, but reliable quantitative estimates of the mitigation potential of improved health will require more research" [16] (p. 111). This paper seeks to respond to this by using a hybrid modelling approach to quantify the production and emissions effects of removing trypanosomiasis from East African cattle systems.

Trypanosomiasis is a disease caused by tsetse-transmitted parasitic protozoans, and is endemic in a tsetse-infested belt that spans across 9 million square kilometres in SSA [17]. With its animal form (called "nagana") and its human form (also known as "sleeping sickness"), trypanosomiasis is widely considered as one of the main threats to human and livestock health and agricultural production, and, as such, a major constraint to rural development and poverty alleviation in SSA [18,19]. It has been identified as "the most economically important disease of livestock in sub-Saharan Africa" [20]. African trypanosomiasis, in its various forms, can cause a wide range of symptoms, including anaemia, wasting, loss of condition, abortion, and reduced milk production. Progression can lead either to death or to a chronic form. For many decades chemotherapy has been the mainstay of trypanosomiasis control, but in recent years resistance has emerged as a growing concern [21].

This paper presents a method for quantifying the effect of trypanosomiasis treatment on the production and EI of a range of cattle production systems. It quantifies the effect of trypanosomiasis treatment on the EI in East Africa. The limitations of the method are discussed and the wider challenges in quantifying the GHG effects of livestock health improvement are explored.

2. Materials and Methods

2.1. Study Area

The study focuses on East Africa, which is one of the most important dairy development zones in SSA. The specific area of study is the Inter-Governmental Authority on Development (IGAD) region, an area of over 5.2 million km^2 that comprises the countries of Djibouti, Eritrea, Ethiopia, Kenya, Somalia, South Sudan, Sudan, and Uganda [22]. Just under half (45%) of the cattle in SSA are reported to be in the IGAD region [23].

2.2. Cattle Production Systems

The cattle in the study area were classified into twelve distinct production systems (Table 1). These refined cattle systems were based on three main livestock systems (i.e., pastoral, agro-pastoral and mixed farming) [24], which were further disaggregated based on the prevalence of work oxen (used for traction) and milk yields of dairy cattle (which have varying degrees of exotic genetic material, usually of the main European dairy breeds that have been widely adopted in high value smallholder dairy production in Eastern Africa).

Table 1. The 12 cattle production systems analysed, table adapted from [25].

Draft Oxen	Production System				
	Pastoral	Agro-Pastoral	Mixed: General [a]	Mixed: Ethiopia [a]	High Milk Yield Dairy
No oxen					×× [b]
Low oxen	×	×	×	×	
Medium oxen		×	×	×	
High oxen		×	×	×	

[a] Two mixed farming systems were analysed, one for Ethiopia and one for the other study countries; [b] High milk yield dairy cows are found in the agro-pastoral and general mixed farming systems.

The high milk yield dairy systems consist of cattle with varying degrees of non-indigenous genetic material derived primarily from the main European dairy breeds, principally Ayrshire, Friesian and Channel Island breeds—Guernsey and Jersey. In the other systems indigenous breeds predominate, i.e., zebu cattle and related crossbreeds [26]. The eastern African indigenous zebu cattle breeds are conventionally divided into three main groups all of which are represented in the countries covered by this study: the Large East African Zebu (including the Boran breeds of Ethiopia, Kenya and Somalia, the Samburu, Karmajong Zebu, Orma Boran and Butana and Kenana of Sudan) the Small East African Zebu (including the Maasai, Abyssinian Shorthorned Zebu, Kamba of Kenya and Serere and Kyoga in the tsetse-infested areas of Uganda) and the Sanga cattle (including the Ankole of Uganda, Danakil cattle of Ethiopia and the Nuer and Dinka of Sudan). Further information is available on the Domestic Animal Genetic Resources Information System (DAGRIS) [27]. More detail on the production systems and how they were defined is available in [28].

Data on the use of work oxen were obtained from in-country informants, reports, census data, livelihood studies [24] and other sources [29–33]. Combining these sources across the region by administrative area, and studying data on herd compositions, enabled three broad categories of oxen use to be distinguished depending on the proportion of cattle used for draught: low ($\leq$10%), medium (>10% and <20%), and high ($\geq$20%).

The proportion of cattle modelled in each system is given in Table 2. The most recent statistics report a total cattle population in the study area of 144.5m in 2016 [23].

Table 2. The proportion of cattle in each system.

Production System	% of Total Cattle Population
Pastoral	12%
Agropastoral, low dairy and oxen	34%
Mixed, low dairy and oxen	11%
Agropastoral, high dairy and low oxen	2%
Agropastoral, low dairy and medium oxen	1%
Agropastoral, low dairy and high oxen	2%
Mixed, high dairy and low oxen	4%
Mixed, low dairy and low oxen	1%
Mixed, low dairy and high oxen	1%
Mixed, low dairy and medium oxen (eth)	7%
Mixed, low dairy and high oxen (eth)	18%
Mixed, low dairy and low oxen (eth)	8%
Total	100%

2.3. Modelling Approach

This study builds on previous work, which quantified the economic benefits of removing trypanosomiasis in East African cattle using the "Mapping the Benefits" (MTB) model [25,28,34]. In this study, the MTB model is combined with an excel version of GLEAM (the UN-FAO's Global Livestock Environmental Assessment Model, [35]) which calculates the GHG emissions arising from production. A schematic diagram of the relationship between the models is provided in Figure 1.

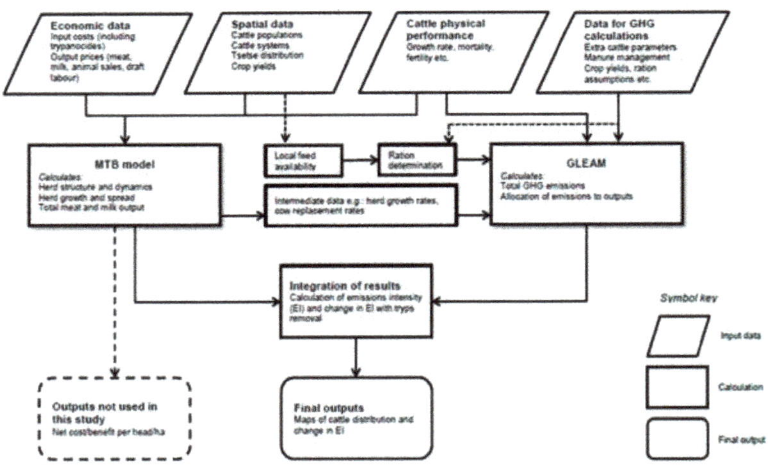

Figure 1. Schematic diagram of the relationship between the two models: Mapping the Benefits (MTB) and Global Livestock Environmental Assessment Model (GLEAM).

2.3.1. Harmonising the GLEAM and MTB Herd Models

At the heart of both GLEAM and MTB are herd models, which determine the herd structure and the number of animals entering and leaving the herd each year. The two herd models perform some of these calculations in different ways. For example, they have different animal cohorts (adult female, replacement female, etc.) and different ways of calculating the number of calves born. Therefore, the herd models had to be harmonised, i.e. adjusted until they produced the same herd structure and dynamics. This was achieved by comparing the results given by the two herd models with the same input data, and adjusting the calculations in GLEAM until the models were producing the same results for all 12 production systems.

2.3.2. Estimating the Effects of Trypanosomiasis Treatment over 20 Years

The herd structure and dynamics, and the output of meat and milk were calculated for the situations with and without trypanosomiasis using the MTB model [34] and applied to cattle population density maps, projected using the herd growth and additional spread models. Regional tsetse maps at 1 km resolution were used [36], complemented by 5km resolution continental datasets [37] where 1 km resolution maps were unavailable. Cattle population densities were available from a series of 1km resolution datasets created within the framework of the IGAD Land Policy Initiative (LPI) using the most recent national statistics available for the period 2000–2005. The mapping methods were those developed to generate the Gridded Livestock of the World [38].

GLEAM was used to calculate the (meat and milk) production, the total greenhouse gas emissions arising from that production and the EI of each commodity for each of the cattle systems with and without trypanosomiasis. The meat production in live weight was calculated by multiplying the number of cattle exiting the herd as offtake, i.e., sold or slaughtered, by the average live weight of each cohort (i.e., adult female, calves etc.). The live weight was then converted to edible protein by multiplying it by (a) the carcass weight: live weight percentage (50–55% depending on the cohort); (b) the bone-free meat: carcass weight percentage (75%) and finally (c) the protein content of bone-free meat (18%). The edible protein from milk was determined by multiplying the net milk production (milk secreted minus milk suckled) by the protein content of milk (3.3%). Emissions are only allocated to edible protein and draft, however cattle perform other functions. It has been estimated that 16% of the emissions should be allocated to the manure, financing and dowry functions of cattle [39]. Doing so would reduce the EI but would have a negligible impact on the change in EI arising from trypanosomiasis removal.

The main GHG-producing activities were included, i.e., the production of feed and fertiliser, the digestion of feed and the excretion of undigested organic matter (Table 3 provides an explanation of the emission categories). The emissions of methane and nitrous oxide were converted to their carbon dioxide equivalent using the Intergovernmental Panel on Climate Change (IPCC) global warming potentials of 25 and 298, respectively. The system boundary of the study is "cradle to farm-gate", i.e., emissions arising from the production of feed and other inputs are included but post-farm gate emissions occurring during processing and distribution are not included; however in the systems studied post-farm gate emissions are likely to represent a small proportion of the total emissions [4]. CO_2 emissions and sequestration in short-term carbon cycles (i.e., respiration and photosynthesis) are not included, in keeping with the IPCC guidelines ([40], Volume 4, p. 10.7). Furthermore, it is assumed that the treatment (chemotherapy) has negligible additional GHG emissions arising (from drug manufacture and distribution) given the small doses involved. Other treatment strategies, such as the removal of tsetse habitats, could lead to significant additional GHG emissions.

Table 3. The emission categories quantified in the study.

Category	Emissions Sources
Manure N_2O	Direct and N_2O arising from the nitrification/denitrification of excreted N during its management and storage.
Manure CH_4	CH_4 arising from the anaerobic decomposition of excreted organic matter during its management and storage.
Enteric fermentation CH_4	CH_4 arising from the microbial decomposition of feed during digestion.
Feed energy CO_2	CO_2 from: fossil fuel use in the cultivation of feed crops; fossil fuel use in the production of non-crop feed materials; processing, transportation and blending of feed materials; fossil fuel use from the production of synthetic fertiliser; land use change arising from soy cultivation.
Feed N_2O	Direct and indirect N_2O from: application of synthetic fertiliser and manure to crops; crop residue management; direct deposition of N by grazing animals; fertiliser manufacture.

2.3.3. Parameterising the Cattle Production Systems

The values for the baseline situation (i.e., with trypanosomiasis) in Table 4 were derived from various sources. A few sources provided information relating to all production systems [32,41], others for the pastoralist systems [42–44], for the agro-pastoral and mixed farming systems [29,31,45–50] and for the high milk yield "grade" dairy cattle kept in a separate system alongside indigenous cattle in the agro-pastoral and mixed farming systems [51–53].

Table 4. Values for selected parameters used in the analysis (T+ refers to the value with trypanosomiasis, T− to the value without trypanosomiasis). The values were estimated based on a review of longitudinal and cross-sectional studies comparing the productivity observed in infected and uninfected individual cattle or whole herds, under conditions of both high and low trypanosomiasis challenge (see Section 2.3.3). Table adapted from [25].

Parameter	Cattle Production Systems									
	Pastoral		Agro-Pastoral		Mixed Farming (General)		Mixed Farming (Ethiopia)		High Milk Yield Dairy	
	T+	T−	T+	T−	T+	T−	T+	T−	T+	T−
Mortality	*% of animals dying each year*									
Female calves	20	17	18	15	16	13	24	20	21	18
Male calves	25	22	20	17	18	15	26	22	26	23
Adult females	7.5	6.5	7.0	6.0	8.0	7.0	9.0	7.5	12.0	10.0
Work oxen	9.0	7.2	8.5	6.8	9.0	7.2	10.0	8.0	-	-
Fertility	*% of cows producing a living calf each year*									
Cow fertility	54	58	52	56	51	55	49	54	53	57
Lactation	*Litres of milk per lactation* [a]									
Milk offtake	275	296	285	306	300	322	280	301	1900	2042

[a] Lactation length varies depending on the system; in this study it is assumed that most of the milk for human consumption is taken off in the 12 months following calving.

The effects of trypanosomiasis treatment on performance were estimated based on a review of longitudinal and cross-sectional studies comparing the productivity observed in infected and uninfected individual cattle or whole herds, under conditions of both high and low trypanosomiasis challenge [33,45,54–61]. The review gave a range of values for the effect of the disease. In this study, it was assumed that disease treatment would have an effect between the middle and lower end of the range reported in studies. These (arguably conservative) assumptions were made in order to correct for the fact that the studies reviewed tended to be conducted in areas where the disease impact was relatively severe.

Similar impacts for trypanosomiasis were applied across the production systems (Table 4). This table also shows that baseline cattle mortality rates observed in Eastern Africa tend to be high. This reflects not just the presence of trypanosomiasis, but also of tick-borne diseases, especially East Coast fever [60] and, locally, very high stocking rates leading to nutritional stress, especially in parts of the Ethiopian highlands [62].

2.3.4. Ration Composition and Nutritional Value

The ration (i.e., the livestock feed) for each system was determined using a combination of approaches. The proportions of the main feed material categories (i.e., forage, crop residues and concentrates) were estimated based on literature, expert knowledge, and unpublished surveys, while the percentage of each crop residue (maize, millet and sorghum stover, wheat, rice and barley straw, and sugar cane tops) were calculated for each cell, based on how much of each crop is produced within the cell. Crop data was extracted for livestock production zones from Spatial Production Allocation Model (SPAM) V3r6 2000 datasets [63]. The nutritional value of the ration was calculated by multiplying the nutritional value of each feed material by its proportion in the ration. The nutritional values of individual feed materials were based on [14,64], the Sub-Saharan Africa Feed Composition Database [65], and Feedipedia [66,67].

2.3.5. Estimating Uncertainty

Calculation of the emission intensities involves parameters that are subject to some degree of uncertainty. A Monte Carlo simulation was undertaken in order to quantify the uncertainty ranges in the results. Parameters were included in the simulation that were (a) likely to have a significant influence on the most important emissions categories (those contributing more than twenty percent of the total emissions, i.e., enteric CH_4 and feed N_2O) and (b) had a high degree of uncertainty or inherent variability. An initial list of parameters was generated and reviewed. Based on the review, some parameters were excluded, e.g., the emission factors (EFs) for indirect N_2O were removed because volatilisation and leaching only account for a small % of feed N_2O emissions, and a previous study had found the contribution to variance of these EFs to be very small (<2%) [14] (p. 81). The distributions and variability of the parameters are given in Table 5. The variability of forage digestibility is based on the standard deviation of the digestibility of Napier grass (Pennisetum purpureum) reported in [67]. The variability of cow milk yield, cow fertility and calf death rates were based on the ranges reported in [32]. The Monte Carlo was performed in Model Risk with 5000 runs for each of 12 systems with and without trypanosomiasis.

Table 5. Parameters included in the Monte Carlo analysis.

Parameter	Distribution	Coefficient of Variance	Minimum-Maximum	Basis
Forage digestibility, DE%	Normal	6%		[67]
Milk yield	Normal	11–39%		[32]
Cow fertility	Normal	8–21%		[32]
Calf death rate	Normal	25–45%		[32]
Enteric CH_4 factor, Ym	Normal	10%		[40], (Vol. 4, p. 10.33)
N_2O Emission factor 1	Lognormal		0.003–0.03	[40], (Vol. 4, p. 11.11)
N_2O Emission factor 3	Lognormal		0.007–0.06	[40], (Vol. 4, p. 11.11)

3. Results

3.1. Baseline Emissions with Trypanosomiasis

The EI by system and emission category is shown in Figure 2. The main sources of emissions are (a) enteric methane and (b) feed nitrous oxide, i.e., the nitrous oxide arising from the deposition of organic N on pasture (either directly via the urine of grazing animals, or via the spreading of the collected manure of housed cattle). The high (milk yield) dairy systems have much lower EI than other systems due to their higher productivity (where productivity is defined as the mass of meat and milk produced by a system each year compared to the live weight mass of the cattle in the system). Differences between the other systems are less marked but also largely driven by productivity. The relationship between productivity and EI for the 12 systems with and without trypanosomiasis is shown in Figure 3. This trend is consistent with other studies [68,69] which found a similar relationship between productivity and EI in cattle systems, i.e., large (but reducing) marginal decreases in EI with increasing productivity in low productivity systems, such as those in East Africa.

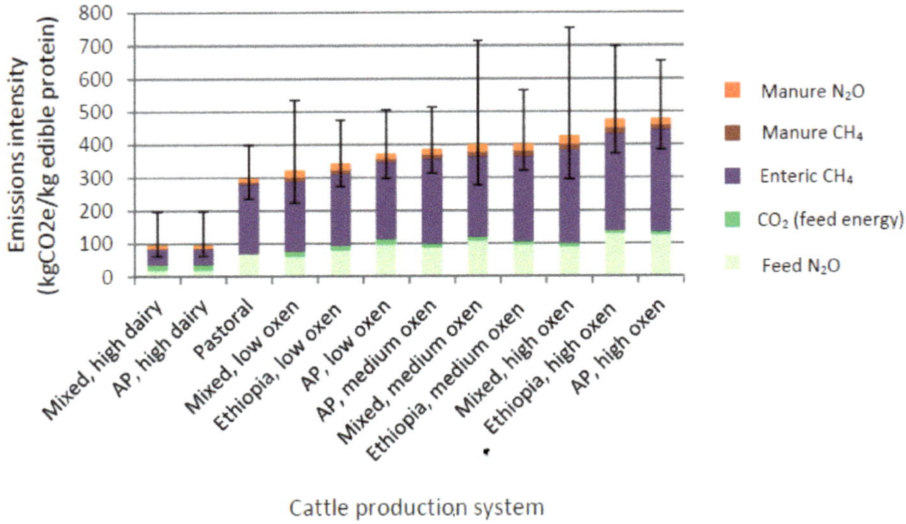

Figure 2. Emissions intensity (EI) by cattle production system and emissions category (kgCO_2e/kg edible protein at farm gate). AP: agro-pastoral. The errors bars show the 95% confidence intervals.

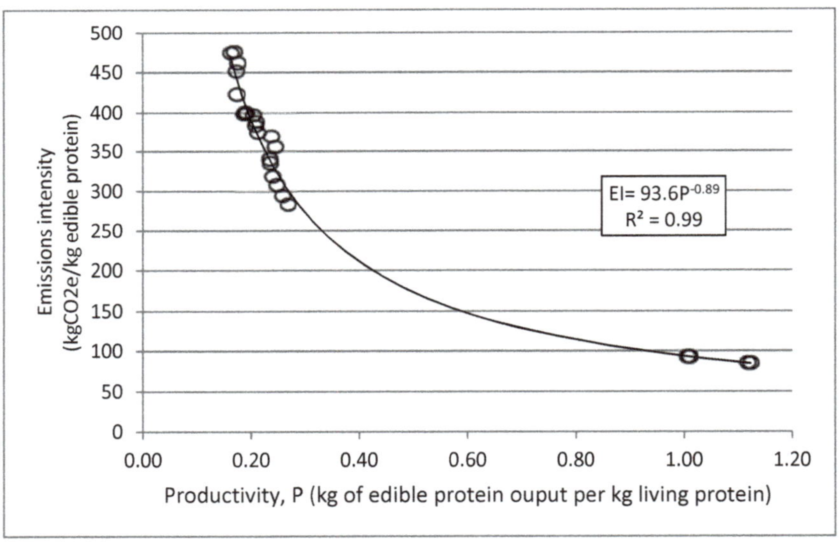

Figure 3. The relationship between productivity and EI for the 12 cattle production systems with and without trypanosomiasis. Productivity is defined as the mass of meat and milk edible protein produced by a system each year compared to the live weight mass of the cattle in the system (measured in terms of mass of living protein).

3.2. Effect of Disease Treatment

Treatment of trypanosomiasis leads to increases in production and emissions across all the systems. Figure 4 shows the increases once the full effects of disease treatment have been achieved. Production increases relatively more than emissions leading to reductions in EI of 0% to 8%.

The reductions in EI are significant in eleven of the twelve systems and are closely (inversely) correlated with the change in productivity (Figure 5).

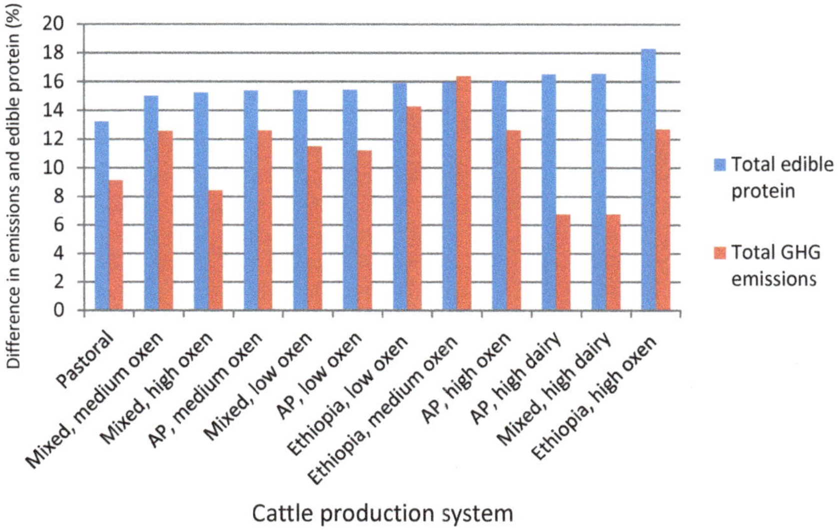

Figure 4. Difference in total edible protein production and GHG emissions with disease treatment. AP: agro-pastoral.

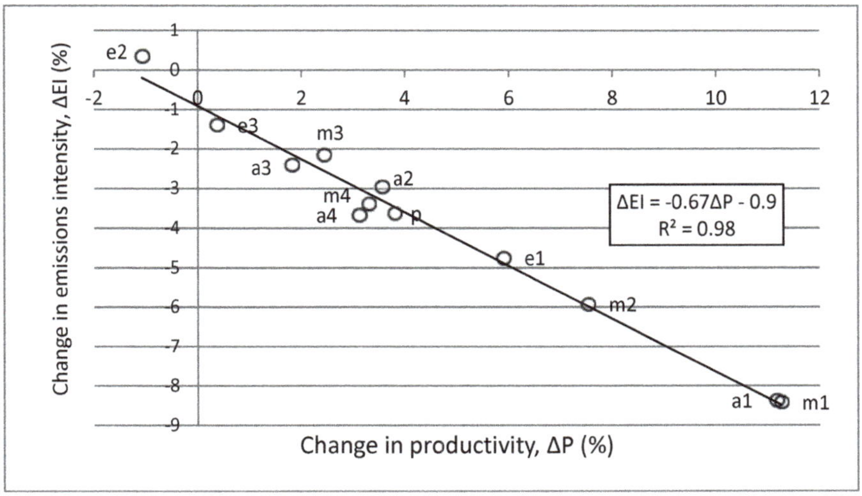

Figure 5. Change in EI and productivity with disease treatment. EI is measured in terms of kgCO_2e per kg of edible protein produced and productivity. Productivity is defined as the mass of meat and milk edible protein produced by a system each year compared to the live weight mass of the cattle in the system (measured in terms of mass of living protein). (Key: a1 = agro-pastoral high dairy; a2 = agro-pastoral low dairy, high oxen; a3 = agro-pastoral low dairy, medium oxen; a4 = agro-pastoral low dairy, low oxen; m1 = mixed high dairy; m2 = mixed low dairy, high oxen; m3 = mixed low dairy, medium oxen; m4 = mixed, low dairy, low oxen; e1 = Ethiopia low dairy, high oxen; e2 = Ethiopia low dairy, medium oxen; e3 = Ethiopia, low dairy, low oxen; p = pastoral.).

3.3. Factors driving the Changes in EI

A sensitivity analysis was undertaken in order to investigate the influence of different parameters on EI (Table 6). The EI is highly sensitive to changes in the digestibility of forage, and to the enteric methane emission factor (Ym), however, both of these remain constant with changes in disease status. Increased milk yield leads to a decrease in EI as a greater percentage of the lactating cow's feed energy requirement is used for the production of edible outputs (i.e., lactation) rather than non-productive activity (e.g., maintenance or locomotion), [68]. The increased cow fertility rate leads to a higher proportion of the adult females producing edible outputs. This is important as while non-lactating adult females produce lower emissions than lactating ones, they produce no edible outputs (i.e., milk or liveweight gain) so their EI is effectively infinite. The increased fertility rate also changes the herd structure, but this has a limited effect on EI. EI is less sensitive to changes in calf mortality than it is to changes in cow milk yield or fertility, however the effect of calf mortality on EI depends on the fate of the dying calves, i.e., what proportion enter the human food chain and are therefore counted as production. Finally, increasing the use of draft oxen leads to increases EI in some systems. Draft animals have a high EI as they produce no/limited edible output, so even when a proportion of the emissions are allocated to labour, they still have a higher EI than younger, faster growing cattle.

Table 6. Summary of the sensitivity analysis, showing the change in EI resulting from a ±10% change in each parameter ("Draft": the % of male cattle aged >3 years old that are used for draft. EF: emission factor; DE: digestible energy).

System	Change in Parameter	Change in EI							
		Constant		*Vary with System*		*Vary with System and Disease Status*			
		N_2O EF1	N_2O EF3	Forage DE%	Enteric CH_4, Ym	Milk Yield	Cow Fertility	Calf Death Rate	Draft
Pastoral	+10%	0.7%	1.2%	−16.4%	7.2%	−4.5%	−3.9%	−0.6%	0.5%
Pastoral	−10%	−0.7%	−1.2%	23.1%	−7.2%	4.9%	4.9%	0.6%	−0.5%
AP, high dairy	+10%	1.6%	0.1%	−15.8%	5.1%	−6.3%	−5.4%	2.2%	0.0%
AP, high dairy	−10%	−1.6%	−0.1%	22.2%	−5.1%	7.5%	6.7%	0.0%	0.0%
AP, low oxen	+10%	1.4%	0.7%	−16.4%	6.3%	−4.4%	−2.9%	−0.7%	1.0%
AP, low oxen	−10%	−1.4%	−0.7%	23.5%	−6.3%	4.8%	3.5%	0.7%	−1.0%
AP, medium oxen	+10%	1.1%	0.7%	−16.5%	6.8%	−4.4%	−3.0%	−0.7%	2.0%
AP, medium oxen	−10%	−1.1%	−0.7%	23.7%	−6.8%	4.8%	3.6%	0.7%	−2.0%
AP, high oxen	+10%	1.3%	0.7%	−16.3%	6.5%	−4.4%	−3.1%	−0.7%	3.8%
AP, high oxen	−10%	−1.3%	−0.7%	23.3%	−6.5%	4.9%	3.9%	0.7%	−3.8%
Mixed, high dairy	+10%	1.6%	0.1%	−15.8%	5.1%	−6.3%	−5.4%	2.3%	0.0%
Mixed, high dairy	−10%	−1.6%	−0.1%	22.2%	−5.1%	7.5%	6.7%	−0.1%	0.0%
Mixed, low oxen	+10%	1.2%	0.3%	−17.1%	6.7%	−4.4%	−1.7%	−0.9%	0.9%
Mixed, low oxen	−10%	−1.2%	−0.3%	24.9%	−6.7%	4.9%	2.0%	0.9%	−0.9%
Mixed, medium oxen	+10%	1.8%	0.3%	−16.7%	6.2%	−4.5%	−2.7%	−0.6%	2.0%
Mixed, medium oxen	−10%	−1.8%	−0.3%	24.1%	−6.2%	4.9%	3.7%	0.6%	−2.0%
Mixed, high oxen	+10%	1.4%	0.3%	−16.8%	6.7%	−4.5%	−3.3%	−0.6%	3.8%
Mixed, high oxen	−10%	−1.4%	−0.3%	24.3%	−6.7%	4.9%	4.1%	0.6%	−3.7%
Ethiopia, low oxen	+10%	1.5%	0.3%	−16.8%	6.4%	−4.5%	−1.7%	−1.5%	0.6%
Ethiopia, low oxen	−10%	−1.5%	−0.3%	24.5%	−6.4%	5.0%	2.0%	1.5%	−0.6%
Ethiopia, medium oxen	+10%	1.6%	0.3%	−16.8%	6.4%	−4.6%	−3.1%	−1.1%	2.0%
Ethiopia, medium oxen	−10%	−1.6%	−0.3%	24.3%	−6.4%	5.1%	3.7%	1.0%	−2.0%
Ethiopia, high oxen	+10%	1.8%	0.3%	−16.6%	6.2%	−4.6%	−3.5%	−1.0%	3.7%
Ethiopia, high oxen	−10%	−1.8%	−0.3%	23.9%	−6.2%	5.1%	4.3%	0.9%	−3.6%

4. Discussion

While the results are consistent with the expectation that improving health is likely to lead to reductions in EI, the validity of the results depends on the input assumptions and the method used to perform the calculations. The approach developed in this study allows multiple effects of a disease to be quantified. The MTB model enables herd growth and livestock movement over time to be quantified. The use of a herd model in GLEAM enables disease impacts on parameters that affect the herd structure to be modelled, e.g. death rates, fertility rates, replacement rates, and offtake rates. The IPCC tier 2 approach [40] to calculating livestock emissions used in GLEAM enables disease impacts on parameters that affect the production and emissions of the individual animal to be modelled, e.g., milk yield, animal growth rates, activity level and ration. However, the approach requires further development in order to capture some potentially important effects, such as the impact of disease status on ration quality.

The results are expressed in terms of the functional unit "kg of CO_2e per kg of edible protein at the farm gate", and therefore don't include post-farm emissions or take into account post-farm losses. While post-farm emissions are likely to represent a small % of the total emissions, losses can make significant differences to the EI, and may vary between different production systems. The functional unit also assumes that: (a) 1 kg of milk protein is equivalent to 1 kg of meat protein and (b) production of edible outputs is the only reason for keeping cattle, which are simplifications. Although some of the emissions are allocated to draught power, none are allocated to other outputs (such as manure) or the less tangible values that cattle may have, such as social value and increased resilience of their owners [39]. While the choice of functional unit means that care needs to be taken when making comparisons of the EI between systems, none of the limitations noted above would affect the change in EI arising from treatment of trypanosomiasis.

As noted in [25] "data on livestock productivity (fertility, mortality, milk yields and output of draught power) were mostly obtained from a limited number of in-depth studies in specific localities. As the large number of references testifies, cattle systems in the study region have been much studied and there is enough information to paint a good general picture, and to cross-check and validate. However, although trypanosomosis has been relatively well researched, there remains a need for more specific studies on this disease's impact". This is particularly the case with draught animals. FAO have developed a geospatial database [70] that should improve understanding of trypanosomiasis prevalence in Africa.

5. Conclusions

The treatment of trypanosomiasis in East African cattle systems could lead to reductions in EI of between 0 and 8%, depending on the system. EI is closely related to productivity, and the largest reductions in EI are in those systems experiencing the largest increases in productivity, i.e., the higher (milk yield) dairy systems. However, when making inter-system comparisons, it should be borne in mind that in some systems cattle perform functions in addition to producing meat and milk.

The approach employed in this study could, in principle, be used to model the impact of other important diseases in SSA (such as foot and mouth disease), if their effects can be adequately described in terms of the model parameters. However, in order to produce valid results, the main direct effects of the disease need to be captured by the model. This is potentially challenging given the lack of data on prevalence and impact for many diseases. While it may be possible to partially fill gaps through the analysis of existing datasets (such as abattoir records), some diseases have important effects that are not routinely recorded or readily measured (for example they may alter feed intake, the efficiency of digestion or energy partitioning) and may require bespoke studies. A second challenge is capturing the indirect effects of disease control. For example, disease treatment would lead to increased livestock populations which could, depending on how this increase is managed, lead to overstocking and thus reduced feed availability and/or quality.

The reduction of EI represents an external benefit of disease control, i.e., it is a benefit not normally factored into farm management or policy decisions. In order to achieve the economically optimal level of disease control, such externalities need to be quantified and included in decision making, alongside other costs and benefits. Given the difficulty of measuring agricultural GHG emissions directly, it is suggested that modelling approaches, such as the one outlined in this paper, have an important contribution to make to the development of disease control strategies.

Author Contributions: Conceptualization, M.M. and T.R.; Data curation, M.M., W.W., and A.S.; Formal analysis, M.M. and A.S.; Funding acquisition, T.R.; Investigation, V.E., W.W., A.S., P.G., G.C., and R.M.; Methodology, M.M.; Writing—original draft, M.M.

Acknowledgments: This project was funded by the International Livestock Research Institute under CCAFS, the CGIAR Research Program "Climate Change, Agriculture and Food Security". The authors are grateful to the UN FAO for permission to use GLEAM in this study. The authors would like to acknowledge the support of the Scottish Government Strategic Research Programme, Theme 2: Productive and Sustainable Land Management and Rural Economies. The support of the Government of Italy through the FAO Project "Improving food security in sub-Saharan Africa by supporting the progressive reduction of tsetse-transmitted trypanosomiasis in the framework of the NEPAD" (codes GTFS/RAF/474/ITA and GCP/RAF/502/ITA) is also acknowledged.

Conflicts of Interest: The authors declare no conflict of interest.

References

1. Robinson, T.P.; Pozzi, F. *Mapping Supply and Demand for Animal-Source Foods to 2030*; Animal Production and Health Working Paper No. 2; Food and Agriculture Organisation of the United Nations: Rome, Italy, 2011.
2. Alexandratos, N.; Bruinsma, J. *World Agriculture towards 2030/2050. The 2012 Revision. Global Perspective Studies Team*; ESA Working Paper No. 12-03; Food and Agriculture Organization of the United Nations: Rome, Italy, 2012.
3. Pingali, P.; McCullough, E. Drivers of change in global agricultural livestock systems. In *Livestock in a Changing Landscape: Drivers, Consequences and Responses*; Steineld, H., Mooney, H.A., Schneider, F., Neville, L.E., Eds.; Island Press: Washington, DC, USA, 2010.
4. Gerber, P.J.; Steinfeld, H.; Henderson, B.; Mottet, A.; Opio, C.; Dijkman, J.; Falcucci, A.; Tempio, G. *Tackling Climate Change through Livestock—A Global Assessment of Emissions and Mitigation Opportunities*; Food and Agriculture Organization of the United Nations: Rome, Italy, 2013.
5. Vallat, B. Feeding the World Better by Controlling Animal Diseases. Available online: http://www.oie.int/for-the-media/editorials/detail/article/feeding-the-world-better-by-controlling-animal-diseases/ (accessed on 23 May 2014).
6. Elliot, J. Animal health and productivity in the EU—Current and emerging issues. In Proceedings of the FACCE–JPI Workshop: Animal Health/Animal Diseases and GHG Mitigation, Madrid, Spain, 21 May 2014.
7. MacLeod, M.; Eory, V.; Gruère, G.; Lankoski, J. *Cost-Effectiveness of Greenhouse Gas Mitigation Measures for Agriculture: A Literature Review, OECD Food*; Agriculture and Fisheries Papers, No. 89; OECD Publishing: Paris, France, 2015.
8. Stocker, T.F.; Qin, D.; Plattner, G.-K.; Tignor, M.; Allen, S.K.; Boschung, J.; Nauels, A.; Xia, Y.; Bex, V.; Midgley, P.M. *IPCC Climate Change 2013: The Physical Science Basis. Contribution of Working Group I to the Fifth Assessment Report of the Intergovernmental Panel on Climate Change*; Cambridge University Press: Cambridge, UK; New York, NY, USA, 2013; p. 1535.
9. Hospido, A.; Sonesson, U. The environmental impact of mastitis: A case study of dairy herds. *Sci. Total Environ.* **2005**, *343*, 71–82. [CrossRef] [PubMed]
10. Stott, A.; MacLeod, M.; Moran, D. *Reducing Greenhouse Gas Emissions through Better Animal Health Rural Policy Centre Policy Briefing July 2010 (RPC PB 2010/01)*; SAC: Edinburgh, UK, 2010.
11. Guelbenzu, M.; Graham, D. *Prevalence of BVD in Northern Ireland Dairy and Suckler Herds*; Agri-Food and Biosciences Institute: Belfast, UK, 2013.
12. ADAS UK Ltd. *Study to Model the Impact of Controlling Endemic Cattle Diseases and Conditions on National Cattle Productivity*; Agricultural Performance and Greenhouse Gas Emissions: Final Report; Defra/AVHLA: London, UK, 2014.
13. Houdijk, J.G.M.; Tolkamp, B.J.; Rooke, J.A.; Hutchings, M.R. Animal health and greenhouse gas intensity: The paradox of periparturient parasitism. *Int. J. Parasitol.* **2017**, *47*, 633–641. [CrossRef] [PubMed]
14. Opio, C.; Gerber, P.; Mottet, A.; Falcucci, A.; Tempio, G.; MacLeod, M.; Vellinga, T.; Henderson, B.; Steinfeld, H. *Greenhouse Gas Emissions from Ruminant Supply Chains—A Global Life Cycle Assessment*; Food and Agriculture Organization of the United Nations: Rome, Italy, 2013.
15. Staal, S.; Poole, J.; Baltenweck, I.; Mwacharo, J.; Notenbaert, A.; Randolph, T.; Thorpe, W.; Nzuma, J.; Herrero, M. *Targeting Strategic Investment in Livestock Development as a Vehicle for Rural Livelihoods. BMGF-ILRI Project on Livestock Knowledge Generation*; International Livestock Research Institute: Nairobi, Kenya, 2009.

16. Hristov, A.N.; Oh, J.; Lee, C.; Meinen, R.; Montes, F.; Ott, T.; Firkins, J.; Rotz, A.; Dell, C.; Adesogan, A.; et al. *Mitigation of Greenhouse Gas Emissions in Livestock Production—A Review of Technical Options for Non-CO_2 Emissions*; Gerber, P.J., Henderson, B., Makkar, H.P.S., Eds.; FAO Animal Production and Health Paper No. 177; Food and Agriculture Organization of the United Nations: Rome, Italy, 2013.
17. Cecchi, G.; Mattioli, R.; Slingenbergh, J.; De La Rocque, S. Land cover and tsetse fly distributions in sub-Saharan Africa. *Med. Vet. Entomol* **2008**, *22*, 364–373. [CrossRef] [PubMed]
18. Cattand, P.; Simarro, P.; Jannin, J.; Ly, C.; Fall, A.; Shaw, A.; Mattioli, R. *Linking Sustainable Human and Animal African Trypanosomiasis Control with Rural Development Strategies*; Programme Against African Trypanosomiasis–Technical and Scientific Series. Vol. 10. 2010; Food and Agriculture Organization of the United Nations: Rome, Italy, 2010.
19. Simarro, P.P.; Cecchi, G.; Franco, J.R.; Paone, M.; Diarra, A.; Ruiz-Postigo, J.A.; Fèvre, E.M.; Mattioli, R.; Jannin, J.G. Estimating and mapping the population at risk of sleeping sickness. *PLoS Negl. Trop. Dis.* **2012**, *6*, 1859. [CrossRef] [PubMed]
20. Taylor, K.; Authié, E.M.-L. Pathogenesis of Animal Trypanosomiasis. In *The Trypanosomiases*; Maudlin, I., Holmes, P.H., Miles, M.A., Eds.; CAB International: Wallingford, UK, 2004.
21. Diall, O.; Cecchi, G.; Wanda, G.; Argilés-Herrero, R.; Vreysen, M.J.B.; Cattoli, G.; Viljoen, G.J.; Mattioli, R.; Bouyer, J. Developing a Progressive Control Pathway for African Animal Trypanosomosis. *Trends Parasitol.* **2017**, *33*, 499–509. [CrossRef] [PubMed]
22. IGAD. The IGAD Region. Available online: https://igad.int/index.php/about-us/the-igad-region (accessed on 24 September 2017).
23. FAOSTAT Live Animals. Available online: http://www.fao.org/faostat/en/#data/QA (accessed on 9 May 2018).
24. Cecchi, G.; Wint, W.; Shaw, A.; Marletta, A.; Mattioli, R.; Robinson, T. Geographic distribution and environmental characterization of livestock production systems in Eastern Africa. *Agric. Ecosyst. Environ.* **2010**, *135*, 98–110. [CrossRef]
25. Shaw, A.P.M.; Cecchi, G.; Wint, G.R.W.; Mattioli, R.C.; Robinson, T.P. Mapping the economic benefits of intervening against bovine trypanosomiasis in Eastern Africa. *Prev. Vet. Med.* **2014**, *113*, 197–210. [CrossRef] [PubMed]
26. Mwai, O.; Hanotte, O.; Kwon, Y.-J.; Cho, S. African Indigenous Cattle: Unique Genetic Resources in a Rapidly Changing World. *Asian-Australas. J. Anim. Sci.* **2015**, *28*, 911–921. [CrossRef] [PubMed]
27. Domestic Animal Genetic Resources Information System (DAGRIS). Available online: http://dagris.info/ (accessed on 9 May 2018).
28. Shaw, A.; Wint, W.; Cecchi, G.; Torr, S.; Waiswa, C.; Alemu, T.; Eregae, M.; Abdi, A.; Muchina, S.; Mugasi, S.; et al. *Intervening against Bovine Trypanosomosis in Eastern Africa: Mapping the Costs and Benefits*; PAAT Technical and Scientific Series No. 11; FAO: Rome, Italy, 2017; Available online: http://www.fao.org/3/a-i7342e.pdf (accessed on 15 May 2018).
29. Itty, P. Economics of Village Cattle Production in Tsetse-affected Areas of Africa. In *A Study of Trypanosomiasis Control Using Trypanotolerant Cattle and Chemotherapy in Ethiopia, Kenya, The Gambia, Côte d'Ivoire, Zaire and Togo*; Hartung-Gorre Verlag: Constance, Germany, 1992.
30. Starkey, P.; Kaumbutho, P. Meeting the Challenges of Animal Traction. In *A Resource Book of the Animal Traction Network for East-ern and Southern Africa (ATNESA), Harare, Zimbabwe*; Intermediate Technology Publications: London, UK, 1999.
31. Rege, J.E.O.; Kahi, A.K.; Okomo-Adhiambo, M.; Mwacharo, J.; Hanotte, O. *Zebu Cattle of Kenya: Uses, Performance, Farmer Preferences, Measures of Genetic Diversity*; ILRI (International Livestock Research Institute) and Kenya Agriculture Research Institute (KARI): Nairobi, Kenya, 2001.
32. Otte, M.J.; Chilonda, P. *Cattle and Small Ruminant Production Systems in Sub-Saharan Africa. A Systematic Review*; FAO (Food and Agriculture Organization of the United Nations): Rome, Italy, 2002.
33. Ocaido, M.; Otim, C.; Okuna, N.; Erume, J.; Ssekitto, C.; Wafula, R.; Kakaire, D.; Walubengo, J.; Monrad, J. Socio-economic and livestock disease survey of agro-pastoral communities in Serere County, Soroti District, Uganda. *Livest. Res. Rural Dev.* **2005**, *17*, 93.
34. Shaw, A.P.M.; Wint, G.R.W.; Cecchi, G.; Torr, S.J.; Mattioli, R.C.; Robinson, T.P. Mapping the benefit-cost ratios of interventions against bovine trypanosomiasis in Eastern Africa. *Prev. Vet. Med.* **2015**, *122*, 406–416. [CrossRef] [PubMed]

35. MacLeod, M.J.; Vellinga, T.; Opio, C.; Falcucci, A.; Tempio, G.; Henderson, B.; Makkar, H.; Mottet, A.; Robinson, T.; Steinfeld, H.; et al. Invited Review: A Position on the Global Livestock Environmental Assessment Model (GLEAM). *Animal* **2018**, *12*, 383–397. [CrossRef] [PubMed]
36. Wint, G.R.W. *Kilometre Resolution Tsetse Fly Distribution Maps for the Lake Victoria Basin and West Africa*; Report to the Joint Food and Agriculture Organization of the United Nations/International Atomic Energy Agency Programme; Food and Agriculture Organization: Rome, Italy, 2001.
37. Wint, W.; Rogers, D. *Predicted Distributions of Tsetse in Africa*; Report of the Environmental Research Group Oxford Ltd and TALA Research Group, Department of Zoology, University of Oxford, for the Animal Health Service of the Animal Production and Health Division of the Food and Agriculture Organisation of the United Nations; FAO: Rome, Italy, 2000.
38. Wint, W.; Robinson, T. *Gridded Livestock of the World*; FAO (Food and Agricultural Organization of the United Nations): Rome, Italy, 2007.
39. Weiler, V.; Udo, H.M.; Viets, T.; Crane, T.A.; De Boer, I.J. Handling multi-functionality of livestock in a life cycle assessment: The case of smallholder dairying in Kenya. *Curr. Opin. Environ. Sustain.* **2014**, *8*, 29–38. [CrossRef]
40. IPCC. *IPCC Guidelines for National Greenhouse Gas Inventories*; Prepared by the National Greenhouse Gas Inventories Programme; Eggleston, H.S., Buendia, L., Miwa, K., Ngara, T., Tanabe, K., Eds.; IPCC: Kanagawa, Japan, 2006.
41. Peeler, E.; Omore, A. *Manual of Livestock Production Systems in Kenya*; KARI (Kenya Agricultural Research Institute): Nairobi, Kenya, 1997.
42. Roderick, S.; Stevenson, P.; Mwendia, C.; Okech, G. The use of trypanocides and antibiotics by Maasai pastoralists. *Trop. Anim. Health Prod.* **2000**, *32*, 361–374. [CrossRef] [PubMed]
43. Roderick, S.; Stevenson, P.; Ndungu, J. The production parameters influencing the composition and structure of pastoral cattle herds in a semi-arid area of Kenya. *Anim. Sci.* **1998**, *66*, 585–594. [CrossRef]
44. Roderick, S.; Stevenson, P.; Ndungu, J. Factors influencing the production of milk from pastoral cattle herds in Kenya. *Anim. Sci.* **1999**, *68*, 201–209.
45. Itty, P.; Swallow, B.; Rowlands, G.; Mulatu, W.; d'Ieteren, G. The economics of village cattle production in a tsetse-infested area of southwest Ethiopia. *Prev. Vet. Med.* **1995**, *22*, 183–196. [CrossRef]
46. Swallow, B. *Impacts of Trypanosomiasis on African Agriculture*; Food and Agriculture Organization of the United Nations (FAO): Rome, Italy, 2000.
47. Muraguri, G.R. *Epidemiological and Financial Impact of Vector-borne Diseases on Productivity of Smallholder Cattle in the Coastal Lowlands of Kenya*; Department of Agriculture, University of Reading: Reading, UK, 2000; p. 219.
48. Machila, N. *Improved Targeting and Appropriate Use of Trypanocidal Drugs for the Control of African Bovine Trypanosomiasis in Tsetse Endemic Areas of Western and Coastal Kenya within the Context of Primary Veterinary Care*; University of Edinburgh: Edinburgh, UK, 2005; p. 304.
49. Musa, L.; Peters, K.; Ahmed, M. On farm characterization of Butana and Kenana cattle breed production systems in Sudan. *Livest. Res. Rural Dev.* **2006**, *18*, 177.
50. Maichomo, M.; Ndungu, J.; Ngare, P.; Ole-Mapenay, I. The performance of Orma Boran and Maasai Zebu crossbreeds in a trypanosomiasis endemic area of Nguruman, south western Kenya. *Onderstepoort J. Vet. Res.* **2010**, *72*, 87–93.
51. Omore, A.; Muriuki, H.; Kenyanjui, M.; Owango, M.; Staal, S. *The Kenyan Dairy Sub-Sector: A Rapid Appraisal*; Smallholder Dairy (Research & Development) Project Report; ILRI (International Live-stock Research Institute): Nairobi, Kenya, 1999.
52. Mudavadi, P.O.; Otieno, K.J.; Wanambacha, W.; Odenya, J.O.; Odendo, M.; Njaro, O.K. *Smallholder Dairy Production and Marketing in Western Kenya: A Review of Literature*; Smallholder Dairy (Research & Development) Project Research Report; ILRI (International Livestock Research Institute): Nairobi, Kenya, 2001.
53. Ongadi, P.; Wakhungu, J.; Wahome, R.; Okitoi, L. Characterization of grade dairy cattle owning households in mixed small scale farming systems of Vihiga, Kenya. *Livest. Res. Rural Dev.* **2007**, *19*, 43.
54. Fox, R.G.R.; Mmbando, S.; Fox, M.; Wilson, A. Effect on herd health and productivity of controlling tsetse and trypanosomiasis by applying deltamethrin to cattle. *Trop. Anim. Health Prod.* **1993**, *25*, 203–214. [CrossRef] [PubMed]
55. Jemal, A.; Hugh-Jones, M.E. Association of tsetse control with health and productivity of cattle in the Didessa Valley, western Ethiopia. *Prev. Vet. Med.* **1995**, *22*, 29–40. [CrossRef]

56. Laker, C.D. *Assessment of the Economic Impact of Bovine Trypanosomiasis and its Control in Dairy Cattle in Mukono County, Uganda*; Makerere University: Kampala, Uganda, 1998.
57. Rowlands, G.; Mulatu, W.; Leak, S.; Nagda, S.; d'Ieteren, G. Estimating the effects of tsetse control on livestock productivity—A case study in southwest Ethiopia. *Trop. Anim. Health Prod.* **1999**, *31*, 279–294. [CrossRef] [PubMed]
58. Shaw, A. Economics of African trypanosomiasis. In *The Trypanosomiases*; Maudlin, I., Holmes, P., Miles, M., Eds.; CABI Publishing: Wallingford, UK, 2004; pp. 369–402.
59. Mugunieri, G.; Matete, G. Association of trypanosomiasis risk with dairy cattle production in western Kenya. *Onderstepoort J. Vet. Res.* **2005**, *72*, 279–284. [CrossRef] [PubMed]
60. Minjauw, B.; McLeod, A. *Tick-borne Diseases and Poverty*; DFID Animal Health Programme: Edinburgh, UK, 2003.
61. Tesfaye, D.; Speybroeck, N.; De Deken, R.; Thys, E. Economic burden of bovine trypanosomiasis in three villages of Metekel zone, Northwest Ethiopia. *Trop. Anim. Health Prod.* **2012**, *44*, 873–879. [CrossRef] [PubMed]
62. Wilson, R.T.; Bech, J.N.; Waveren, E. Draught animals in the livelihoods of mixed smallholder farmers in North Wollo, Ethiopia. *Draught Anim. News* **2002**, *37*, 2–8.
63. You, L.; Guo, Z.; Koo, J.; Sebastian, K.; Wood-Sichra, U. *Spatial Production Allocation Model (SPAM) 2005 Version 1*; Provided through Harvest Choice; International Food Policy Research Institute: Washington, DC, USA; University of Minnesota: St. Paul, MN, USA, 2014; Available online: http://harvestchoice.org (accessed on 8 January 2015).
64. Lukuyu, B.; Gachuiri, C.K.; Lukuyu, M.N.; Lusweti, C.; Mwendia, S. *Feeding Dairy Cattle in East Africa*; East Africa Dairy Development Project: Nairobi, Kenya, 2012.
65. CGIAR/ILRI Sub-Saharan Africa Feed Composition Database. Available online: http://192.156.137.110/ssafeed/ (accessed on 1 December 2015).
66. Feedipedia. Feedipedia Animal Feed Resources Information System, a Programme by INRA, CIRAD, AFZ and FAO. Available online: http://www.feedipedia.org (accessed on 1 August 2015).
67. Feedipedia Napier Grass. Available online: https://www.feedipedia.org/node/395 (accessed on 13 December 2017).
68. Capper, J.L.; Cady, R.A.; Bauman, D.E. The environmental impact of dairy production: 1944 compared with 2007. *J. Anim. Sci.* **2009**, *87*, 2160–2167. [CrossRef] [PubMed]
69. Gerber, P.; Vellinga, T.; Opio, C.; Steinfeld, H. Productivity gains and greenhouse gas EI in dairy systems. *Livest. Sci.* **2011**, *139*, 100–108. [CrossRef]
70. Cecchi, G.; Paone, M.; Feldmann, U.; Vreysen, M.J.B.; Diall, O.; Mattioli, R.C. Assembling a geospatial database of tsetse-transmitted animal trypanosomiasis for Africa. *Parasites Vectors* **2014**, *7*, 39. [CrossRef] [PubMed]

sustainability

MDPI

Review

Reservoirs and Transmission Pathways of Resistant Indicator Bacteria in the Biotope Pig Stable and along the Food Chain: A Review from a One Health Perspective

Ricarda Maria Schmithausen [1,*], Sophia Veronika Schulze-Geisthoevel [2], Céline Heinemann [2], Gabriele Bierbaum [3], Martin Exner [1], Brigitte Petersen [2] and Julia Steinhoff-Wagner [2]

1. Institute for Hygiene and Public Health, University Hospital Bonn, 53127 Bonn, Germany; martin.exner@ukbonn.de
2. Institute of Animal Science, Preventive Health Management Group, University of Bonn, Katzenburgweg 7-9, 53115 Bonn, Germany; Sophia.Schulze-Geisthoevel@lwk.nrw.de (S.V.S.-G.); celine.heinemann@uni-bonn.de (C.H.); b-petersen@uni-bonn.de (B.P.); jste@itw.uni-bonn.de (J.S.-W.)
3. Institute for Medical Microbiology, Immunology and Parasitology, University Hospital of Bonn, 53127 Bonn, Germany; g.bierbaum@uni-bonn.de

* Correspondence: ricarda.schmithausen@ukbonn.de; Tel.: +49-228-287-13452

Received: 1 September 2018; Accepted: 22 October 2018; Published: 31 October 2018

Abstract: The holistic approach of "One Health" includes the consideration of possible links between animals, humans, and the environment. In this review, an effort was made to highlight knowledge gaps and various factors that contribute to the transmission of antibiotic-resistant bacteria between these three reservoirs. Due to the broad scope of this topic, we focused on pig production and selected "indicator bacteria". In this context, the role of the bacteria livestock-associated methicillin-resistant *Staphylococcus aureus* (LA-MRSA) and extended spectrum beta-lactamases carrying *Escherichia coli* (ESBL-E) along the pig production was particularly addressed. Hotspots of their prevalence and transmission are, for example, pig stable air for MRSA, or wastewater and manure for ESBL-E, or even humans as vectors in close contact to pigs (farmers and veterinarians). Thus, this review focuses on the biotope "stable environment" where humans and animals are both affected, but also where the end of the food chain is not neglected. We provide basic background information about antibiotics in livestock, MRSA, and ESBL-bacteria. We further present studies (predominantly European studies) in tabular form regarding the risk potentials for the transmission of resistant bacteria for humans, animals, and meat differentiated according to biotopes. However, we cannot guarantee completeness as this was only intended to give a broad superficial overview. We point out sustainable biotope approaches to try to contribute to policy management as critical assessment points in pig housing conditions, environmental care, animal health, and food product safety and quality as well as consumer acceptance have already been defined.

Keywords: MRSA; ESBL-E; pigs; transmission; antibiotic resistance; one health

1. Introduction

The use of antibiotics in human and veterinary medicine and their dissemination in the environment have favored the emergence and spread of antibiotic-resistant microorganisms [1–5]. For example, both extended-spectrum betalactamase-producing *Escherichia coli* (ESBL-E) [6,7] and livestock-associated methicillin-resistant *Staphylococcus aureus* (LA-MRSA) have already been separately isolated at different stages of the pork production chain [8–11]. Hence, there has been an ongoing debate regarding the potential sources, transmission routes, and risk factors of the continuous spread of MRSA and ESBL-E between animals, humans, and the environment [12–16]. Thus, they are also perceived as a potential threat to public health [17–23]. Therefore, in this review, methicillin-resistant *Staphylococcus aureus* (MRSA) and ESBL-E have been used as "indicator" bacteria with possible zoonotic potential and occurrence in both the stable biotopes and healthcare sector and community [24]. Consequently, farmers have been confronted with several consequences of this problem: The potential danger of animal colonization with drug-resistant bacteria for (i) humans living on or in the vicinity of farms [25–27], and (ii) for consumers of animal products [3,28–31]. The intended audience of this review is farmers of intensive pig production systems worldwide as well as healthcare workers, as humans living and working in close contact with pigs and patients in rural areas have been found to be colonized with LA-MRSA [25,32–38]. On the other hand, in the food chain, meat products can serve as potential transmission factors of ESBL-E from animals to humans [29,39–41].

Furthermore, the global spread of multidrug-resistant *Enterobacteriaceae* may be linked to wastewater from hotspots like hospitals and/or intensive livestock production settings in Germany and worldwide [15,16]. Consequently, the stable environment (air, wastewater, etc.) needs to be considered as a reservoir and source of the dissemination of multidrug-resistant bacteria. Notably, bacteria not only persists on/in the living animal, but also on surfaces that are in contact with the animals (compartment walls and equipment) [42]. In correlation with this, LA-MRSA bacteria were detected by Friese et al. [43] in dust samples from the investigated breeding farms in Germany and in the stable air of a fattening farm, while ESBL-*Escherichia coli* isolates have been isolated from pig livestock production sites [44]. Furthermore, in the farm environment, both commensal and environmental bacteria serve as reservoirs for the transfer of antimicrobial resistance genes to pathogenic bacteria [45,46]. Pietsch et al. [47] found isolates of distinct *E. coli* clonal lineages in all three reservoirs: Human, animal, and food in Germany. Thus, it is assumed that the contribution of the animal biotope in terms of antimicrobial resistance in humans is not negligible [48].

Therefore, a holistic approach in the sense of "One Health" needs to be integrated into interdisciplinary research at the interface between humans and animals and their common environment [49–52]. In the framework of sustainable biotope infection control measures in pig housing conditions, hygienic stable environment, animal health, and food product safety to reduce the prevalence of resistant bacteria remains challenging in pig production [53].

2. Background Information

2.1. Antibiotic Use in Livestock Production

The first cases of antimicrobial resistance occurred in the late 1930's [54]. From that point until now, the production of food animals has been associated with large farms, a high density of animals, and in this respect, an improvement in disease management [2].

In a report on antibiotic consumption and resistance in Germany in 2015 (GERMAP), tetracycline, sulfonamide, and betalactam antibiotics (penicillin, first to fourth generation cephalosporin, and the betalactamase inhibitors) [55] were found to be the most commonly used antibiotics in veterinary medicine [56]. The total consumption of around 1706 tons of antibiotics used in the veterinary field in 2011 decreased to 733 tons in 2017 [57,58]. It can be assumed that the reduction was not only based on the EU ban, but on the basis of governmental and private antibiotic monitoring programs. However, the causal relationship between the decrease in resistance rates and the decrease in antibiotic use, observed on behalf of the patient on the basis of the quantities administered and/or the frequency of treatment, cannot be examined on the basis of the available data. Metaphylactic therapy (in which high doses of antibiotics are usually given for a short period) is still preferred in the standard management of some highly organized and efficient management systems due to the optimized disease prevention rates [59,60]. The adaptation of bacterial metabolism to antimicrobial agents can be forced by under-dosing and the non-achievement of required blood and tissue levels over several days. To avoid the selection of drug-resistant bacteria, it is necessary to relate the dosage of antibiotics to body mass and perform several repeat applications [61]. Further data has shown a correlation between the frequency of treatment and the occurrence of multidrug-resistant bacteria. The higher the treatment frequency (the average number of days each animal in the herd is treated with antibiotics), the higher the rate of resistance identified in isolates from animals and animal products [62,63]. Fundamentally, animal protection laws demand that animals with infections have the right to (medical) treatment [64]. Furthermore, the use of antibiotics is indicated for the maintenance of a balanced pig's health and its physical condition [65,66].

2.2. Characteristics of Multidrug-Resistant Bacteria (MRSA and ESBL-E)

2.2.1. MRSA

S. aureus is a Gram-positive, coagulase-positive bacterium. *S. aureus* demonstrates a high robustness of months for desiccation, heat, UV radiation, and various disinfectants [67]. *S. aureus* can transiently or persistently colonize the skin and mucous membranes of the respiratory tract of humans and animals [68]. Colonization alone does not cause symptoms, but can lead to an increased risk for secondary infections: superficial skin lesions and soft tissue infections, invasive life-threatening bloodstream infections, and sepsis [67]. Methicillin-resistant *Staphylococcus aureus* (MRSA) strains acquire their resistance by transferring the mobile genetic element, SCC*mec* [69]. The resistance is based on the *mec*A gene on the SCC*mec* coding for the 78 kDa alternative penicillin binding protein (PBP2a) [70]. The PBP2a has a very low binding affinity for all betalactam antibiotics. This allows the PBP2a to synthesize the cross-links of the peptidoglycan of the bacterial cell wall without being inhibited by methicillin and any traditional beta-lactam antibiotics [71]. However, new "5th generation" cephalosporins (ceftaroline and ceftobiprole) have a high affinity for this PBP, resulting in enhanced activity against methicillin-resistant *S. aureus* [72–74]. MRSA in humans was first isolated in the United Kingdom and Denmark after penicillinase-stable betalactam methicillin was introduced in 1959 [75,76]. In animals, MRSA was identified in 1972 [77]. Three different MRSA types linked with human infection can be classified based on their occurrence and distribution patterns: hospital-associated MRSA (HA-MRSA), health care-associated community MRSA (HCA-MRSA), and community-associated MRSA (CA-MRSA) [78,79]. The fourth category is called livestock-associated MRSA (LA-MRSA) and is associated with livestock as well as the humans who are in close contact with the livestock [78]. LA-MRSA has retained a pathogenic potency for humans [80,81] although some virulence factors have been lost [78]. This underlines the potential of "bidirectional zoonotic exchange", which adds risks to public health [78]. Most LA-MRSA strains are classified as clonal lineage CC398 [79,82]. Price et al. [78] assumed that clonal lineages of LA-MRSA CC398 originated from human methicillin-sensitive *Staphylococcus aureus* (MSSA), which has adapted to livestock animals [78,83].

2.2.2. ESBL and Resistance Genes

Extended-spectrum betalactamases (ESBLs) and the less common AmpC enzyme are able to hydrolyze a broader spectrum of betalactam antibiotics than the original betalactamases from which they are derived. ESBLs inactivate betalactam antibiotics with an oxyimino group such as oxyimino-cephalosporin (e.g., ceftazidime and cefotaxime) and the oxyimino-monobactam aztreonam [84,85]. ESBL and AmpC enzymes can combine to expand the resistance profile to include betalactamase inhibitors and even carbapenems [86]. Paterson et al. [87] also described ciprofloxacin resistance as being highly associated with ESBL-producing strains [87]. A further increase in resistance is provided by carbapenemases (e.g., VIM [Verona integron-encoded metallo-betalactamase], NDM [New Delhi metallo-betalactamase], OXA-48 [oxacillin resistant]) [88], which are betalactamases that are able to hydrolyze the carbapenem reserve antibiotics (imipenem, meropenem, and ertapenem) [89,90]. The most common types of ESBL enzymes detected in *Enterobacteriaceae* are the following: Temoneira (TEM)-type betalactamase, cefotaximase-Munich

(CTX-M)-type betalactamase, sulfhydryl variable (SHV)-type betalactamase, and oxacillin resistant (OXA)-type betalactamase [91]. Currently, the most common worldwide CTX-M-type enzymes can be classified into five main groups depending on their amino acid sequence: CTX-M-1, CTX-M-2, CTX-M-8, CTX-M-9, and CTX-M-25 [13,92–94]. There are more than 40 different CTX-M types assigned to the five main groups [95]. The enzyme that is most widely distributed between human *Enterobacteriaceae* is CTX-M-15, which has also been detected in *E. coli* from pigs and poultry [96,97]. The genes for ESBL (bla_{CTX-M} genes) [98,99] are transferred by both vertical and horizontal transfer [100]. Therefore, apart from direct transmission in *Enterobacteriaceae*, it is of particular interest that resistance genes can be transferred with the help of plasmids to other pathogenic and apathogenic microorganisms. They encode a variety of resistance genes [101]. MGEs appear in a huge variety: Transposons, phages, plasmids, chromosomal resistance cassettes, and others, which can be taken up by bacteria in different biotopes [102]. However, the role of antibiotic resistance genes and their exchange of resistance genes between the human and animal microflora and between the environment is not completely known [103].

2.3. *Bioavailability and Transmission Potential of Antibiotic-Resistant Bacteria*

The following aspects can be discussed to influence the spread and bioavailability of drug-resistant bacteria in different reservoirs in Table S1 in the Supplementary Material.

3. Reservoirs and Transmission Pathways of LA-MRSA and ESBL-E

3.1. *Transmission Pathways*

The epidemiology and mechanisms of emergence and spread of antimicrobial resistance and antibiotic-resistant bacteria are diverse and several possible pathways in different systems exist for their transmission. Valentin et al. [104] indicated that the same subtypes of isolates were detected in isolates from human and livestock and companion animal populations, suggesting a possible exchange of bacteria or bacterial genes. Figure 1 presents some possible transmission pathways of antibiotic-resistant bacteria between human and animal vectors and between reservoirs (air, dust, water, manure, food, etc.) as well as exposure routes in their biotopes (stable, abattoir, etc.). Here, the term "biotope" is defined according to Nehring and Albrecht [105] and Lötzli [106] as the living space of a community of species.

Nevertheless, many transmission routes are still unclear. Therefore, in the following passages, a selection of possible transmission pathways in different biotopes and reservoirs will be demonstrated using the selected bacteria MRSA and ESBL-E. We used both MRSA and ESBL-producing *E. coli* (ESBL-E) as indicators to evaluate the movement of antibiotic-resistant bacteria in the environment [24] as *Staphylococcus aureus* occurs in both the mucous membrane of humans and animals and is very resistant to the environment. The MRSA strain CC398 is known to be transferred via direct contact between animals [107] and humans [25,108–110], as well as through the environment [111], particularly airborne transmission [43,112]. On the other hand, *Escherichia coli* is frequently used as indicator bacteria in the animal sector and in environmental hygiene. Additionally, due to its association with swine and pork, we decided to use the resistant variant ESBL-E [113,114]. In the following sections, the different sources and reservoirs, especially within the stable environment, are presented and discussed in more detail regarding the different transmission potentials.

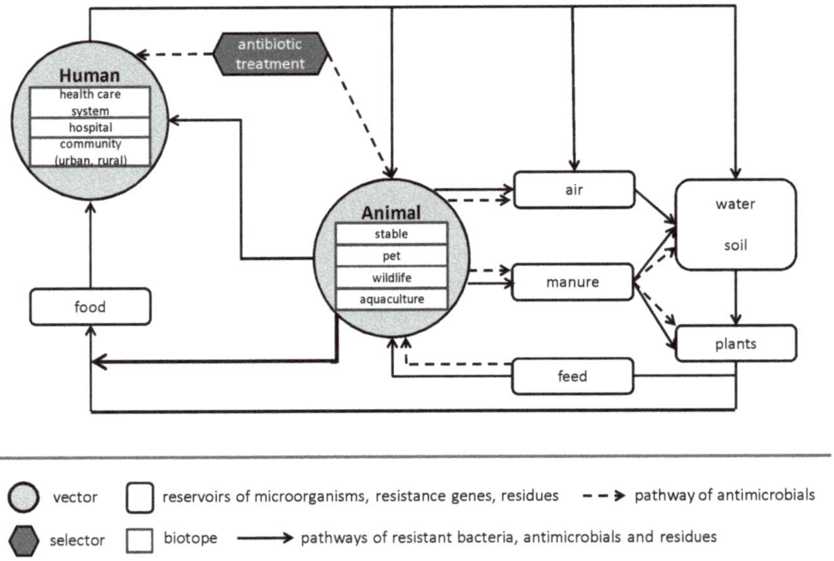

Figure 1. Some possible routes of exposure and transmission pathways of antibiotic-resistant bacteria and antimicrobial resistance between different reservoirs and biotopes.

3.2. LA-MRSA and ESBL-E in Humans in the Stable Biotope

The administration of antibiotics in humans and animals selects resistant strains within the microflora and enables colonization of the mucosa with new isolates. Animal strains may be able to transiently colonize human indigenous microflora (and vice versa) and transfer resistance genes [115]. Thus, the human respiratory and gastrointestinal tract can be reservoirs for bacteria with antibiotic resistance [116] (Table S2). Humans working in the stable environment make contact with different resistant strains which can act as vectors in different biotopes [117]: Farmers and veterinarians belong to the stable's biotope [118]. Pig farmers and veterinarians belong to this defined risk group [119]. Table S2 gives an overview over the occurrence of MRSA and ESBL-E rates in farmers and veterinarians in Europe and possible transmission pathways. Nevertheless, the importance of drug-resistant *Staphylococcus aureus* and/or *E. coli* strains isolated from the animal biotope has not yet been quantified or evaluated. However, it is assumed that the contribution of the animal biotope in terms of antimicrobial resistance in humans is not negligible [48]. Kraemer et al. [120] even showed associations between antibiotic use and resistant bacteria carriage.

3.3. LA-MRSA and ESBL-E in Pigs

Possible sources for the transmission of drug-resistant bacteria and/or antimicrobial resistances include direct contact between animals on the farm and within compartments [121,122] or via pig trading [9,10] in livestock vehicles [123] within or between regions, and the introduction of new animals in herds [124,125]. Pigs can also be colonized during transportation between pig production steps or on their way to the abattoir [8,126]. Until now, the LA-MRSA CC398 strain has been found in all stages of the food chain, varying between European countries (Table S3). Other risk factors are reported to be associated with the occurrence of multidrug-resistant bacteria are for antimicrobial use, purchase of gilts, and hygiene measures [127,128]. Generally, the effectiveness of the cleaning and disinfection methods used on commercial pig farms needs to be evaluated in more detail [129,130].

Table S3 gives an overview of several studies describing the different possibilities of the transmission of antibiotic-resistant bacteria (MRSA and ESBL-E) between pigs.

3.4. Reservoir Stable Environment

Studies on the agricultural environment and the food chain have revealed that pathogen reservoirs also exist in the environment outside their animal host [129,131–134]. In farm environments, commensal, and environmental bacteria may be reservoirs for the transfer of antimicrobial resistance genes to pathogenic bacteria [97,135–138].

3.4.1. LA-MRSA and ESBL-E in Air and Dust

Dynamic processes influencing the spread of bacteria in the biotope of the animal stable are presented in Figure 2.

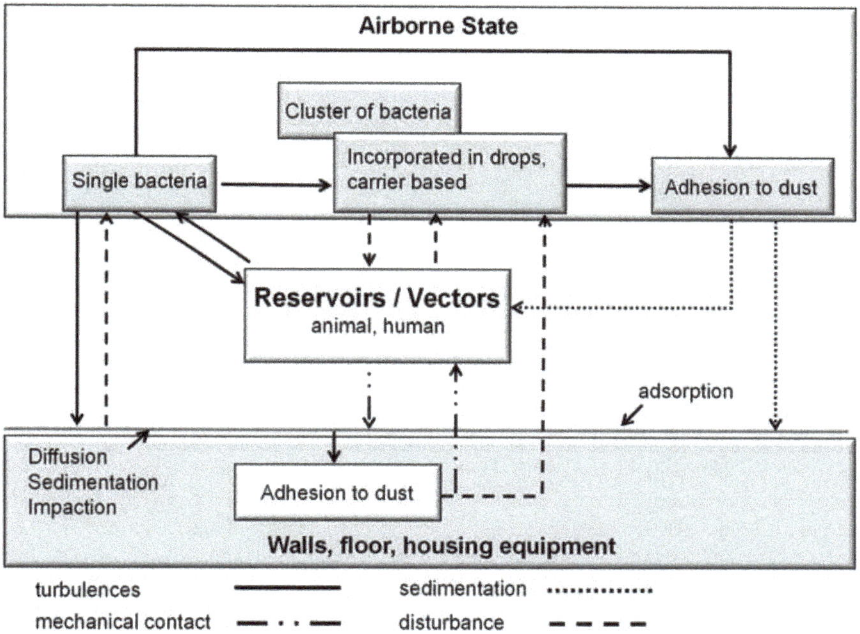

Figure 2. Dynamic processes of bacteria in the biotope of the animal stable (modified according to Müller et al. [139]). Bacteria can spread from their source via different dynamic processes:

- *Turbulence*: Spread of bacteria due to air turbulence;
- *Mechanical contact*: Transmission of bacteria by direct contact (passive via stable equipment or active via persons/animals);
- *Sedimentation*: Reduction of circulation from the airborne state and subsequent deposit;
- *Re-suspension*: Bacteria not primarily and actively transported by air flow; and
- *Re-entrainment*: Re-return of already sedimented bacteria to the air [139].

In the pig stable, about 80% of airborne bacteria are staphylococci (Gram positive) [140] and 0.5% [141] or even 5% [142] are coliform bacteria such as *E. coli* (Gram negative). The imbalance between Gram-positive and Gram-negative bacteria is attributed to the lower survival time of Gram-negative bacteria in their airborne condition [141]. *S. aureus* can disperse in the air as directly suspendable airborne particles. However, they are most often attached to and carried by dust particles [141]. Once the microorganisms get into the air, they prefer to merge into larger clusters or adhere to larger dust particles. Airborne bacteria return via air turbulence or by sedimentation to their reservoirs on stable surfaces or are acquired by animals or humans [126,128,143]. In contrast, coliform bacteria, which are excreted in feces, are mechanically bound to stable surfaces and/or litter and feed [144]. Only after the animal feces have dried out, and due to air turbulence by animal activities, does the whirling dust and dirt particles raise them in the stable air. The level of contamination of stable air by both airborne and coliform bacteria depends on both the animal and stable environment including the density, age, and activity of animals, the ventilation system, the quantity of dust [145], the humidity of the surfaces, and air as well as air turbulence. Survival of intestinal bacteria depends on the environmental conditions outside the animal organism. Survival rates of coliforms are best in the immediate vicinity of the animal and, therefore, in feces and fecal contaminated surfaces, whereas the detection rate in air immediately decreases [146–148]. The bacterial concentration in the air is based on physical processes (e.g., air flow and sedimentation) and on the organisms' biological viability. Influencing factors are temperature, humidity, UV radiation, air change rate, noxious gas concentration, and the organisms' aerodynamic shape [63,149].

Environmental contamination with MRSA was identified in the study by Weese et al. [150]. Friese et al. [43] determined that 85.2% of pig stables in Germany were LA-MRSA positive in the air. They also identified dust as an important factor for the occurrence of MRSA in the air. Drug-resistant *S. aureus* could be identified inside and outside the pig stables and originated from contaminated dust [151]. The dissemination of MRSA between pigs within a farm by air was based on a positive association of samples from pigs and the environment [43]. Agersø et al. [112] indicated a high sensitivity of air samples equal to the within-herd prevalence. Therefore, they recommended air sampling for initial testing or even screening of herds. Bos et al. [152] and Gilbert et al. [111] even confirmed a strong association between nasal ST398 MRSA carriage in people working on the farms for >20 h per week and MRSA air levels. In people working in the barns <20 h per week, there was a strong association between nasal carriage and the number of working hours. This study showed that working in the lairage area or scalding and dehairing area were the major risk factors for MRSA carriage in pig slaughterhouse workers, while the overall prevalence of MRSA carriage was low. Occupational exposure to MRSA decreased along the slaughter line, and the risk of carriage showed a parallel decrease. Heinemann et al. [153] showed that a working time of three-to-six hours could be enough for positive findings of nasal colonization with MRSA. However, it is unknown whether the presence of LA-MRSA is a result of the carriage or retention of MRSA-contaminated dust. Nevertheless, the persistence of LA-MRSA CC398 in humans depends on the intensity of animal contact [109]. Several studies have investigated dust and air samples from pig farms. An overview of some of these studies and their data is given in Table S4.

The main message of all the presented studies (in Table S4) is that exposure to the stable environment and air in pig farms is often contaminated with LA-MRSA, and can act as a transmission source for humans, which is an important determinant for nasal carriage, especially in this highly exposed group of farmers. This is next to the duration of the contact with animals. Intervention measures should therefore also target the reduction of ST398 MRSA air levels including the improvement of environmental and operating parameters of air quality and pig performance [154,155]. In contrast, only a few studies have been performed on the spread of ESBL-E in the stable environment such as by dust and air in pig settings [126,156–158]. Laube et al. [159] assumed dust to be a major source for the transmission and spread of ESBL-E within stables and during the release of animals (especially poultry) from stables. They found a high prevalence of ESBL-E in the pooled feces and dust samples obtained on broiler chicken farms [159]. Von Salviati et al. [157] identified the transmission potential between ESBL-E between pig farmers and their surroundings (surfaces, barn, and ambient air). However, the detection of ESBL/AmpC–*E. coli* in stable air and ambient environment was low and also found ESBL/AmpC–*E. coli* on surfaces in the vicinity (see details in Table S4). Furthermore, they proved emission via slurry and transmission via flies. Hoffmann [160] hypothesized a possible transmission of CTX-M-1 subtypes to humans via the inhalation of contaminated dust particles during exposure in the stable environment. This assumption was confirmed by the study of Dohmen et al. [158], who found that CTX-M-1 carriage in pig farmers and the presence of CTX-M-1 in dust were associated, indicating that air transmission of CTX-M 1 might be possible on pig farms.

In summary, the transmission pathway of ESBL-*Enterobacteriaceae* via air and/or dust spread via the airborne route or via different vectors seems possible [127]. Otherwise, it is not possible to distinguish between the two transmission pathways of direct contact between humans and animals and the indirect airborne transmission pathway [160]. Contaminated manure presented the major emission source for ESBL/AmpC-producing *E. coli* on pig farms [157].

3.4.2. LA-MRSA and ESBL-E in Water, Wastewater and Manure

Water contaminated with antibiotic-resistant bacteria is an important reservoir for the emergence and spread of resistance mechanisms and mobile genetic elements [161,162]. Ingestion and dealing with contaminated water can result in the colonization of the gastrointestinal tract of humans and animals [163]. Thus, water constitutes not only a way for the dissemination of antibiotic-resistant organisms among human and animal populations as drinking water is produced from surface water, but also a major route by which resistance genes are introduced to natural bacterial ecosystems. In such systems, nonpathogenic bacteria can serve as a reservoir of resistance genes [162]. Lupo et al. [161] and Schwartz et al. [164] highlighted horizontal gene transfer by transduction as the main mechanism conferring drug resistance in drinking water, surface water, and wastewater. Several other studies have reported the presence of drug-resistant bacteria and resistance genes in fresh drinking water, rivers, and sewage in Europe and all over the world [16,44,165–171]. Nevertheless, to date, more and more information about the risks of contamination of livestock drinking water with antibiotic-resistant bacteria have been provided [128,172,173]. Pletinckx et al. [174] did not find any evidence of MRSA-positive isolates in animal drinking water on pig and poultry farms, while environmental samples (dust, animal feed, and manure) and samples from pigs and farmers were contaminated with MRSA. However, MRSA has been found in contaminated feed and trough water [130,172]. Thus, water sources might facilitate the transmission of MRSA between different animals [175]. Feed and troughs also represent possible sources of exposure to foodborne coliform bacteria like *E. coli* [176]. Animals potentially face daily exposure to bacterial contamination from these sources. Nevertheless, LeJeune et al. [172] indicated that the degree of bacterial contamination was associated with potentially controllable management factors. Water systems in stables represent a complex ecosystem that is affected by multiple factors. These factors influence the persistence and the effects of bacteria in drinking water (pipelines and troughs) [172,177]. The multiple factors are diverse: entry of biomass into the system promotes biofilm growth as a medium for microorganisms, installation defects, backward spread of bacteria, high temperature, etc. Therefore, it is essential to control the quality of the animals' drinking water at regular intervals (at least once a year) [178]. Hartung and Kamphues [179] determined basic practical and species-specific aspects in housing conditions including the water collection and distribution system as basic requirements for sustainable livestock and housing management [178,179]. Heinemann et al. [130] found high bacterial loads in animal drinkers after cleaning and disinfection, which could lead to a vertical transfer of pathogens to newly arriving pigs. In this context, Heinemann et al. [130] indicated that decontamination strategies such as intensive cleaning and disinfection were effective at reducing the levels of *Enterobacteriaceae* on stable floors. Nevertheless, residual contamination remains in the environment and on the surfaces of troughs and drinkers [129,130]. The findings of Jutkina et al. [180] therefore indicate that exposure to sub-lethal concentrations of certain antimicrobials may contribute to the emerging problem of antibiotic resistance not only through selection of certain resistance phenotypes, but also by means of stimulating transfer of antimicrobial resistance traits directly. Schwartz et al. [164] assumed a possible transfer from wastewater and surface water to drinking water distribution systems. Hölzel et al. [181] expected that antibiotic-resistant bacteria could also reach the food chain via human and animal manure applications. Thus, the possible risk for the transmission of antibiotic-resistant bacteria, resistance genes, and antibiotics from human wastewater or animal manure to the environment, including drinking waters and the food chain, cannot be neglected. Some studies have reported multidrug-resistant pathogens in the soil and water of farm environments. The ESBL-variants detected corresponded to those previously found in animals or humans living in a farm environment [182,183] (Table S5). To date, several non-European studies have reported multidrug-resistant pathogens in the vegetables, soil, and water of farm environments. The ESBL-variants detected corresponded to those previously found in animals or humans living in a farm environment [182,183]. Vital et al. [184] proved that multidrug-resistant isolates were observed in irrigation water, soil, and vegetables in urban farms, indicating that water serves as a possible route for a wide distribution across all types of borders. All of the presented studies commonly demonstrate (molecular) homology analysis of CTX-M-producing *E. coli* isolates collected from water and/or swine, and/or human, implying that multidrug-resistant pathogens in the aquatic environment might derive from both humans and animals [44,185]. However, contaminated slurry was presented as the major emission source for ESBL/AmpC-producing *E. coli* in pig fattening farms [158]. Several other studies identified ESBL strains in manure samples from pig farms (Table S5). However, a possible link between the prevalence of ESBL-E in hospitals and other sources such as local food, water, or animal sources has not been identified [186].

3.5. LA-MRSA and ESBL-E in the Food Chain: Abattoir Biotope

Abattoirs can be possible sources of drug-resistant bacteria [187]: Carry-over of resistant bacteria at the time of slaughter, the hygiene related to the slaughter processes, meat processing hygiene, and retail handling [188,189] (Table S6). Pietsch et al. [47] found isolates of distinct *E. coli* clonal lineages in all three reservoirs of human, animal, and food in Germany. However, a direct clonal relationship of isolates from food animals and humans was only noticeable for a few cases. All foodborne outbreaks have been caused by enteropathogens. The role of the ESBL-producing *Enterobacteriaceae* in these persons was considered as commensal flora. Thus, in comparison to the general population, humans involved in food processing need to be considered as very important intermediate reservoirs and vectors for ESBL genes remaining in food retailers contaminated with ESBL-producing bacteria [190]. Normanno et al. [191] clearly demonstrated the need for improved hygiene standards to reduce the risk of occupational and foodborne infection linked to the handling/consumption of raw pork containing resistant bacteria like MRSA and ESBL-*Enterobacteriaceae*.

4. Sustainable Biotope Approaches

4.1. Animal Health in the Livestock Biotope

The origin of many health problems in pig production is very complex in nature [192]. Diseases that occur in a multi-factor system such as pig production are not only defined by one factor, but by the interplay of a variety of processes that take place in the animate and inanimate environments of the animal and within the animal [192]. Künneken [143] described the interaction between several factors in the stable environment in his model (Figure 3).

In modern intensive livestock, the following infection-promoting and infection-inhibiting factors can influence the balance of the stable biotope:

1. A high number of passages of host pathogens with potentially increased virulence within herds;
2. Higher rates of the adhesion of pathogens;
3. The rapid spread of pathogens via direct host contact, food, water, air, and living vectors;
4. The use of high-performance animals that are more sensitive to environmental stressors;
5. The neglect of hygiene principles, e.g., sufficient drying times during cleaning and disinfection; and
6. The reduced possibility of individual health control and animal observation [143].

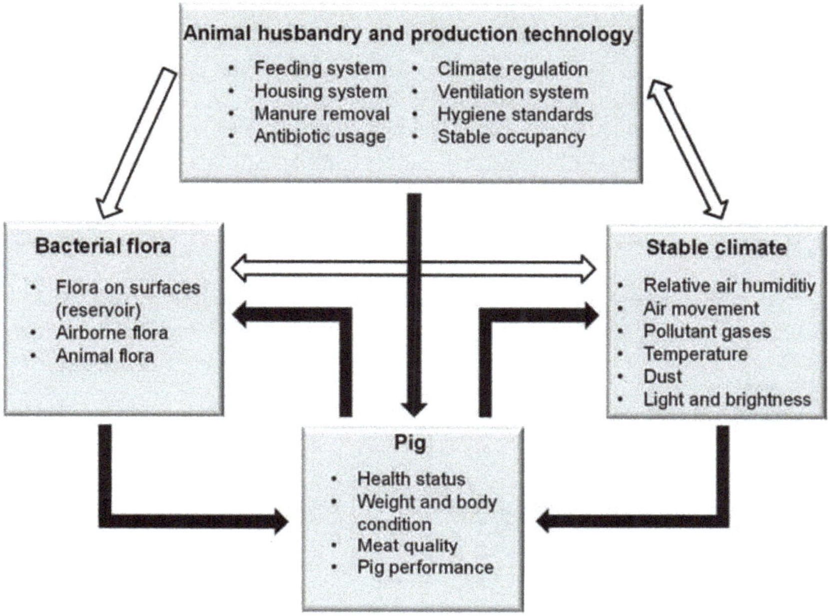

Figure 3. Relationships between animate and inanimate factors in environmental reservoirs (modified according to Künneken [143]).

These infection-promoting and infection-inhibiting factors are involved in constantly changing processes in the stable biotope. If there is a summation of many infection-promoting factors in one or all of the three systems, a chain of events is initiated with the possible outcome of an infectious disease [192]. Walker et al. [193] identified three aspects in the problem of increasing the resistance of bacteria in, for example, a pig producing system: Pigs have difficulties developing and stabilizing their individual natural barrier and immune response (stable-specific immunity) against the increase and spread of pathogens caused by a high pig density or when pigs are introduced into a new stable with a foreign pig population. These and other actions can trigger chronic stress responses with the pigs becoming more susceptible to infections that are caused by potentially infectious commensal bacteria [194]. Thus, closed systems show lower colonization rates with multidrug-resistant bacteria than those farms with a high rate of incoming pigs (maybe even from one or even more than two different suppliers), which interferes with stable immunity [10,195]. Thielen et al. [196] referred to animal health in livestock as a balance between infection pressure and the animals' immune systems.

4.2. Preventive Health Management: Part of Quality Management

At the beginning of the last century, animals were only treated when they were clearly suffering from a disease or if their lives were acutely threatened. Whereas in recent decades, the concept of "integrated veterinary herd health care" has been established, focusing on the combination of systematic animal health management of the entire animal population for the prevention and early detection of diseases. Today, the focus is on the implementation to react to any deviation from the normal state as fast as possible given that diseases, even if they occur without visible symptoms, impair the well-being of the animals, and thus have effects on eating and drinking behavior [197].

From the perspective of protecting consumer health, it is important to prevent the entry of pathogens from animal origins that are likely to cause human zoonosis [198]. Strategies for the maintenance of animal health afford a direct contribution towards optimizing consumer health protection [199]. In recent years, research projects have focused on the implementation and organization of intra and inter-farm health management systems in meat-production chains [200–210]. The main task of health management is the maintenance of animal health at all stages of animal production. Animal health not only has special significance for the production of healthy food, but also for the economical optimization of the various production stages, namely farrowing, nursery, and finishing [201,204,205]. According to Berns [200] and Welz [208], diseases are considered as process errors and quality-reducing effects in terms of quality management. Therefore, preventive quality management focuses on health precautions on livestock farms [211].

Similar to human medicine, animal health management can be divided into three levels (Table 1) [212].

Table 1. Definition of the three levels of prevention (according to Schulze Althoff, [212]).

Level	Definition
Primary prevention	Structural, group-based and individual measures to prevent the occurrence of disturbances
Secondary prevention	Preventive measures initiated once the pathogen has been identified in order to prevent progression of the disturbance
Tertiary prevention	Measures to prevent aggravation of the disturbance and mitigate the effects of the disturbance

According to Berns [200], the main priorities of preventive animal health management are the following:

- Detect diseases while still in the subclinical stage;
- Prevent infections from progressing to a clinical stage and stop the spread from a single animal to the whole herd; and
- Identify and promptly eliminate stressors and risks to animal health from the environment.

Consequently, to ensure food safety and quality, the traditional quality control at the end of the production process needs to be supplemented by control inspections at an incoming and at an intermediate level—when selling and purchasing piglets, finishing pigs, and at slaughter [204]. To assure food safety and consumer protection, only healthy pigs that originate from farms with a certified health status should be slaughtered [212]. Health and quality assurance in pork production encompasses:

- Pig supplying farms;
- Networks of primary producers; and
- Slaughtering, processing, and distribution [205,213].

Thus, animal health management is deemed to be an integral part of quality management in a food supply chain, all the way to grocery retail stores [198,214]. Coordinating these measures with several enterprises along the value-added pork chain defines the inter-farm chain-oriented quality management [215]. The concept of a quality management system is defined as systematic planning, implementation, and documentation with an influence on the quality of a product [206,209]. The driving force is each participating enterprise's awareness that by improving its own quality management, it can make the quality management efforts of other enterprises in the value chain more efficient. This enables the maximization of the profits of the enterprises that compose the value chain [216,217].

4.3. *"One Health" Crossing Biotopes*

The interdisciplinary cooperation in solving complex health problems is intended to create incentives for system innovation [50]. The resistance problem is one of many examples that can be perfectly addressed within the "One Health" approach [218–221]. In the case of pork production, this means considering humans and animals in the "livestock" biotope, as well as the "food processing and transport" and "health care system" biotopes, are all interconnected via the environment.

Every pig production farm that produces animals for sale constitutes a separate biotope. Pigs, throughout their life, are often moved from one biotope to another. The production of pigs usually includes four transfers and changes of stables. Each farm involved constitutes a geographical and organizational unit within the chain [209]. Nevertheless, it should not be underestimated that on finishing farms, the source of the colonization of pigs, whether acquired at the current stage or during farrowing or nursery stages, often cannot be determined [222]. De Neeling et al. [222] and other authors [223,224] have reported that pigs, colonized at the farrowing or nursery farm, carry their microbial load over to the finishing stage. In this context, the livestock biotope greatly benefits from inter-farm measures that detect and promptly reduce the risk of infections spreading from one stage of production to the next [201,214]. The establishment of monitoring systems and the certification of farms according to their health status use this approach. Hereby, single and intra and inter-farm audits as well as monitoring for specific pathogens (e.g., for LA-MRSA and ESBL-E) have been implemented [201,204,205]. With the renewal of the German Drug Act (16th AMG-Novelle) in 2014, antibiotic consumption in farm animal production is monitored to reduce its usage. Exceeding the farm-specific biannual therapeutic frequency beyond the 75% quantile obligates livestock owners to submit a written plan of measures to the public veterinary authority. Four years after its implementation, an in-depth scientific evaluation of the antibiotic monitoring is still due [225]. Epidemiological trends within a value-added chain can, according to Schulze-Geisthövel et al. (2015), be recognized, and farm comparisons such as industry marketing can be made by the common use of investigation data. *Salmonella* monitoring, for instance, is regularly included in the coordinated investigations between supplier farms and abattoirs [226,227]. Schulze Althoff [209] and Düsseldorf [201] stressed that it would make sense along the pig production chain for a receiving inspection to be referred to the supplier, an intermediate checkup referred to the farm, and a final inspection referred to the customer. This is because the carcass can be contaminated through the slaughtering process as well as by the intestinal contents of the slaughtered animals [7]. Therefore, all E.U. member states are obligated by guideline 2003/99/EG to monitor for zoonosis and zoonotic bacteria. In the context of zoonosis monitoring, representative data regarding the occurrence of zoonotic bacteria has for years been acquired in the most important food supplying animal species and products to measure the infection risk for consumers from consuming food.

Furthermore, many decision-makers in politics and science have demanded and promoted from agricultural science, veterinarian and human medicine, and environmental sciences, a common collaborative holistic strategy (the "One Health" approach) against the spread of antibiotic-resistant bacteria [51,220].

5. Final Remarks, Recommendations and Future Directions

The holistic collaborative "One Health" approach between human medicine and agricultural sciences has advanced the risk assessment of the dynamics of MRSA and ESBL-E transmission pathways as well as that of human exposure related to livestock production (particularly pigs) and processing. The impact of the importance of these "indicator bacteria" is still under debate; especially as new resistance problems and transferable resistance genes are emerging. For example, the global rise of Carbapenemase-producing *Enterobacteriaceae* has resulted in the increased use of Colistin resistance and is associated with the risk of emerging resistance. This concerns agriculture just as much as human medicine and requires round table discussions on the prudent and sustainable use of antibiotics on both sides. As this review has demonstrated, these discussions will no longer only focus on humans and animals and their food products. The focus of "One Health" will expand and focus on the environmental aspect. On the other hand, in environmental hotspots with their own dynamics and links to the reservoirs and vectors, and "animal" and "human" in different biotopes are currently crystallizing out.

This review tried to make a small contribution to the better understanding of possible transmission pathways between the presented reservoirs and highlights the following aspects:

- Healthy animals do not need antibiotic medication, thus further suppressing the risk of the occurrence of non-pathogenic resistant bacteria. The constant administration of antibiotics to animals will destroy potential antibiotics. Therefore, to save existing potential antibiotics, the government, physicians, and farm industries should limit the prescription of antibiotics to prevent antibiotic resistance, and people should not easily obtain access to self-medicated antibiotics, especially in developing countries.
- Farmers and veterinarians come into contact with antimicrobial-resistant bacteria from pigs within the stable environment. Strong associations between the isolation of resistant commensal bacteria (both MRSA and ESLB-E) and contact with pigs and even the working hours in the stable could be made.
- However, not only do animals present potential reservoirs or vectors for transferring resistance genes and resistant bacteria, but even farmers and farm workers themselves should also be considered for their transmission potential.
- Thus, aside from organic factors, the inanimate environment such as the stable climate has a substantial influence on the well-being and health status of pigs and the tenacity of bacteria. Air and dust were clearly determined as sources for the contamination of humans and animals mainly with MRSA.
- Wastewater in general (municipal, urban, with clinical and/or agroindustrial influence) serves as a melting pot for the possible horizontal transfer between resistance genes and multidrug-resistant bacteria. Whereas the impact of animal wastewater on surface water has yet to be investigated.

Thus far, resistant bacteria could be isolated with a higher percentage from swine and slaughterhouse wastewater, only indicating the potential role of agricultural wastewater within the context of environmental resistance pollution.

- Many resistant bacteria in animal drinkers could be identified after cleaning and disinfection, which could lead to a vertical transfer of pathogens to newly arriving pigs. Therefore, methods for cleaning performances, especially regarding the water systems in pig stables, should be evaluated.
- A high prevalence of ESBL-E was found in pig manure, indicating a high emission and transmission potential into the stable environment and their surroundings.
- Contamination of meat with ESBL-producing *E. coli* and MRSA is no longer surprising. However, the growing diversity of ESBL-E indicates a growing dissemination of ESBL-genes in *E. coli* in meat products from porcine origin.

Some aspects could improve our ability to mitigate the spread of resistances and would be a useful supplement to the already existing health management initiatives. Possible strategies for the enhancement of individual defense mechanisms and control of the resistance status could be the following:

- Limit the purchase of new pigs to those that are accompanied with health certificates from the supply farms.
- Determine the MRSA/ESBL status (similar to the Salmonella monitoring), take part in a continued health-monitoring program and create financial incentives for reduction measures.
- Use workshops and training to transfer scientific knowledge and sensitize for reduction measures.

However, permanent colonization should be distinguished from transient colonization. Interventions—as already implemented in the health care system—may prevent transient colonization and may hence be a useful control of MRSA not only in the hospital, but also in the stable environment.

In general, knowledge and information about the biotope-crossing potential of multidrug-resistant bacteria and antibiotic resistances should be the most important approach within the "One Health" concept. Several articles in this review demonstrated that these aspects are not yet anchored in the teaching and education of farmers, (medical and agriculture) students, multipliers, and policy makers. In summary, these approaches contribute to a larger goal of achieving the early recognition of antimicrobial resistance in bacterial livestock pathogens. Therefore, future scientific efforts should clarify the persistence and spread of highly resistant bacteria in the environment, especially in the stable and aquatic environment and their interface.

Within this framework, the implications of the "One Health" approach could have a positive impact on all sustainable future strategies.

Supplementary Materials: The following are available online at http://www.mdpi.com/2071-1050/10/11/3967/s1. Table S1: Discussion of eight factors possibly influencing transmission and bioavailability of drug-resistant bacteria. Table S2: Overview of LA-MRSA and ESBL-E in humans in the stable biotope. Abbreviations for the transmission pathways: HH = Human to Human, AA = Animal to Animal, HA = Human to Animal, AH = Animal to Human, HE = Human to Environment, EH = Environment to Human, AE = Animal to Environment, EA = Environment to Animal, EE = Environment to Environment, FH = Food to Human, HF = Human to Food, EF = Environment to Food, AF = Animal to Food. Abbreviations for the countries: NL = Netherlands, ESP = Spain, IT = Italy, EU = Europe, DEN = Denmark, UK = United Kingdom, GER = Germany, BEL = Belgium, FR = France, USA = United States of America, CHE = Switzerland, HKG = Hong Kong. Table S3: Overview over several studies describing the different possibilities of the transmission of antibiotic-resistant bacteria (MRSA and ESBL-E) between pigs. Abbreviations for the transmission pathways: HH = Human to Human, AA = Animal to Animal, HA = Human to Animal, AH = Animal to Human, HE = Human to Environment, EH = Environment to Human, AE = Animal to Environment, EA = Environment to Animal, EE = Environment to Environment, FH = Food to Human, HF = Human to Food, EF = Environment to Food, AF = Animal to Food. Abbreviations for the countries: NL = Netherlands, ESP = Spain, IT = Italy, EU = Europe, DEN = Denmark, UK = United Kingdom, GER = Germany, BEL = Belgium, FR = France, USA = United States of America, CHE = Switzerland, INT = international (worldwide). Table S4: Overview of LA-MRSA and ESBL-E in air and dust in the stable environment. Abbreviations for the transmission pathways: HH = Human to Human, AA = Animal to Animal, HA = Human to Animal, AH = Animal to Human, HE = Human to Environment, EH = Environment to Human, AE = Animal to Environment, EA = Environment to Animal, EE = Environment to Environment, FH = Food to Human, HF = Human to Food, EF = Environment to Food, AF = Animal to Food. Abbreviations for the countries: HUN = Hungary, NL = Netherlands, ESP = Spain, IT = Italy, EU = Europe, CHN = China, NOR =Norway, DEN = Denmark, UK = United Kingdom, GER = Germany, BEL = Belgium, FR = France, USA = United States of America, CHE = Switzerland. Table S5: Table overview over LA-MRSA and ESBL-E in Water, Wastewater and Manure of pig production facilities. Abbreviations for the transmission pathways: HH = Human to Human, AA = Animal to Animal, HA = Human to Animal, AH = Animal to Human, HE = Human to Environment, EH = Environment to Human, AE = Animal to Environment, EA = Environment to Animal, EE = Environment to Environment, FH = Food to Human, HF = Human to Food, EF = Environment to Food, AF = Animal to Food. Abbreviations for the countries: NL = Netherlands, ESP = Spain, IT = Italy, EU = Europe, CHN = China, NOR = Norway, DEN = Denmark, UK = United Kingdom, GER = Germany, BEL = Belgium, FR = France, USA = United States of America, CHE = Switzerland, PHL = Philippines. Table S6: Table overview over LA-MRSA and ESBL-E in the Food Chain: Abattoir Biotope and Human vectors in Pig Meat Processing. Abbreviations for the transmission pathways: HH = Human to Human, AA = Animal to Animal, HA = Human to Animal, AH = Animal to Human, HE = Human to Environment, EH = Environment to Human, AE = Animal to Environment, EA = Environment to Animal, EE = Environment to Environment, FH = Food to Human, HF = Human to Food, EF = Environment to Food, AF = Animal to Food. Abbreviations for the countries: CAN = Canada, NL = Netherlands, ESP = Spain, IT = Italy, EU = Europe, CHN = China, NOR = Norway, DEN = Denmark, UK = United Kingdom, GER = Germany, BEL = Belgium, FR = France, USA = United States of America, CHE = Switzerland, JAP = Japan, AUS = Austria.

Author Contributions: Each author made substantial contributions to: Conceptualisation: R.M.S., S.V.S.-G., G.B., M.E. and B.P.; Formal analysis: R.M.S. and J.S.-W.; Investigation: R.M.S. and S.V.S.-G.; Methodology: R.M.S, S.V.S.-G. and C.H.; Project administration: R.M.S. and B.P.; Supervision: M.E.; Validation: R.M.S and J.S.-W.; Visualization: R.M.S. and S.V.S.-G.; Writing—original draft: R.M.S.; Writing—review & editing: R.M.S., C.H., G.B. and J.S.-W.

Funding: "This research received no external funding" at the Institute of Animal Sciences at the University of Bonn and the Institutes of Medical Microbiology, Immunology and Parasitology, Hygiene and Public Health at the University Hospital Bonn.

Conflicts of Interest: The authors declare no conflict of interest.

References

1. Gilchrist, M.J.; Greko, C.; Wallinga, D.B.; Beran, G.W.; Riley, D.G.; Thorne, P.S. The potential role of concentrated animal feeding operations in infectious disease epidemics and antibiotic resistance. *Environ. Health Perspect.* **2007**, *115*, 313–316. [CrossRef] [PubMed]
2. Mathew, A.G.; Cissell, R.; Liamthong, S. Antibiotic resistance in bacteria associated with food animals: A United States perspective of livestock production. *Foodborne Pathog. Dis.* **2007**, *4*, 115–133. [CrossRef] [PubMed]
3. Threlfall, E.J.; Ward, L.R.; Frost, J.A.; Willshaw, G.A. The emergence and spread of antibiotic resistance in food-borne bacteria. *Int. J. Food Microbiol.* **2000**, *62*, 1–5. [CrossRef]
4. Pruden, A.; Larsson, D.G.; Amézquita, A.; Collignon, P.; Brandt, K.K.; Graham, D.W.; Lazorchak, J.M.; Suzuki, S.; Silley, P.; Snape, J.R.; et al. Management options for reducing the release of antibiotics and antibiotic resistance genes to the environment. *Environ. Health Perspect.* **2013**, *121*, 878–885. [CrossRef] [PubMed]
5. Landers, T.F.; Cohen, B.; Wittum, T.E.; Larson, E.L. A review of antibiotic use in food animals: Perspective, policy, and potential. *Public Health Rep.* **2012**, *127*, 4–22. [CrossRef] [PubMed]
6. Escudero, E.; Vinué, L.; Teshager, T.; Torres, C.; Moreno, M.A. Resistance mechanisms and farm-level distribution of fecal Escherichia coli isolates resistant to extended-spectrum cephalosporins in pigs in Spain. *Res. Vet. Sci.* **2010**, *88*, 83–87. [CrossRef] [PubMed]
7. Hansen, K.H.; Damborg, P.; Andreasen, M.; Nielsen, S.S.; Guardabassi, L. Carriage and fecal counts of cefotaxime M-producing Escherichia coli in pigs: A longitudinal study. *Appl. Environ. Microbiol.* **2013**, *79*, 794–798. [CrossRef] [PubMed]
8. Broens, E.M.; Graat, E.A.; Van der Wolf, P.J.; Van de Giessen, A.W.; De Jong, M.C. Transmission of methicillin resistant Staphylococcus aureus among pigs during transportation from farm to abattoir. *Vet. J.* **2011**, *189*, 302–305. [CrossRef] [PubMed]
9. Broens, E.M.; Graat, E.A.; van der Wolf, P.J.; van de Giessen, A.W.; van Duijkeren, E.; Wagenaar, J.A.; van Nes, A.; Mevius, D.J.; de Jong, M.C. MRSA CC398 in the pig production chain. *Prev. Vet. Med.* **2011**, *98*, 182–189. [CrossRef] [PubMed]
10. van Duijkeren, E.; Ikawaty, R.; Broekhuizen-Stins, M.J.; Jansen, M.D.; Spalburg, E.C.; de Neeling, A.J.; Allaart, J.G.; van Nes, A.; Wagenaar, J.A.; Fluit, A.C. Transmission of methicillin-resistant Staphylococcus aureus strains between different kinds of pig farms. *Vet. Microbiol.* **2008**, *126*, 383–389. [CrossRef] [PubMed]
11. Schulze-Geisthövel, S.; Schmithausen, R.; Stemmer, F.; El-Sade, M.; Hoerauf, A.; Exner, M.; Bierbaum, G.; Bekeredjian-Ding, I.; Petersen, B. Occurrence of MRSA and ESBL in different habitats and different stages of the pork production chain. In Proceedings of the DACH Epidemiologietagung "Tiergesundheit und Ökonomie", Zürich, Schweiz, 3–5 September 2014.
12. Hunter, P.A.; Dawson, S.; French, G.L.; Goossens, H.; Hawkey, P.M.; Kuijper, E.J.; Nathwani, D.; Taylor, D.J.; Teale, C.J.; Warren, R.E.; et al. Antimicrobial-resistant pathogens in animals and man: Prescribing, practices and policies. *J. Antimicrob. Chemother.* **2010**, *65* (Suppl. 1), i3–i17. [CrossRef] [PubMed]
13. Hawkey, P.M.; Jones, A.M. The changing epidemiology of resistance. *J. Antimicrob. Chemother.* **2009**, *64* (Suppl. 1), i3–i10. [CrossRef] [PubMed]
14. Li, X.Z.; Mehrotra, M.; Ghimire, S.; Adewoye, L. Beta-Lactam resistance and beta-lactamases in bacteria of animal origin. *Vet. Microbiol.* **2007**, *121*, 197–214. [CrossRef] [PubMed]
15. Müller, A.; Stephan, R.; Nüesch-Inderbinen, M. Distribution of virulence factors in ESBL-producing Escherichia coli isolated from the environment, livestock, food and humans. *Sci. Total Environ.* **2016**, *541*, 667–672. [CrossRef] [PubMed]
16. Westphal-Settele, K.; Konradi, S.; Balzer, F.; Schönfeld, J.; Schmithausen, R. The environment as a reservoir for antimicrobial resistance: A growing problem for public health? *Bundesgesundheitsblatt Gesundheitsforschung Gesundheitsschutz* **2018**, *61*, 533–542. [CrossRef] [PubMed]
17. Mehndiratta, P.L.; Bhalla, P. Use of antibiotics in animal agriculture & emergence of methicillin-resistant Staphylococcus aureus (MRSA) clones: Need to assess the impact on public health. *Indian J. Med. Res.* **2014**, *140*, 339–344. [PubMed]

18. Liverani, M.; Waage, J.; Barnett, T.; Pfeiffer, D.U.; Rushton, J.; Rudge, J.W.; Loevinsohn, M.E.; Scoones, I.; Smith, R.D.; Cooper, B.S.; et al. Understanding and managing zoonotic risk in the new livestock industries. *Environ. Health Perspect.* **2013**, *121*, 873–877. [CrossRef] [PubMed]
19. Mather, A.E.; Matthews, L.; Mellor, D.J.; Reeve, R.; Denwood, M.J.; Boerlin, P.; Reid-Smith, R.J.; Brown, D.J.; Coia, J.E.; Browning, L.M.; et al. An ecological approach to assessing the epidemiology of antimicrobial resistance in animal and human populations. *Proc. Biol. Sci.* **2012**, *279*, 1630–1639. [CrossRef] [PubMed]
20. EFSA. *Assessment of the Public Health Significance of Methicillin Resistant Staphylococcus Aureus (MRSA) in Animals and Foods*; EFSA: Parma, Italy, 2009.
21. Da Costa, P.M.; Loureiro, L.; Mator, A.J.F. Transfer of multidrug-resistant bacteria between intermingled ecological niches: The interface between humans, animals and the environment. *Int. J. Environ. Res. Public Health* **2013**, *10*, 278–294. [CrossRef] [PubMed]
22. BfR. *ESBL-Bildende Bakterien in Lebensmitteln und Deren Übertragbarkeit auf den Menschen*; Bundesinstitut für Risikobewertung (BfR): Berlin, Germany, 2010.
23. Willyard, C. The drug-resistant bacteria that pose the greatest health threats. *Nature* **2017**, *543*, 15. [CrossRef] [PubMed]
24. Guenther, S.; Ewers, C.; Wieler, L.H. Extended-spectrum beta-lactamases producing *E. coli* in wildlife, yet another form of environmental pollution? *Front. Microbiol.* **2011**, *2*, 246. [CrossRef] [PubMed]
25. Cuny, C.; Nathaus, R.; Layer, F.; Strommenger, B.; Altmann, D.; Witte, W. Nasal colonization of humans with methicillin-resistant Staphylococcus aureus (MRSA) CC398 with and without exposure to pigs. *PLoS ONE* **2009**, *4*, e6800. [CrossRef] [PubMed]
26. Deiters, C.; Günnewig, V.; Friedrich, A.W.; Mellmann, A.; Köck, R. Are cases of Methicillin-resistant Staphylococcus aureus clonal complex (CC) 398 among humans still livestock-associated? *Int. J. Med. Microbiol.* **2015**, *305*, 110–113. [CrossRef] [PubMed]
27. Van den Bogaard, A.E.; Stobberingh, E.E. Epidemiology of resistance to antibiotics. Links between animals and humans. *Int. J. Antimicrob. Agents* **2000**, *14*, 327–335. [CrossRef]
28. Marshall, B.M.; Levy, S.B. Food animals and antimicrobials: Impacts on human health. *Clin. Microbiol. Rev.* **2011**, *24*, 718–733. [CrossRef] [PubMed]
29. Aarestrup, F.M.; Hasman, H.; Agersø, Y.; Jensen, L.B.; Harksen, S.; Svensmark, B. First description of blaCTX-M-1-carrying Escherichia coli isolates in Danish primary food production. *J. Antimicrob. Chemother.* **2006**, *57*, 1258–1259. [CrossRef] [PubMed]
30. Liebana, E.; Batchelor, M.; Hopkins, K.L.; Clifton-Hadley, F.A.; Teale, C.J.; Foster, A.; Barker, L.; Threlfall, E.J.; Davies, R.H. Longitudinal farm study of extended-spectrum beta-lactamase-mediated resistance. *J. Clin. Microbiol.* **2006**, *44*, 1630–1634. [CrossRef] [PubMed]
31. De Boer, E.; Zwartkruis-Nahuis, J.T.; Wit, B.; Huijsdens, X.W.; de Neeling, A.J.; Bosch, T.; van Oosterom, R.A.; Vila, A.; Heuvelink, A.E. Prevalence of methicillin-resistant Staphylococcus aureus in meat. *Int. J. Food Microbiol.* **2009**, *134*, 52–56. [CrossRef] [PubMed]
32. Wulf, M.W.; Verduin, C.M.; van Nes, A.; Huijsdens, X.; Voss, A. Infection and colonization with methicillin resistant Staphylococcus aureus ST398 versus other MRSA in an area with a high density of pig farms. *Eur. J. Clin. Microbiol. Infect. Dis.* **2012**, *31*, 61–65. [CrossRef] [PubMed]
33. Van Cleef, B.A.; Verkade, E.J.; Wulf, M.W.; Buiting, A.G.; Voss, A.; Huijsdens, X.W.; van Pelt, W.; Mulders, M.N.; Kluytmans, J.A. Prevalence of livestock-associated MRSA in communities with high pig-densities in The Netherlands. *PLoS ONE* **2010**, *5*, e9385. [CrossRef] [PubMed]
34. Monaco, M.; Pedroni, P.; Sanchini, A.; Bonomini, A.; Indelicato, A.; Pantosti, A. Livestock-associated methicillin-resistant Staphylococcus aureus responsible for human colonization and infection in an area of Italy with high density of pig farming. *BMC Infect. Dis.* **2013**, *13*, 258. [CrossRef] [PubMed]
35. Aubry-Damon, H.; Grenet, K.; Sall-Ndiaye, P.; Che, D.; Cordeiro, E.; Bougnoux, M.E.; Rigaud, E.; Le Strat, Y.; Lemanissier, V.; Armand-Lefèvre, L.; et al. Antimicrobial resistance in commensal flora of pig farmers. *Emerg. Infect. Dis.* **2004**, *10*, 873–879. [CrossRef] [PubMed]
36. Köck, R.; Schaumburg, F.; Mellmann, A.; Köksal, M.; Jurke, A.; Becker, K.; Friedrich, A.W. Livestock-associated methicillin-resistant Staphylococcus aureus (MRSA) as causes of human infection and colonization in Germany. *PLoS ONE* **2013**, *8*, e55040. [CrossRef] [PubMed]

37. Köck, R.; Bischoff, M.; Cuny, C.; Eckmanns, T.; Fetsch, A.; Harmsen, D.; Goerge, T.; Oberheitmann, B.; Schwarz, S.; Selhorst, T.; et al. The burden of zoonotic MRSA colonization and infection in Germany. *Berl. Munch. Tierarztl. Wochenschr.* **2014**, *127*, 384–398. [PubMed]
38. Bisdorff, B.; Scholhölter, J.L.; Claußen, K.; Pulz, M.; Nowak, D.; Radon, K. MRSA-ST398 in livestock farmers and neighbouring residents in a rural area in Germany. *Epidemiol. Infect.* **2012**, *140*, 1800–1808. [CrossRef] [PubMed]
39. Johnson, J.R.; Kuskowski, M.A.; Smith, K.; O'Bryan, T.T.; Tatini, S. Antimicrobial-resistant and extraintestinal pathogenic Escherichia coli in retail foods. *J. Infect. Dis.* **2005**, *191*, 1040–1049. [CrossRef] [PubMed]
40. Overdevest, I.; Willemsen, I.; Rijnsburger, M.; Eustace, A.; Xu, L.; Hawkey, P.; Heck, M.; Savelkoul, P.; Vandenbroucke-Grauls, C.; van der Zwaluw, K.; et al. Extended-spectrum β-lactamase genes of Escherichia coli in chicken meat and humans, The Netherlands. *Emerg. Infect. Dis.* **2011**, *17*, 1216–1222. [CrossRef] [PubMed]
41. Jakobsen, L.; Spangholm, D.J.; Pedersen, K.; Jensen, L.B.; Emborg, H.D.; Agersø, Y.; Aarestrup, F.M.; Hammerum, A.M.; Frimodt-Møller, N. Broiler chickens, broiler chicken meat, pigs and pork as sources of ExPEC related virulence genes and resistance in Escherichia coli isolates from community-dwelling humans and UTI patients. *Int. J. Food Microbiol.* **2010**, *142*, 264–272. [CrossRef] [PubMed]
42. Weese, J.S.; Rousseau, J. Attempted eradication of methicillin-resistant staphylococcus aureus colonisation in horses on two farms. *Equine Vet. J.* **2005**, *37*, 510–514. [CrossRef] [PubMed]
43. Friese, A.; Schulz, J.; Hoehle, L.; Fetsch, A.; Tenhagen, B.A.; Hartung, J.; Roesler, U. Occurrence of MRSA in air and housing environment of pig barns. *Vet. Microbiol.* **2012**, *158*, 129–135. [CrossRef] [PubMed]
44. Hu, Y.Y.; Cai, J.C.; Zhou, H.W.; Chi, D.; Zhang, X.F.; Chen, W.L.; Zhang, R.; Chen, G.X. Molecular typing of CTX-M-producing escherichia coli isolates from environmental water, swine feces, specimens from healthy humans, and human patients. *Appl. Environ. Microbiol.* **2013**, *79*, 5988–5996. [CrossRef] [PubMed]
45. Blake, D.P.; Hillman, K.; Fenlon, D.R.; Low, J.C. Transfer of antibiotic resistance between commensal and pathogenic members of the Enterobacteriaceae under ileal conditions. *J. Appl. Microbiol.* **2003**, *95*, 428–436. [CrossRef] [PubMed]
46. Martinez, J.L. The role of natural environments in the evolution of resistance traits in pathogenic bacteria. *Proc. Biol. Sci.* **2009**, *276*, 2521–2530. [CrossRef] [PubMed]
47. Pietsch, M.; Irrgang, A.; Roschanski, N.; Brenner Michael, G.; Hamprecht, A.; Rieber, H.; Käsbohrer, A.; Schwarz, S.; Rösler, U.; Kreienbrock, L.; et al. Whole genome analyses of CMY-2-producing Escherichia coli isolates from humans, animals and food in Germany. *BMC Genom.* **2018**, *19*, 601. [CrossRef] [PubMed]
48. Hammerum, A.M.; Heuer, O.E. Human health hazards from antimicrobial-resistant *Escherichia coli* of animal origin. *Clin. Infect. Dis.* **2009**, *48*, 916–921. [CrossRef] [PubMed]
49. Wilcox, B.; Kueffer, C. Transdisciplinarity in EcoHealth: Status and future prospects. *Ecohealth* **2008**, *5*, 1–3. [CrossRef] [PubMed]
50. Zinsstag, J.; Schelling, E.; Waltner-Toews, D.; Tanner, M. From "one medicine" to "one health" and systemic approaches to health and well-being. *Prev. Vet. Med.* **2011**, *101*, 148–156. [CrossRef] [PubMed]
51. Haxton, E.; Sinigoj, S.; Betson, M.; Riviere-Cinnamond, A. Evidence-based added value of One Health: The Network for Evaluation of One Health (NEOH). *One Health Sweden* **2015**. Available online: http://neoh.onehealthglobal.net/wp-content/uploads/sites/2/2015/01/NEOH-UPDATES-NOV-2015.pdf (accessed on 23 October 2018). [CrossRef] [PubMed]
52. Köck, R.; Kreienbrock, L.; van Duijkeren, E.; Schwarz, S. Antimicrobial resistance at the interface of human and veterinary medicine. *Vet. Microbiol.* **2017**, *200*, 1–5. [CrossRef] [PubMed]
53. Denis, O.; Suetens, C.; Hallin, M.; Catry, B.; Ramboer, I.; Dispas, M.; Willems, G.; Gordts, B.; Butaye, P.; Struelens, M.J. Methicillin-resistant Staphylococcus aureus ST398 in swine farm personnel, Belgium. *Emerg. Infect. Dis.* **2009**, *15*, 1098–1101. [CrossRef] [PubMed]
54. Alanis, A.J. Resistance to antibiotics: Are we in the post-antibiotic era? *Arch. Med. Res.* **2005**, *36*, 697–705. [CrossRef] [PubMed]
55. Smet, A.; Martel, A.; Persoons, D.; Dewulf, J.; Heyndrickx, M.; Herman, L.; Haesebrouck, F.; Butaye, P. Broad-spectrum β-lactamases among Enterobacteriaceae of animal origin: Molecular aspects, mobility and impact on public health. *FEMS Microbiol. Rev.* **2010**, *34*, 295–316. [CrossRef] [PubMed]

56. GERMAP. *GERMAPAntibiotikaresistenz und -verbrauchBericht über den Antibiotikaverbrauch und die Verbreitung von Antibiotikaresistenzen in der Human- und VeterinÄrmedizin in Deutschland*; GERMAP/PEG: Langen, Germany, 2015.
57. BMEL. *Lagebild zum Antibiotikaeinsatz*; BMELV: Bonn, Germany, 2017.
58. BVL. *Zweite Datenerhebung zur Antibiotikagabe in der Tiermedizin*; BVL: Berlin, Germany, 2017.
59. Callens, B.; Persoons, D.; Maes, D.; Laanen, M.; Postma, M.; Boyen, F.; Haesebrouck, F.; Butaye, P.; Catry, B.; Dewulf, J. Prophylactic and metaphylactic antimicrobial use in Belgian fattening pig herds. *Prev. Vet. Med.* **2012**, *106*, 53–62. [CrossRef] [PubMed]
60. Economou, V.; Gousia, P. Agriculture and food animals as a source of antimicrobial-resistant bacteria. *Infect. Drug Resist.* **2015**, *8*, 49–61. [CrossRef] [PubMed]
61. Trolldernier, H. Zu Grundlagen der Antibiotikaresistenz. *Mh Vet. Med.* **1975**, *41*, 234–237.
62. Hering, J.; Hille, K.; Frömke, C.; von Münchhausen, C.; Hartmann, M.; Schneider, B.; Friese, A.; Roesler, U.; Merle, R.; Kreienbrock, L. Prevalence and potential risk factors for the occurrence of cefotaxime resistant Escherichia coli in German fattening pig farms—A cross-sectional study. *Prev. Vet. Med.* **2014**, *116*, 129–137. [CrossRef] [PubMed]
63. Gothe, C.; Petersen, B. *Qualitätsmerkmal Tierwohl*; Rhenish Friedrich-Wilhelms University: Bonn, Germany, 2018.
64. TierSchG. Tierschutzgesetz in der Fassung der Bekanntmachung vom 18. Mai 2006. (BGBl. I S. 1206, 1313), das Zuletzt Durch Artikel 141 des Gesetzes vom. März 2017 (BGBl. I S. 626) Geändert Worden ist. 1972. Available online: https://www.tierhyg.vetmed.uni-muenchen.de/tierschutz/tierschutzlinks/ts_downloads/tschg.pdf (accessed on 23 October 2018).
65. Bollwahn, W. Grauer Arzneimittelmarkt. Aufklären, keine neuen Gesetze. *Top Agrar* **1981**, *9*, 63–67.
66. Aarestrup, F. Sustainable farming: Get pigs off antibiotics. *Nature* **2012**, *486*, 465–466. [CrossRef] [PubMed]
67. Quinn, P.; Markey, B.; ME, C.; Donnelly, W.; Leonard, C. *Veterinary Microbiology and Microbial Disease*; Blackwell Sciences Ltd.: Oxford, UK, 2002.
68. Kenner, J.; O'Connor, T.; Piantanida, N.; Fishbain, J.; Eberly, B.; Viscount, H.; Uyehara, C.; Hospenthal, D. Rates of carriage of methicillin-resistant and methicillin-susceptible Staphylococcus aureus in an outpatient population. *Infect. Control Hosp. Epidemiol.* **2003**, *24*, 439–444. [CrossRef] [PubMed]
69. Ito, T.; Okuma, K.; Ma, X.X.; Yuzawa, H.; Hiramatsu, K. Insights on antibiotic resistance of Staphylococcus aureus from its whole genome: Genomic island SCC. *Drug Resist. Updat.* **2003**, *6*, 41–52. [CrossRef]
70. Hiramatsu, K.; Cui, L.; Kuroda, M.; Ito, T. The emergence and evolution of methicillin-resistant Staphylococcus aureus. *Trends Microbiol.* **2001**, *9*, 486–493. [CrossRef]
71. Pinho, M.G.; Filipe, S.R.; de Lencastre, H.; Tomasz, A. Complementation of the essential peptidoglycan transpeptidase function of penicillin-binding protein 2 (PBP2) by the drug resistance protein PBP2A in Staphylococcus aureus. *J. Bacteriol.* **2001**, *183*, 6525–6531. [CrossRef] [PubMed]
72. Alm, R.A.; McLaughlin, R.E.; Kos, V.N.; Sader, H.S.; Iaconis, J.P.; Lahiri, S.D. Analysis of Staphylococcus aureus clinical isolates with reduced susceptibility to ceftaroline: An epidemiological and structural perspective. *J. Antimicrob. Chemother.* **2014**, *69*, 2065–2075. [CrossRef] [PubMed]
73. Moisan, H.; Pruneau, M.; Malouin, F. Binding of ceftaroline to penicillin-binding proteins of Staphylococcus aureus and Streptococcus pneumoniae. *J. Antimicrob. Chemother.* **2010**, *65*, 713–716. [CrossRef] [PubMed]
74. Lovering, A.L.; Gretes, M.C.; Safadi, S.S.; Danel, F.; de Castro, L.; Page, M.G.; Strynadka, N.C. Structural insights into the anti-methicillin-resistant Staphylococcus aureus (MRSA) activity of ceftobiprole. *J. Biol. Chem.* **2012**, *287*, 32096–32102. [CrossRef] [PubMed]
75. De Lencastre, H.; Chung, M.; Westh, H. Archaic strains of methicillin-resistant Staphylococcus aureus: Molecular and microbiological properties of isolates from the 1960s in Denmark. *Microb. Drug Resist.* **2000**, *6*, 1–10. [CrossRef] [PubMed]
76. Jevons, M.P. "Celbenin" resistant staphylococci. *Br. Med. J.* **1961**, *1*, 124–125. [CrossRef]
77. Devriese, L.A.; Van Damme, L.R.; Fameree, L. Methicillin (cloxacillin)-resistant Staphylococcus aureus strains isolated from bovine mastitis cases. *Zentralbl. Veterinarmed. B* **1972**, *19*, 598–605. [CrossRef] [PubMed]
78. Price, L.B.; Stegger, M.; Hasman, H.; Aziz, M.; Larsen, J.; Andersen, P.S.; Pearson, T.; Waters, A.E.; Foster, J.T.; Schupp, J.; et al. Staphylococcus aureus CC398: Host adaptation and emergence of methicillin resistance in livestock. *MBio* **2012**, *3*. [CrossRef] [PubMed]

79. Catry, B.; Van Duijkeren, E.; Pomba, M.C.; Greko, C.; Moreno, M.A.; Pyörälä, S.; Ruzauskas, M.; Sanders, P.; Threlfall, E.J.; Ungemach, F.; et al. Reflection paper on MRSA in food-producing and companion animals: Epidemiology and control options for human and animal health. *Epidemiol. Infect.* **2010**, *138*, 626–644. [CrossRef] [PubMed]
80. Witte, W.; Strommenger, B.; Stanek, C.; Cuny, C. Methicillin-resistant Staphylococcus aureus ST398 in humans and animals, Central Europe. *Emerg. Infect. Dis.* **2007**, *13*, 255–258. [CrossRef] [PubMed]
81. Fitzgerald, J.R. Livestock-associated Staphylococcus aureus: Origin, evolution and public health threat. *Trends Microbiol.* **2012**, *20*, 192–198. [CrossRef] [PubMed]
82. Cuny, C.; Friedrich, A.; Kozytska, S.; Layer, F.; Nübel, U.; Ohlsen, K.; Strommenger, B.; Walther, B.; Wieler, L.; Witte, W. Emergence of methicillin-resistant Staphylococcus aureus (MRSA) in different animal species. *Int. J. Med. Microbiol.* **2010**, *300*, 109–117. [CrossRef] [PubMed]
83. Lowder, B.V.; Guinane, C.M.; Ben Zakour, N.L.; Weinert, L.A.; Conway-Morris, A.; Cartwright, R.A.; Simpson, A.J.; Rambaut, A.; Nübel, U.; Fitzgerald, J.R. Recent human-to-poultry host jump, adaptation, and pandemic spread of Staphylococcus aureus. *Proc. Natl. Acad. Sci. USA* **2009**, *106*, 19545–19550. [CrossRef] [PubMed]
84. Bradford, P.A. Extended-spectrum beta-lactamases in the 21st century: Characterization, epidemiology, and detection of this important resistance threat. *Clin. Microbiol. Rev.* **2001**, *14*, 933–951. [CrossRef] [PubMed]
85. Bradford, P.A.; Petersen, P.J.; Fingerman, I.M.; White, D.G. Characterization of expanded-spectrum cephalosporin resistance in E. coli isolates associated with bovine calf diarrhoeal disease. *J. Antimicrob. Chemother.* **1999**, *44*, 607–610. [CrossRef] [PubMed]
86. Bradford, P.A.; Urban, C.; Mariano, N.; Projan, S.J.; Rahal, J.J.; Bush, K. Imipenem resistance in Klebsiella pneumoniae is associated with the combination of ACT-1, a plasmid-mediated AmpC beta-lactamase, and the foss of an outer membrane protein. *Antimicrob. Agents Chemother.* **1997**, *41*, 563–569. [CrossRef] [PubMed]
87. Paterson, D.L.; Mulazimoglu, L.; Casellas, J.M.; Ko, W.C.; Goossens, H.; Von Gottberg, A.; Mohapatra, S.; Trenholme, G.M.; Klugman, K.P.; McCormack, J.G.; et al. Epidemiology of ciprofloxacin resistance and its relationship to extended-spectrum beta-lactamase production in Klebsiella pneumoniae isolates causing bacteremia. *Clin. Infect. Dis.* **2000**, *30*, 473–478. [CrossRef] [PubMed]
88. Kaase, M. Carbapenemases in gram-negative bacteria. Current data and trends of resistance resulting from the work of national reference centres. *Bundesgesundheitsblatt Gesundheitsforschung Gesundheitsschutz* **2012**, *55*, 1401–1404. [CrossRef] [PubMed]
89. Walsh, T.R. Emerging carbapenemases: A global perspective. *Int. J. Antimicrob. Agents* **2010**, *36* (Suppl. 3), S8–S14. [CrossRef]
90. Patel, G.; Bonomo, R.A. Status report on carbapenemases: Challenges and prospects. *Expert Rev. Anti Infect. Ther.* **2011**, *9*, 555–570. [CrossRef] [PubMed]
91. Livermore, D.M. Beta-Lactamases in laboratory and clinical resistance. *Clin. Microbiol. Rev.* **1995**, *8*, 557–584. [CrossRef] [PubMed]
92. Tham, J.; Walder, M.; Melander, E.; Odenholt, I. Prevalence of extended-spectrum beta-lactamase-producing bacteria in food. *Infect. Drug Resist.* **2012**, *5*, 143–147. [CrossRef] [PubMed]
93. Pitout, J.D. Infections with extended-spectrum beta-lactamase-producing enterobacteriaceae: Changing epidemiology and drug treatment choices. *Drugs* **2010**, *70*, 313–333. [CrossRef] [PubMed]
94. Naseer, U.; Sundsfjord, A. The CTX-M conundrum: Dissemination of plasmids and *Escherichia coli* clones. *Microb. Drug Resist.* **2011**, *17*, 83–97. [CrossRef] [PubMed]
95. Pitout, J.D.; Gregson, D.B.; Church, D.L.; Elsayed, S.; Laupland, K.B. Community-wide outbreaks of clonally related CTX-M-14 beta-lactamase-producing Escherichia coli strains in the Calgary health region. *J. Clin. Microbiol.* **2005**, *43*, 2844–2849. [CrossRef] [PubMed]
96. Smet, A.; Martel, A.; Persoons, D.; Dewulf, J.; Heyndrickx, M.; Catry, B.; Herman, L.; Haesebrouck, F.; Butaye, P. Diversity of extended-spectrum beta-lactamases and class C beta-lactamases among cloacal Escherichia coli Isolates in Belgian broiler farms. *Antimicrob. Agents Chemother.* **2008**, *52*, 1238–1243. [CrossRef] [PubMed]

97. Tian, G.B.; Wang, H.N.; Zou, L.K.; Tang, J.N.; Zhao, Y.W.; Ye, M.Y.; Tang, J.Y.; Zhang, Y.; Zhang, A.Y.; Yang, X.; et al. Detection of CTX-M-15, CTX-M-22, and SHV-2 extended-spectrum beta-lactamases (ESBLs) in Escherichia coli fecal-sample isolates from pig farms in China. *Foodborne Pathog. Dis.* **2009**, *6*, 297–304. [CrossRef] [PubMed]
98. Geser, N.; Stephan, R.; Korczak, B.M.; Beutin, L.; Hächler, H. Molecular identification of extended-spectrum-β-lactamase genes from Enterobacteriaceae isolated from healthy human carriers in Switzerland. *Antimicrob. Agents Chemother.* **2012**, *56*, 1609–1612. [CrossRef] [PubMed]
99. Kiiru, J.; Kariuki, S.; Goddeeris, B.M.; Butaye, P. Analysis of β-lactamase phenotypes and carriage of selected β-lactamase genes among *Escherichia coli* strains obtained from Kenyan patients during an 18-year period. *BMC Microbiol.* **2012**, *12*, 155. [CrossRef] [PubMed]
100. Shah, A.A.; Hasan, F.; Ahmed, S.; Hameed, A. Characteristics, epidemiology and clinical importance of emerging strains of Gram-negative bacilli producing extended-spectrum beta-lactamases. *Res. Microbiol.* **2004**, *155*, 409–421. [CrossRef] [PubMed]
101. Frost, L.S.; Leplae, R.; Summers, A.O.; Toussaint, A. Mobile genetic elements: The agents of open source evolution. *Nat. Rev. Microbiol.* **2005**, *3*, 722–732. [CrossRef] [PubMed]
102. Malachowa, N.; DeLeo, F.R. Mobile genetic elements of Staphylococcus aureus. *Cell. Mol. Life Sci.* **2010**, *67*, 3057–3071. [CrossRef] [PubMed]
103. Allen, H.K.; Donato, J.; Wang, H.H.; Cloud-Hansen, K.A.; Davies, J.; Handelsman, J. Call of the wild: Antibiotic resistance genes in natural environments. *Nat. Rev. Microbiol.* **2010**, *8*, 251–259. [CrossRef] [PubMed]
104. Valentin, L.; Sharp, H.; Hille, K.; Seibt, U.; Fischer, J.; Pfeifer, Y.; Michael, G.B.; Nickel, S.; Schmiedel, J.; Falgenhauer, L.; et al. Subgrouping of ESBL-producing Escherichia coli from animal and human sources: An approach to quantify the distribution of ESBL types between different reservoirs. *Int. J. Med. Microbiol.* **2014**, *304*, 805–816. [CrossRef] [PubMed]
105. Nehring, S.; Albrecht, U. Biotop, Habitat, Mikrohabitat—Ein Diskussionsbeitrag zur Begriffsdefinition. *Lauterbonia* **2000**, *38*, 75–84.
106. Lötzli, K. Ökosystem. In *Handbuch zur Ökologie*; Kuttler: Berlin, Germany, 1993; pp. 288–295. Available online: https://www.zobodat.at/pdf/Lauterbornia_2000_38_0075-0084.pdf (accessed on 30 October 2018).
107. Broens, E.M. *Livestock-Associated Methicillin Resistant Staphylococcus Aureus in Pigs—Prevalence, Risk Factors and Transmission Dynamics*; Wageningen University: Wageningen, The Netherlands, 2011.
108. Armand-Lefevre, L.; Ruimy, R.; Andremont, A. Clonal comparison of Staphylococcus aureus isolates from healthy pig farmers, human controls, and pigs. *Emerg. Infect. Dis.* **2005**, *11*, 711–714. [CrossRef] [PubMed]
109. Graveland, H.; Wagenaar, J.A.; Bergs, K.; Heesterbeek, H.; Heederik, D. Persistence of livestock associated MRSA CC398 in humans is dependent on intensity of animal contact. *PLoS ONE* **2011**, *6*, e16830. [CrossRef] [PubMed]
110. Lozano, C.; Aspiroz, C.; Charlez, L.; Gómez-Sanz, E.; Toledo, M.; Zarazaga, M.; Torres, C. Skin lesion by methicillin-resistant Staphylococcus aureus ST398-t1451 in a Spanish pig farmer: Possible transmission from animals to humans. *Vector Borne Zoonotic Dis.* **2011**, *11*, 605–607. [CrossRef] [PubMed]
111. Gilbert, M.J.; Bos, M.E.; Duim, B.; Urlings, B.A.; Heres, L.; Wagenaar, J.A.; Heederik, D.J. Livestock-associated MRSA ST398 carriage in pig slaughterhouse workers related to quantitative environmental exposure. *Occup. Environ. Med.* **2012**, *69*, 472–478. [CrossRef] [PubMed]
112. Agersø, Y.; Vigre, H.; Cavaco, L.M.; Josefsen, M.H. Comparison of air samples, nasal swabs, ear-skin swabs and environmental dust samples for detection of methicillin-resistant Staphylococcus aureus (MRSA) in pig herds. *Epidemiol. Infect.* **2014**, *142*, 1727–1736. [CrossRef] [PubMed]
113. Agersø, Y.; Aarestrup, F.M.; Pedersen, K.; Seyfarth, A.M.; Struve, T.; Hasman, H. Prevalence of extended-spectrum cephalosporinase (ESC)-producing Escherichia coli in Danish slaughter pigs and retail meat identified by selective enrichment and association with cephalosporin usage. *J. Antimicrob. Chemother.* **2012**, *67*, 582–588. [CrossRef] [PubMed]
114. Jakobsen, L.; Kurbasic, A.; Skjøt-Rasmussen, L.; Ejrnaes, K.; Porsbo, L.J.; Pedersen, K.; Jensen, L.B.; Emborg, H.D.; Agersø, Y.; Olsen, K.E.; et al. Escherichia coli isolates from broiler chicken meat, broiler chickens, pork, and pigs share phylogroups and antimicrobial resistance with community-dwelling humans and patients with urinary tract infection. *Foodborne Pathog. Dis.* **2010**, *7*, 537–547. [CrossRef] [PubMed]

115. Guardabassi, L. Sixty years of antimicrobial use in animals: what is next? *Vet. Rec.* **2013**, *173*, 599–603. [CrossRef] [PubMed]
116. Smet, A.; Martel, A.; Persoons, D.; Dewulf, J.; Heyndrickx, M.; Cloeckaert, A.; Praud, K.; Claeys, G.; Catry, B.; Herman, L.; et al. Comparative analysis of extended-spectrum-β-lactamase-carrying plasmids from different members of Enterobacteriaceae isolated from poultry, pigs and humans: Evidence for a shared β-lactam resistance gene pool? *J. Antimicrob. Chemother.* **2009**, *63*, 1286–1288. [CrossRef] [PubMed]
117. Wulf, M.W.; Tiemersma, E.; Kluytmans, J.; Bogaers, D.; Leenders, A.C.; Jansen, M.W.; Berkhout, J.; Ruijters, E.; Haverkate, D.; Isken, M.; et al. MRSA carriage in healthcare personnel in contact with farm animals. *J. Hosp. Infect.* **2008**, *70*, 186–190. [CrossRef] [PubMed]
118. Vanderhaeghen, W.; Hermans, K.; Haesebrouck, F.; Butaye, P. Methicillin-resistant Staphylococcus aureus (MRSA) in food production animals. *Epidemiol. Infect.* **2010**, *138*, 606–625. [CrossRef] [PubMed]
119. Köck, R.; Harlizius, J.; Bressan, N.; Laerberg, R.; Wieler, L.H.; Witte, W.; Deurenberg, R.H.; Voss, A.; Becker, K.; Friedrich, A.W. Prevalence and molecular characteristics of methicillin-resistant Staphylococcus aureus (MRSA) among pigs on German farms and import of livestock-related MRSA into hospitals. *Eur. J. Clin. Microbiol. Infect. Dis.* **2009**, *28*, 1375–1382. [CrossRef] [PubMed]
120. Kraemer, J.G.; Pires, J.; Kueffer, M.; Semaani, E.; Endimiani, A.; Hilty, M.; Oppliger, A. Prevalence of extended-spectrum β-lactamase-producing Enterobacteriaceae and Methicillin-Resistant Staphylococcus aureus in pig farms in Switzerland. *Sci. Total Environ.* **2017**, *603–604*, 401–405. [CrossRef] [PubMed]
121. Schouten, J.M.; Bouwknegt, M.; van de Giessen, A.W.; Frankena, K.; De Jong, M.C.; Graat, E.A. Prevalence estimation and risk factors for *Escherichia coli* O157 on Dutch dairy farms. *Prev. Vet. Med.* **2004**, *64*, 49–61. [CrossRef] [PubMed]
122. Correira-Gomes, C.; Economou, T.; Bailey, T.; Brazdil, P.; Vieira-Pinto, M.; Niza-Ribiero, J. Simulation Model for *Salmonella* Typhimurium on a Farrow-to-Finish herd. In Proceedings of the 10th International Conference on the Epidemiology and Control of Biological, Chemical and Physical Hazards in Pigs and Pork, Lisbon, Portugal, 9–12 September 2013.
123. Grøntvedt, C.A.; Elstrøm, P.; Stegger, M.; Skov, R.L.; Skytt Andersen, P.; Larssen, K.W.; Urdahl, A.M.; Angen, Ø.; Larsen, J.; Åmdal, S.; et al. Methicillin-Resistant Staphylococcus aureus CC398 in humans and pigs in Norway: A "One Health" perspective on introduction and transmission. *Clin. Infect. Dis.* **2016**, *63*, 1431–1438. [CrossRef] [PubMed]
124. Cernicchiaro, N.; Pearl, D.L.; Ghimire, S.; Gyles, C.L.; Johnson, R.P.; LeJeune, J.T.; Ziebell, K.; McEwen, S.A. Risk factors associated with Escherichia coli O157:H7 in Ontario beef cow-calf operations. *Prev. Vet. Med.* **2009**, *92*, 106–115. [CrossRef] [PubMed]
125. Snow, L.C.; Warner, R.G.; Cheney, T.; Wearing, H.; Stokes, M.; Harris, K.; Teale, C.J.; Coldham, N.G. Risk factors associated with extended spectrum beta-lactamase Escherichia coli (CTX-M) on dairy farms in North West England and North Wales. *Prev. Vet. Med.* **2012**, *106*, 225–234. [CrossRef] [PubMed]
126. Schmithausen, R.M.; Schulze-Geisthoevel, S.V.; Stemmer, F.; El-Jade, M.; Reif, M.; Hack, S.; Meilaender, A.; Montabauer, G.; Fimmers, R.; Parcina, M.; et al. Analysis of transmission of MRSA and ESBL-E among pigs and farm personnel. *PLoS ONE* **2015**, *10*, e0138173. [CrossRef] [PubMed]
127. Broens, E.M.; Graat, E.A.; Van der Wolf, P.J.; Van de Giessen, A.W.; De Jong, M.C. Prevalence and risk factor analysis of livestock associated MRSA-positive pig herds in The Netherlands. *Prev. Vet. Med.* **2011**, *102*, 41–49. [CrossRef] [PubMed]
128. Schmithausen, R.M.; Kellner, S.R.; Schulze-Geisthoevel, S.V.; Hack, S.; Engelhart, S.; Bodenstein, I.; Al-Sabti, N.; Reif, M.; Fimmers, R.; Körber-Irrgang, B.; et al. Eradication of methicillin-resistant Staphylococcus aureus and of Enterobacteriaceae expressing extended-spectrum beta-lactamases on a model pig farm. *Appl. Environ. Microbiol.* **2015**, *81*, 7633–7643. [CrossRef] [PubMed]
129. Mannion, C.; Leonard, F.C.; Lynch, P.B.; Egan, J. Efficacy of cleaning and disinfection on pig farms in Ireland. *Vet. Rec.* **2007**, *161*, 371–375. [CrossRef] [PubMed]
130. Heinemann, C.; Steinhoff-Wagner, J.; Petersen, B. Evaluation of methods for determining cleaning performances in pig stables. *J. Anim. Sci.* **2017**, *95*, 48. [CrossRef]
131. Wellington, E.M.; Boxall, A.B.; Cross, P.; Feil, E.J.; Gaze, W.H.; Hawkey, P.M.; Johnson-Rollings, A.S.; Jones, D.L.; Lee, N.M.; Otten, W.; et al. The role of the natural environment in the emergence of antibiotic resistance in gram-negative bacteria. *Lancet Infect. Dis.* **2013**, *13*, 155–165. [CrossRef]

132. Young, J.S.; Gormley, E.; Wellington, E.M. Molecular detection of Mycobacterium bovis and Mycobacterium bovis BCG (Pasteur) in soil. *Appl. Environ. Microbiol.* **2005**, *71*, 1946–1952. [CrossRef] [PubMed]
133. Gaze, W.H.; Burroughs, N.; Gallagher, M.P.; Wellington, E.M. Interactions between Salmonella typhimurium and Acanthamoeba polyphaga, and observation of a new mode of intracellular growth within contractile vacuoles. *Microb. Ecol.* **2003**, *46*, 358–369. [CrossRef] [PubMed]
134. Beuchat, L.R. Ecological factors influencing survival and growth of human pathogens on raw fruits and vegetables. *Microbes Infect.* **2002**, *4*, 413–423. [CrossRef]
135. Hiroi, M.; Harada, T.; Kawamori, F.; Takahashi, N.; Kanda, T.; Sugiyama, K.; Masuda, T.; Yoshikawa, Y.; Ohashi, N. A survey of β-lactamase-producing Escherichia coli in farm animals and raw retail meat in Shizuoka Prefecture, Japan. *Jpn. J. Infect. Dis.* **2011**, *64*, 153–155. [PubMed]
136. Blanc, V.; Mesa, R.; Saco, M.; Lavilla, S.; Prats, G.; Miró, E.; Navarro, F.; Cortés, P.; Llagostera, M. ESBL- and plasmidic class C beta-lactamase-producing E. coli strains isolated from poultry, pig and rabbit farms. *Vet. Microbiol.* **2006**, *118*, 299–304. [CrossRef] [PubMed]
137. Mesa, R.J.; Blanc, V.; Blanch, A.R.; Cortés, P.; González, J.J.; Lavilla, S.; Miró, E.; Muniesa, M.; Saco, M.; Tórtola, M.T.; et al. Extended-spectrum beta-lactamase-producing Enterobacteriaceae in different environments (humans, food, animal farms and sewage). *J. Antimicrob. Chemother.* **2006**, *58*, 211–215. [CrossRef] [PubMed]
138. Heuer, H.; Schmitt, H.; Smalla, K. Antibiotic resistance gene spread due to manure application on agricultural fields. *Curr. Opin. Microbiol.* **2011**, *14*, 236–243. [CrossRef] [PubMed]
139. Müller, W.; Wieser, P.; Woiwode, J. Zur Größe koloniebildender Einheiten in der Stallluft. *Berl. Münch. Tierärztl. Wschr.* **1977**, *90*, 6–11.
140. Green, C.F.; Gibbs, S.G.; Tarwater, P.M.; Mota, L.C.; Scarpino, P.V. Bacterial plume emanating from the air surrounding swine confinement operations. *J. Occup. Environ. Hyg.* **2006**, *3*, 9–15. [CrossRef] [PubMed]
141. Hartung, J.; Seedorf, J.; Trickl, T.; Gronauer, H. Emission of particulates from a pig farm with central air exhaust in the pig stall. *Deutsche Tierarztliche Wochenschrift* **1998**, *105*, 244–245. [PubMed]
142. Zucker, B.A.; Trojan, S.; Müller, W. Airborne gram-negative bacterial flora in animal houses. *J. Vet. Med. B Infect. Dis. Vet. Public Health* **2000**, *47*, 37–46. [CrossRef] [PubMed]
143. Künneken, J. *Simulation der Tier-Bakterium-Umwelt-Interaktion im Biotop "Abferkelstall"*; University of Bonn: Bonn, Germany, 1991.
144. Hartung, J.; Whyte, R. Erfassung und Bewertung von Luftverunreinigungen in der Nutztierhaltung. *Atemwegs-Lungenkr.* **1994**, *20*, 17–25.
145. Venglovský, J.; Gregová, G.; Kmeť, V.; Sasáková, N. Detection of airborne microorganisms and antibiotic resistance from animal housing facilities. In Proceedings of the XVth International Congress of the International Society for Animal Hygiene, Vienna, Austria, 12–16 August 2011; pp. 813–815.
146. Hilliger, H.G. Determination of bacterial flora in stall air. *Zentralbl Veterinarmed B* **1984**, *31*, 493–504. [CrossRef] [PubMed]
147. Benbough, J.E. Death mechanisms in airborne Escherichia coli. *J. Gen. Microbiol.* **1967**, *47*, 325–333. [CrossRef] [PubMed]
148. Gundermann, K.O. Life-span of bacterial strains in dust as influenced by various degrees of air humidity. *Zentralbl Bakteriol Orig B* **1972**, *156*, 422–429. [PubMed]
149. Hoffmann, D.H. Experimental endogenous inflammation of the internal eye by Candida albicans. Ophthalmoscopic, histological and microbiological studies on the course of the infection in rabbits. *Ophthalmologica* **1966**, *151*, 1–84. [PubMed]
150. Weese, J.S.; DaCosta, T.; Button, L.; Goth, K.; Ethier, M.; Boehnke, K. Isolation of methicillin-resistant Staphylococcus aureus from the environment in a veterinary teaching hospital. *J. Vet. Intern. Med.* **2004**, *18*, 468–470. [CrossRef] [PubMed]
151. Gibbs, S.G.; Green, C.F.; Tarwater, P.M.; Scarpino, P.V. Airborne antibiotic resistant and nonresistant bacteria and fungi recovered from two swine herd confined animal feeding operations. *J. Occup. Environ. Hyg.* **2004**, *1*, 699–706. [CrossRef] [PubMed]
152. Bos, M.E.; Verstappen, K.M.; van Cleef, B.A.; Dohmen, W.; Dorado-García, A.; Graveland, H.; Duim, B.; Wagenaar, J.A.; Kluytmans, J.A.; Heederik, D.J. Transmission through air as a possible route of exposure for MRSA. *J. Expo. Sci. Environ. Epidemiol.* **2016**, *26*, 263–269. [CrossRef] [PubMed]

153. Heinemann, C.; Schmithausen, R.M.; Sib, E.; Meyer, I.; Petersen, B.; Steinhoff-Wagner, J. Preventive effect of nasal lavage with physiologic saline on the colonization with MRSA after working in porcine stable. *J. Anim. Sci.* **2017**, *95*, 250. [CrossRef]
154. Bos, M.E.; Taverne, F.J.; van Geijlswijk, I.M.; Mouton, J.W.; Mevius, D.J.; Heederik, D.J.; SDa, N.V.M.A. Consumption of antimicrobials in pigs, veal calves, and broilers in the Netherlands: Quantitative results of nationwide collection of data in 2011. *PLoS ONE* **2013**, *8*, e77525. [CrossRef] [PubMed]
155. Wenke, C.; Pospiech, J.; Reutter, T.; Altmann, B.; Truyen, U.; Speck, S. Impact of different supply air and recirculating air filtration systems on stable climate, animal health, and performance of fattening pigs in a commercial pig farm. *PLoS ONE* **2018**, *13*, e0194641. [CrossRef] [PubMed]
156. Laube, H.; Friese, A.; von Salviati, C.; Guerra, B.; Rösler, U. Transmission of ESBL/AmpC-producing Escherichia coli from broiler chicken farms to surrounding areas. *Vet. Microbiol.* **2014**, *172*, 519–527. [CrossRef] [PubMed]
157. Von Salviati, C.; Laube, H.; Guerra, B.; Roesler, U.; Friese, A. Emission of ESBL/AmpC-producing Escherichia coli from pig fattening farms to surrounding areas. *Vet. Microbiol.* **2015**, *175*, 77–84. [CrossRef] [PubMed]
158. Dohmen, W.; Schmitt, H.; Bonten, M.; Heederik, D. Air exposure as a possible route for ESBL in pig farmers. *Environ. Res.* **2017**, *155*, 359–364. [CrossRef] [PubMed]
159. Laube, H.; Friese, A.; von Salviati, C.; Guerra, B.; Käsbohrer, A.; Kreienbrock, L.; Roesler, U. Longitudinal monitoring of extended-spectrum-beta-lactamase/AmpC-producing *Escherichia coli* at German broiler chicken fattening farms. *Appl. Environ. Microbiol.* **2013**, *79*, 4815–4820. [CrossRef] [PubMed]
160. Hoffmann, A. *Presence of CTX-M-1 Group Extended-Spectrum-ß-Lactamase in Dust from Dutch Pig Farms*; Utrecht University: Utrecht, The Netherlands, 2014.
161. Lupo, A.; Coyne, S.; Berendonk, T.U. Origin and evolution of antibiotic resistance: The common mechanisms of emergence and spread in water bodies. *Front. Microbiol.* **2012**, *3*, 18. [CrossRef] [PubMed]
162. Baquero, F.; Martínez, J.L.; Cantón, R. Antibiotics and antibiotic resistance in water environments. Curr. Opin. Biotechnol. **2008**, *19*, 260–265. [CrossRef] [PubMed]
163. Coleman, B.L.; Salvadori, M.I.; McGeer, A.J.; Sibley, K.A.; Neumann, N.F.; Bondy, S.J.; Gutmanis, I.A.; McEwen, S.A.; Lavoie, M.; Strong, D.; et al. The role of drinking water in the transmission of antimicrobial-resistant *E. coli*. *Epidemiol. Infect.* **2012**, *140*, 633–642. [CrossRef] [PubMed]
164. Schwartz, T.; Kohnen, W.; Jansen, B.; Obst, U. Detection of antibiotic-resistant bacteria and their resistance genes in wastewater, surface water, and drinking water biofilms. *FEMS Microbiol. Ecol.* **2003**, *43*, 325–335. [CrossRef] [PubMed]
165. Mulamattathil, S.G.; Esterhuysen, H.A.; Pretorius, P.J. Antibiotic-resistant gram-negative bacteria in a virtually closed water reticulation system. *J. Appl. Microbiol.* **2000**, *88*, 930–937. [PubMed]
166. Kaspar, C.W.; Burgess, J.L.; Knight, I.T.; Colwell, R.R. Antibiotic resistance indexing of Escherichia coli to identify sources of fecal contamination in water. *Can. J. Microbiol.* **1990**, *36*, 891–894. [CrossRef] [PubMed]
167. Ozgumus, O.B.; Celik-Sevim, E.; Alpay-Karaoglu, S.; Sandalli, C.; Sevim, A. Molecular characterization of antibiotic resistant Escherichia coli strains isolated from tap and spring waters in a coastal region in Turkey. *J. Microbiol.* **2007**, *45*, 379–387. [PubMed]
168. Gao, P.; Munir, M.; Xagoraraki, I. Correlation of tetracycline and sulfonamide antibiotics with corresponding resistance genes and resistant bacteria in a conventional municipal wastewater treatment plant. *Sci. Total Environ.* **2012**, *421–422*, 173–183. [CrossRef] [PubMed]
169. Madec, J.Y.; Haenni, M.; Ponsin, C.; Kieffer, N.; Rion, E.; Gassilloud, B. Sequence Type 48 Escherichia coli Carrying the blaCTX-M-1 IncI1/ST3 Plasmid in Drinking Water in France. *Antimicrob. Agents Chemother.* **2016**, *60*, 6430–6432. [CrossRef] [PubMed]
170. Zarfel, G.; Lipp, M.; Gürtl, E.; Folli, B.; Baumert, R.; Kittinger, C. Troubled water under the bridge: Screening of River Mur water reveals dominance of CTX-M harboring Escherichia coli and for the first time an environmental VIM-1 producer in Austria. *Sci. Total Environ.* **2017**, *593–594*, 399–405. [CrossRef] [PubMed]
171. Kittinger, C.; Lipp, M.; Baumert, R.; Folli, B.; Koraimann, G.; Toplitsch, D.; Liebmann, A.; Grisold, A.J.; Farnleitner, A.H.; Kirschner, A.; et al. Antibiotic resistance patterns of pseudomonas spp. isolated from the River Danube. *Front. Microbiol.* **2016**, *7*, 586. [CrossRef] [PubMed]
172. LeJeune, J.T.; Besser, T.E.; Merrill, N.L.; Rice, D.H.; Hancock, D.D. Livestock drinking water microbiology and the factors influencing the quality of drinking water offered to cattle. *J. Dairy Sci.* **2001**, *84*, 1856–1862. [CrossRef]

173. LeJeune, J.T.; Besser, T.E.; Hancock, D.D. Cattle water troughs as reservoirs of Escherichia coli O157. *Appl. Environ. Microbiol.* **2001**, *67*, 3053–3057. [CrossRef] [PubMed]
174. Pletinckx, L.J.; Verhegghe, M.; Dewulf, J.; Crombé, F.; De Bleecker, Y.; Rasschaert, G.; Goddeeris, B.M.; De Man, I. Screening of poultry-pig farms for methicillin-resistant Staphylococcus aureus: Sampling methodology and within herd prevalence in broiler flocks and pigs. *Infect. Genet. Evol.* **2011**, *11*, 2133–2137. [CrossRef] [PubMed]
175. Porrero, M.C.; Mentaberre, G.; Sánchez, S.; Fernández-Llario, P.; Casas-Díaz, E.; Mateos, A.; Vidal, D.; Lavín, S.; Fernández-Garayzábal, J.F.; Domínguez, L. Carriage of Staphylococcus aureus in free-living wild animals in Spain. *Appl. Environ. Microbiol.* **2014**. [CrossRef] [PubMed]
176. van der Wolf, P.J.; Rothkamp, A.; Junker, K.; de Neeling, A.J. Staphylococcus aureus (MSSA) and MRSA (CC398) isolated from post-mortem samples from pigs. *Vet. Microbiol.* **2012**, *158*, 136–141. [CrossRef] [PubMed]
177. Exner, M.; Kramer, A.; Lajoie, L.; Gebel, J.; Engelhart, S.; Hartemann, P. Prevention and control of health care-associated waterborne infections in health care facilities. *Am. J. Infect. Control* **2005**, *33*, S26–S40. [CrossRef] [PubMed]
178. Hartung, J. Rules to control drinking water supply systems in livestock farming. *Deutsche Tierarztliche Wochenschrift* **2000**, *107*, 302–304. [PubMed]
179. Hartung, J.; Kamphues, J. Do we need regulation of drinking water for animals? Recommendations for the water supply of farm animals and pets. *Deutsche Tierarztliche Wochenschrift* **2000**, *107*, 343–345. [PubMed]
180. Jutkina, J.; Marathe, N.P.; Flach, C.F.; Larsson, D.G.J. Antibiotics and common antibacterial biocides stimulate horizontal transfer of resistance at low concentrations. *Sci. Total Environ.* **2018**, *616–617*, 172–178. [CrossRef] [PubMed]
181. Hölzel, C.S.; Schwaiger, K.; Harms, K.; Küchenhoff, H.; Kunz, A.; Meyer, K.; Müller, C.; Bauer, J. Sewage sludge and liquid pig manure as possible sources of antibiotic resistant bacteria. *Environ. Res.* **2010**, *110*, 318–326. [CrossRef] [PubMed]
182. Ben Said, L.; Jouini, A.; Klibi, N.; Dziri, R.; Alonso, C.A.; Boudabous, A.; Ben Slama, K.; Torres, C. Detection of extended-spectrum beta-lactamase (ESBL)-producing Enterobacteriaceae in vegetables, soil and water of the farm environment in Tunisia. *Int. J. Food Microbiol.* **2015**, *203*, 86–92. [CrossRef] [PubMed]
183. Purohit, M.R.; Chandran, S.; Shah, H.; Diwan, V.; Tamhankar, A.J.; Stålsby Lundborg, C. Antibiotic resistance in an Indian rural community: A 'One-Health' observational study on commensal coliform from humans, animals, and water. *Int. J. Environ. Res. Public Health* **2017**, *14*, 386. [CrossRef] [PubMed]
184. Vital, P.G.; Zara, E.S.; Paraoan, C.E.M.; Dimasupil, A.Z.; Abello, J.J.M.; Santos, I.T.G.; Rivera, W.L. Antibiotic resistance and extended-spectrum Beta-Lactamase production of Escherichia coli isolated from irrigation waters in selected urban farms in metro Manila, Philippines. *Water* **2018**, *10*, 548. [CrossRef]
185. Zurfluh, K.; Power, K.A.; Klumpp, J.; Wang, J.; Fanning, S.; Stephan, R. A novel Tn3-like composite transposon harboring blaVIM-1 in Klebsiella pneumoniae spp. pneumoniae isolated from river water. *Microb. Drug Resist.* **2015**, *21*, 43–49. [CrossRef] [PubMed]
186. Moore, J.E.; Watabe, M.; Millar, B.C.; Loughrey, A.; McCalmont, M.; Goldsmith, C.E.; Heaney, J.C.; Buckley, T.; Egan, C.; McDowell, D.A.; et al. Screening of clinical, food, water and animal isolates of Escherichia coli for the presence of blaCTX-M extended spectrum beta-lactamase (ESBL) antibiotic resistance gene loci. *Ulster Med. J.* **2010**, *79*, 85–88. [PubMed]
187. Ikegbunam, M.N.; Anagu, L.O.; Iroha, I.R.; Ejikeugwu, C.E.; Esimone, C.O. Abattoirs as non-hospital source of extended spectrum beta lactamase producers: Confirmed by the double disc synergy test and characterized by matrix-assisted laser desorption/ionization time of flight mass spectrometry. *PLoS ONE* **2014**, *9*, e94461. [CrossRef] [PubMed]
188. Johnson, J.R.; Sannes, M.R.; Croy, C.; Johnston, B.; Clabots, C.; Kuskowski, M.A.; Bender, J.; Smith, K.E.; Winokur, P.L.; Belongia, E.A. Antimicrobial drug-resistant Escherichia coli from humans and poultry products, Minnesota and Wisconsin, 2002–2004. *Emerg. Infect. Dis.* **2007**, *13*, 838–846. [CrossRef] [PubMed]
189. Vincent, C.; Boerlin, P.; Daignault, D.; Dozois, C.M.; Dutil, L.; Galanakis, C.; Reid-Smith, R.J.; Tellier, P.P.; Tellis, P.A.; Ziebell, K.; et al. Food reservoir for Escherichia coli causing urinary tract infections. *Emerg. Infect. Dis.* **2010**, *16*, 88–95. [CrossRef] [PubMed]

190. Lavilla, S.; González-López, J.J.; Miró, E.; Domínguez, A.; Llagostera, M.; Bartolomé, R.M.; Mirelis, B.; Navarro, F.; Prats, G. Dissemination of extended-spectrum beta-lactamase-producing bacteria: The food-borne outbreak lesson. *J. Antimicrob. Chemother.* **2008**, *61*, 1244–1251. [CrossRef] [PubMed]
191. Normanno, G.; Dambrosio, A.; Lorusso, V.; Samoilis, G.; Di Taranto, P.; Parisi, A. Methicillin-resistant Staphylococcus aureus (MRSA) in slaughtered pigs and abattoir workers in Italy. *Food Microbiol.* **2015**, *51*, 51–56. [CrossRef] [PubMed]
192. Sommer, M. *Epidemiologische Untersuchungen zur Tiergesundheit in Schweinezuchtbeständen unter besonderer Berücksichtigung von Managementfaktoren und des Einsatzes von Antibiotika und Homöopathika*; Hanover University: Hanover, Germany, 2009.
193. Walker, B.; Holling, C.S.; Carpenter, S.; Kinzig, A. Resilience, adaptability and transformability in social-ecological systems. *Ecol. Soc.* **2004**, *5*, 9. [CrossRef]
194. Rauw, W.M.; Kanis, E.; Noordhuizen-Stassen, E.N.; Grommers, F.J. Undesirable side effects of selection for high production efficiency in farm animals: A review. *Livest. Prod. Sci.* **1998**, *56*, 15–33. [CrossRef]
195. Tenhagen, B.-A.; Fetsch, A.; Alt, K.; Käsbohrer, A.; Bräunig, A.; Appel, B. MRSA in herds of fattening pigs in Germany—Associated risk factors. In Proceedings of the Safe Pork 2009, Québec, QC, Canada, 30 September–2 October 2009.
196. Thielen, T.L.; Castle, S.S.; Terry, J.E. Anterior ocular infections: An overview of pathophysiology and treatment. *Ann. Pharmacother.* **2000**, *34*, 235–246. [CrossRef] [PubMed]
197. Steinhoff-Wagner, J. Ethische Zielkonflikte aus agrarwissenschaftlicher Perspektive. 2018; unpublished data.
198. Hörügel, K. Tiergesundheitsmanagement in der Schweinehaltung. *Schriftenreihe der Sächsischen Landesanstalt für Landwirtschaft* **2001**, *7*, 63–67.
199. Böckel, V. *Untersuchungen zur quantitativen Bewertung der Tiergesundheit von Schweinebeständen*; Hanover University: Hanover, Germany, 2008.
200. Berns, G. *Einbindung von Check-Listen und mobilen Analyselabor in Beratungskonzepte zur Erweiterung von Gesundheitsvorsorge- und Qualitätsmanagementsystemen in der Schweinefleischerzeugung*; University of Bonn: Bonn, Germany, 1996.
201. Düsseldorf, S. *Concept of key Performance Indicators Controlling Consumer Oriented Quality and Herd Health Management in a Bavarian Pork Chain*; Univertiy of Bonn: Bonn, Germany, 2013.
202. Ellebrecht, A. *Nutzenbetrachtung Internetbasierter Informationssysteme im Einzel- und Überbetrieblichen Gesundheitsmanagement*; University of Bonn: Bonn, Germany, 2008.
203. Ellebrecht, S. *Application Concept for Combined Preventive Quality Management methods in inter-Enterprise Health Management in Pork Chains*; University of Bonn: Bonn, Germany, 2012.
204. Gymnich, S. *Haptoglobin als Screeningparameter im Gesundheitsmanagement von Ferkelaufzuchtbetrieben*; University of Bonn: Bonn, Germany, 2001.
205. Klauke, T. *Risk Based Approach towards More Sustainablilty in European Pig Production*; University of Bonn: Bonn, Germany, 2012.
206. Mack, A. *Konzeption für Die Anwendung Computergestützter Techniken im Gesundheitsmanagement Fleischerzeugender Ketten*; University of Bonn: Bonn, Germany, 2009.
207. Schütz, V. *Modell zur Planung von Dienstleistungen für das Überbetriebliche Gesundheitsmanagement in der Fleischwirtschaft*; University of Bonn: Bonn, Germany, 2009.
208. Welz, M. *Bewertung von Erkrankungen als Qualitätshemmende Faktoren mit Hilfe der Fehler- Möglichkeits- und Einfluss-Analyse (FMEA) im Rahmen der Erzeugung von Qualitätsfleisch*; Bonn University: Bonn, Germany, 1993.
209. Schulze Althoff, G. *Konzept und Erprobung von Werkzeugen für Grenzüberschreitende, Kettenübergreifende Qualitätsmanagementsysteme in der Schweinefleischerzeugung*; University of Bonn: Bonn, Germany, 2007.
210. Petersen, B.; Nüssel, M.; Hamer, M. Quality and risk management in agri-food chains. In *European Review of Agricultural Economics*; Wageningen Academic Publishers: Wageningen, The Netherlands, 2014; pp. 1–3.
211. Schulze Althoff, G.; Breuer, O.; Jaeger, F.; Petersen, B. Working together to improve early warning—Cross border approaches to advancing prevention concepts. *Fleischwirtsch. Int.* **2007**, *2*, 55–58.
212. Petersen, J.V.; Andersen, J.K.; Sørensen, F.; Knudsen, H. Food safety on the slaughterline: Inspection of pig heads. *Vet. Rec.* **2002**, *150*, 782–784. [CrossRef] [PubMed]
213. Petersen, B.; Spiller, A.; Theuvsen, L. *Vom Viehmarkter zum Dienstleistungsprofi*; GIQS: Bonn, Germany, 2010.
214. Lehnert, H. Animal health monitoring. *TopAgrar* **2009**, *11*, 12–17.

215. Schiefer, G. Tracking, Tracing and Business Process Interests: A Framework for Analysis. In Proceedings of the 26th Conference of the International Association of Agricultural Economists, Queensland, Australia, 12–16 August 2006.
216. Van der Vorst, J. *Effective Food Supply Chains. Generating, Modelling and Evaluating Supply Chain Scenarios*; University Wageningen: Wageningen, The Netherlands, 2000.
217. Petersen, B.; Lehnert, S. *Qualitätsmanagement Maßgeschneidert für die Agrar- und Ernährungswirtschaft*; Rheinische Friedrich-Wilhelms-Universität Bonn, International FoodNetCenter: Bonn, Germany, 2017; Volume 1.
218. Ewers, C.; Bethe, A.; Semmler, T.; Guenther, S.; Wieler, L.H. Extended-spectrum β-lactamase-producing and AmpC-producing Escherichia coli from livestock and companion animals, and their putative impact on public health: A global perspective. *Clin. Microbiol. Infect.* **2012**, *18*, 646–655. [CrossRef] [PubMed]
219. Ferreira, J.P.; Anderson, K.L.; Correa, M.T.; Lyman, R.; Ruffin, F.; Reller, L.B.; Fowler, V.G. Transmission of MRSA between companion animals and infected human patients presenting to outpatient medical care facilities. *PLoS ONE* **2011**, *6*, e26978. [CrossRef] [PubMed]
220. Häsler, B. (NEOH) Network for Evaluation of One Health. Available online: http://www.accelopment.com/de/projects/neoh (accessed on 23 October 2018).
221. Steinhoff-Wagner, J.; Ilg, Y.; Tappe, E.-V.; Lehnert, S.; Lehnert, H.; Petersen, B. *One Health-Konzepte und ihre Umsetzung—Forschungsprojekte, Hochschulausbildung und Netzwerkinitiativen*; ATD: Bonn, Germany, 2017; Volume 2.
222. De Neeling, A.J.; van den Broek, M.J.; Spalburg, E.C.; van Santen-Verheuvel, M.G.; Dam-Deisz, W.D.; Boshuizen, H.C.; van de Giessen, A.W.; van Duijkeren, E.; Huijsdens, X.W. High prevalence of methicillin resistant Staphylococcus aureus in pigs. *Vet. Microbiol.* **2007**, *122*, 366–372. [CrossRef] [PubMed]
223. Dickhöfer, D. *Haptoglobin als Screeningparameter für Atemwegserkrankungen des Schweines*; University of Bonn and University of Veterinary Medicine Hannover: Hannover, Germany, 2002.
224. Lipperheide, C.; Knura, S.; Petersen, B. *Sicherung der Regionalen Vermarktung von Ferkeln für Nordrhein-Westfälische Qualitätsfleischprogramme Durch die Einführung Überbetrieblicher Gesundheitsmanagement- und Frühwarnsysteme*; Lehr- und Forschungsschwerpunkt "Umweltverträgliche und Standortgerechte Landwirtschaft": Bonn, Germany, 1999.
225. BMEL. *16. AMG-Novelle des Arzneimittelgesetzes*; BMELV: Bonn, Germany, 2014.
226. QS. Wenn es um Fleisch geht. Qualität. Warenkunde. Küchentechniken. Rezeptideen. In *Qualitätssicherung. Vom Landwirt bis zur Ladentheke*; QS: Bonn, Germany, 2011; Available online: https://www.q-s.de/flip/ratgeber_fleisch/files/assets/downloads/publication.pdf (accessed on 23 October 2018).
227. Merle, R.; Kösters, S.; May, T.; Portsch, U.; Blaha, T.; Kreienbrock, L. Serological Salmonella monitoring in German pig herds: Results of the years 2003–2008. *Prev. Vet. Med.* **2011**, *99*, 229–233. [CrossRef] [PubMed]

sustainability

MDPI

Article

The Valorization of Ammonia in Manure Digestate by Means of Alternative Stripping Reactors

Marco Baldi [1,*], Maria Cristina Collivignarelli [2], Alessandro Abbà [3] and Ilaria Benigna [2]

1 Department of Chemistry, University of Pavia, viale Taramelli 10, 27100 Pavia, Italy
2 Department of Civil Engineering and Architecture, University of Pavia, via Ferrata 1, 27100 Pavia, Italy; mcristina.collivignarelli@unipv.it (M.C.C.); ilaria.benigna01@universitadipavia.it (I.B.)
3 Department of Civil, Environmental, Architectural Engineering and Mathematics, University of Brescia, via Branze 43, 25123 Brescia, Italy; alessandro.abba@unibs.it
* Correspondence: marco.baldi@unipv.it; Tel.: +39-03-8298-7270

Received: 12 July 2018; Accepted: 27 August 2018; Published: 29 August 2018

Abstract: The proper recovery of resources such as nitrogen and phosphorus present in the manure from intensive livestock farming is essential in order to allow environmental sustainable zootechny especially in densely populated areas where these activities are historically prevalent. The experiences at full-scale established that the ammonia stripping allows recovery from 35% to 50% of nitrogen depending on the type of substrate treated with anaerobic digestion and on the nitrogen content/form in the digestate. This study focuses on the ammonia stripping on digestate derived from anaerobic digestion of livestock manure and corn silage. Two different full-scale plants are studied including a packed column and an air bubble reactor without filling material with the aim to reduce fouling issues due to the content of suspended solids in digestate. The main results suggest that the use of an air bubble reactor could treat digestate with high concentration of suspended solids. A deeper study based on a two-level factorial experiment highlights that the temperature is an important parameter that influences the ammonia removal yields. Thus, a proper management of available thermal energy is very important.

Keywords: anaerobic digestion; livestock manure; digestate; ammonia recovery; fouling issue

1. Introduction

The management of agricultural waste, animal manure (bovine, swine, and poultry), and food residues by means of anaerobic digestion (AD) is becoming more important [1,2]. In anaerobic digestion (AD), the microorganisms in the absence of oxygen break down the biodegradable matter and produce a valuable biogas consisting of mostly methane [3]. The biogas can either be used directly or it can be converted into electrical energy. The other residue of anaerobic digestion is digestate [4]. The digestate represents the AD effluent (i.e., the digested substrate), which is removed from the reactor. The nature and composition of treated substrate and the operating parameters of the AD process have a significant influence on the physical and chemical characteristics of digestate [3]. The ammonia content of digestate depends on the total nitrogen content of feedstock. Cereal grains, poultry, and pig manures show high ammonia content. However, the bovine manure and corn silage present a lower ammonia content [5].

The main issues concerning the landspreading of digestate on agricultural soils that represents the main route are related to the high content of nutrients especially nitrogen [6] and represent the increasing number of AD plants especially in the region with intensive livestock farming [7]. Therefore, in order to comply with the European Nitrates Directive (91/676/EEC) [8], suitable treatments for nitrogen reduction in digestate must be applied. Moreover, the recent Italian legislation has restricted the landspreading of digestate derived from AD of different materials and/or residues [9].

The first step of digestate treatments usually carried out at farm scale concerns the solid/liquid separation. The solid fraction with high content of organic nitrogen can be spread on agricultural soils directly or after a composting process [10]. In the liquid phase, the high contents of nitrogen generally with high ammonia percentage, phosphorus, and other macro and micronutrients [11] represent a serious problem for its recovery/disposal.

Furthermore, some authors [12] have studied the feasibility to increase the methane productions in AD up to 104% [13] by using the pretreatment of animal manure fibers with aqueous ammonia soaking. Despite this treatment being very interesting, the proposed solution could increase the ammonia content in digestate involving, especially in Italy, additional issues for its recovery on agricultural soils.

As concerns grow regarding the ammonia removal from liquid digestate, several treatments are proposed in the scientific literature. The methods based on physical-chemical processes include struvite precipitation: the ammonia removal yields from the digestate of poultry manure reached 86.4% (with Mg:N:P molar ratio of 1:1:1) and 97.4% (adjusting Mg:N:P molar ratio to 1.5:1:1) [14]. However, the contents of heavy metals precipitated together with struvite [15] must be controlled.

The ammonia stripping on livestock manure digestates allowed to obtain ammonia removal yields up to 85%, which operates at 40 °C with a pH adjustment [16]. However, Serna-Maza et al. [17] showed that the ammonia removal from fresh digestate was more difficult than digestate stored for long periods.

Moreover, the evaporation is an interesting solution to remove nitrogen from pig slurry with a previous anaerobic digestion to increase the economic feasibility [18]. Guercini et al. [19] proposed to use the cogeneration heat to support both the AD plant and the subsequent evaporation of the digestate.

The membrane technology for ammonia removal from digestate could be applied. The use of hollow fiber membranes, submerged into the digestate from the anaerobic reactor that treats slaughterhouse wastes, led to a reduction of free ammonia by 70% [20]. However, the membrane parameters (such as bubble point, breakthrough pressure, airflow, pore size, contact angle, thickness, etc.) must be evaluated to predict the performance process [21]. As concerns the membrane processes, an innovative solution based on a thermophilic aerobic membrane reactor for the treatment of sewage sludge and high strength aqueous waste [22] is able to increase the ammonia content in permeate and the precipitation of nutrients (nitrogen and phosphorus) in the form of salts. Moreover, the proposed solution can be conveniently integrated in traditional wastewater treatment plants (based on a conventional activated sludge process) with the aim to reduce the content of nutrients in the discharged effluent and to recover the heat from the thermophilic aerobic reactor [23].

The choice of suitable technology and the required pretreatments depend on different aspects mainly related to the qualitative characteristics of digestate. The ammonia stripping and chemical precipitation require simple pretreatments (usually screw press or belt filter press) in order to reduce the content of suspended solid. The other proposed technologies instead require pretreatments with ultrafiltration (UF) followed by cleaning/regeneration cycles [24]. The farmers prefer to adopt simple and robust treatment plants with low investment and maintenance costs.

The ammonia stripping technology is a process where the nitrogen, in the form of ammonia, is removed from a liquid by gas flow through the liquid. The volatility of ammonia in an aqueous solution can be enhanced by increasing the temperature and pH (usually obtained through chemical addition).

The ammonia stripping process includes the following operations [25]: (i) conversion of ammonium ions (NH_4^+) to ammonia gas (NH_3) (ammonia dissociation equilibrium), (ii) diffusion of NH_3 to the air-water interface (water-side mass transfer), (iii) release of NH_3 to the air at the interface (volatilization), and (iv) diffusion of NH_3 from the air-water interface into the air above (air-side mass transfer). The whole process depends on pH, temperature, and mass transfer area.

Ammonia gas (NH_3) is thereafter absorbed in an acid solution (sulphuric or nitric acid) in order to obtain fertilizer for agricultural use (ammonium sulphate or nitrate respectively) with low organic contamination. The cleaned gas can be reused in the stripping column.

The efficiency of air stripping mainly depends on the following parameters: feed pH, feed temperature, ratio of air to feed, and feed characteristics.

The high thermal energy requirement has often restricted the applicability of ammonia stripping. Therefore, this technology is used when there is heat available, which results from the production of electrical power by endothermic motors in excess of the intrinsic needs arising from the need to control the temperature of the plant for the biogas production.

Concerning the stripping reactor, the design of the contacting system between the digestate and the gas used to strip out the NH_3 is very important. The aim is to maximize the extent of contact (maximum rate of mixing and highest specific surface area) while minimizing energy costs due to the equipment design. The most common mass transfer design for air stripping systems uses packed towers: columns filled with packing material in order to increase the surface area available for the ammonia mass transfer.

Despite the fact that most common reactors used for ammonia stripping are based on packed columns, the suspended solids can clog the column. This aspect is called fouling. The high content of suspended solids reduces the stripping performance. This reduction is due both to the fouling that involves problems on stripping management and to high content of organic acids and colloidal compounds that reduce the process efficiency [26]. As a consequence, efficient solid–liquid separation is necessary beforehand. In addition, a high maintenance and cleaning effort may be necessary.

Regarding the nitrogen partition in livestock manure, Fabbri and Piccinini (2012) [27] show that, in swine, manure ammonia represents about 60% of total nitrogen, which increases by 10% due to AD. Even in the case of bovine manure, the AD involves an increase of ammonia content from 43% to 52%. Furthermore, concerning the solid/liquid separation process, they show that the ammonia nitrogen with respect to total nitrogen increases: (i) in the case of swine manure, from 70% to 90% and (ii) in the case of bovine manure, from 50% to 70%.

In the present study, the ammonia recovery from digestate derived from the AD of swine, bovine, poultry manure, and corn silage was analyzed. Experimental activities were carried out on two different air stripping full-scale plants: the first is a packed column reactor and the second is an air bubble reactor without filling material in order to reduce the issues concerning fouling. In this study, the advantages and disadvantages of the alternative solutions are reported.

2. Material and Methods

In this section, the characteristics of the ammonia stripping plants at full-scale are reported. The main characteristics of liquid digestate (treated in these plants) are also shown. Moreover, the operative conditions of the tests carried out during the experimental activities are reported.

2.1. Description of Plants and Digestate Characteristics

The Figure 1 reports a schematic diagram of the process for the anaerobic treatment of livestock manure and agricultural residues in order to recover ammonia as fertilizer. The treatment consists of an anaerobic digestion where bovine, swine, poultry manure, and corn silage are fed as well as a solid/liquid separation for the obtaining of liquid digestate and the subsequent ammonia recovery section. Three different flows come from the treatment including: (i) the solid fraction, which is sent to a composting process, (ii) the liquid fraction, which is sent to ammonia stripping, and (iii) the air contaminated by ammonia, which is sent to an absorption tower for fertilizer production.

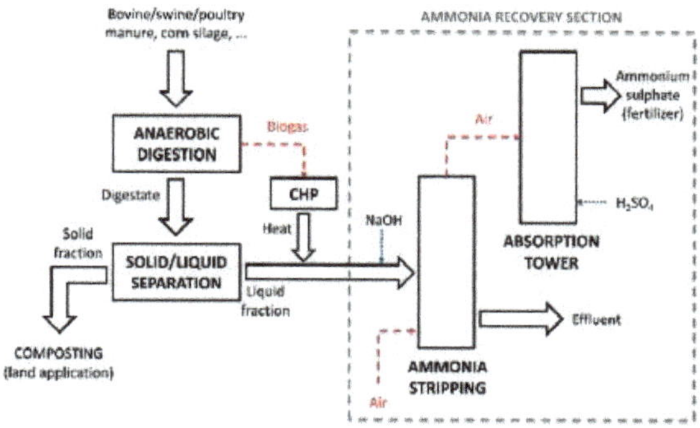

Figure 1. Schematic diagram of the process.

During the experimental activities, two different kinds of plants were used. The stripping reactor of plant #1 consists of a packed column that receives the liquid digestate obtained from a rotary drum with a mesh of 350 μm followed in a second phase of experimentation by a vibrating screen with a mesh of 150 μm. The ammonia stripping section of plant #2 is an air bubble reactor that treats the liquid digestate derived from a screw press separator with a mesh of 5 mm. Both plants are equipped with a similar packing absorption tower for ammonium sulphate production (see Table 1).

Table 1. Geometrical characteristic of the plants.

Plant Section	Parameter	Plant #1	Plant #2
Ammonia stripping	Diameter [m]	1.2	2.8
	Surface [m^2]	1.1	6
	Packed height [m]	3	-
	Liquid height [m]	-	2
	Specific surface of packing material [m^2 m^{-3}]	200	-
	Recirculation flowrate [m^3 h^{-1}]	16	22
	Air flowrate [Nm^3 h^{-1}]	4000–4500	500–700
Absorption tower	Diameter [m]	0.8	0.55
	Packed height [m]	3.0	2.5
	Specific surface of packing material [m^2 m^{-3}]	245	245
	Absorbent	H_2SO_4 (50% *v/v* solution)	H_2SO_4 (50% *v/v* solution)

Regarding plant #1 (Figure 2a), the ammonia stripping reactor is a vertical column filled with packing material. The specific surface is 200 m^2 m^{-3}. The flow treated is 2.5 m^3 h^{-1} and the air flowrate is about 4500 m^3 h^{-1}. In the first part of the experiment, plant #1 operated continuously while, in a second period in a semi-batch mode (i.e., liquid phase in batch with air flow). In semi-batch conditions, the volume of liquid digestate (5 m^3 and 13.2 m^3 for plant#1 and #2, respectively) was maintained into the stripping reactor for a suitable time (2 h for plant #1 and 6 h for plant #2). The air flowrate is supplied continuously.

In the stripping column of plant #2 (Figure 2b), the air is supplied with medium/fine bubble diffusers. The plant worked both in continuous (with complete mixing) and in a semi-batch mode.

In both plants, the ammonia recovery is obtained in an acid scrubber with a pH of 4.3. In these conditions, even if the concentration of ammonia in the absorbent solution reaches the 10% *w*/*w* (weight to weight), the concentration of NH_3 in the flu-gas is maintained around 3 to 5 mgN m^{-3} of dry air. For the absorption process, a 50% *v*/*v* (volume/volume) sulphuric acid (H_2SO_4) solution is used, which allows the production of di-ammonium sulphate solution with a nitrogen content of 8% to 10%.

The geometrical characteristics of the plants used in the experimental activities are reported in Table 1. In Table 2, the qualitative characteristics of the liquid fraction of digestate fed to the different plants are shown. It can be observed that plant #2 is able to treat liquid fraction with suspended solids content up to 5%. This value cannot be reached in the case of plant #1 due to the issue concerning the fouling of packing material. In fact, the configuration of plant #1 does not allow the treatment of the substrate with a suspended solid content higher than 2%.

(a) Plant#1 (b) Plant#2

Figure 2. Ammonia recovery sections of plant #1 (**a**) and plant #2 (**b**).

Table 2. Qualitative characteristics of liquid digestate.

Parameter	Plant #1	Plant #2
pH	7.7–8.2	7.7–8.2
Suspended solids [%]	<2	2–5
Dissolved solids [%]	0.8–1.2	0.8–1.2
Ammoniacal nitrogen [g L^{-1}]	2.5–3.5	2.0–3.0

2.2. *Operative Conditions of Experimental Tests*

The operative conditions of the tests carried out during the experiment are shown in Table 3. Concerning plant #1, at the beginning of the experiment, it worked in a continuous mode (phase I) with pH and temperature ranging from 9.5 to 10 pH and from 35 to 40 °C, respectively. Then the plant worked in semi-batch conditions (phase II) with pH ranges from 9.2 to 9.5. In this case, the alkaline reagent (NaOH) consumption was significantly lower than in phase I.

The semi-batch conditions, due to the high rates of ammonia removal, reached the same performance obtained in the continuous mode, but with pH values significantly lower, which indicates an important reduction of alkaline reagent consumption.

The vibrating screen with a mesh of 150 μm works at the end of phase I.

At the beginning of experimental activities with plant #2, a preliminary test was performed in semi-batch conditions using an aqueous solution of ammonium sulphate with ammonia concentration equal to 2500 mgN L^{-1}. Then plant #2 worked in a continuous mode with a temperature of 48 °C and pH ranging from 9.5 to 10.1 (obtained with NaOH dosage). This test showed a foaming issue due to the dosage of an alkaline reagent and the air supply with medium/fine bubble diffusers. Thus, 6 additional tests, always in a continuous mode, were performed without the use of an alkaline reagent (4 tests) and with the presence of NaOH and a defoamer agent (2 tests).

Table 3. Operative conditions of experimental tests on plant #1 and #2.

Plant	Work Condition	Day	Test #	Digestate Flowrate [$m^3\ h^{-1}$]	HRT [h]	Temperature [°C]	pH	Air Flowrate [$Nm^3\ h^{-1}$]
Plant#1	Phase I continuous	0	1	2.5	2	38.5	9.5	4000–4500
		24	2			37.5	9.6	
		91	3			35	9.5	
		225	4			39	9.2	
		372	5			40.5	9.2	
		518	6			37.5	10.1	
		771	7			35.5	10.0	
		954	8			40.5	10.2	
	Phase II semi-batch	1086	9	(*)	2	50	9.5	4000–4500
		1092	10			50.5	9.5	
		1185	11			46.5	9.4	
		1316	12			50.5	9.5	
		1428	13			52	9.5	
		1548	14			49	9.5	
		1676	15			50.5	9.3	
		1807	16			52.5	9.4	
		2004	17			51	9.5	
		2215	18			49	9.3	
		2354	19			51.5	9.2	
		2563	20			50	9.3	
		2739	21			49.5	9.3	
		2927	22			49.5	9.3	
		3081	23			48.5	9.2	
Plant#2	Semi-batch *	0	1	(**)	6	52	9.4–10.1	500
	Continuous	3	2	2.2	6	48	9.5–10.1	500–700
		18	3			53	9.4 **	
		59	4			46	9.2 **	
		116	5			45	9.2 **	
		232	6			55	9.5 **	
		234	7			56	10.6	
		280	8			58	10.5	

HRT: Hydraulic Retention Time, (*) treated volume: 5 m^3, (**) treated volume: 13.2 m^3, * preliminary test, ** no dosage of an alkaline reagent.

In order to evaluate the influence of operating parameters on the ammonia removal in plant #2, a deeper study based on a two-level factorial experiment is developed. The development of the factorial experimentation was focused on three operating parameters: temperature, digestate flowrate, and air flowrate. The pH was not modified with an alkaline reagent dosage. In this case, the stabilization of pH (at 9.2 to 9.3) is due to the equilibrium obtained from CO_2 and ammonia removal. This choice is due to the high concentration of bicarbonate ion in the solution fed to the plants. In fact, the high concentration of bicarbonate ion (0.32 N) involves that the increase (with NaOH solution) of digestate pH more than 9.2 is too expensive and the stabilization of pH (without a reagent dosage) to a pH of 9.2 to 9.3 (from 7.7 to 7.8), due to the CO_2 removal obtained from the bicarbonate decomposition, is favored.

The variation of operating parameters in the factorial experimentation is reported in Table 4. All tests were carried out in a continuous mode with an HRT (Hydraulic Retention Time) equal to 6 h for a higher digestate flowrate (2.2 $m^3\ h^{-1}$) and 9.5 h for a lower flowrate (1.4 $m^3\ h^{-1}$).

Table 4. Operative conditions of factorial experimentation (plant #2).

Temperature [°C]	Air Flowrate [$Nm^3\ h^{-1}$]	Digestate Flowrate [$m^3\ h^{-1}$]
50	500	1.4
		2.2
	700	1.4
		2.2
60	500	1.4
		2.2
	700	1.4
		2.2

With regard to the deepening study based on a two-level factorial experiment, the correlation between the real removal yields of ammonia (R_R) and the values estimated (R_S) with two linear models is considered.

The aim of this study is to evaluate the effects of the operating parameters tested (temperature, digestate flowrate, and air flowrate) on ammoniacal nitrogen removal yields by the use of a simple linear model in which no constant is present: the hypothesis is $R_S = R_R$ for $R_R = 0$.

The following linear model was applied below.

$$R_S = a\,T + b\,Q_a + c\,Q_d \tag{1}$$

where T is the temperature [°C], Q_a is the air flowrate [$Nm^3\ h^{-1}$], Q_d is the digestate flowrate [$m^3\ h^{-1}$], and a, b, and c are the weights of the operating parameters reported above.

In addition, a more complex linear model is proposed. In this case, the correlation between the operating parameters at two and three levels is taken into account.

$$R_S = a\,T + b\,Q_a + c\,Q_d + d\,T\,Q_a + e\,T\,Q_d + f\,Q_a\,Q_d + g\,T\,Q_a\,Q_d \tag{2}$$

3. Results

The results of the tests carried out on plant #1 and plant #2 are reported in this section. In particular, the effects of different operative conditions on the ammonia removal yields are shown. Moreover, the results of a two-level factorial experiment (carried out on plant #2) are reported in order to study the influence of operating parameters on the ammonia removal in plant #2.

3.1. Plant #1 Results

Figure 3 reports pH and temperature in the stripping reactor and the trend of both ammoniacal nitrogen input and output concentrations during the working of plant #1 (both in a continuous and a semi-batch mode). Moreover, in Figure 4, the ammonia removal yields are reported.

As shown in these figures, the removal yields on plant #1 that works in a continuous mode (like a rear-mixed plug-flow column) varied from 22% to 37%. In these conditions (phase I), the system has a low efficiency. Low enhancement of ammonia removal yields (mean values from 25% up to 34%) can be obtained only by increasing the dosages of an alkaline reagent by 2 to 2.5 times (as shown by tests #6, #7, and #8 in which the pH rises from mean values of 9.4—in tests #1 to #5—up to 10.1).

When plant #1 worked in semi-batch conditions (phase II) and a pH similar to the first period of phase I (mean value of 9.4) performance increased, the ammonia removal yields raised from 25% (mean value of the first period of phase I) to 66% (tests #14). Moreover, in these conditions, the operative pH has been obtained with a lower dosage of an alkaline reagent (less than about 30%) with respect to a continuous mode (phase I).

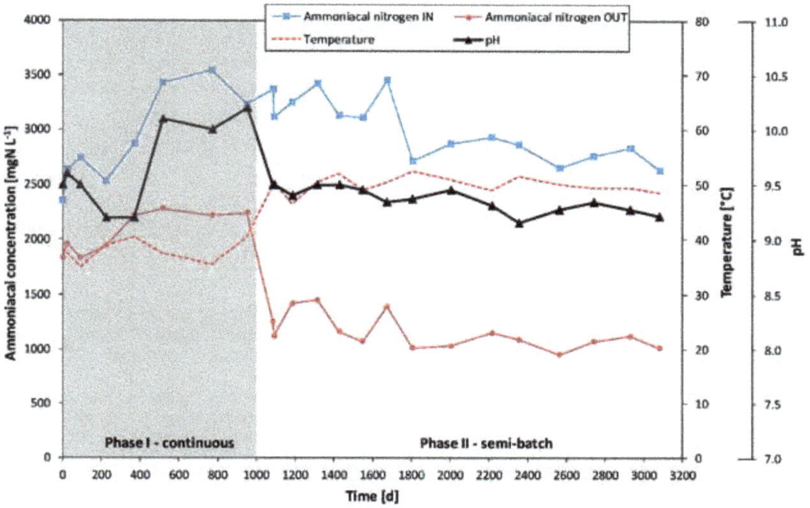

Figure 3. Plant #1 results: input and output ammoniacal nitrogen concentrations and pH and temperature in the stripping reactor.

According to the Henry law on gas/liquid equilibrium of ammonia/water system, in the same conditions, the ammonia removal yields increase with a rising temperature. As expected, in Figure 4b, the performance enhancement of the system vs. temperature can be observed. The low value of the correlation index (R^2) equal to 0.19 point out that, in addition to the temperature, several operative parameters (that are controlled such as pH or uncontrolled such as fouling) have an important effect on the results.

The influence of these parameters is more evident in Figure 4a in which the decrease of ammonia removal yields seems be due to the temperature increasing.

The results reported in Figure 4 show that an increase in the ammonia removal yields from 25% to 62% (mean values) can be observed when the temperature increases from the mean values of 38 °C (Figure 4a) to 50 °C (Figure 4b). These results are consistent with the Henry law and with the work conditions that changed from a continuous to a semi-batch mode.

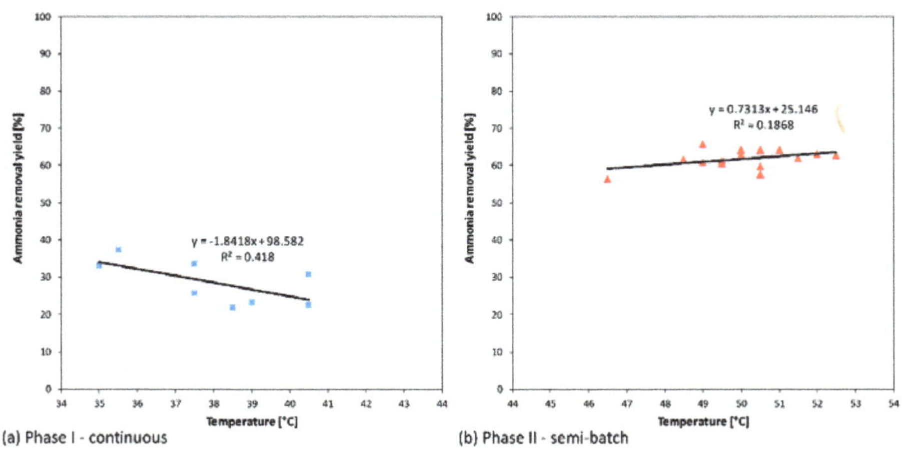

Figure 4. Plant #1 results: ammonia removal yields.

3.2. *Plant #2 Results*

Regarding plant #2, Table 5 shows the trend of the input and output ammoniacal nitrogen concentrations and the removal yields obtained during continuous and semi-batch (for preliminary test only) conditions. The operating temperature and pH are also reported.

Table 5. Plant #2 results.

Work Condition	Day	Test #	Temperature [°C]	pH	Type of Alkaline Reagent	Ammoniacal Concentration [g L^{-1}]		Removal Yield [%]
						IN	OUT	
Semi-batch	0	1	52	9.4–10.1	NaOH	2.75	0.57	79
Continuous	3	2	48	9.5–10.1	NaOH	2.12	0.81	62
	18	3	53	9.4	-	2.18	1.32	39
	59	4	46	9.2	-	2.53	1.94	23
	116	5	45	9.2	-	2.61	1.82	30
	232	6	55	9.5	-	2.65	1.61	39
	234	7	56	10.6	NaOH	2.71	1.12	59
	280	8	58	10.5	NaOH	2.64	0.91	66

In the preliminary test (in the semi-batch mode), the ammonia removal yield is equal to 79%.

The lower ammonia removal yields (from 23% to 39%) are obtained in a continuous mode with a temperature ranging from 45 to 55 °C and with pH ranging from 9.2 to 9.5 (obtained without an alkaline reagent dosage). Moreover, the increase of temperature and pH improves the performance. An ammonia removal yield of 66% was reached when operating at 58 °C and a pH equal to 10.5.

It can be noted that the high bicarbonate concentrations in the digestate fed to the stripping reactor with the temperature increasing by 45 to 55 °C allow us to work with a pH from 9.2 to 9.5 without the dosage of a strong alkaline reagent. This behavior is due to the thermal decomposition of bicarbonates in carbonates with the development of CO_2, which is subsequently removed from the aqueous phase.

When the pH value is 9.3 to 9.5, the concentration of un-dissociated ammonia (also known as free ammonia nitrogen) is about 40% while the ammonium ion content (NH_4^+) is 60%. The increase of pH up to the value of 10.5 significantly enhances the performance by raising the free ammonia nitrogen percentage up to values close to 100%.

Regarding the study based on a two-level factorial experimentation, the results (in terms of real removal yields of ammonia—R_R) are reported in Table 6.

Table 6. Results of two-level factorial experimentation (plant #2).

Temperature [°C]	Air Flowrate [Nm^3 h^{-1}]	Digestate Flowrate [m^3 h^{-1}]	Real Removal Yields of Ammonia (R_R) [%]
50	500	1.4	32.7
		1.4	30.9
		2.2	25.4
		2.2	27.2
	700	1.4	42.1
		1.4	43.0
		2.2	45.8
		2.2	44.1
60	500	1.4	48.9
		1.4	48.6
		2.2	35.3
		2.2	37.3
	700	1.4	44.8
		1.4	44.9
		2.2	40.7
		2.2	38.2

The results of the simple linear model (Equation (1)), obtained by the minimization of the squared error sum $(R_R - R_S)^2$, are the following: a = 0.711, b = 0.0228, and c = −7.88.

Regarding the more complex linear model (Equation (2)), the results are: a = 1.898, b = −0.2387, c = −8.4630, d = 0.00253, e = −0.6221, f = 0.1297, and g = −0.0012.

In Figure 5, the correlations between R_R and R_S for both models are reported. Concerning the linear model described with Equation (1), the slope of the interpolation line is near 0.99. This result suggests that the operating parameters taken into account play an important role in process performance (Figure 5a). Despite the fact that the correlation coefficient (R^2) is not very high due to the variation of digestate characteristics during the experimentation, the model is able to obtain a suitable estimation of ammoniacal nitrogen removal yields.

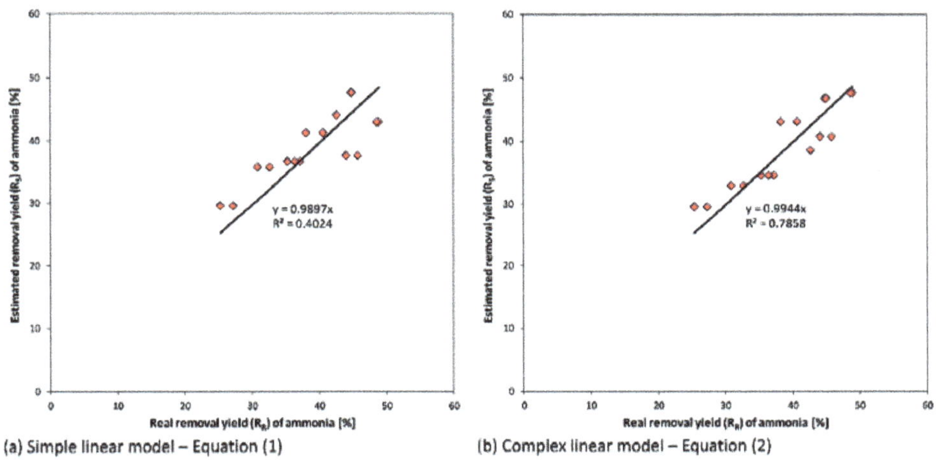

Figure 5. Plant #2 results: correlations between real removal yields (R_R) and the estimated values (R_S) for the models studied.

Regarding the more complex model (Equation (2)), the results obtained (Figure 5b) in terms of a correlation coefficient are better than the first model. However, the physical meaning of the calculated weights is very difficult to explain. Therefore, the simple linear model (Equation (1)) is recommended. The calculated values of weights (a, b, and c) revealed the positive effects of temperature and air flowrate on the ammonia removal yields while it is observed that the increase of digestate flowrate (with a reduction of the mean HRT in the stripping reactor) involves a performance reduction.

Since the weights are inversely related to the values of respective operating parameters (T, Q_a, and Q_d) in order to define the real weight of the variables on the ammoniacal nitrogen removal yield, which are the coefficients obtained by using the square error sum minimization reported above, these have been normalized. The coefficients a, b, and c have been multiplied by the 10% of the average value of the corresponding parameters. The results obtained are: $a_{normalized}$ = 3.91, $b_{normalized}$ = 1.37, and $c_{normalized}$ = −1.42. Therefore, the normalization (to the range 0–100) of the mean values for each parameter (T = 55 °C, Q_a = 600 $Nm^3\ h^{-1}$, Qd = 1.8 $m^3\ h^{-1}$) was carried out.

4. Discussion

Concerning plant #1, the experimental results show that the main parameters that play an important role on ammonia removal are work conditions (i.e., semi-batch or continuous mode), pH, and temperature.

Concerning pH, an increase from 9.4 to 10 involves the enhancement of ammonia removal yields from 25% to 34%. As confirmed by Guštin and Marinšek-Logar [28], the increase of pH values promote the ammonia removal but only up to a pH of 10.

The effects of temperature and work conditions are not evaluable individually because the change from continuous to semi-batch conditions falls with the main increasing of the temperature (from 38 to 50 °C).

The results point out that the work conditions of the plant (i.e., semi-batch or continuous mode) should be the main parameters that affect the performance of a stripping reactor. We have observed that, in semi-batch conditions, the ammonia removal yields are higher than 50% with respect to a continuous mode operation.

The management of plant #1 highlighted the following main issues: high-energy costs and fouling. The high-energy costs are mainly due to the electrical power required from the radial fan (12 to 14 kW) and the thermal power (120 kW in winter conditions) for the preheated air supply in the stripping reactor (4000 to 4500 $Nm^3\ h^{-1}$ at 45 °C) by using an air/water exchanger. Furthermore, regarding the fouling issue, a more efficient digestate pretreatment (with a mesh of 150 μm) with respect to the conventional solid/liquid separators (screw press or rotary drum) must be applied. Therefore, a better removal of suspended solids especially with a fibrous matrix can be obtained. The concentration of suspended solids in the pretreated digestate inlet to stripping column is lower than 1% and the fouling of packing material is reduced with a significant performance improvement.

Plant #2 worked only in a continuous mode (with the exception of the preliminary test—see Table 3). The results obtained during the nine months of management (HRT equal to 6 h) with a temperature variation from 48 to 58 °C (that depends on external weather conditions) showed that the pH seems to be the main parameter that affects the performance. Therefore, the dosage of NaOH is the key factor regarding (i) the ammonia removal yields and (ii) the management costs of the plant.

An important result obtained from this work concerns the equilibrium pH, which, as already pointed out, at a temperature between 45 to 55 °C, autonomously (without the dosage of alkaline reagent) reaches values from 9.2 to 9.5, i.e., in a suitable range to obtain an efficient ammonia removal from the aqueous phase. These results are confirmed by previous experimental works [29,30]. However, this aspect is strictly related to the characteristics of the substrate treated with AD especially to its alkalinity that depends both on the different feeding substrate and on the anaerobic degradation process of the same substrate.

In any case, in order to verify this operational feasibility on the digestate treated during this work, the two-levels of factorial experimentation are based on three parameters (temperature, air flowrate, and digestate flowrate) was developed. The two-level factorial test has proven that the temperature is a main parameter when no alkaline reagent is added. In this case, despite the fact that no statistical analyses have been carried out, the temperature seems to significantly affect the process performance. In fact, the temperature influences the reaction kinetics and the equilibrium of phase transfer by an exponential law. However, the other two parameters (air and digestate flowrates) must be taken into account in order to obtain ammonia removal yields that allow the correct management of nitrogen content in the digestate. The experiment results suggest the need for a proper management of available thermal energy with the aim to operate always in optimal conditions with respect to the process temperature.

The study of plant #2 allows us to confirm the feasibility to work on digestate with high concentrations of suspended solids without the occurrence of a fouling problem. Moreover, it was observed that, despite the fact that the high concentration of suspended solids (values higher than 1%) has a negative effect on the rheology, this aspect does not affect the alkaline reagent dosage to obtain pH values from 10.0 to 10.5. The amount of alkaline reagent significantly influences the digestate alkalinity.

Concerning economic aspects, the operating cost, in terms of euro per m^3 of digestate treated, related to plant #2 (without the dosage of alkaline reagent) is about one-third of that in plant #1.

The avoided dosage of the alkaline reagent involves a significant cost saving (estimated in 2–3 € per m^3 of digestate). Moreover, regarding the air supply, the electrical power of the rotary lobe compressor in plant #2 (7–8 kW) is lower than the value required for the radial fan of plant #1 (12–14 kW). Lastly, regarding plant #1, the semi-batch operative mode seems to be more efficient than the continuous mode with complete mixing.

Future developments regarding the study of a stripping reactor allowed to treat (with a lower thermal energy input) livestock manure (or other substrate) with low energy content cannot suitable for biogas production with anaerobic digestion. In this case, the thermal energy required for the process could be obtained from the protonation/dissolution reaction of ammonia gas in sulphuric acid.

5. Conclusions

In this work, two different ammonia stripping full-scale plants are studied. The substrate tested is liquid digestate derived from the anaerobic digestion of livestock manure and corn silage.

Plant #1, based on a packed column, shows low ammonia removal yields (from 22% to 37%) when it works in a continuous mode. Instead, the semi-batch conditions increase the performance (up to 66%) and reduce the alkaline reagent dosage due to the application of pH significantly lower than the values maintained in the continuous mode tests (9.2–9.5 with respect to 10–10.2). Despite these results, the main issues are related to high-energy costs due to the preheated air supply and to the significant digestate pretreatment required to avoid fouling.

Therefore, plant #2, based on an air bubble reactor without filling material, was studied. The lower performance (ammonia removal yields from 23% to 39%) were obtained in a continuous mode with temperatures ranging from 45 to 55 °C and, without an alkaline reagent dosage, pH autonomously raised from 9.2 to 9.5. When the temperature and pH increase, the performance enhances and reaches an ammonia removal yield of 66%.

The use of plant #2 allows us to obtain ammonia removal yields up to 50% (with HRT equal to 9.5 h) in a continuous mode. This value is higher than the performance obtained with the use of plant #1 in a continuous operation. Moreover, the fouling issue is avoided, the air flowrate required is one order of magnitude lower, and no dosage of NaOH is provided. This aspect is very important because the digestate with low sodium content can be recovered on agricultural soil avoiding issues concerning a salinity increase.

Author Contributions: M.B. supervised the experimental activities and data analysis. M.C.C., A.A., and I.B. performed the results interpretation and the paper drafting.

Funding: This research received no external funding.

Conflicts of Interest: The authors declare no conflict of interest.

References

1. Bousek, J.; Scroccaro, D.; Sima, J.; Weissenbacher, N.; Fuchs, W. Influence of the gas composition on the efficiency of ammonia stripping of biogas digestate. *Bioresour. Technol.* **2016**, *203*, 259–266. [CrossRef] [PubMed]
2. Panepinto, D.; Genon, G.; Brizio, E.; Russolillo, D. Production of green energy from co-digestion: Perspectives for the Province of Cuneo, energetic balance and environmental sustainability. *Clean Technol. Environ. Policy* **2013**, *15*, 1055–1062. [CrossRef]
3. Provenzano, M.R.; Iannuzzi, G.; Fabbri, C.; Senesi, N. Qualitative Characterization and Differentiation of Digestates from Different Biowastes Using FTIR and Fluorescence Spectroscopies. *J. Environ. Prot.* **2011**, *2*, 83–89. [CrossRef]
4. Collivignarelli, M.C.; Bertanza, G.; Abbà, A.; Sordi, M.; Pedrazzani, R. Synergy between Anaerobic Digestion and a Post-Treatment Based on Thermophilic Aerobic Membrane Reactor (TAMR). *Environ. Prog. Sustain. Energy* **2017**, *36*, 1802–1809. [CrossRef]
5. Möller, K.; Müller, T. Effects of anaerobic digestion on digestate nutrient availability and crop growth: A review. *Eng. Life Sci.* **2012**, *12*, 242–257. [CrossRef]

6. Cornalba, V. Energy Recovery from Biomass: Technical Aspects and Environmental Impact. Ph.D. Thesis, University of Pavia, Pavia, Italy, 2011.
7. Blumenstein, B.; Siegmeier, T.; Bruckhaus, C.; Anspach, V.; Möller, D. Integrated Bioenergy and Food Production—A German Survey on Structure and Developments of Anaerobic Digestion in Organic Farming Systems. *Sustainability* **2015**, *7*, 10709–10732. [CrossRef]
8. European Parliament and Council. Council Directive 91/676/EEC of 12 December 1991 Concerning the Protection of Waters against Pollution Caused by Nitrates from Agricultural Sources. 1991. Available online: https://eur-lex.europa.eu/legal-content/EN/TXT/?uri=CELEX%3A31991L0676 (accessed on 10 July 2018).
9. Italian Parliament. General Criteria and Technical Standards for Regional Regulating Agricultural Use of Livestock Manure and Wastewater, as Well as for the Production and the Agronomic Use of the Digestate, Ministerial Decree 25 February 2016. Available online: http://www.gazzettaufficiale.it/eli/id/2016/04/18/16A02762/sg (accessed on 10 July 2018).
10. Drosg, B.; Fuchs, W.; Al Seadi, T.; Madsen, M.; Linke, B. *Nutrient Recovery by Biogas Digestate Processing*; BAXTER, D., Ed.; IEA Bioenergy: Paris, France, 2015; ISBN 978-1-910154-16-8.
11. Gong, H.; Yan, Z.; Liang, K.Q.; Jin, Z.Y.; Wang, K.J. Concentrating process of liquid digestate by disk tube-reverse osmosis system. *Desalination* **2013**, *326*, 30–36. [CrossRef]
12. Jurado, E.; Antonopoulou, G.; Lyberatos, G.; Gavala, H.N.; Skiadas, I.V. Continuous anaerobic digestion of swine manure: ADM1-based modelling and effect of addition of swine manure fibers pretreated with aqueous ammonia soaking. *Appl. Energy* **2016**, *172*, 190–198. [CrossRef]
13. Mirtsou-Xanthopoulou, C.; Jurado, E.; Skiadas, I.V.; Gavala, H.N. Effect of Aqueous Ammonia Soaking in the Methane Yield and Composition of Digested Manure Fibers Applying Different Ammonia Concentrations and Treatment Durations. *Energies* **2014**, *7*, 4157–4158. [CrossRef]
14. Yilmazel, Y.D.; Demirer, G.N. Removal and recovery of nutrients as struvite from anaerobic digestion residues of poultry manure. *Environ. Technol.* **2011**, *32*, 783–794. [CrossRef] [PubMed]
15. Uysal, A.; Yilmazel, Y.D.; Demirer, G.N. The determination of fertilizer quality of the formed struvite from effluent of a sewage sludge anaerobic digester. *J. Hazard. Mater.* **2010**, *181*, 248–254. [CrossRef] [PubMed]
16. Provolo, G.; Perazzolo, F.; Mattacchini, G.; Finzi, A.; Naldi, E.; Riva, E. Nitrogen removal from digested slurries using a simplified ammonia stripping technique. *Waste Manag.* **2017**, *69*, 154–161. [CrossRef] [PubMed]
17. Serna-Maza, A.; Heaven, S.; Banks, C.J. Biogas stripping of ammonia from fresh digestate from a food waste digester. *Bioresour. Technol.* **2015**, *190*, 66–75. [CrossRef] [PubMed]
18. Bonmatì, A.; Campos, E.; Flotats, X. Concentration of pig slurry by evaporation: Anaerobic digestion as the key process. *Water Sci. Technol.* **2003**, *48*, 189–194. [CrossRef] [PubMed]
19. Guercini, S.; Castelli, G.; Rumor, C. Vacuum evaporation treatment of digestate: Full exploitation of cogeneration heat to process the whole digestate production. *Water Sci. Technol.* **2014**, *70*, 479–485. [CrossRef] [PubMed]
20. Lauterbock, B.; Ortner, M.; Haider, R.; Fuchs, W. Counteracting ammonia inhibition in anaerobic digestion by removal with a hollow fiber membrane contactor. *Water Res.* **2012**, *46*, 4861–4869. [CrossRef] [PubMed]
21. Lauterbock, B.; Moder, K.; Germ, T.; Fuchs, W. Impact of characteristic membrane parameters on the transfer rate of ammonia in membrane contactor application. *Sep. Purif. Technol.* **2013**, *116*, 327–334. [CrossRef]
22. Collivignarelli, M.C.; Abbà, A.; Castagnola, F.; Bertanza, G. Minimization of municipal sewage sludge by means of a thermophilic membrane bioreactor with intermittent aeration. *J. Clean. Prod.* **2017**, *143*, 369–376. [CrossRef]
23. Collivignarelli, M.C.; Abbà, A.; Bertanza, G.; Setti, M.; Barbieri, G.; Frattarola, A. Integrating novel (thermophilic aerobic membrane reactor-TAMR) and conventional (conventional activated sludge-CAS) biological processes for the treatment of high strength aqueous wastes. *Bioresour. Technol.* **2018**, *255*, 213–219. [CrossRef] [PubMed]
24. Zarebska, A.; Romero Nieto, D.; Christensen, K.V.; Søtof Fjerbæk, L.; Norddhal, B. Ammonium Fertilizers Production from Manure—A Critical Review. *Crit. Rev. Environ. Sci. Technol.* **2015**, *45*, 1469–1521. [CrossRef]
25. Limoli, A.; Langone, M.; Andreottola, G. Ammonia removal from raw manure digestate by means of a turbulent mixing stripping process. *J. Environ. Manag.* **2016**, *176*, 1–10. [CrossRef] [PubMed]

26. Piccinini, S.; Moscatelli, G.; Fabbri, C. Nitrogen recovery from digestate with ammonia stripping: Results of 27 tests. In Proceedings of the ManuREsource 2013—Manure Management and Valorization, Bruges, Belgium, 5–7 December 2013; Available online: http://www.crpa.it/media/documents/crpa_www/Convegni/20131206_manureResource_BELGIO/Piccinini_Ammonia_Stripping.pdf (accessed on 10 July 2018).
27. Fabbri, C.; Piccinini, S. Dairy Cattle and Biogas: Guidelines for the Construction and Management of Plants. Centro Ricerche Produzioni Animali, S.p.A.: Reggio Emilia, Italy, 2012. (In Italian). Available online: http://www.crpa.it/media/documents/crpa_www/Pubblicazi/E-book/lineebiogas/Bovini_da_latte_e_biogas.pdf (accessed on 10 July 2018).
28. Guštin, S.; Marinšek-Logar, R. Effect of pH, temperature and air flow rate on the continuous ammonia stripping of the anaerobic digestion effluent. *Process Saf. Environ. Prot.* **2011**, *89*, 61–66. [CrossRef]
29. Collivignarelli, C.; Baldi, M.; Gorini, R. Ammonia removal from MSW landfill leachate by means of air stripping. *Solid Waste* **1992**, *VI*, 1–6. (In Italian)
30. Collivignarelli, C.; Bertanza, G.; Baldi, M.; Avezzù, F. Ammonia stripping from MSW landfill leachate in bubble reactors: Process modeling and optimization. *Waste Manag. Res.* **1998**, *16*, 455–466. [CrossRef]

MDPI
St. Alban-Anlage 66
4052 Basel
Switzerland
Tel. +41 61 683 77 34
Fax +41 61 302 89 18
www.mdpi.com

Sustainability Editorial Office
E-mail: sustainability@mdpi.com
www.mdpi.com/journal/sustainability

www.ingramcontent.com/pod-product-compliance
Lightning Source LLC
Chambersburg PA
CBHW041753220926
51847CB00014B/185